Space Sciences Series of ISSI

Volume 65

More information about this series at http://www.springer.com/series/6592

Robert Pincus · David Winker
Sandrine Bony · Bjorn Stevens
Editors

Shallow Clouds, Water Vapor, Circulation, and Climate Sensitivity

Previously published in *Surveys in Geophysics*, Volume 38, Issue 6, 2017

Springer

Editors
Robert Pincus
University of Colorado
Boulder, CO
USA

and

NOAA Earth System Research Lab
Boulder, CO
USA

Sandrine Bony
Institute Pierre Simon Laplace
Laboratoire de Météorologie Dynamique
Paris
France

Bjorn Stevens
Max Planck Institute for Meteorology
Hamburg
Germany

David Winker
Atmospheric Sciences
NASA Langley Research Center
Hampton, VA
USA

ISSN 1385-7525
Space Sciences Series of ISSI
ISBN 978-3-030-08415-8 ISBN978-3-319-77273-8 (eBook)
https://doi.org/10.1007/978-3-319-77273-8

Softcover re-print of the Hardcover 1st edition 2018

Cover illustration: Organization of shallow clouds pictured over the tropical Atlantic near Barbados during the NARVAL2 airborne field campaign (photo by Sandrine Bony).

Printed on acid-free paper

This Springer imprint is published by the registered company Springer International Publishing AG part of Springer Nature
The registered company address is: Gewerbestrasse 11, 6330 Cham, Switzerland

Contents

Convective Self-Aggregation in Numerical Simulations: A Review 1
A. A. Wing, K. Emanuel, C. E. Holloway and C. Muller

Observing Convective Aggregation 27
C. E. Holloway, A. A. Wing, S. Bony, C. Muller, H. Masunaga, T. S. L'Ecuyer, D. D. Turner and P. Zuidema

An Observational View of Relationships Between Moisture Aggregation, Cloud, and Radiative Heating Profiles 65
M. D. Lebsock, T. S. L'Ecuyer and R. Pincus

Correction to: An Observational View of Relationships Between Moisture Aggregation, Cloud, and Radiative Heating Profiles 83
M. D. Lebsock, T. S. L'Ecuyer and R. Pincus

Implications of Warm Rain in Shallow Cumulus and Congestus Clouds for Large-Scale Circulations 85
L. Nuijens, K. Emanuel, H. Masunaga and T. L'Ecuyer

A Survey of Precipitation-Induced Atmospheric Cold Pools over Oceans and Their Interactions with the Larger-Scale Environment 111
P. Zuidema, G. Torri, C. Muller and A. Chandra

Low-Cloud Feedbacks from Cloud-Controlling Factors: A Review 135
S. A. Klein, A. Hall, J. R. Norris and R. Pincus

Mechanisms and Model Diversity of Trade-Wind Shallow Cumulus Cloud Feedbacks: A Review 159
J. Vial, S. Bony, B. Stevens and R. Vogel

Importance Profiles for Water Vapor 183
B. Mapes, A. S. Chandra, Z. Kuang and P. Zuidema

Structure and Dynamical Influence of Water Vapor in the Lower Tropical Troposphere 199
B. Stevens, H. Brogniez, C. Kiemle, J.-L. Lacour, C. Crevoisier and J. Kiliani

The Representation of Tropospheric Water Vapor Over Low-Latitude Oceans in (Re-)analysis: Errors, Impacts, and the Ability to Exploit Current and Prospective Observations 227
R. Pincus, A. Beljaars, S. A. Buehler, G. Kirchengast, F. Ladstaedter and J. S. Whitaker

Airborne Lidar Observations of Water Vapor Variability in Tropical Shallow Convective Environment 253
C. Kiemle, S. Groß, M. Wirth and L. Bugliaro

Emerging Technologies and Synergies for Airborne and Space-Based Measurements of Water Vapor Profiles 273
A. R. Nehrir, C. Kiemle, M. D. Lebsock, G. Kirchengast, S. A. Buehler, U. Löhnert, C.-L. Liu, P. C. Hargrave, M. Barrera-Verdejo and D. M. Winker

Observational Constraints on Cloud Feedbacks: The Role of Active Satellite Sensors 311
D. Winker, H. Chepfer, V. Noel and X. Cai

Shallow Circulations: Relevance and Strategies for Satellite Observation 337
G. Bellon, O. Reitebuch and A. K. Naumann

EUREC^{4}A: A Field Campaign to Elucidate the Couplings Between Clouds, Convection and Circulation 357
S. Bony, B. Stevens, F. Ament, S. Bigorre, P. Chazette, S. Crewell, J. Delanoë, K. Emanuel, D. Farrell, C. Flamant, S. Gross, L. Hirsch, J. Karstensen, B. Mayer, L. Nuijens, J. H. Ruppert, Jr., I. Sandu, P. Siebesma, S. Speich, F. Szczap, J. Totems, R. Vogel, M. Wendisch and M. Wirth

Surv Geophys (2017) 38:1171–1172
https://doi.org/10.1007/s10712-017-9441-3

Preface to the Special Issue "ISSI Workshop on Shallow Clouds and Water Vapor, Circulation and Climate Sensitivity"

Robert Pincus[1] · **David Winker**[2] · **Sandrine Bony**[3] · **Bjorn Stevens**[4]

Received: 16 October 2017 / Accepted: 19 October 2017 / Published online: 31 October 2017

Each of the 15 chapters in this work presents a paper gleaned from presentations at an International Space Science Institute Workshop on *Shallow Clouds, Water vapor, Circulation and Climate Sensitivity* organized as part of the World Climate Research Programme's Grand Science Challenge on Clouds, Circulation, and Climate Sensitivity. The workshop's somewhat awkward title reflects the practice of treating each subject separately—a state of affairs that the workshop sought to address. As the roughly 40 participating experts from around the world emphasized, the coupling of clouds and water vapor to circulation helps determine the nature of circulation systems in the past and present as well as the climate sensitivity that characterizes the response of the Earth's surface temperature to radiative forcing.

It has been known for more than a decade that an understanding of factors controlling the distribution and amount of the low-level, fair-weather, clouds over the tropical oceans is critical for determining Earth's climate sensitivity. What has become clear only recently is that these clouds do not simply respond passively to the large-scale circulations in which they form. Studies of clouds and circulations across a range of scales, enabled by increasing computational power, have shown that clouds help set these circulations through their interactions with radiation. Radiative cooling from low clouds drives low-level temperature and pressure gradients that reinforce the regions of gentle subsidence in which they prevail. This pathway is also thought to be responsible for the clustering—or self-

✉ Robert Pincus
Robert.Pincus@colorado.edu

[1] University of Colorado and NOAA/Earth System Research Lab, Physical Sciences Division, Boulder, CO 80305, USA

[2] NASA/Langley Research Center, Hampton, VA 23681, USA

[3] CNRS, LMD/IPSL, Université Pierre et Marie Curie, 75252 Paris, France

[4] Max Planck Institute for Meteorology, Bundesstr. 53, 20146 Hamburg, Germany

aggregation—of deep convection seen in simulations with spatially uniform forcing. The relevance of self-aggregation behavior to clouds on Earth was one theme emerging from the workshop.

Water vapor, like clouds, interacts powerfully with radiant energy, and there is a longstanding appreciation that water vapor influences Earth's climate sensitivity, especially through changes in water near the tropopause. But as several papers in this collection highlight, small departures in the relative humidity of the lower atmosphere can be just as important in influencing Earth's radiative balance. Moreover, perhaps no other quantity is as important for patterning the distribution of deep convection. Despite the importance of lower tropospheric humidity for a vast array of climate relevant processes it is poorly characterized in the absence of field campaign measurements, largely because current remote sensing techniques have difficulty unambiguously detecting the structure of water vapor in the tropical boundary layer.

The 15 chapters in this volume expand on these themes. The first set of papers focuses on convective self-aggregation, or the tendency of convection to organize even in the absence of external influences. Allison Wing and her co-authors provide a timely review of a rapidly expanding literature. Following this lead two groups, led by Christopher Holloway and Matthew Lebsock, respectively, explore the observational record for signatures of convective self-aggregation. A team consisting of Louise Nuijens, Kerry Emanuel, Hirohiko Masunaga and Tristan L'Ecuyer asks to what extent radiative forcing from somewhat deeper clouds, cumulus congestus, also influence the pattern of large-scale circulation. Paquita Zuidema and Giuseppe Torri consider the disaggregating aspect of deep convection, namely the cold pools that propagate away from convecting areas and initiate deeper convection well away from aggregated clusters.

Two articles address the question as to how shallow clouds influence climate sensitivity. One, a concise summary by Stephen Klein et al., synthesizes five recent studies leveraging cloud controlling factors to try and infer how low clouds may change in a warming climate. The other, by Jessica Vial, Sandrine Bony and their co-authors, explores why climate models are so sensitive to the representation of low clouds, and to what extent more fundamental modeling approaches, like large-eddy simulation, may help to constrain their behavior.

The importance of water vapor is reviewed in two articles, one led by Brian Mapes and one by Bjorn Stevens, Helene Brogniez and co-authors. The latter article, and one by Robert Pincus and his co-authors, assesses the ability of our present observing system to characterize the distribution of water in the lower troposphere.

The book's final section looks to the future. New technologies for observing lower tropospheric water vapor are becoming more widely available. An article by Christopher Kiemle demonstrates the use of one such technology in airborne field campaigns, while the contribution led by Amin Nehrir surveys the range of emerging technologies being developed for air- and space-borne measurements. The prospects of using related space-based observations of clouds to constrain the temperature-mediated dependence of clouds and circulation is reviewed by Dave Winker, Helene Chepfer and Xia Cai. Observational strategies for directly exploring how satellite observations might be used to measure shallow circulations are explored in an article by Gilles Bellon, Oliver Reitenbuch and Ann Kristin Naumann—an especially timely contribution given that the first satellite-based wind lidar will be launched just as this work appears. Finally, Sandrine Bony, Bjorn Stevens and a large international team of co-authors outline an exciting new field campaign in which they plan to measure the interaction of low clouds, water vapor and circulation with the goal of putting better bounds on Earth's climate sensitivity.

Surv Geophys (2017) 38:1173–1197
https://doi.org/10.1007/s10712-017-9408-4

Convective Self-Aggregation in Numerical Simulations: A Review

Allison A. Wing[1,2] · **Kerry Emanuel**[3] · **Christopher E. Holloway**[4] · **Caroline Muller**[5]

Received: 16 October 2016 / Accepted: 8 February 2017 / Published online: 22 February 2017

Abstract Organized convection in the tropics occurs across a range of spatial and temporal scales and strongly influences cloud cover and humidity. One mode of organization found is "self-aggregation," in which moist convection spontaneously organizes into one or several isolated clusters despite spatially homogeneous boundary conditions and forcing. Self-aggregation is driven by interactions between clouds, moisture, radiation, surface fluxes, and circulation, and occurs in a wide variety of idealized simulations of radiative–convective equilibrium. Here we provide a review of convective self-aggregation in numerical simulations, including its character, causes, and effects. We describe the evolution of self-aggregation including its time and length scales and the physical mechanisms leading to its triggering and maintenance, and we also discuss possible links to climate and climate change.

Keywords Self-aggregation · Convective organization · Radiative–convective equilibrium · Convective processes · Tropical convection · Idealized modeling

1 Introduction

Radiative–convective equilibrium (RCE) is the statistical equilibrium state that the atmosphere and surface would reach in the absence of lateral energy transport, in which there is a balance between net radiative cooling and convective heating. It has long been

✉ Allison A. Wing
awing@fsu.edu

1 Lamont-Doherty Earth Observatory, Columbia University, Palisades, NY, USA

2 Present Address: Florida State University, Tallahassee, FL, USA

3 Massachusetts Institute of Technology, Cambridge, MA, USA

4 University of Reading, Reading, UK

5 Laboratoire de Météorologie Dynamique, École Normale Supérieure, Paris, France

used as an idealization of the tropical atmosphere in simulations with single-column models (e.g., Manabe and Strickler 1964; Renno et al. 1994), two- and three- dimensional cloud-resolving models (e.g., Held et al. 1993; Bretherton et al. 2005), and regional/global models with parameterized convection (e.g., Held et al. 2007).

Convective self-aggregation is the spontaneous spatial organization of convection in numerical simulations of radiative–convective equilibrium despite homogeneous boundary conditions and forcing. This instability of the RCE state arises due to interactions among convection, radiation, environmental moisture, and surface fluxes.

Aggregation has been found to occur in simulations with two-dimensional cloud-resolving models (Held et al. 1993; Grabowski and Moncrieff 2001, 2002; Stephens et al. 2008), small-domain square three-dimensional cloud-resolving models (Tompkins and Craig 1998; Bretherton et al. 2005; Khairoutdinov and Emanuel 2010; Muller and Held 2012; Jeevanjee and Romps 2013; Wing and Emanuel 2014; Abbot 2014; Muller and Bony 2015; Holloway and Woolnough 2016; Hohenegger and Stevens 2016), elongated channel three-dimensional cloud-resolving models (Tompkins 2001; Posselt et al. 2008, 2012; Stephens et al. 2008; Wing and Cronin 2016), regional/global models with parameterized clouds and convection (Su et al. 2000; Held et al. 2007; Popke et al. 2013; Becker and Stevens 2014; Reed et al. 2015; Arnold and Randall 2015; Reed and Medeiros 2016; Coppin and Bony 2015; Silvers et al. 2016; Hohenegger and Stevens 2016; Bony et al. 2016) or super-parameterized clouds and convection (Arnold and Randall 2015), and global models with explicit convection (Satoh and Matsuda 2009; Satoh et al. 2016).

An example of self-aggregation in a cloud-resolving model (CRM) with no rotation is found in Fig. 1, which shows snapshots of outgoing longwave radiation, where low values indicate the presence of high, deep convective clouds. Initially, convection is quasi-randomly distributed across the domain (Fig. 1a), but after tens of days, the convection has aggregated into a single, intensely precipitating moist cluster surrounded by dry, subsiding air (Fig. 1b).

An example of self-aggregation in aquaplanet simulations of non-rotating RCE with several general circulation models (GCMs) with parameterized convection is found in Fig. 2. In GCMs, self-aggregation entails the development of a few isolated clusters of deep convection within a large area of subsidence, and the development of a large-scale circulation.

Spontaneous symmetry breaking in two-dimensional and two-column models (Raymond 2000; Nilsson and Emanuel 1999) and multiple equilibria in weak temperature gradient simulations with single column (Sobel et al. 2007; Emanuel et al. 2014; Daleu et al. 2015) and two-dimensional cloud-resolving models (Sessions et al. 2010, 2015, 2016) have also been interpreted as analogous to/consistent with convective self-aggregation. In single column or small-domain CRM weak temperature gradient simulations, in which the large-scale circulation is parameterized, the dry equilibrium corresponds to the dry, non-convecting regions of a self-aggregated state and the equilibrium with precipitation corresponds to the moist cluster in a self-aggregated state. Theory and simple models of aggregation (or its root cause, instability of the RCE state) have been presented by Bretherton et al. (2005), Craig and Mack (2013), Emanuel et al. (2014) and Beucler and Cronin (2016).

The above refers to aggregation under conditions of non-rotating RCE. Under conditions of *rotating* RCE, self-aggregation takes the form of spontaneous genesis of tropical cyclones (Bretherton et al. 2005; Nolan et al. 2007; Held and Zhao 2008; Khairoutdinov and Emanuel 2013; Shi and Bretherton 2014; Zhou et al. 2014; Boos et al. 2015; Reed and Chavas 2015; Davis 2015; Wing et al. 2016; Merlis et al. 2016; Zhou et al. 2017). While aggregation occurs across a wide variety of different models, with different dynamical

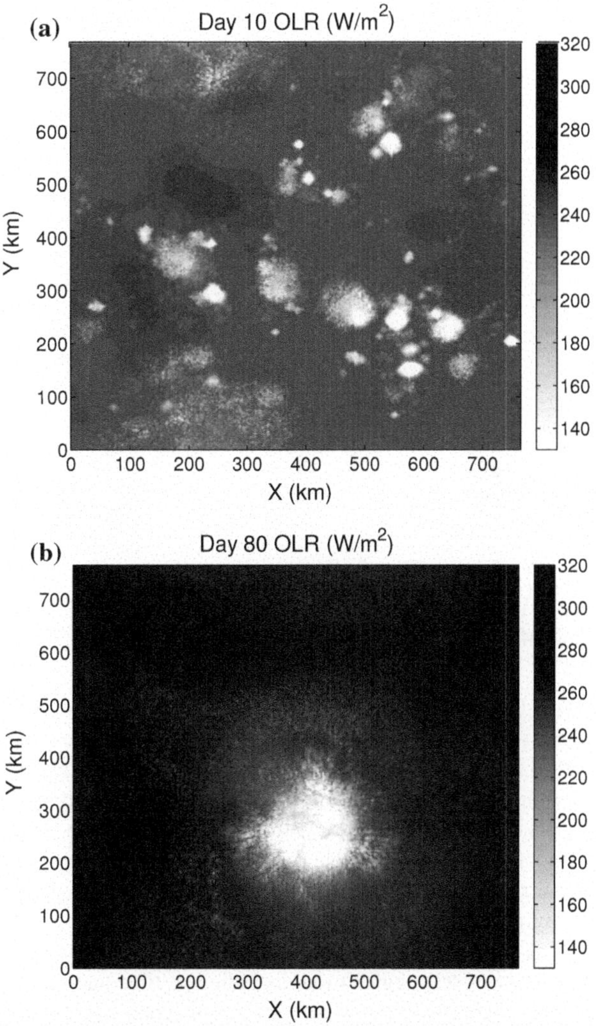

Fig. 1 Snapshot of outgoing longwave radiation (OLR) at **a** day 10 and **b** day 80 of a radiative–convective equilibrium simulation at 305 K. Reprinted from Wing and Emanuel (2014).

cores, different convective parameterizations (in the case of GCMs), and different radiation and microphysics schemes, the exact nature and sensitivities of aggregation vary and depend on model details. This reflects the fact that multiple processes contribute to aggregation, involving intricate interactions between clouds, moisture, radiation, and circulation. In CRMs, self-aggregation occurs more readily with large domains and coarser resolution (Muller and Held 2012), although the domain-size dependence disappears if the evaporation of rain at low levels (which causes downdrafts and cold pools) is eliminated (Jeevanjee and Romps 2013; Muller and Bony 2015). Interactive radiation and surface fluxes are generally necessary for self-aggregation to occur, a result supported by mechanism denial experiments carried out by many different authors; the details are described in Sect. 3. Some studies find that self-aggregation is favored by high sea surface temperatures (SSTs) (Held et al. 1993; Khairoutdinov and Emanuel 2010; Wing and Emanuel 2014; Emanuel et al. 2014; Coppin and Bony 2015), although it can also occur at temperatures far below current tropical SSTs (Abbot 2014; Coppin and Bony 2015; Wing and Cronin

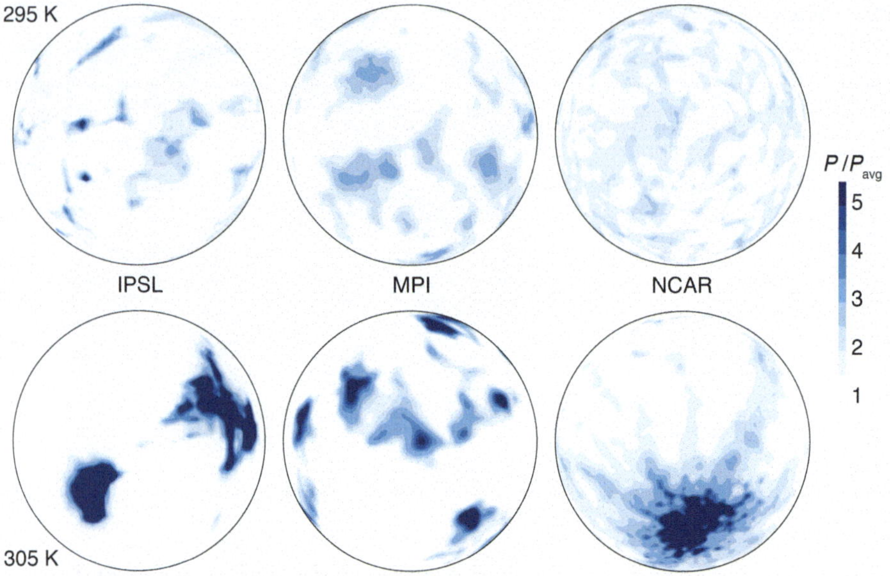

Fig. 2 Hemispheric view of monthly precipitation (normalized by its global mean value) predicted by the IPSL, MPI, and NCAR GCMs in RCE simulations forced by an SST of (*top*) 295 K and (*bottom*) 305 K. Reprinted from Bony et al. (2016)

2016; Holloway and Woolnough 2016). Vertical wind shear and/or strong mean winds make self-aggregation less likely (Held et al. 1993; Bretherton et al. 2005; Wing 2014; Abbot 2014), although much remains to be understood about the sensitivity of aggregation to vertical wind shear. Self-aggregation also exhibits hysteresis; that is, once convection occurs, it is difficult for it to disaggregate, and the aggregated state can be maintained without the feedbacks that are necessary to trigger it from homogeneous conditions (Khairoutdinov and Emanuel 2010; Muller and Held 2012; Muller and Bony 2015; Holloway and Woolnough 2016).

The structure of the remainder of this paper is as follows: First, we review the characteristics of self-aggregation, including its time and length scales and impacts on modeled climate. Next, we review the physical mechanisms leading to self-aggregation, including longwave radiation, shortwave radiation, surface fluxes, moisture feedbacks, and advective processes. We then discuss the importance of self-aggregation: Why studies of aggregation in idealized simulations might be important for understanding the tropical atmosphere and climate. Finally, we conclude with a synthesis of self-aggregation in idealized numerical models and its applicability to the real world, and what needs to be explored further. We will focus our discussion on non-rotating self-aggregation, but will note several areas in which the rotating case behaves differently.

2 Characteristics of Self-Aggregation

2.1 General Evolution of Aggregation

Simulations of convection in non-rotating RCE using three-dimensional cloud-resolving models initially produce distributions of convection that are nearly random in space and time. Aggregation generally begins with the emergence of one or several dry regions in which convection is suppressed. The dry regions have strong radiative cooling, weaker surface enthalpy fluxes, and subsidence, which yields further drying and suppression of convection. Over time, these persistent dry regions amplify and expand such that convection, clouds, and precipitation are increasingly confined to one (or several, depending on the domain setup) moist area. As in observations (e.g., Bretherton et al. 2004), there is a strong correlation between moister columns and more active deep convection. Throughout the evolution of aggregation, the dry regions get drier and the moist regions get moister, such that the distribution of precipitable water widens considerably. Usually the dry regions amplify first, but there are at least a few simulations in which several dry and moist regions amplify roughly at the same time during the early stages of aggregation (Holloway and Woolnough 2016). This general picture of the evolution of aggregation is notably different in simulations of rotating RCE, in which a large moist region is formed which then spins up into a tropical cyclone (Wing et al. 2016) or multiple moist vortices merge (Davis 2015), rather than dry regions amplifying and expanding (the online supplemental videos[1] from Wing et al. (2016) show this distinction nicely). Non-rotating RCE simulations in which rain evaporation is suppressed also are characterized by growth and merger of moist regions (e.g., Wing 2014).

2.2 Identifying Metrics

To study self-aggregation, we need to identify it and quantify its strength. As suggested by the large number of changes that occur during the evolution of self-aggregation in idealized simulations, there are a number of different metrics that have been used to characterize the aggregation state.

One category of metrics includes measurements of the drying of the non-convective environment and the associated increase in the variance of precipitable water (PW) and in column-integrated moist static energy (MSE). The non-convective environment grows and dries, while the convective region stays moist and often becomes moister, but the domain-mean PW decreases with aggregation. Many studies use the decrease in domain-mean PW and, in particular, the increase in the spread in the PW distribution as measured by, for instance, the interquartile range (IQR) of PW to quantify self-aggregation (e.g., Bretherton et al. 2005; Muller and Held 2012; Arnold and Randall 2015; Holloway and Woolnough 2016). Since self-aggregation in non-rotating RCE should have very weak horizontal temperature gradients, the horizontal variability in PW is also the dominant source of horizontal variability in column-integrated MSE, $\widehat{h}$. Wing and Emanuel (2014) use the horizontal variance of $\widehat{h}$ as a measure of aggregation state (Fig. 3a), and they further derive an equation for the budget of this quantity that allows for the quantification of the contribution of different processes to the growth of the total variance. Craig and Mack (2013) use the distribution of free tropospheric PW instead of total PW, and they are also one of several studies which normalize PW distributions (or measures of spread such as IQR) by saturation PW to allow for fairer comparisons between simulations with different SSTs. An example of one of these normalized metrics, the spatial variance of column relative humidity, is shown in Fig. 3b. Note that while the horizontal variance of $\widehat{h}$ increases with SST because of the Clausius–Clapeyron exponential dependence of water vapor on

[1] Video S1 and Video S2 from Wing et al. (2016) can be found at http://dx.doi.org/10.1175/JAS-D-15-0380.1.

temperature (Fig. 3a), the horizontal variance of column relative humidity is roughly the same across simulations with different SSTs (Fig. 3b).

The increase in domain-mean outgoing longwave radiation has also been used as a measure of aggregation (e.g., Khairoutdinov and Emanuel 2010; Wing and Emanuel 2014). Global RCE studies have also used the increase in "subsiding fraction," the fractional area of the domain covered by large-scale mid-tropospheric subsidence, as a measure of aggregation (Coppin and Bony 2015). A limitation of the metrics mentioned above is that they do not quantify the horizontal scale of the aggregated convective (or non-convective) regions. Methods of quantifying the horizontal length scales for convective aggregation and what may define them are discussed more in Sect. 2.4 below.

For most smaller square domains, aggregation in non-rotating RCE appears to be binary, either on or off (e.g., Muller and Held 2012; Wing and Emanuel 2014) although this is not true for at least one study with smaller square domains (Holloway and Woolnough 2016). For long-channel experiments or larger domains, aggregation appears to exist more on a gradual continuum as boundary conditions are varied or mechanisms are suppressed (e.g., Wing and Cronin 2016; Coppin and Bony 2015). This suggests that small domain size, at least for some models, may prevent key aggregation feedbacks from occurring at all for certain setups (or enhance the influence of negative feedbacks opposing aggregation (Jeevanjee and Romps 2013)). Even for simulations that do show strong aggregation, the time scale for aggregation to fully develop can vary greatly for different simulations, as discussed in the next section.

2.3 Time Scale

The time scale for idealized self-aggregation from homogeneous initial conditions to reach a relatively stable state can vary from 15 to 100 days or more. For instance, Bretherton et al. (2005) found this time scale to be approximately 40 days for a $(576\ \text{km})^2$ domain, 301 K SST, and a 3-km grid. Holloway and Woolnough (2016) had 16 days for a fairly similar setup but a different model. Muller and Held (2012) found it took 20–25 days or more, with less time for coarser grid spacing. Wing and Emanuel (2014) found a time scale of 60 days for a $(768\ \text{km})^2$ domain, 305 K SST, and 3-km grid. Holloway and Woolnough (2016) found a time scale of only 8 days when rain evaporation and hence downdrafts and cold pools were suppressed, suggesting that downdrafts and/or cold pools slow or suppress aggregation as proposed by Jeevanjee and Romps (2013) and Muller and Bony (2015). Muller and Held (2012) and Holloway and Woolnough (2016) both found a disaggregation time scale (which is the time needed to return to a less aggregated equilibrium) as small as 10 days when simulations were initialized with an aggregated state and then interactive radiation was suppressed.

Over a range of SSTs in the square-domain setup of Wing and Emanuel (2014), the e-folding time for growth of spatial variance of column-integrated MSE was $\sim$11–13 days (Wing 2014). Wing and Cronin (2016) found the times to reach a stable aggregated state ranged from 15 days for 280 K to 50 days for 310 K, though rapid growth and most of the organization occurred in the first 10–20 days for all temperatures and the initial e-folding growth time scale from logistic fits was 2–6 days (Fig. 3); they used long-channel domains of 12,228 km $\times$ 192 km with a 3-km grid. This kind of exponential growth will lead to much larger horizontal scales in a given amount of time when starting from larger initial clustering, as is typically found in nature. The horizontal scale of aggregation is addressed in the next section.

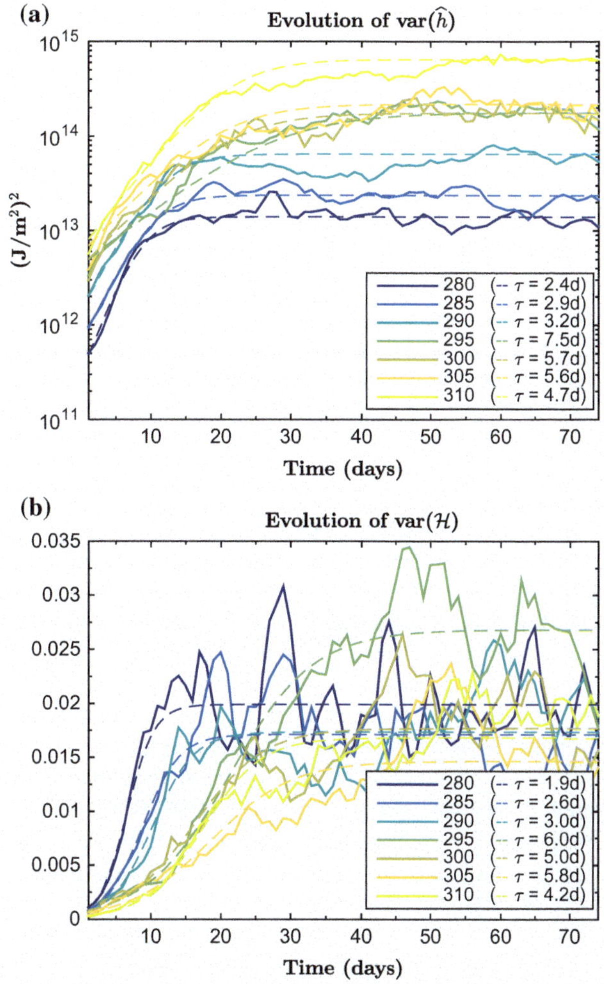

Fig. 3 Evolution of **a** spatial variance of column frozen moist static energy, var($\widehat{h}$) (J^2 m^{-4}), and **b** spatial variance of column relative humidity, var($\mathcal{H}$), in *solid lines*, and logistic fits to each, in *thin dashed lines*, for the channel RCE simulations at different values of SST (K). The vertical scale is **a** logarithmic and **b** linear, and the *legend* indicates the initial e-folding growth time-scale from each logistic fit. Reprinted from Wing and Cronin (2016). ©2015 Royal Meteorological Society

The time scale for self-aggregation results from the growth rates associated with different feedbacks that favor or oppose aggregation. Bretherton et al. (2005) developed a simple semi-empirical model to predict the initial e-folding rate of self-aggregation. They used physically motivated curve fits of the advective, surface flux, and radiative forcing from the initial stages of aggregation in their CRM simulations as parameters in an ordinary differential equation for column relative humidity. Using this semi-empirical model, Bretherton et al. (2005) found an e-folding time of the instability of 9 days. The contribution of different processes to the growth rate of column moist static energy variance is also quantified by the budget introduced by Wing and Emanuel (2014).

2.4 Length Scale

One of the limitations of square-domain simulations of non-rotating aggregation conducted thus far is that, when such simulations reach a fully aggregated state, they only contain one moist, precipitating cluster. This suggests that the size of the domain constrains the size of the cluster, and so it is difficult to define the length scale of aggregation or develop a theory for it. In square-domain simulations, the absolute size of the aggregated area increases with domain size, such that the aggregated area is 20–25% of the total domain area (Muller and Held 2012). There is no correlation between the area of the aggregated region and resolution (Muller and Held 2012) or SST (Wing 2014). The precise size of the cluster is somewhat sensitive to the metric used to define it; for example, Muller and Held (2012) used various threshold values for precipitable water. Defining the cluster size as the area where the precipitable water is greater than one standard deviation above the domain-mean value, Wing (2014) found that, across a range of SSTs at a given domain size, the cluster covered 15–17% of the horizontal area of the domain. In rotating RCE, where the nonzero Coriolis parameter introduces the Rossby radius of deformation and the ratio of tropical cyclone potential intensity over the Coriolis parameter as important horizontal scales, large enough domains allow for multiple cyclonic features with measurable average size and separation (e.g., Held and Zhao 2008; Khairoutdinov and Emanuel 2013).

Simulations using an elongated channel domain geometry in non-rotating RCE (e.g., Posselt et al. 2012; Wing and Cronin 2016) have the advantage of containing multiple aggregated areas, allowing a spatial scale to be more easily defined. In such simulations, the average spacing between convective regions in the longer horizontal dimension is a simple measure of scale (e.g., Stephens et al. 2008). The autocorrelation length scale of PW, which is the largest horizontal scale at which the average horizontal autocorrelation coefficient is $\geq e^{-1}$, is a more objective measure of this scale and has been shown to grow with self-aggregation in idealized models (Craig and Mack 2013; Wing and Cronin 2016; Holloway and Woolnough 2016). The power spectrum can also be calculated to characterize dominant scales of horizontal variability of PW or $\widehat{h}$ (Bretherton and Khairoutdinov 2015; Wing and Cronin 2016).

Over a 30 K range of SSTs, Wing and Cronin (2016) found that, using average wavenumber and correlation length metrics, the spatial scale of the aggregation varied from $\sim$1000 to $\sim$4000 km, with simulations at higher SSTs having smaller spatial scales. They presented a theory for the separation distance between convectively active regions based on boundary layer remoistening. A length scale resulting from this theory, proportional to the boundary layer height divided by the surface enthalpy exchange coefficient, was highly correlated with the spatial scale of aggregation across the main set of Wing and Cronin (2016)'s simulations; however, attempts to confirm the scaling were inconclusive. Further, this theory related to the maximum size of a dry region, and it is unknown whether the size of a moist region scales with that of a dry region or is controlled by other mechanisms. In addition, there could be a (perhaps temperature-dependent) minimum length scale of aggregation, below which the instability does not emerge. Bretherton and Khairoutdinov (2015) examined the scale dependence of self-aggregation feedbacks in near-global aquaplanet simulations of realistic tropical variability; similar analysis in idealized simulations may lead to insights on what controls the intrinsic length scale of self-aggregation. As of now, though, the question of what sets the spatial scale of self-aggregation in non-rotating RCE remains largely unsolved.

2.5 Impacts

As alluded to above, self-aggregation is not solely a spatial reorganization of the convection, but has dramatic impacts on the domain-mean climate. Figure 4 shows horizontally averaged profiles of humidity, MSE, and saturation MSE averaged over day 1 (disorganized convection) and day 50 (aggregated convection) from a non-rotating RCE simulation from Bretherton et al. (2005). The entire depth of the free troposphere is substantially drier, with relative humidities near 20%, when convection is aggregated. This is because the domain mean is dominated by the dry, non-convecting areas. Mean drying is found in all numerical simulations of self-aggregation, to our knowledge, although some simulations exhibit a less extreme form of aggregation, and consequently, a more muted (though still substantial) amount of drying (e.g., Wing and Cronin (2016), their Figure 2). Associated with this decrease in humidity, the domain-mean outgoing longwave radiation is ~10–30 Wm^{-2} larger when convection is aggregated (Wing and Cronin 2016), which is comparable to the increase of ~20–30 Wm^{-2} found in observations of aggregated regimes (Tobin et al. 2013).

Aggregation is also associated with domain-mean warming in the free troposphere, as indicated by the increase in saturation MSE in Fig. 4, which corresponds to several degrees of temperature increase (also see Wing and Cronin (2016), their Figure 1). The temperature increase is consistent with the fact that, when aggregated, the convecting environment is moister, which reduces the influence of entrainment and drives the troposphere closer to a

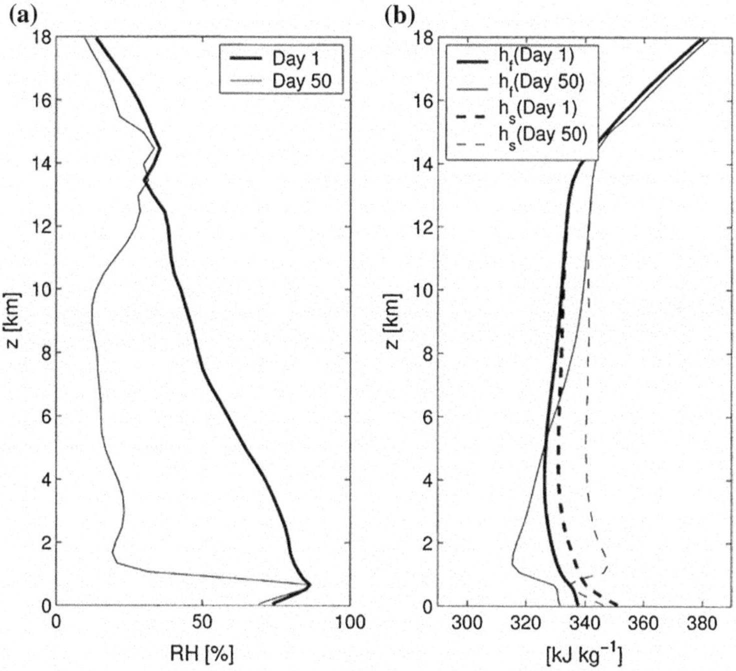

Fig. 4 Horizontally averaged profiles of **a** relative humidity and **b** moist static energy (h_f) and saturation moist static energy (h_s) averaged over days 1 and 50 from an RCE simulation. Reprinted from Bretherton et al. (2005). ©2005 American Meteorological Society

moist adiabat. It is also consistent with an increase in boundary layer moisture and MSE in the convective region, shifting buoyant parcels toward a warmer moist adiabat.

Changes in radiative fluxes by aggregation are strongly influenced by changes in cloudiness. A decrease in high clouds with aggregation is found in CRM simulations (Fig. 5), GCM simulations of RCE with parameterized convection (Bony et al. 2016), and in observations (Tobin et al. 2012, 2013). Bony et al. (2016) argued that the decrease in anvil cloud fraction with aggregation was a result of an increase in upper tropospheric stability due to the increase in temperature. By mass conservation, the amount of mass divergence in the convecting regions corresponds to the maximum clear-sky radiatively driven divergence in the upper troposphere. As the upper troposphere warms with aggregation, it becomes more stable, reducing the amount of divergence necessary to balance the same clear-sky radiative cooling. This mechanism is similar to arguments based on subsidence regions which can explain the weakening of the overturning circulations in the tropics as a response to greenhouse warming (Knutson and Manabe 1995; Held and Soden 2006). The reduction in anvil cloud fraction is then linked to the reduction in convective outflow. If the frequency or degree of aggregation changes with warming, the reduction in high clouds (and the increased drying) could affect cloud feedbacks on surface warming and climate sensitivity (Mauritsen and Stevens 2015).

In CRM simulations, this decrease in high clouds is largely offset by an increase in low clouds, such that the reflected shortwave radiation changes little (Fig. 5, Wing 2014; Wing and Cronin 2016). This result is uncertain, however, because the horizontal resolution of 3 km used in those studies is too coarse to model low clouds accurately. The response of the top-of-atmosphere net radiation budget in CRM simulations differs from observations of aggregated convection, which find that the reflected shortwave radiation is reduced due to a reduced total cloud fraction, which largely cancels the increase in outgoing longwave radiation (Tobin et al. 2012, 2013). Both numerical simulations and observations agree that the domain-mean tropospheric radiative cooling increases with aggregation, due to the drier troposphere.

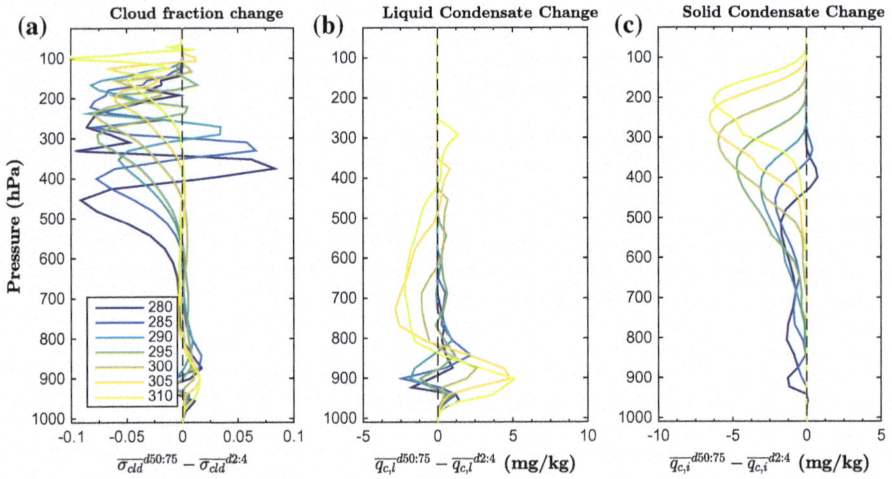

Fig. 5 Domain-average profiles of change in **a** cloud fraction, **b** liquid condensate $q_{c,l}$ and **c** solid condensate $q_{c,i}$ between days 2 and 4 and 50 and 75 of RCE channel simulations. *Colors* indicate the sea-surface temperature (K) of the channel simulation, and an *overbar* indicates a mean over the time range indicated. Reprinted from Wing and Cronin (2016). ©2015 Royal Meteorological Society

3 Mechanisms of Self-Aggregation

Numerical studies of self-aggregation have identified multiple processes involving convection-moisture-radiation feedbacks that are capable of creating an area around clouds that is more favorable for future convection than areas further away. These processes have been identified using both mechanism denial experiments and diagnostic frameworks. One diagnostic framework that has been employed is a budget for the spatial variance of column-integrated moist static energy (Wing and Emanuel 2014), which is given by:

$$\frac{1}{2}\frac{d\widehat{h'}^2}{dt} = \widehat{h'}F'_K + \widehat{h'}N'_S + \widehat{h'}N'_L - \widehat{h'}\nabla_h \cdot \widehat{\mathbf{u}h}. \tag{1}$$

where h is the moist static energy (or frozen moist static energy), F_K is the surface enthalpy flux, N_S is the column shortwave flux convergence, N_L is the column longwave flux convergence, and $-\nabla_h \cdot \widehat{\mathbf{u}h}$ is the "advective term," the horizontal convergence of the density-weighted vertical integral of the flux of frozen moist static energy. A primed quantity, $(\cdot)'$, denotes the spatial anomaly from the horizontal mean, $\langle(\cdot)\rangle$, and $\widehat{\cdot}$ indicates a density-weighted vertical integral.

The advantage of this (or similar) frameworks is that, since self-aggregation is associated with an increase in the spatial variance of MSE, the budget enables the quantification of the each feedback associated with a process that is a source or sink of MSE, hence contributing or opposing self-aggregation. Each feedback can be quantified across the entire evolution of the simulation. The magnitude of these feedbacks can be compared to each other within a given simulation and across simulations using different boundary conditions, parameters, and models. However, while this budget diagnoses the direct effect of radiative processes [the second and third terms on the right-hand side of Eq. (1)], it does not explicitly diagnose the indirect effect of a radiatively driven circulation [this, among other dynamical contributions, is a part of the fourth term on the right-hand side of Eq. (1)]. Since it is a vertically integrated budget, the impact of shallow circulations or the sensitivity to the profile of radiative heating anomalies is also not explicitly included (they are indirectly included, insofar as they determine the circulation response which is a part of the advective term). On the other hand, mechanism denial experiments, in which a feedback process is disallowed, test the sensitivity of self-aggregation to all aspects of that feedback. For example, horizontally homogenizing the radiative heating rates removes both the direct and indirect feedbacks associated with radiation. If self-aggregation still occurs without a particular feedback enabled, this indicates that that feedback is not necessary for aggregation. However, since many different processes can contribute to aggregation, caution must be taken to interpret the results of such sensitivity tests, as the importance of a particular feedback could vary depending on what parameters are used and what other feedback processes are active. In addition, if there is a critical SST for aggregation to occur, aggregation will be sensitive to virtually everything when one is near the critical point.

One fundamental aspect of moist convection in the tropics that underpins the mechanisms of aggregation discussed here is that, in a weak temperature gradient environment, deep convection is more active in moister tropospheric columns, as shown by precipitation observations, for example (Bretherton et al. 2004). This should be kept in mind when interpreting the feedbacks on aggregation identified in this section.

In this section, we review the various processes leading to the self-aggregation of convection in RCE simulations. This includes longwave radiation (Bretherton et al. 2005;

Muller and Held 2012; Posselt et al. 2012; Wing and Emanuel 2014; Emanuel et al. 2014; Muller and Bony 2015; Coppin and Bony 2015; Arnold and Randall 2015; Wing and Cronin 2016; Holloway and Woolnough 2016), shortwave radiation (Wing and Emanuel 2014; Wing and Cronin 2016), surface fluxes (Bretherton et al. 2005; Wing and Emanuel 2014; Coppin and Bony 2015; Wing and Cronin 2016), moisture feedbacks (Tompkins 2001; Craig and Mack 2013; Muller and Bony 2015; Holloway and Woolnough 2016), and advective processes (Bretherton et al. 2005; Muller and Held 2012; Muller and Bony 2015). At the end of this section, we address the relative importance of some of these processes for the maintenance of an aggregated state as opposed to the initial stages of self-aggregation. We focus on non-rotating simulations, although we note a few instances where the mechanisms differ if instead rotating RCE is simulated.

3.1 Surface Fluxes

Surface flux feedbacks favor self-aggregation, though they are not always necessary for aggregation to occur. Tompkins and Craig (1998) and Bretherton et al. (2005) both found that sensitivity runs without radiative feedbacks or without surface flux feedbacks did not aggregate. Muller and Held (2012) extended these sensitivity runs to cover a large range of domains and resolutions. They also found that surface flux feedbacks favor self-aggregation, but convection could still aggregate without them as long as radiative feedbacks are active and the domain is large enough. The converse is not true. Holloway and Woolnough (2016) confirmed that sensitivity runs with homogeneous surface fluxes can self-aggregate or not depending on the strength of the surface fluxes imposed. Non-rotating RCE simulations without radiative feedbacks do not aggregate (unless rain evaporation is artificially removed; see below section on moisture feedbacks). Therefore, at current temperatures, it seems that surface flux feedbacks are not sufficient on their own, without longwave radiative feedbacks, for non-rotating aggregation to occur. Rotating RCE simulations, on the other hand, have stronger surface flux feedbacks and can aggregate without radiative feedbacks (Wing et al. 2016).

Physically, there are two opposing contributions to the surface flux feedback (Wing and Emanuel 2014). The air-sea enthalpy disequilibrium is smaller in the moist region than in the dry region, which would tend to suppress surface fluxes in the moist region, a negative feedback on aggregation. On the other hand, the surface winds are stronger in the moist, convecting region, which would tend to enhance surface fluxes there, a positive feedback on aggregation. The latter dominates in initial triggering of aggregation, yielding an overall positive surface flux feedback.

3.1.1 Sensitivity to SST

In the GCM simulations of Coppin and Bony (2015), the surface flux-wind feedback was the leading mechanism of aggregation at high temperatures. In the high-temperature regime, strong surface winds in the convective region yield strong surface fluxes, moistening the high-MSE convective region, thereby enhancing the MSE gradient and favoring self-aggregation. In contrast, Wing and Cronin (2016) found that the total surface flux feedback, as diagnosed from the MSE variance budget, was approximately constant in magnitude across a wide range of temperatures (280–310 K). Both of these studies used fixed SSTs and thus do not have surface energy balance. Therefore, caution must be taken in interpreting these results; with fixed SST, there is no guarantee that the relationship between surface evaporation and SST is correct and so the behavior of the surface flux feedback should not be taken as general.

3.2 Longwave Radiation

Mechanism denial experiments have shown that the longwave radiative feedback is essential for non-rotating aggregation to occur (Bretherton et al. 2005; Muller and Held 2012; Wing 2014; Holloway and Woolnough 2016). Consistent with this, interactive radiation drastically increases the parameter range that supports multiple equilibria (the analog to self-aggregation) in CRM simulations employing the weak temperature gradient approximation, although there are some limited conditions under which multiple equilibria can exist with fixed radiation (Sessions et al. 2016). There is also broad agreement that the formation of one or several dry areas that are driven by enhanced longwave radiative cooling, termed “radiatively driven cold pool” by Coppin and Bony (2015), is important in driving non-rotating aggregation.

Longwave radiative feedbacks also contribute substantially to spontaneous tropical cyclone genesis in simulations of rotating RCE, but are not strictly necessary for it to occur (Wing et al. 2016).

Enhanced longwave radiative cooling in the dry regions triggers aggregation in two ways: the direct diabatic effect, where the enhanced cooling relative to the moist regions decreases the MSE in the dry regions; and the indirect effect mediated by a circulation, where the enhanced longwave cooling in the dry regions drives a shallow circulation between the dry and moist regions and this circulation transports MSE upgradient. Both effects act to suppress convection in the dry regions and enhance convection in the moist regions. Note that there is no guarantee that there will be enhanced longwave cooling in the drier regions; this depends on temperature (Emanuel et al. 2014), clouds, and the vertical structure of the moisture perturbation (Beucler and Cronin 2016). In some circumstances, the opposite may occur (perhaps at cold temperatures), which would yield a negative feedback on aggregation (Emanuel et al. 2014; Wing and Cronin 2016). There is also no guarantee that the circulation driven by radiative heating anomalies transports moist static energy upgradient (this depends on the vertical structure of the heating anomalies (Muller and Bony 2015)).

In simulations that aggregate, the direct diabatic effect of the longwave feedback in the dry regions at the beginning of the aggregation process, as diagnosed with the MSE variance budget, is large and positive and results from both clear sky and cloud effects (Wing and Emanuel 2014; Holloway and Woolnough 2016; Wing and Cronin 2016). However, while the direct longwave feedback is important at amplifying the MSE anomalies in the early stages of aggregation, as aggregation proceeds, it switches to become a negative feedback in the dry regions, at least at temperatures near current tropical SSTs (Wing and Emanuel 2014). The partitioning of the enhanced longwave cooling in the dry regions between clear sky and cloud effects is sensitive to the choice of radiation scheme (Wing and Cronin 2016), and it would not be surprising if this was also sensitive to the cloud microphysics or, in the case of GCMs, the cloud parameterization. Cloud amounts and hence cloud radiative effects can also be sensitive to resolution and domain size (Muller and Held 2012; Muller and Bony 2015).

The shallow radiatively driven circulation is largely induced by strong longwave cooling from low-level clouds in the dry region (Muller and Held 2012; Muller and Bony 2015; Coppin and Bony 2015; Holloway and Woolnough 2016). This low-level cooling in the dry region yields low-level subsidence and outflow from dry to moist regions near the surface where the MSE is large. This results in a divergence of moist static energy out of the dry region, causing upgradient MSE transport that reinforces MSE gradients and further drives aggregation. The role of advective processes in aggregation is discussed further below.

3.2.1 Sensitivity to SST

In the GCM simulations of Coppin and Bony (2015), the enhanced low-level cooling in the dry region and its associated circulation were found to be most efficient in driving aggregation at temperatures near current tropical SSTs and colder. On the other hand, instability driven by the direct clear-sky longwave feedback is favored by higher temperatures, because the rapid increase in water vapor concentration with temperature causes the lower troposphere to be very opaque in the longwave at high temperature (Emanuel et al. 2014). This dependence was interpreted as the cause of the temperature dependence of non-rotating aggregation in the square simulations of Wing and Emanuel (2014), but subsequent results have cast doubt on that conclusion. Although the direct longwave feedback is initially large and negative in the cold (SST $<$ 295 K) simulations of Wing and Cronin (2016), this is not sufficient to prevent aggregation and, moreover, the negative longwave feedback is nearly entirely a result of clouds, not clear-sky processes (the clear-sky longwave feedback is near zero). Wing and Cronin (2016) hypothesized that this is due to the fact that a low-temperature atmosphere is so optically thin that the presence of clouds (in the moist regions) would actually increase the longwave cooling of the atmosphere by increasing the number of longwave emitters. We note, though, that the initial negative longwave cloud feedback in cold simulations does not persist; after a few days, the longwave cloud feedback is positive (Wing and Cronin 2016; Holloway and Woolnough 2016).

Nearly all simulations of self-aggregation have used fixed sea-surface temperature, but a few studies that have employed a slab ocean have found that coupling between the SST and the net surface energy may disrupt self-aggregation or delay its onset, or even prevent it if the slab is thin enough (Bretherton et al. 2005; Reed et al. 2015; Hohenegger and Stevens 2016). Hohenegger and Stevens (2016) found that, in a coupled simulation, SST gradients develop which tend to oppose the development of the radiatively driven low-level circulation, therefore delaying self-aggregation. However, air-sea coupling could also allow other instabilities to be realized (Beucler and Emanuel 2016); more work is needed to fully understand the behavior of self-aggregation with an interactive surface.

3.3 Shortwave Radiation

Shortwave feedbacks can contribute to self-aggregation, but do not appear to be essential for it to occur. The direct, diabatic effect is positive as measured by its contribution to the MSE variance budget (only accounting for diabatic warming/cooling in moist/dry regions), although its magnitude is smaller than the longwave and surface flux terms at current temperatures (Wing and Emanuel 2014; Holloway and Woolnough 2016; Wing and Cronin 2016). This direct positive shortwave feedback is due to a reduction in shortwave heating from clear-sky atmospheric absorption in dry regions. The shortwave cloud feedback can be either positive or negative; it is generally negative in the moist regions after convection has aggregated, where deep clouds reflect shortwave radiation before it can penetrate into the column and be absorbed.

In sensitivity experiments including both direct (diabatic cooling/heating) and indirect (dynamic response to this diabatic forcing) effects, shortwave radiative feedbacks slightly oppose aggregation (Muller and Held 2012). This was clarified in Holloway and Woolnough (2016) as resulting from the positive anomalous heating at high levels by high clouds in the moist regions, favoring upward motion and yielding higher MSE divergence at high levels from the moist region. This transports MSE down-gradient and damps the convective aggregation. Overall, the impact of shortwave radiation is controlled by this MSE transport, at least in the simulations of Muller and Held (2012) and Holloway and Woolnough (2016).

3.3.1 Sensitivity to SST

The direct shortwave feedback was found to be much stronger at low (SST < 295 K) temperatures in the channel CRM simulations of Wing and Cronin (2016), and, along with the surface flux feedback, is the dominant initial driver of aggregation in those low temperature simulations. They found that the positive shortwave feedback at low temperatures was a result of clouds. They proposed several hypotheses for a positive shortwave cloud feedback, including direct shortwave absorption by cloud water and ice, reflected shortwave by low clouds in the moist regions (back to the atmosphere, where it could be absorbed by water vapor), and increased atmospheric absorption because of a higher fraction of diffuse radiation in cloud regions. It is not known which, if any, of these effects dominates, or why they should be stronger at lower temperature. Holloway and Woolnough (2016) found that the clear-sky shortwave feedback was smaller than but of comparable magnitude to the total shortwave feedback in a simulation at 290 K. This further indicates that the relative importance of clouds and clear-sky process to radiative feedbacks depends on the model and radiation package used.

3.4 Advective Processes

As alluded to in Sect. 3.2, advective processes may also contribute to self-aggregation. Bretherton et al. (2005) first showed that upgradient transport of MSE by the circulation consistent with a negative gross moist stability occurred during self-aggregation, as diagnosed from the MSE budget and visualized with a moisture-sorted streamfunction. Muller and Held (2012) and Muller and Bony (2015) further emphasized the importance of upgradient advection and specified that strong radiative cooling at the top of low clouds in the dry region was responsible for driving a shallow circulation that transported MSE upgradient (an indirect effect of radiation on aggregation). This is shown in Fig. 6, which displays the moisture-sorted stream function introduced by Bretherton et al. (2005), radiative cooling rates, MSE, and clouds. The bottom two panels show a simulation without low clouds and the resulting difference in the circulation. The total contribution of all advection processes, integrated over the entire column, as quantified with the MSE variance budget is a positive feedback during the intermediate stages of aggregation, but a negative feedback during other times (Wing and Emanuel 2014; Holloway and Woolnough 2016). Using the same metric, Wing and Cronin (2016) found that, in an elongated channel geometry, advective processes always damped the MSE variance tendency, exporting MSE from the moist regions.

However, even when the total column-integrated advective feedback across the domain, as expressed by the MSE variance budget, is negative, there could still be local upgradient transport (Coppin and Bony 2015), and the shallow component of the circulation could still be acting to further aggregation. In addition, this metric includes all advective processes, so it is possible that the indirect effect of radiative anomalies on the circulation could have a positive influence on aggregation, but is counteracted by other processes. Indeed, Holloway and Woolnough (2016) found that a low-level circulation did appear to transport MSE from drier to moister regions, but that this circulation was mostly balanced by other advective effects of the opposite sign and was forced primarily by horizontal anomalies of convective heating (leading to low-level upward motion in the moist region), rather than radiation. Note that Holloway and Woolnough (2016) used the weak temperature gradient approximation to diagnose circulation components caused by different diabatic processes,

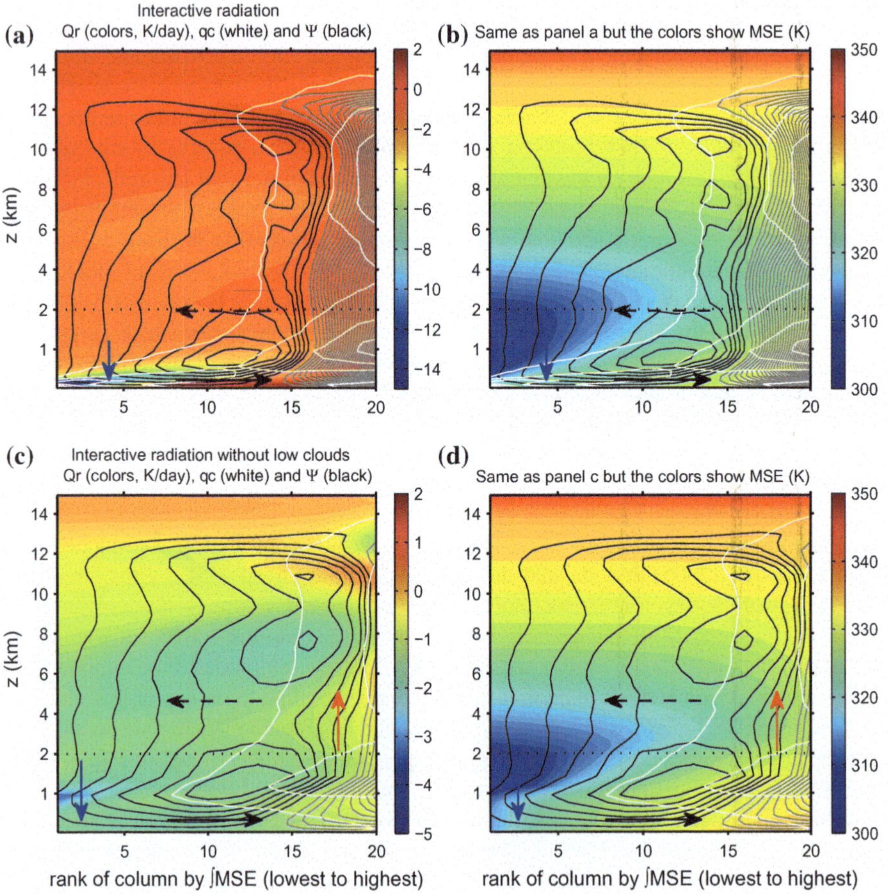

Fig. 6 Radiative cooling rates (*colors* in **a** and **c**), moist static energy (*colors* in **b** and **d**), cloud water content (liquid + ice, white contours every 5×10^{-2} g/kg, starting at 5×10^{-3} g/kg), and stream function (*black contours* for counterclockwise, *gray contours* otherwise, every 8×10^{-3} kg m^{-2} s^{-1}), averaged over the last 20 days of aggregated RCE simulations, plotted as a function of height z and vertically integrated MSE. Note the stretched vertical coordinate z below 2 km. **a**, **b** Simulations with fully interactive radiation; **c**, **d** similar simulation but without the low-cloud radiative effects. The *arrows* schematically represent the subsidence generated by the radiative cooling (*blue*) and rising motion by the warming (*red*), as well as the low-level (*solid black*) and midlevel (*dashed black*) flows induced. Reprinted from Muller and Bony (2015). ©2015. American Geophysical Union. All Rights Reserved

and so they neglected the potential effects of radiative cooling within the boundary layer on the circulation.

Overall, it is clear that advective processes contribute to non-rotating self-aggregation, but there remains some disagreement in the literature as to whether they trigger aggregation on their own or amplify it once direct diabatic feedbacks have started the process.

3.5 Moisture Feedbacks

Moisture feedbacks, which result from the interaction between convection and humidity, are known to organize convection (Tompkins 2001; Grabowski and Moncrieff 2004; Mapes and Neale 2011). Perhaps surprisingly, it was recently found that those feedbacks could lead to the full convective aggregated state, even in the absence of radiative feedbacks (Muller and Bony 2015; Holloway and Woolnough 2016). In that case, the aggregation process is different from the radiatively driven dry cold pool expansion discussed earlier. Instead, it develops similarly to the coarsening process described in the theoretical model of Craig and Mack (2013), with moist areas growing and merging or dying out with little horizontal drift of moist and dry regions. This occurs when the evaporation of rain is artificially suppressed, hence when evaporation-driven downdrafts and cold pools below clouds are weak. This implies that the rain falls without evaporation, which is not realistic in standard conditions, but may occur when the boundary layer is nearly saturated and the precipitation efficiency approaches 100%.

The positive moisture feedback is one in which more moisture favors convection, which in turn yields more moisture. The details of the physical process are still unclear, though several processes have been proposed. First, convection is favored where the boundary layer is anomalously moist (hence boundary layer parcels are more buoyant). Without downdrafts advecting dry air into the boundary layer, it remains moist and the upward motion remains above the boundary layer moisture anomaly. Second, the absence of cold pools in this case may also be important (both downdrafts and cold pools are absent when the evaporation of rain is suppressed), consistent with evidence from Jeevanjee and Romps (2013) that cold pools actually slow down the aggregation process by increasing low-level mixing between moist and dry regions. Third, the moisture feedback could be due to entrainment, since a parcel ascending in a moister environment will be less cooled by entrainment, leading to a larger parcel buoyancy and stronger convection (Tompkins 2001; Holloway and Neelin 2009; Mapes and Neale 2011). In a version of the simple model of Emanuel et al. (2014), the sensitivity of convection to free tropospheric water vapor can boost the otherwise radiatively driven instability. Emanuel et al. (2014) also found that aggregation is favored by increased precipitation efficiency; in fact, when the lower troposphere is opaque in the infrared (i.e., at high temperature) and the precipitation efficiency is unity, their two-layer model is always unstable.

3.6 Triggering Versus Maintenance

Several studies have now confirmed that some feedbacks which are not sufficient to trigger self-aggregation from homogeneous conditions may still be able to maintain aggregation once it is established (Khairoutdinov and Emanuel 2010; Muller and Held 2012; Holloway and Woolnough 2016). This is consistent with the feedback analysis from the MSE variance budget, which shows a strong time evolution of the leading feedback throughout the aggregation process (Wing and Emanuel 2014; Wing and Cronin 2016). The strongest positive feedbacks are typically found in the dry region at early times, while at later times strong positive feedbacks are found in the moist region.

Although strong longwave cooling in the dry regions, at least partially due to low clouds, was found to be crucial for the onset of aggregation, for maintenance, low-cloud longwave cooling is not necessary (Muller and Held 2012; Muller and Bony 2015). Instead, high-level clouds in the moist regions and clear-sky longwave cooling in the dry regions can maintain aggregation. The direct diabatic effect of high-level clouds is a strong longwave positive feedback in the very moistest regions where all the deep clouds are concentrated, primarily because the column longwave cooling is strongly reduced by the longwave opacity and low temperature of high clouds. This is the strongest positive feedback that maintains the high-MSE region during the mature phase of self-aggregation (Wing and Emanuel 2014).

Surface flux feedbacks are not sufficient to maintain aggregation (Holloway and Woolnough 2016), at least not at current climate temperatures. While the surface flux feedback is positive during the early stages of aggregation, later in the evolution of aggregation, as the boundary layer in the dry regions gets drier, the total surface flux feedback becomes negative (Wing and Emanuel 2014). This is not the case in simulations of rotating RCE, in which the surface flux feedback remains positive throughout and in fact dominates over the radiative feedbacks once a broad vortex has formed (Wing et al. 2016).

4 Importance of Self-Aggregation

Self-aggregation of moist convection represents an important phase transition in moist convective systems, at least those that have been modeled in cloud system permitting models and in aquaplanet GCMs. In some simulations (e.g., Wing and Emanuel 2014), the phase transition is discrete, occurring above some threshold temperature, while in others (e.g., Bony et al. 2016) it is gradual. Either way, the transition is accompanied by a substantial drying of the free troposphere (Bretherton et al. 2005), an effect which, if the surface temperature was allowed to vary, would cool the system by reducing the greenhouse effect of water vapor. Khairoutdinov and Emanuel (2010) proposed that this drying, combined with the temperature dependence of self-aggregation, could strongly regulate tropical climate. In a simple model, they proposed that this feedback would result in a self-organized critical state in which the system is attracted to the critical temperature for aggregation. The general idea that aggregation can act as a kind of thermostat was extended to the rotating case by Khairoutdinov and Emanuel (2013) and has been described as a kind of "iris" effect by Mauritsen and Stevens (2015). The temperature dependence of self-aggregation remains uncertain, however, as it can occur at SSTs far below current tropical values (e.g., Wing and Cronin 2016) and it is not obvious how or whether the degree of aggregation depends on temperature in those or other simulations.

When aggregation takes the form of tropical cyclones, an additional set of feedbacks comes into play, involving turbulent mixing of the upper ocean (e.g., Bender et al. 1993). The mixing cools the surface waters and warms deeper waters, conserving the ocean column enthalpy. But the surface cold wakes recover over a period of weeks, and this represents a net warming of the column. So, ironically, although tropical cyclones operate by extracting heat from the ocean, their net effect, after a few weeks, is to transport enthalpy from the atmosphere to the ocean. This may have effects on ocean circulation (Emanuel 2001), although the magnitude of this effect is disputed (Jansen and Ferrari

2009; Jansen et al. 2010). Mixing of nutrients and dissolved CO_2 to the surface may also affect marine biology and the carbon cycle (Lin et al. 2003).

Whether and to what degree all these feedbacks operate in the natural world remains an open question, although attempts to bridge the gap between self-aggregation in idealized simulations and real organized convection are beginning to be made. For example, Bretherton and Khairoutdinov (2015) investigated feedbacks related to self-aggregation in near-global aquaplanet cloud-resolving simulations of realistic tropical variability, finding that radiative feedbacks amplify humidity variance at all scales, consistent with idealized CRM simulations. Vertical shear of the large-scale horizontal wind is known to be destructive to the formation of tropical cyclones and, given what we know about the physics of non-rotating self-aggregation, it seems likely that shear would inhibit this as well. The modeling work described here leaves little doubt that the character and perhaps even the existence of self-aggregation depends on how clouds, radiation, convection, and the boundary layer are modeled. This casts into some doubt whether current climate models can simulate aggregation or, if they do, whether it is simulated accurately. Given that aggregation physics may be important for such phenomena as tropical cyclones (Bretherton et al. 2005; Nolan et al. 2007; Davis 2015; Wing et al. 2016) and the Madden-Julian Oscillation (Bretherton et al. 2005; Arnold and Randall 2015), the sensitivity to physics may help explain why such phenomena have been notoriously difficult to simulate with global models. If aggregation does indeed have an important negative feedback on climate change, it is not clear how well this is handled by current GCMs.

Precipitation efficiency is much higher in aggregated convection, because rain falls through humid air and loses less mass to evaporation. Evaporation tends to concentrate heavier isotopes of hydrogen and oxygen in the raindrops. Thus, we expect to find smaller ratios of the heavier to lighter isotopes in rain from aggregated convection. This is a well-known signal of tropical cyclone rainfall (Lawrence and Gedzelman 1996) and may provide a proxy for aggregation that could be used as an aggregation metric by measuring the isotopic composition of rain. Since this composition is recorded in, for example, tree rings (Miller et al. 2006) and cave deposits (Frappier et al. 2007), there is some hope that one could detect past variations in aggregation in past climates. This might help test the hypothesis that aggregation of moist convection serves as a brake on tropical climate change.

5 Conclusions

5.1 What Aspects of Self-Aggregation do Modeling Studies Agree on?

In the 20+ years since self-aggregation was first described by Held et al. (1993), a growing body of literature has investigated its characteristics, mechanisms, and impacts. In particular, a great deal of progress has been made in the last $\sim$5 years, as there has been a resurgence of interest in radiative–convective equilibrium as an idealization of the tropical atmosphere which, despite its simplicity, exhibits a rich spectrum of behavior that is yet to be completely understood. Several aspects of non-rotating self-aggregation have emerged as robust across these modeling studies; these common features are noted here.

1. Moist static energy variance is dominated by the variance in humidity above the boundary layer, a consequence of the maintenance of weak temperature gradients in the tropical atmosphere.

2. Convection preferentially occurs in humid, high moist static energy regions.
3. As convection aggregates, there is an increase in humidity variance and, in most cases, not only do the dry regions become drier, but the moist regions also become moister.
4. Self-aggregation is not merely a spatial reorganization of convection, it entails large changes to the domain-mean climate. Most dramatically, there is a decrease in the domain-mean humidity.
5. Feedbacks between longwave radiation and water vapor and/or clouds are essential for triggering and maintaining aggregation.
6. Surface flux feedbacks favor the development of aggregation. In the rotating case, surface flux feedbacks dominate.
7. The amplification and expansion of dry regions, in which convection is suppressed, is important in the triggering of self-aggregation.
8. The self-aggregated state exhibits strong hysteresis.

5.2 What Remains Uncertain?

Although the fundamentals of self-aggregation have been established, there are many details that remain uncertain. In particular, there is disagreement in the literature regarding the following issues:

1. The relative importance of cloud versus clear-sky radiative processes. This is likely dependent on the treatment of radiative transfer and cloud microphysics, and the fact that shallow convection is not well represented at typical CRM resolutions.
2. The relative contributions of the direct (diabatic) and indirect (circulation mediated) effects of radiative forcing on the growth of moist static energy anomalies and evolution of self-aggregation.
3. The role of advective processes. Is advective transport of MSE by the circulation essential for triggering self-aggregation, or does it only contribute after diabatic processes have started the process?
4. The temperature dependence of self-aggregation. Some studies find it to be favored by high temperatures, while others find that it occurs across a wide range of temperatures including those much colder than current tropical SSTs.

5.3 What Could be Explored More?

In addition to reconciling the disagreements between studies noted above, there are many aspects of self-aggregation that need to be explored further to achieve a complete understanding of its physics and importance for climate. Several of them are noted here:

1. How does self-aggregation operate when subjected to mean winds and/or vertical wind shear? Does wind shear affect the initiation and maintenance of aggregation differently? Does unidirectional shear simply change the form of aggregation to be more squall line-like, or does it prevent it?
2. How are the mechanisms of self-aggregation altered when the sea-surface temperature is interactive (i.e., calculated from surface energy balance over a slab ocean) versus fixed?
3. Does self-aggregation occur over land surfaces? If it does, how are its behavior and dependencies altered?
4. What controls the spatial scale of self-aggregation?

5. How and why does the degree of aggregation depend on temperature?
6. How does self-aggregation impact climate and climate sensitivity?
7. What is the sensitivity of self-aggregation to boundary layer processes? Most CRM simulations of self-aggregation are too coarse to fully resolve boundary layer turbulence yet lack a boundary layer parameterization. Does this have a fundamental effect on the aggregation of convection?
8. What is the sensitivity of self-aggregation to the dynamical model? A model intercomparison study in which the simulation design and model configuration is controlled would enable a better understanding of the robustness of self-aggregation. "RCEMIP," a recently proposed model intercomparison of radiative–convective equilibrium involving both cloud-resolving models and GCMs with convective parameterizations, may be able to answer this question.
9. How does the self-aggregation found in idealized simulations of radiative–convective equilibrium relate to organized convection in the real world? For what observed convective phenomenon is the self-aggregation of convection in RCE the best simple starting point for understanding? Which aspects of self-aggregation are found in nature, and which are unrealistic? An overview of observational work on self-aggregation and ways forward in this area is presented in Holloway et al. (2017).

5.4 Synthesis

Self-aggregation of moist convection represents a new frontier in meteorology and climate, not simply because a new phenomenon has been added to the panoply of atmospheric processes, but because it also represents a novel intellectual endeavor, breaking the classical stove pipes of, e.g., dynamics vs. radiation physics vs. cloud microphysics. Rapid progress is being made largely by a new generation of atmospheric scientists who are well versed in traditional dynamics, convective and cloud physics, thermodynamics, and radiative transfer.

The novelty of self-aggregation is reflected by the many remaining unanswered questions about its character, causes and effects. It is clear that interactions between longwave radiation and water vapor and/or clouds are critical: non-rotating aggregation does not occur when they are omitted. Beyond this, the field is in play, with the relative roles of surface fluxes, rain evaporation, cloud versus water vapor interactions with radiation, wind shear, convective sensitivity to free atmosphere water vapor, and the effects of an interactive surface yet to be firmly characterized and understood. The sensitivity of simulated aggregation not only to model physics but to the size and shape of the numerical domain and resolution remains a source of concern about whether we have even robustly characterized and simulated the phenomenon. While aggregation has been observed in models (e.g., global models) in which moist convection is parameterized, it is not yet clear whether such models simulate aggregation with any real fidelity. The ability to simulate self-aggregation using models with parameterized convection and clouds will no doubt become an important test of the quality of such schemes.

Understanding self-aggregation may hold the key to solving a number of obstinate problems in meteorology and climate. There is, for example, growing optimism that understanding the interplay among radiation, surface fluxes, clouds, and water vapor may lead to robust accounts of the Madden Julian oscillation and tropical cyclogenesis, two long-standing problems in atmospheric science. Indeed, the difficulty of modeling these phenomena may be owing in part to the challenges of simulating them using representations of clouds and convection that were not designed or tested with self-aggregation in mind. Perhaps most exciting is the prospect that understanding self-aggregation may lead to an improved understanding of climate. The strong hysteresis observed in many simulations of aggregation—once a cluster is formed it tends to be robust to changing environmental conditions—points to the possibility of intransitive or almost intransitive behavior of tropical climate. The strong drying that accompanies aggregation, by cooling the system, may act as a kind of thermostat, if indeed the existence or degree of aggregation depends on temperature. Whether or how well this regulation is simulated in current climate models depends on how well such models can simulate aggregation, given the imperfections of their convection and cloud parameterizations.

Clearly, there is much exciting work to be done on aggregation of moist convection.

Acknowledgements This paper arises from the International Space Science Institute (ISSI) workshop on "Shallow clouds and water vapor, circulation and climate sensitivity." AAW acknowledges support from an National Science Foundation AGS Postdoctoral Research Fellowship under Award 1433251.

References

Abbot D (2014) Resolved snowball Earth clouds. J Climate 27:4391–4402. doi:10.1175/JCLI-D-13-00738.1

Arnold NP, Randall DA (2015) Global-scale convective aggregation: implications for the Madden-Julian oscillation. J Adv Model Earth Syst. doi:10.1002/2015MS000498

Becker T, Stevens B (2014) Climate and climate sensitivity to changing CO_2 on an idealized land planet. J Adv Model Earth Syst 6:1205–1223. doi:10.1002/2014MS000369

Bender MA, Ginis I, Kurihara YY (1993) Numerical simulations of tropical cyclone-ocean interaction with a high resolution coupled model. J Geophys Res 98:23245–23263

Beucler T, Cronin T (2016) Moisture-radiative cooling instability. J Adv Model Earth Syst. doi:10.1002/2016MS000763

Beucler T, Emanuel KA (2016) Instabilities of radiative convective equilibrium with an interactive surface. In: Preprints 32nd Conference on hurricanes and tropical meteorology, San Juan, PR. American Meteorological Society, Tucson, AZ

Bony S, Stevens B, Coppin D, Becker T, Reed KA, Voigt A, Medeiros B (2016) Thermodynamic control of anvil cloud amount. Proc Nat Acad Sci 113(32):8927–8932. doi:10.1073/pnas.161472113

Boos WR, Fedorov AV, Muir L (2015) Convective self-aggregation and tropical cyclogenesis under the hypohydrostatic rescaling. J Atmos Sci. doi:10.1175/JAS-D-15-0049.1

Bretherton CS, Khairoutdinov MF (2015) Convective self-aggregation feedbacks in near-global cloud-resolving simulations of an aquaplanet. J Adv Model Earth Syst 7(4):1765–1787. doi:10.1002/2015MS000499

Bretherton CS, Peters ME, Back LE (2004) Relationships between water vapor path and precipitation over the tropical oceans. J Climate 17:1517–1528

Bretherton CS, Blossey PN, Khairoutdinov M (2005) An energy-balance analysis of deep convective self-aggregation above uniform SST. J Atmos Sci 62:4237–4292. doi:10.1175/JAS3614.1

Coppin D, Bony S (2015) Physical mechanisms controlling the initiation of convective self-aggregation in a general circulation model. J Adv Model Earth Syst 7(4):2060–2078. doi:10.1002/2015MS000571

Craig GC, Mack JM (2013) A coarsening model for self-organization of tropical convection. J Geophys Res Atmos 118:8761–8769. doi:10.1002/jgrd.50674

Daleu CL, Plant RS, Woolnough SJ, Sessions S, Herman MJ, Sobel A, Wang S, Kim D, Cheng A, Bellon G, Peyrille P, Ferry F, Siebesma P, van Ulft L (2015) Intercomparison of methods of coupling between convection and large-scale circulation: I. Comparison over uniform surface conditions. J Adv Model Earth Syst. doi:10.1002/2015MS000468

Davis CA (2015) The formation of moist vortices and tropical cyclones in idealized simulations. J Atmos Sci 72:3499–3516. doi:10.1175/JAS-D-15-0027.1

Emanuel K, Wing AA, Vincent EM (2014) Radiative-convective instability. J Adv Model Earth Syst 6:75–90. doi:10.1002/2013MS000270

Emanuel KA (2001) The contribution of tropical cyclones to the oceans' meridional heat transport. J Geophys Res 106:14771–714782

Frappier AB, Sahagian D, Carpenter SJ, Gonzlez LA, Frappier BR (2007) Stalagmite stable isotope record of recent tropical cyclone events. Geology 35:111–114
Grabowski W, Moncrieff M (2001) Large-scale organization of tropical convection in two-dimensional explicit numerical simulations. Q J R Meteorol Soc 127:445–468
Grabowski W, Moncrieff M (2002) Large-scale organization of tropical convection in two-dimensional explicit numerical simulations: effects of interactive radiation. Q J R Meteorl Soc 128:2349–2375. doi:10.1256/qj.01.104
Grabowski WW, Moncrieff MW (2004) Moisture-convection feedback in the tropics. Q J R Meteorol Soc 130:3081–3104. doi:10.1256/qj.03.135
Held IM, Soden BJ (2006) Robust responses of the hydrological cycle to global warming. J Climate 19:5686–5699. doi:10.1175/JCLI3990.1
Held IM, Zhao M (2008) Horizontally homogeneous rotating radiative-convective equilibrium at GCM resolution. J Atmos Sci 65:2003–2013. doi:10.1175/2007JAS2604.1
Held IM, Hemler RS, Ramaswamy V (1993) Radiative–convective equilibrium with explicit two-dimensional moist convection. J Atmos Sci 50:3909–3927
Held IM, Zhao M, Wyman B (2007) Dynamic radiative–convective equilibria using GCM column physics. J Atmos Sci 64:228–238. doi:10.1175/JAS3825.11
Hohenegger C, Stevens B (2016) Coupled radiative convective equilibrium simulations with explicit and parameterized convection. J Adv Model Earth Syst. doi:10.1002/2016MS000666
Holloway CE, Neelin JD (2009) Moisture vertical structure, column water vapor, and tropical deep convection. J Atmos Sci 66:1665–1683
Holloway CE, Woolnough SJ (2016) The sensitivity of convective aggregation to diabatic processes in idealized radiative–convective equilibrium simulations. J Adv Model Earth Syst 8(1):166–195. doi:10.1002/2015MS000511
Holloway CE, Wing AA, Bony S, Muller C, Masunaga H, L'Ecuyer TS, Turner DD, Zuidema P (2017) Observing convective aggregation. Surv Geophys (submitted)
Jansen M, Ferrari R (2009) Impact of the latitudinal distribution of tropical cyclones on ocean heat transport. Geophys Res Lett 36(L06):604. doi:10.1029/2008GL036796
Jansen MF, Ferrari R, Mooring TA (2010) Seasonal versus permanent thermocline warming by tropical cyclones. Geophys Res Lett 37(L03):602. doi:10.1029/2009GL041808
Jeevanjee N, Romps DM (2013) Convective self-aggregation, cold pools, and domain size. Geophys Res Lett 40:1–5. doi:10.1002/grl/50204
Khairoutdinov MF, Emanuel K (2013) Rotating radiative–convective equilibrium simulated by a cloud-resolving model. J Adv Model Earth Syst. doi:10.1002/2013MS000253
Khairoutdinov MF, Emanuel KA (2010) Aggregation of convection and the regulation of tropical climate. In: Preprints. 29th Conference on hurricanes and tropical meteorology. American Meteorological Society, Tucson, AZ
Knutson TR, Manabe S (1995) Time-mean response over the tropical pacific to increased CO_2 in a coupled ocean-atmosphere model. J Climate 8:2181–2199
Lawrence JR, Gedzelman SD (1996) Low stable isotope ratios of tropical cyclone rains. Geophys Res Lett 23:527–530
Lin I et al (2003) New evidence for enhanced ocean primary production triggered by tropical cyclone. Geophys Res Lett. doi:10.1029/2003GL017,141
Manabe S, Strickler RF (1964) Thermal equilibrium of the atmosphere with a convective adjustment. J Atmos Sci 21:361–385
Mapes B, Neale R (2011) Parameterizing convective organization to escape the entrainment dilemma. J Adv Model Earth Syst 3(M06):004. doi:10.1029/2011MS000042
Mauritsen T, Stevens B (2015) Missing iris effect as a possible cause of muted hydrological change and high climate sensitivity in models. Nat Geosci 8:346–351. doi:10.1038/ngeo2414
Merlis TM, Zhou W, Held IM, Zhao M (2016) Surface temperature dependence of tropical cyclone-permitting simulations in a spherical model with uniform thermal forcing. Geophys Res Lett 43:2859–2865. doi:10.1002/2016GL067730
Miller DL, Mora CI, Grissino-Mayer HD, Mock CJ, Uhle ME, Sharp Z (2006) Tree-ring isotope records of tropical cyclone activity. Proc Nat Acad Sci 103:14,294–14,297
Muller C, Bony S (2015) What favors convective aggregation and why? Geophys Res Lett 42:5626–5643. doi:10.1002/2015GL064260
Muller CJ, Held IM (2012) Detailed investigation of the self-aggregation of convection in cloud resovling simulations. J Atmos Sci 69:2551–2565. doi:10.1175/JAS-D-11-0257.1
Nilsson J, Emanuel K (1999) Equilibrium atmospheres of a two-column radiative–convective model. Q J R Meteorol Soc 125:2239–2264

Nolan DS, Rappin ED, Emanuel KE (2007) Tropical cyclogenesis sensitivity to environmental parameters in radiative–convective equilibrium. Q J R Meteorol Soc 133:2085–2107. doi:10.1002/qj.170

Popke D, Stevens B, Voigt A (2013) Climate and climate change in a radiative–convective equilibrium version of ECHAM6. J Adv Model Earth Syst 5:1–14. doi:10.1029/2012MS000191

Posselt D, van den Heever S, Stephens G (2008) Trimodal cloudiness and tropical stable layers in simulations of radiative–convective equilibrium. Geophys Res Lett 35(L08):802. doi:10.1029/2007GL033029

Posselt D, van den Heever S, Stephens G, Igel M (2012) Changes in the interaction between tropical convection, radiation, and the large-scale circulation in a warming environment. J Climate 25:557–571. doi:10.1175/2011JCLI4167.1

Raymond D (2000) The Hadley circulation as a radiative–convective instability. J Atmos Sci 57:1286–1297

Reed K, Medeiros B (2016) A reduced complexity framework to bridge the gap between AGCMs and cloud-resolving models. Geophys Res Lett 43:860–866. doi:10.1002/2015GL066713

Reed KA, Chavas DR (2015) Uniformly rotating global radiative–convective equilibrium in the community atmosphere model, version 5. J Adv Model Earth Syst. doi:10.1002/2015MS000519

Reed KA, Medeiros B, Bacmeister JT, Lauritzen PH (2015) Global radiative–convective equilibrium in the community atmosphere model 5. J Atmos Sci. doi:10.1175/JAS-D-14-0268.1

Renno NO, Emanuel KA, Stone PH (1994) Radiative–convective model with an explicit hydrological cycle. Part i: formulation and sensitivity to model parameters. J Geophys Res. 99:14,429–14,441

Satoh M, Matsuda Y (2009) Statistics on high-cloud areas and their sensitivities to cloud microphysics using single-cloud experiments. J Atmos Sci 66:2659–2677. doi:10.1175/2009JAS2948.1

Satoh M, Aramaki K, Sawada M (2016) Structure of tropical convective systems in aquaplanet experiments: radiative–convective equilibrium versus the Earth-like experiment. SOLA 12:220–224. doi:10.2151/sola.2016-044

Sessions SL, Sugaya S, Raymond DJ, Sobel AH (2010) Multiple equilibria in a cloud-resolving model using the weak temperature gradient approximation. J Geophys Res. doi:10.1029/2009JD013376

Sessions SL, Herman MJ, Sentic S (2015) Convective response to changes in the thermodynamic environment in idealized weak temperature gradient simulations. J Adv Model Earth Syst 7:712–738. doi:10.1002/2015/MS000446

Sessions SL, Sentic S, Herman MJ (2016) The role of radiation in organizing convection in weak temperature gradient simulations. J Adv Model Earth Syst 8:244–271. doi:10.1002/2015MS000587

Shi X, Bretherton CS (2014) Large-scale character of an atmosphere in rotating radiative–convective equilibrium. J Adv Model Earth Syst 6:616–629. doi:10.1002/2014MS000342

Silvers LG, Stevens B, Mauritsen T, Giorgetta M (2016) Radiative convective equilibrium as a framework for studying the interaction between convection and its large-scale environment. J Adv Model Earth Syst. doi:10.1002/2016MS000629

Sobel AH, Bellon G, Bacmeister J (2007) Multiple equilibria in a single-column model of the tropical atmosphere. Geophys Res Lett 34(L22):804. doi:10.1029/2007GL031320

Stephens GL, van den Heever S, Pakula L (2008) Radiative–convective feedbacks in idealized states of radiative–convective equilibrium. J Atmos Sci 65:3899–3916. doi:10.1175/2008JAS2524.1

Su H, Bretherton CS, Chen SS (2000) Self-aggregation and large-scale control of tropical deep convection: a modeling study. J Atmos Sci 57:1797–1816

Tobin I, Bony S, Roca R (2012) Observational evidence for relationships between the degree of aggregation of deep convection, water vapor, surface fluxes, and radiation. J Climate 25:6885–6904

Tobin I, Bony S, Holloway CE, Grandpeix JY, Seze G, Coppin D, Woolnough SJ, Roca R (2013) Does convective aggregation need to be represented in cumulus parameterizations? J Adv Model Earth Syst 5:692–703. doi:10.1002/jame.20047

Tompkins A, Craig G (1998) Radiative–convective equilibrium in a three-dimensional cloud-ensemble model. Q J R Meteorol Soc 124:2073–2097

Tompkins AM (2001) Organization of tropical convection in low vertical wind shears: the role of water vapor. J Atmos Sci 58:529–545

Wing AA (2014) Physical mechanisms controlling self-aggregation of convection in idealized numerical modeling simulations. PhD thesis, MIT, Cambridge, 146 pp

Wing AA, Cronin TW (2016) Self-aggregation of convection in long channel geometry. Q J R Meteorol Soc 142:1–15. doi:10.1002/qj.2628

Wing AA, Emanuel KA (2014) Physical mechanisms controlling self-aggregation of convection in idealized numerical modeling simulations. J Adv Model Earth Syst 6:59–74. doi:10.1002/2013MS000269

Wing AA, Camargo SJ, Sobel AH (2016) Role of radiative–convective feedbacks in spontaneous tropical cyclogenesis in idealized numerical simulations. J Atmos Sci 73:2633–2642. doi:10.1175/JAS-D-15-0380.1

Zhou W, Held IM, Garner ST (2014) Parameter study of tropical cyclones in rotating radiative–convective equilibrium with column physics and resolution of a 25 km GCM. J Atmos Sci 71:1058–1068. doi:10.1175/JAS-D-13-0190.1

Zhou W, Held I, Garner S (2017) Tropical cyclones in rotating radiative–convective equilibrium with coupled SST. J Atmos Sci. doi:10.1175/JAS-D-16-0195.1

Surv Geophys (2017) 38:1199–1236
https://doi.org/10.1007/s10712-017-9419-1

Observing Convective Aggregation

Christopher E. Holloway[1] · **Allison A. Wing**[2,3] · **Sandrine Bony**[4] · **Caroline Muller**[5] · **Hirohiko Masunaga**[6] · **Tristan S. L'Ecuyer**[7] · **David D. Turner**[8] · **Paquita Zuidema**[9]

Received: 23 November 2016 / Accepted: 14 June 2017 / Published online: 28 June 2017

Abstract Convective self-aggregation, the spontaneous organization of initially scattered convection into isolated convective clusters despite spatially homogeneous boundary conditions and forcing, was first recognized and studied in idealized numerical simulations. While there is a rich history of observational work on convective clustering and organization, there have been only a few studies that have analyzed observations to look specifically for processes related to self-aggregation in models. Here we review observational work in both of these categories and motivate the need for more of this work. We acknowledge that self-aggregation may appear to be far-removed from observed convective organization in terms of time scales, initial conditions, initiation processes, and mean state extremes, but we argue that these differences vary greatly across the diverse range of

✉ Christopher E. Holloway
c.e.holloway@reading.ac.uk

1 Department of Meteorology, University of Reading, Reading RG6 6BB, UK

2 Lamont-Doherty Earth Observatory, Columbia University, PO Box 1000, 61 Route 9W, Palisades, NY 10964-1000, USA

3 Department of Earth, Ocean and Atmospheric Science, Florida State University, Mail Code 4520, PO Box 3064520, Tallahassee, FL 32306-4520, USA

4 Sorbonne University, LMD/IPSL, CNRS, Univ Paris 06, mailbox 99, 4 Place Jussieu, 75252 Paris cedex 05, France

5 LMD/IPSL, CNRS, École Normale Supérieure, Paris Sciences Et Lettres, 24 rue Lhomond, 75230 Paris cedex 05, France

6 Institute for Space-Earth Environmental Research, Nagoya University, Furo-cho, Chikusa-ku, Nagoya 464-8601, Japan

7 Department of Atmospheric and Oceanic Sciences, University of Wisconsin-Madison, 1225 West Dayton Street, Madison, WI 53706, USA

8 NOAA/Earth System Research Laboratory, Global Systems Division, 325 Broadway, Boulder, CO 80305-3337, USA

9 Rosenstiel School of Marine and Atmospheric Science, University of Miami, 4600 Rickenbacker Causeway, Miami, FL 33149, USA

model simulations in the literature and that these comparisons are already offering important insights into real tropical phenomena. Some preliminary new findings are presented, including results showing that a self-aggregation simulation with square geometry has too broad distribution of humidity and is too dry in the driest regions when compared with radiosonde records from Nauru, while an elongated channel simulation has realistic representations of atmospheric humidity and its variability. We discuss recent work increasing our understanding of how organized convection and climate change may interact, and how model discrepancies related to this question are prompting interest in observational comparisons. We also propose possible future directions for observational work related to convective aggregation, including novel satellite approaches and a ground-based observational network.

Keywords Self-aggregation · Tropical convection · Convective organization · Climate sensitivity · Cloud feedback

1 Introduction

From the very first studies describing convective self-aggregation (e.g., Held et al. 1993; Tompkins 2001; Bretherton et al. 2005), the spontaneous clustering of convection, cloud, and moisture in idealized numerical simulations of radiative–convective equilibrium (RCE) despite homogeneous initial conditions, boundary conditions, and forcing (cf. Wing et al. 2017), there has been a recurring question: Is this "real"? In other words, is the intriguing clumping behavior representative of actual convective organization in nature, or is it just a model artifact? And, to the extent that the behavior is relevant for understanding real atmospheric convection, what does it tell us about the role of convective organization in weather and climate?

Here we argue that this behavior in models *does* appear to be relevant to real-world convection and climate. Certainly, the study of convective self-aggregation is leading to exciting new insights into processes that allow convection to interact with its environment in models. There are encouraging signs that these processes may operate in nature too, as we discuss below. There are also some aspects of self-aggregation in models that conflict with observations, and many aspects that need more observational study.

This paper is organized as follows. In the remainder of this section, we motivate the study of aggregation as a means of understanding real-world climate and review the literature on observations of organized convection and convective aggregation. Section 2 presents a fairly brief review of processes important for self-aggregation and the maintenance of aggregated convection in idealized simulations, with a focus on aspects of these processes that could be targeted in observational studies. We then discuss observational pathways toward assessing the relevance of the idealized framework for real-world applications, including some new results comparing humidity profiles from radiosondes with humidity profiles from idealized self-aggregation, in Sect. 3. Section 4 provides observational perspectives on the possible interaction between convective aggregation and climate change, while Sect. 5 proposes novel approaches to observing convective aggregation, including ideas for new satellite studies and ground-based networks; this is followed by our conclusions.

Fig. 1 A visible satellite image showing an active Madden–Julian Oscillation (MJO) event on 8 April 2009 and convection organized over a wide range of scales. Image taken from the NERC Satellite Receiving Station, Dundee University, Scotland http://www.sat.dundee.ac.uk/

1.1 Importance of Aggregation

Convective clouds exhibit a very large diversity of spatial organization, ranging from spatially random distributions to coherent structures such as mesoscale cloud clusters, cloud streets, and squall lines up to cloud envelopes of planetary scale (Fig. 1). For many decades, studies of convective organization have been developed by mesoscale meteorologists and weather forecasters, motivated by the wish to understand why convection would organize in one form rather than another, and by the evidence that the organization of convection matters for the prediction of severe weather. Over the last decade, however, the ability to study the organization of convection with models running at increasingly fine resolution over increasingly large domains has led to new perspectives and to a new line of questioning: Does it make any difference *for climate* whether convection organizes in one form or another?

It has long been recognized that convective organization influences the diabatic heating profile of the atmosphere and thus affects the mean large-scale atmospheric circulation (e.g., Hartmann et al. 1984). More recent numerical studies show that the clumping of convection can occur spontaneously even in the absence of external drivers such as inhomogeneous surface boundary conditions or equatorial wave dynamics (e.g., Held et al. 1993; Bretherton et al. 2005; Muller and Held 2012; Wing and Emanuel 2014) and that this behavior, referred to as convective self-aggregation, may be considered as a fundamental instability of radiative–convective equilibrium (Emanuel et al. 2014). Could tropical phenomena such as tropical cyclones or Madden–Julian Oscillation (MJO) events represent manifestations of this self-aggregation behavior at different spatial scales (Khairoutdinov and Emanuel 2010, 2013; Arnold and Randall 2015)? Answering this question would provide new opportunities to understand and to predict these phenomena through completely novel approaches.

Numerical studies of convective aggregation also show that the clumping of convection is associated with changes in the large-scale state, including a drying of the atmosphere, a shrinking of upper-tropospheric clouds, and an enhanced ability of the atmosphere to lose heat to space (e.g., Wing and Emanuel 2014; Wing and Cronin 2016; Holloway and Woolnough 2016; Bony et al. 2016). Self-aggregation in numerical models also exhibits some temperature dependence (Khairoutdinov and Emanuel 2010; Wing and Emanuel 2014; Emanuel et al. 2014; Coppin and Bony 2015; Wing and Cronin 2016). The combination of these different findings implies that changes in convective organization could occur under climate change, potentially affecting the water vapor and cloud feedbacks. These numerical results shed new light on the role that convective aggregation might play in climate (Mapes 2016): Could a sensitivity of convective aggregation to temperature modulate climate sensitivity and hydrological sensitivity (Khairoutdinov and Emanuel 2010; Mauritsen and Stevens 2015; Bony et al. 2015)? In a warmer climate, could it play a role in the intensification of the MJO (Arnold and Randall 2015; Arnold et al. 2015) or in the narrowing of tropical rain belts (Bony et al. 2016)?

Many of these exciting scientific questions primarily stem from numerical investigations. However, numerous studies (e.g., Stephens et al. 2008; Muller and Bony 2015; Wing and Cronin 2016; Holloway and Woolnough 2016; Silvers et al. 2016; Tompkins and Semie 2017) demonstrate that the behavior of convective aggregation in models can be sensitive to aspects of the experimental setup (such as the size of the domain) and/or to the models themselves (e.g., horizontal resolution, the representation of diabatic processes or the parameterization of subgrid-scale mixing).

To move forward, we must therefore expand our study of the aggregation of convection using observations. We must probe links between processes in idealized self-aggregation and observed convective organization and also confront differences between idealized frameworks and the real world. We first present a review of relevant literature below before addressing these topics in the following sections.

1.2 Literature Review: Observational Studies of Convective Organization

There is a rich history of observational work on convective clustering and organization, much of which details the climatology and life cycles of these systems. The primary source of data for this observational work is infrared and visible images from geostationary satellites, dating back to at least Arkin (1979) and encompassing Velasco and Fritsch (1987), Miller and Fritsch (1990), Laing and Fritsch (1993a, b), Machado and Rossow (1993), Mapes and Houze (1993), Laing and Fritsch (1997), Zuidema (2003), and Hennon et al. (2012), but some more recent studies have also used other types of satellite data such as precipitation radar (Nesbitt et al. 2000; Schumacher and Houze 2003; Futyan and Genio 2007; Peters et al. 2009), microwave measurements of column water vapor (CWV) (Mapes et al. 2009), and scatterometer winds (Mapes et al. 2009). While cloud clusters are often identified by searching for large, contiguous cold cloud shields, more advanced techniques search for the combined signature of deep convection and extensive stratiform cloud and precipitation area. For example, higher stratiform rain fractions are associated with organized convection, which can be diagnosed from satellite precipitation radar data (Schumacher and Houze 2003), as are large optical thicknesses and low-cloud top pressures, which can distinguish a particular cloud regime (Tselioudis et al. 2010; Tan et al. 2013).

A significant fraction of the observational work on organized convection has focused on mesoscale convective systems (MCSs), or a subset of them known as mesoscale convective complexes (MCCs), which occur in both the tropics and mid-latitudes. A global climatology of MCCs, which are identified by a large (>10^5 km^2), long-lasting (>6 h), quasi-circular cold cloud shield, was compiled by Laing and Fritsch (1997) based on previous regional studies (Miller and Fritsch 1990; Laing and Fritsch 1993a, b; Velasco and Fritsch 1987).

Other studies have detailed the properties of, more generally, tropical cloud clusters and deep convective systems. This includes studies on the structural characteristics and radiative properties of tropical high cloud systems (Machado and Rossow 1993), the life cycles of deep convective systems (Futyan and Genio 2007; Mapes et al. 2009), the size distribution of cloud clusters (Mapes and Houze 1993; Roca and Ramanathan 2000; Zuidema 2003; Peters et al. 2009), and the spatial and temporal variability in cloud clusters and their efficiency at producing tropical cyclones (Hennon et al. 2012). Studies have also pointed out significant self-similarity between MCSs and convectively coupled equatorial waves (Mapes et al. 2006; Kiladis et al. 2009).

Despite the fact that the occurrence of mesoscale organized convection makes up a small fraction of the total frequency of cloud/precipitation features in the tropics (<6%, Mapes and Houze 1993; Nesbitt et al. 2000; Tan et al. 2013), it contributes a significant proportion of total tropical cloudiness[1] and about half of total tropical precipitation.[2] Tropical cloud clusters therefore may modulate the radiative heating of the surface and atmosphere (e.g., Machado and Rossow 1993) and strongly influence the large-scale circulation, moisture distribution, and hydrological cycle. There is observational evidence that the frequency of organized convection has increased across the tropics over the past ~30 years (Tselioudis et al. 2010) and that most of the regional increases in tropical precipitation over that period are associated with this increase (Tan et al. 2015). In addition to their contribution to tropical cloudiness and precipitation, tropical cloud clusters also play an important role as precursors to tropical cyclones, with globally 6.4% of tropical cloud clusters developing into tropical cyclones each year (Hennon et al. 2012).

Another observational finding which may be relevant to self-aggregation is the evidence that tropical precipitation has properties like those of a critical phenomenon. Peters and Neelin (2006) found that there is a power law increase in precipitation with CWV above a critical CWV value, and a sharp peak in the variance of precipitation at the critical value. Holloway and Neelin (2009) further found that free-tropospheric moisture plays a key role in the transition to deep convection and linked the increase in precipitation with CWV to an increase in the buoyancy of entraining plumes, which relates to the proposed moisture–convection feedback in self-aggregation. Neelin et al. (2009) also noted that the atmosphere is near criticality a larger fraction of the time when it is over warm sea surface temperatures (SSTs). Peters et al. (2009) found that precipitation clusters exhibited scale-

[1] Mapes and Houze (1993) found that half of the very cold cloudiness was contributed by cloud clusters greater than 2×10^4 km^2 in size and half of the moderately cold cloudiness was contributed by cloud clusters greater than 10^5 km^2 in size. Mapes (1993) found that 43% of the total cold cloud coverage in the tropics was associated with large superclusters.

[2] Tan et al. (2013) found that the cloud regime associated with organized convection contributes 45% of total tropical rainfall. Nesbitt et al. (2000) found that precipitation features that include an MCS contribute 38–55% of total rainfall in various regions of the tropics.

free size distributions including much larger clusters near-critical CWV than below it, suggesting a possible link between clustering within the moist convective regions in idealized self-aggregation and near-critical CWV values.

While the literature on observations of tropical cloud clusters is extensive (only a small segment of which was reviewed here), only a few studies have specifically looked for processes related to modeled self-aggregation using observations. The first such paper, Tobin et al. (2012), used geostationary satellite infrared brightness temperature in snapshots of large tropical latitude-longitude boxes (10° × 10°) to categorize observations by their degree of convective organization. To do this, they devised the Simple Convective Aggregation Index (SCAI) as a combined measure of cluster number and inter-cluster distance, with cluster pixels defined as having brightness temperature below 240 K and with larger SCAI corresponding to a less aggregated state. They found that cluster number was statistically sufficient to discriminate between different levels of aggregation, so they often used the number of clusters as a metric for the degree of aggregation, with fewer clusters corresponding to a more aggregated state. By controlling for measures of box-mean convective intensity and large-scale forcing, including rainfall from microwave satellite data, SST from infrared satellite data, and vertical velocity from reanalyses, they could compare atmospheric conditions for varying amounts of convective organization in a way that was analogous to comparing different stages of aggregation in idealized models.

Tobin et al. (2012) found several similarities between their observational analyses and idealized simulations of self-aggregation. Holding large-scale SST and rainfall constant, they found that more aggregated states had a drier free troposphere in the non-convective environment and, consequently, in the domain as a whole. They also found an increase in outgoing longwave radiation (OLR) at the top of the atmosphere (by as much as 30 W m^{-2}) with aggregation (Fig. 2a), mainly because of a reduction of mid-level and upper level cloudiness. These main conclusions, in agreement with all studies of idealized self-aggregation in models, were also supported by a related paper, Tobin et al. (2013), which looked at smaller (3° × 3°) domains using higher-resolution satellite brightness

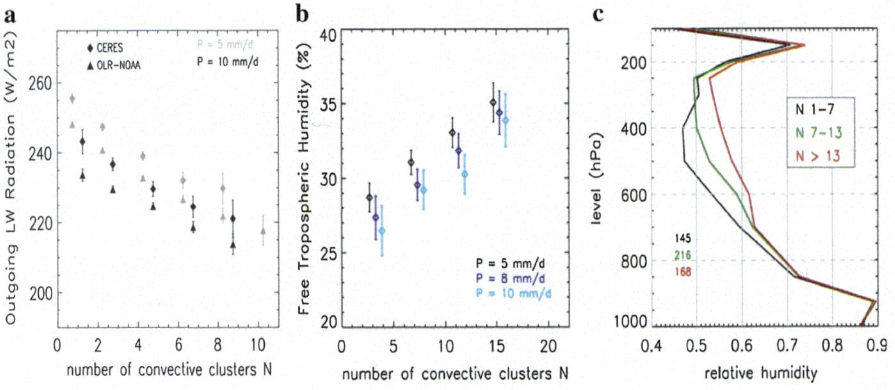

Fig. 2 **a** Composites over many 10° × 10° snapshots of domain-averaged OLR from CERES and NOAA for two different average rain rates for different satellite-derived cluster numbers, with fewer clusters representing more aggregated convection. **b** Similar analysis for 3° × 3° snapshots of domain-averaged free-tropospheric humidity derived from Meteosat Tb in the WV channel, for three different average rain rates. **c** Domain-averaged AIRS relative humidity composited on the same 3° × 3° snapshots as in **b** for three cluster number bins for a precipitation rate of 8 mm day^{-1}. Figures from Tobin et al. (2012) (panel a) and Tobin et al. (2013) (panels b, c)

temperature data (Fig. 2b, c), as well as Stein et al. (2017), which analyzed the vertical cloud structure for different SCAI values using CloudSat–CALIPSO data.

On the other hand, Tobin et al. (2012, 2013) found some results that were inconclusive, mixed, or contradictory when compared with modeling studies. For instance, Tobin et al. (2012) found that surface turbulent heat fluxes increased both inside and outside convective regions when aggregation increased, whereas Tobin et al. (2013) found little sensitivity of these fluxes to aggregation at the smaller scales they investigated (although this discrepancy could be due to limitations in satellite retrievals of surface fluxes). In idealized simulations, surface fluxes generally increase with self-aggregation (e.g., Bretherton et al. 2005; Wing 2014; Holloway and Woolnough 2016), with the increase due to larger wind speeds in general and larger air–sea enthalpy disequilibrium in the dry environment (Wing and Emanuel 2014). (Note that this modest increase in surface fluxes for idealized models is also consistent with slightly larger atmospheric radiative cooling rates and precipitation rates in radiative–convective equilibrium after aggregation has occurred.) Tobin et al. (2012, 2013) also found that the top-of-atmosphere net radiation budget was not significantly affected by aggregation because increased OLR was offset by decreased reflected shortwave radiation. This differs from idealized simulations discussed by Wing and Cronin (2016), in which an increase in low-cloud fraction with aggregation left reflected shortwave largely unchanged, leading to a net loss of radiation at the top of atmosphere for aggregated conditions. Tobin et al. (2013) and Stein et al. (2017) both found evidence for an increase in low-cloud fraction with aggregation, while Tobin et al. (2012) found the opposite, so the models are supported by at least some observational studies regarding low-cloud changes.

In the next section, we briefly review processes found to be important for self-aggregation in models with a focus on links to observed convective organization.

2 Observational Perspectives on Processes Important for Idealized Convective Aggregation

There are longstanding attempts to reconcile the well-observed clumping of tropical convection with simple theory (e.g., Mapes 1993). Randall and Huffman (1980) proposed that clumping occurs when clouds can create an area around themselves that is more favorable for future convection than areas further away. Numerical studies of self-aggregation have identified multiple processes involving convection–moisture–radiation feedbacks that are capable of doing exactly that. The diversity of processes that can lead to convective aggregation may explain why it has been observed by multiple different modeling groups using very different models, from high-resolution cloud-resolving models to global climate models (GCMs) with parameterized convection. Additionally, different feedbacks that lead to aggregation may be excited by different initial conditions.

We will mostly discuss self-aggregation in idealized settings: radiative convective equilibrium (RCE) over constant uniform SST in non-rotating, three-dimensional, doubly periodic square domains, though some rectangular and aquaplanet simulations will occasionally be discussed as well. It is worth noting that self-aggregation has been shown to be robust to the presence of rotation (Bretherton et al. 2005; Khairoutdinov and Emanuel 2013; Bretherton and Khairoutdinov 2015; Davis 2015; Wing et al. 2016), vertical shear (Bretherton et al. 2005), diurnal cycle (Wing and Cronin 2016), two-dimensional or three-dimensional settings (Held et al. 1993; Jeevanjee and Romps 2013), and an interactive ocean mixed layer (Bretherton et al. 2005; Hohenegger and Stevens 2016), and to occur as well in global climate simulations with parameterized convection in aquaplanet non-rotating settings (Coppin and Bony 2015; Popke et al. 2013; Reed et al. 2015).

In this section, we briefly review the various processes leading to the self-aggregation of convection in RCE simulations and the metrics used to quantify them, including the physical processes that lead to aggregation from homogeneous initial conditions as well as those which can maintain convective aggregation once it is established. We focus on those which could be targeted in observations; a more complete review can be found in Wing et al. (2017).

2.1 Metrics to Quantify Feedbacks

Several methods have been proposed to analyze the leading order feedbacks in simulations (and also possibly in observations). They all share the methodology of stratifying the data by vertically integrated moist static energy (MSE). In the tropics, weak temperature gradients imply that horizontal variability of MSE is largely dictated by variability in CWV. Using this methodology, different variables can be moisture-ranked.

Wing and Emanuel (2014) introduced an analysis framework employing a budget for the spatial variance of MSE. Self-aggregation is associated with a very strong increase in MSE variance. The equation for the time evolution of MSE variance allows one to estimate the various contributions to the enhanced MSE variability. The terms of this budget include the horizontal convergence or divergence of MSE, as well as the direct diabatic contributions from radiative and surface fluxes, i.e., whether a heating/moistening diabatic tendency reinforces (positive feedback) or smoothes (negative feedback) MSE gradients. The potential use of observations to calculate the diabatic terms in the MSE spatial variance budget is discussed more in Sect. 2.4.

Note that these diabatic terms include the direct diabatic effects of radiative and surface flux feedbacks, not the circulation that the diabatic terms generate. For instance a positive shortwave (SW) feedback means anomalous SW heating in the high-MSE region and/or anomalous SW cooling in the low-MSE region, thereby enhancing the MSE gradient. The diabatic feedback term does not account for the dynamical response to this SW heating distribution, which can also transport MSE up- or down-gradient. This transport is a component of the horizontal convergence term, but is not explicitly diagnosed separately from the other dynamical contributions in this framework. Another related issue is that these diagnostics are based on vertical integrals and hence do not explicitly capture the sensitivity to the vertical distribution of diabatic forcings found in Muller and Bony (2015). Indeed diabatic tendencies applied at different heights can yield different MSE transports since MSE varies strongly with height.

An assessment of both the direct diabatic effect and the indirect circulation and MSE transport corresponding to a heating anomaly is achieved in model simulations with sensitivity runs in which diabatic terms are horizontally homogenized, removing both direct and indirect effects, as done in Muller and Held (2012). This is obviously not possible with observations. The remaining option is to analyze the circulation generated by diabatic forcing and infer the MSE transport as done in Holloway and Woolnough (2016). The visualization of the MSE transport is usually done with a stream function in moisture and height space (Bretherton et al. 2005; Muller and Held 2012; Holloway and Woolnough 2016). This quantifies the energy transport between the dry region and the moist region and hence determines whether it is up-gradient, which is typical of aggregation in idealized model studies (e.g., Bretherton et al. 2005; Muller and Held 2012) though the total vertically integrated transport is not always up-gradient (e.g., Coppin and Bony 2015). This visualization method is useful in simulations, where vertical profiles of vertical velocity as a function of MSE are available, but it is not clear whether it is applicable to observations. Also, quantifying the role played by radiation in the circulation requires vertical profiles of radiation as a function of MSE (cf. Muller and Bony 2015), which are only beginning to become available in observations (e.g., Haynes et al. 2013). Section 5.3 explores possible approaches to observing these profiles using ground-based instruments.

Bretherton et al. (2005) showed that they could capture the aggregation instability in a semiempirical toy model accounting for the sensitivity of radiative and surface fluxes, as well as MSE convergence, to humidity. In their theoretical paper of convective aggregation, Craig and Mack (2013) take a somewhat similar approach, although the physical processes are modeled differently (in particular the MSE convergence is modeled as a diffusive process). The end result is an expression of the rate of change of humidity as a function of humidity itself $\partial I/\partial t = f(I) = -\delta V/\delta I + T$, where I is the order parameter (in this case column-integrated free-tropospheric water vapor), $V(I)$ is a potential function, and T is a diffusive transport term. The minima of the functional $V(I)$ are equilibrium values of humidity. The structure of the functional $V(I)$ therefore highlights the appearance of multiple equilibria typical of self-aggregation, with the two minima corresponding to the moist and dry solutions. Although this framework allows for the identification of aggregation, it is unclear if it can be used to identify feedbacks involved in the aggregation process. Aggregation from different feedbacks may have different signatures in the functional dependence $V(I)$.

More work using theory, as well as idealized (and perhaps more realistic) simulations, is desirable to compare conceptual frameworks and metrics of aggregation and determine how these could be applied to observations.

2.2 Initiation Processes

At SSTs close to our current tropical climate (300 K or so), the leading physical process behind the spontaneous self-aggregation of convection seems to be a "radiatively driven cold pool" outside deep convection, as seen in a schematic from Coppin and Bony (2015) (Fig. 3). One or several dry regions appear and expand, with strong longwave radiative cooling and subsidence yielding further drying. Moisture and convection are confined to the rest of the domain, as the dry convection-free region expands. In the following, we briefly review the various physical processes contributing to the formation of this radiatively driven cold pool for temperatures close to current tropical atmospheric temperatures,

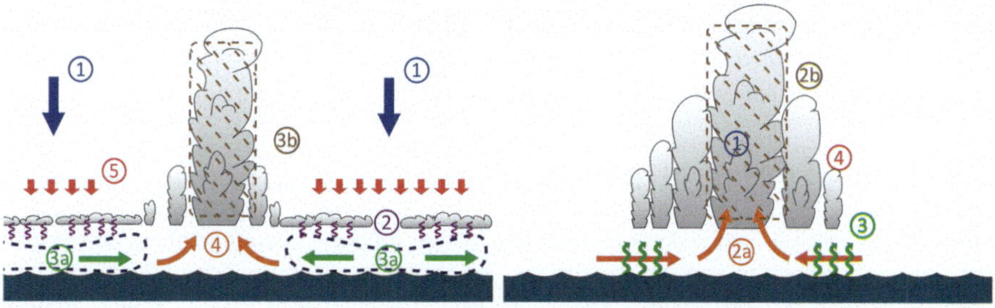

Fig. 3 Aggregation processes for: (*left*) cold SSTs with radiatively driven cold pools, and (*right*) warm SSTs with surface flux feedbacks. Figure adapted from Coppin and Bony (2015)

and then we discuss the sensitivity of these processes to SST. A more complete review can be found in Wing et al. (2017).

2.2.1 Longwave Radiation

As mentioned above, aggregation generally begins with the formation of a dry region with strong radiative cooling. The strong longwave cooling in the dry region is largely induced by low-level clouds (Muller and Held 2012; Muller and Bony 2015; Coppin and Bony 2015; Holloway and Woolnough 2016), although clear-sky cooling also contributes (Wing and Emanuel 2014; Wing and Cronin 2016).

Strong subsidence in dry regions, theoretically predicted by the RCE instability study of Emanuel et al. (2014), further promotes the formation of low-level clouds. These in turn enhance the radiative cooling, forming radiatively driven cold pools in dry regions responsible for the clumping of convection in the rest of the domain. Note that these radiatively driven cold pools are colocated with the dry regions and therefore do not mix boundary layer air between moist and dry regions, whereas "conventional cold pools" (defined here as cold pools resulting from downdrafts caused by rain evaporation and/or condensate loading) can propagate from moist to dry regions and tend to slow or weaken aggregation (cf. Jeevanjee and Romps 2013).

2.2.2 Surface Fluxes

Feedbacks involving surface enthalpy fluxes favor the initiation of self-aggregation due to larger surface winds in the moist, convecting area, which enhance the up-gradient MSE transport associated with the radiatively driven cold pool discussed above. However, while sensitivity runs with homogenized surface fluxes (no feedback) sometimes do not aggregate (Tompkins and Craig 1998; Bretherton et al. 2005; Wing 2014), they *can* aggregate depending on the domain size, strength of the surface fluxes imposed, and availability of radiative feedbacks (Muller and Held 2012; Holloway and Woolnough 2016). Therefore, surface fluxes feedbacks are not critical for aggregation to occur, at least at current temperatures.

2.2.3 Shortwave Radiation

The direct, diabatic effect of shortwave radiation is a positive feedback on aggregation due to variations in the absorption of shortwave radiation by water vapor (Wing and Emanuel 2014), but it is weaker than the longwave and surface flux feedbacks. In sensitivity experiments that include both direct and indirect (dynamic response to the diabatic forcing) effects, shortwave feedbacks slightly oppose aggregation. Either way, the impact of shortwave radiation appears to be secondary, at least at current temperatures.

2.2.4 Moisture–Convection Feedbacks

Moisture–convection feedbacks, in which convection moistens the atmosphere and is also more likely to occur in moister conditions, amplify the instabilities leading to self-aggregation (Tompkins 2001; Mapes and Neale 2011; Emanuel et al. 2014). When radiation feedbacks (which are normally required for self-aggregation) are suppressed while rain evaporation is also suppressed (preventing conventional cold pools which can otherwise weaken aggregation), these moisture–convection feedbacks are strong enough to cause aggregation on their own (Muller and Bony 2015; Holloway and Woolnough 2016). This appears to occur through a process similar to the coarsening process in Craig and Mack (2013) in which initial perturbations of a bistable system grow over time. These feedbacks are difficult to quantify directly, even in models, and they would also be difficult to target in observations. A place to start (perhaps using field campaign data) would be to correlate convective activity with moist (high MSE) locations and then to estimate the transport of MSE (part of the convergence term in the MSE spatial variance budget) due to circulations forced by this anomalous convective heating.

2.3 Sensitivity to SST

Several aspects of self-aggregation are sensitive to SST, such as its initiation mechanisms, spatial scale, and perhaps degree of organization (Khairoutdinov and Emanuel 2010; Wing and Emanuel 2014; Emanuel et al. 2014; Wing and Cronin 2016; Coppin and Bony 2015; Abbot 2014; Holloway and Woolnough 2016). It is worth noting that self-aggregation is found at temperatures much colder than our current climate, including 243 K in snowball Earth simulations (Abbot 2014) and 280 K in long-channel experiments (Wing and Cronin 2016), as well as much warmer (e.g., 310 K, Wing and Cronin 2016). The radiatively driven cold pools discussed above seem to be most efficient at cold and current temperatures (Fig. 1, Coppin and Bony 2015), possibly because climate models with strong positive low-cloud feedback (like the model used in that study) do not have any low clouds at high temperatures. However, cloud radiative feedbacks may behave differently at much colder temperatures (Wing and Cronin 2016) and clear-sky longwave feedbacks are favored by warm temperatures (Emanuel et al. 2014). At warm temperatures, surface-flux-wind feedbacks in the high-MSE convective region are the leading mechanism for self-aggregation in GCM simulations (Fig. 1, Coppin and Bony 2015).

In their semiempirical model of self-aggregation based on cloud-permitting simulations at present-day temperatures, Bretherton et al. (2005) found a slightly stronger sensitivity of radiative fluxes to moisture than that of surface fluxes. These sensitivities are likely to be different at different temperatures.

2.4 Maintenance Processes

Given that the real tropical atmosphere is never starting from a homogeneous background state, as in the idealized simulations, the processes that maintain existing convective aggregation may be easier to observe than those initiating it. While the strongest positive feedbacks in the early stages of idealized self-aggregation are usually found in the dry region, at later times strong consistently positive feedbacks are found only in the moist region. Muller and Held (2012) and Muller and Bony (2015), which find low clouds to be necessary for the initiation of aggregation using mechanism denial experiments, find that low clouds are not necessary to maintain self-aggregation in their simulations. Instead, high clouds in the moist regions and clear-sky longwave feedbacks can maintain aggregation (Muller and Held 2012; Wing and Emanuel 2014; Muller and Bony 2015; Wing and Cronin 2016). Possible sensitivity of these maintenance processes to SST is discussed in Sect. 4 below.

Surface flux feedbacks are neither necessary nor sufficient to maintain non-rotating aggregation (Holloway and Woolnough 2016), at least at current climate temperatures. Indeed, the surface flux feedback becomes negative in later stages of non-rotating aggregation, due to the opposing influences of surface winds and air–sea enthalpy disequilibrium (Wing and Emanuel 2014). However, surface flux feedbacks could behave differently in simulations with an interactive SST calculated from surface energy balance.

Quantifying the strength of these maintenance feedbacks in observations would be desirable. As in the simulations discussed earlier, the radiative and surface flux feedbacks could be diagnosed by their contributions to the MSE spatial variance budget. These require simultaneous measurements over a large area of the top-of-atmosphere and surface radiative fluxes, as well as observed surface enthalpy fluxes and vertically integrated MSE. Alternatively, an MSE temporal variance budget could be computed at a given location, assuming that with time, both the dry and moist regions of aggregated convection would pass over the station. Methods for estimating quantities needed to calculate these terms using satellite data are explored in Sect. 5.1. The strength of the radiative and surface flux feedbacks could also be correlated with the degree of aggregation as measured by SCAI (defined in Tobin et al. 2012). Tobin et al. (2013) used SCAI calculated from observations to suggest that intraseasonal variations of aggregation tend to amplify dynamical anomalies. Similarly, recent work compositing on MJO events during the DYNAMO field campaign has shown that radiation and, to a lesser extent, surface heat fluxes play an important role in amplifying MJO variability (Sobel et al. 2014), revealing potential links to the aggregation work proposed here.

3 Comparing the Idealized World to the Natural World

In addition to process-oriented studies, observations can also be used to test the realism of the mean state, variability, and convective characteristics of the idealized models. Here we explore similarities and differences between these aspects of idealized simulations of self-aggregation and observations. We also discuss processes that are not usually captured by idealized models, such as ocean interaction. Linking self-aggregation processes to environments with further complexity, such as non-uniform SST or the effects of land and orography, is not addressed here but deserves future investigation. The motivation for this section is that, in order to have confidence in the relevance of self-aggregation processes found in idealized simulations for observed convective organization, we need to be able to understand and explain differences between the idealized world and the natural world.

3.1 Time Scales of Self-Aggregation

One common critique of idealized self-aggregation is that the time scale of the aggregation process is much longer than typical time scales for observed convective organization. This is a valid concern, but there are several rebuttals which are discussed below. First, there is a broad range of time scales for self-aggregation and disaggregation in the literature, and these appear to depend on model, domain size, resolution, initial conditions, SST, and the inclusion or suppression of processes such as conventional cold pools. Second, self-aggregation from homogeneous initial conditions includes the spin-up of small-scale convective activity and clustering without pre-existing large-scale features, and while we can learn a lot from these early stages they are not likely to occur simultaneously across a large region in the real world where asymmetries are always present. Third, while it is likely that the processes important for idealized self-aggregation are not important for all types of convective organization in nature, and may be less important for rapidly organizing convective systems, some types of organized convection (particularly on longer time scales) do show intriguing links to self-aggregation.

As mentioned in Wing et al. (2017), self-aggregation in idealized models can take 15–100 days or more to reach a relatively stable aggregated state when starting from homogeneous initial conditions (though the longer time scales likely relate to an initial gestation period in some simulations which occurs before aggregation has started at all). There is some sensitivity of this to domain size and grid scale (Muller and Held 2012). When rain evaporation and conventional cold pools were suppressed, Holloway and Woolnough (2016) found that the time scale decreased to only 8 days as opposed to 16 days in their control run, supporting the idea proposed in Jeevanjee and Romps (2013) that conventional cold pools slow or suppress aggregation in idealized simulations. Perhaps also relevant to understanding processes that keep convective clusters organized in nature, Muller and Held (2012) and Holloway and Woolnough (2016) both found a disaggregation time scale (which is the time needed to return to a less aggregated equilibrium) as small as 10 days when simulations were initialized with an aggregated state and then interactive radiation was suppressed.

Wing (2014) found that the spatial MSE variance grew with an e-folding time of $\approx$11–13 days. As mentioned in Wing et al. (2017), this kind of exponential growth will lead to much larger scales in a given amount of time when starting from larger initial clustering, as is typically found in nature. In other words, much of the time scale for self-aggregation from homogeneous initial conditions may not be especially relevant to comparisons with nature because these time periods involve spinning up mesoscale activity from extremely small initial length scales (and may also involve gestation periods before aggregation begins at all). In fact, convective cluster growth across scales (but especially at larger scales) was found to be linked to radiative feedbacks in near-global RCE channel runs (including rotation) in Bretherton and Khairoutdinov (2015), with e-folding time scales of 6–14 days. Those authors suggest that diabatic feedbacks (mainly longwave radiation feedback) may be especially important for large-scale convective organization such as the MJO.

3.2 Mean Wind and Wind Shear

Most idealized RCE studies have no imposed mean wind or wind shear. While wind shear can act to enhance some kinds of mesoscale organization such as squall lines (e.g., Houze 2004; Muller 2013), it has also been shown to slow or prevent self-aggregation in idealized simulations such as those in Bretherton et al. (2005) and Khairoutdinov and Emanuel (2010), although the latter found that there was hysteresis, since an already aggregated state did not disaggregate with some levels of imposed shear.

Nonzero mean vertical velocity due to large-scale circulations is common in regions containing organized tropical systems in nature but cannot occur for the domain mean in a typical RCE setup. Global-scale simulations, however, do represent these circulations (e.g., Coppin and Bony 2015), and smaller RCE simulations can impose them (e.g., Su et al. 2000) or parameterize them using reference profiles and assumptions of weak temperature or pressure gradients (e.g., Sessions et al. 2016). As these kinds of modeling studies progress, there will be more opportunities to evaluate their simulated relationships between aggregation and large-scale circulations using observations.

3.3 Humidity Profiles

One of the potentially unrealistic aspects of self-aggregation as seen in idealized models that needs to be reconciled with observations is the presence of very dry humidity profiles that occur in the non-convecting areas of the domain. Since humidity plays a key role in all of the feedbacks important for self-aggregation, it is especially important to investigate this issue. To that end, we include here some examples of humidity profiles (and the related radiative heating profiles) in moist and dry areas of simulated self-aggregation. These profiles are from the simulations presented by Wing and Cronin (2016). We show profiles from two simulations: one with a square domain that is 1536 km × 1536 km in the horizontal (*sq*) and one that is an elongated channel with dimensions of 12,288 km × 192 km in the horizontal (*ch*). The *sq* simulation has one circular, intensely precipitating moist cluster while the *ch* simulation has multiple moist and dry bands. All other aspects of the simulations are identical.[3] Fig. 4 shows water vapor mixing ratio and relative humidity in the moist and dry areas, averaged over the last 10 days of the two simulations. In Fig. 4a, the "moist" area is defined as the area where the CWV is greater than 80% of the maximum CWV found in the last 10 days of simulation. The rest of the domain is classified as the "dry" area. Profiles using an alternate definition of moist and dry areas are shown in Fig. 4b, in which the "dry" and "moist" areas are the driest 10% and moistest 10% of the domain according to CWV. Here, we show profiles from both the developing and mature stage of aggregation, using 5-day averages centered at day 10 and day 70, respectively. Figure 4c, d show similar plots to Fig. 4a, b but for relative humidity, with ranking done according to column relative humidity (CRH, defined as CWV divided by column-integrated saturation specific humidity) instead of CWV.

As shown in Fig. 4, the water vapor mixing ratio and relative humidity are substantially reduced in the dry regions relative to the moist regions at all levels (including the boundary layer), but most strongly in the mid-troposphere. The difference between the dry and moist regions is stronger for the *sq* simulation than the *ch* simulation, reflecting the more extreme (and arguably less realistic) aggregation that occurs in square domain simulations. There is significantly more radiative cooling in the dry regions than the moist regions, especially in the lower troposphere (Fig. 5), which further amplifies the anomalies.

These results naturally lead to several questions about how representative these idealized simulations are of humidity variability in the real tropics. The behavior of the humidity profiles across the different evolutionary states of aggregation in the simulations (from developing to mature aggregation) is interesting; substantial drying is present in the upper troposphere as early as day 10, but drying of the middle–lower troposphere and boundary layer does not appear until later in the simulation (Fig. 4b, d). One potential avenue of research to link this to observations is to relate the evolution of humidity in the dry regions and the stage of aggregation to the altitude depth of the bimodality of water vapor (Mapes 2001, 2016; Zhang et al. 2003). However, a more basic starting point is to determine whether humidity in the tropics exhibits a similar range of variability between dry and moist conditions: do humidity profiles as dry as the ones in simulated aggregation exist in the real tropics?

[3] SST = 305 K, no rotation, diurnal cycle of insolation at 19.45 N at perpetual Julian day 80.5, Rapid Radiative Transfer Model (RRTM) radiation scheme, 3 km horizontal resolution, 64 vertical levels, rigid lid at 28 km, doubly periodic lateral boundaries, initialized with white noise in boundary layer temperature field.

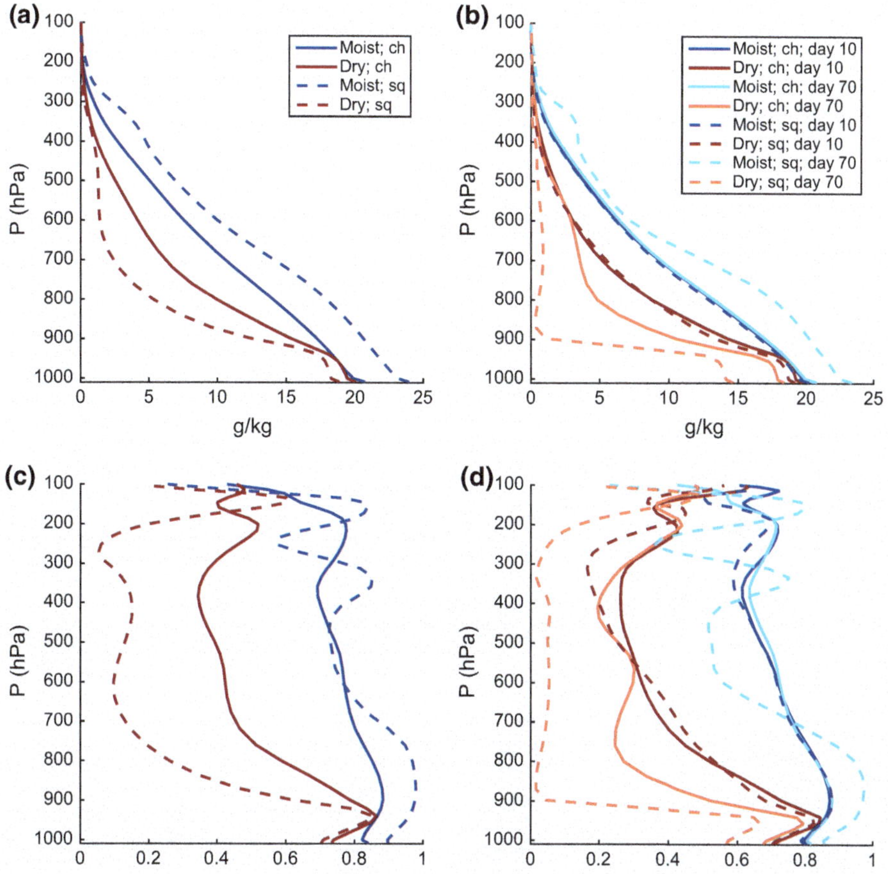

Fig. 4 Profiles of water vapor mixing ratio (*top panels*) and relative humidity (*bottom panels*) in simulations in elongated channel (*solid lines*) and square (*dashed lines*) domains. The *left panels* define moist regions (*blue*) as area where CWV ≥ 0.8 max(CWV) [or CRH ≥ 0.8 max(CRH) in panel **c**], dry regions (*red*) defined as the rest of the domain; the profiles are averaged over the last 10 days of the simulation. The *right panels* show profiles from the moistest (*shades of blue*) and driest (*shades of red*) 10% of the domain, according to CWV [or CRH in **d**]. Profiles from both the developing (5-day average centered at day 10; *lighter colors*) and mature (5-day average centered at day 70; *darker colors*) stages of aggregation are plotted. **a** q where CWV >/< 0.8 max (CWV), **b** q in moistest/driest 10% of domain, **c** RH where CRH >/< 0.8 max (CRH) and **d** RH where CRH >/< 0.8 max (CRH)

As a first step toward answering this question, we compare the humidity data from the idealized simulations in Wing and Cronin (2016) to twice-daily radiosondes from Nauru in the Pacific warm pool. Figure 6 shows humidity profiles from 5 years of the Nauru radiosondes. These data span the period from April 1, 2001 to August 16, 2006 and are from the former Atmospheric Radiation Measurement (ARM) site (Mather et al. 1998; Long et al. 2016). The data, which include 3491 retained sondes, are described in more detail in Holloway and Neelin (2009). Figure 6a shows mean specific humidity profiles of two subsets of sondes divided by a CWV threshold of 0.8 times the 99th percentile of CWV (55 mm, the 63rd percentile). Despite coming from a range of SSTs which are generally a few degrees cooler than 305 K, these profiles look quite similar to the mixing ratio profiles from the *ch* simulation shown in Fig. 4a, while the *sq* simulation in that

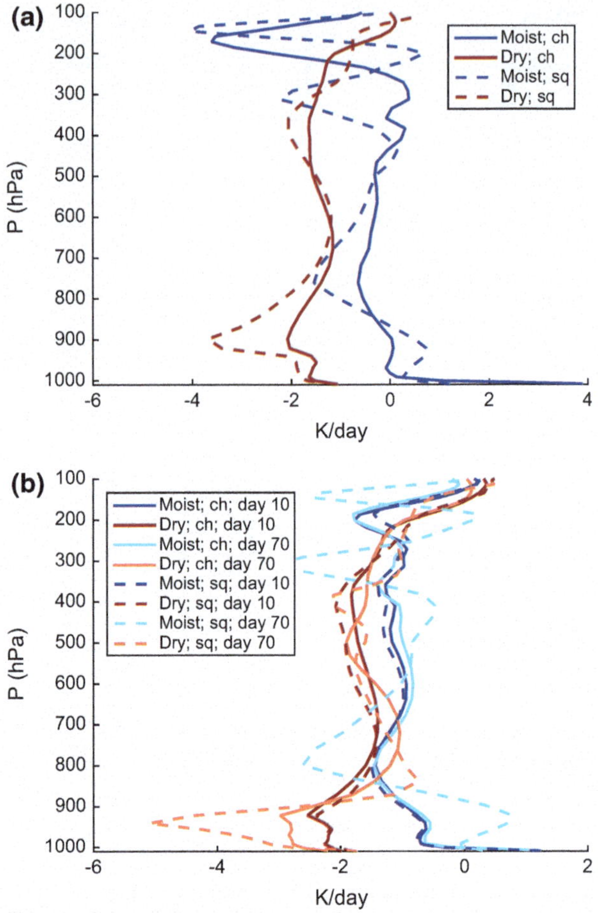

Fig. 5 Profiles of radiative heating rate in simulations in elongated channel (*solid lines*) and square (*dashed lines*) domains. **a** Moist regions (*blue*) as area where $CWV \geq 0.8\ \max(CWV)$, dry regions (*red*) defined as the rest of the domain; the profiles are averaged over the last 10 days of the simulation. **b** Profiles from the moistest (*shades of blue*) and driest (*shades of red*) 10% of the domain, according to CWV. Profiles from both the developing (5-day average centered at day 10; *lighter colors*) and mature (5-day average centered at day 70; *darker colors*) stages of aggregation are plotted. **a** Q_{rad} where CWV >/< 0.8 max (CWV) and **b** Q_{rad} in moistest/driest 10% of domain

figure shows much more spread between moist and dry profiles. Similarly, the extreme moistest 10% and driest 10% of sondes in Fig. 6b are much more similar to those for day 70 of the *ch* simulation than for day 70 of the *sq* simulation in Fig. 4b; indeed, the extreme 10% quantiles in the *sq* simulation at day 70 within the lower and middle free troposphere are much more extreme than even the extreme 1% quantiles for the sondes. The driest 10% quantile in the *sq* simulation at day 70 suggests that air is subsiding from the upper troposphere down to almost 900 hPa without encountering significant moistening by mixing or convection, something not seen in the observations.

Figure 6c, d shows similar analysis to Fig. 6a, b but for relative humidity (defined with respect to ice for temperatures below 0°C) and CRH. Note that, even for the driest 1% of sondes, relative humidity in the boundary layer is always above 65% on average. The corresponding profiles for the simulations in Fig. 4c, d are consistent with the *ch* simulation being more realistic than the *sq* simulation, at least regarding humidity variability. Note that the near-surface relative humidity averaged for the driest 10% of the domain in the *sq* simulation at day 70 is about 10% drier (in relative humidity units) than the average for the driest 1% of sondes at Nauru.

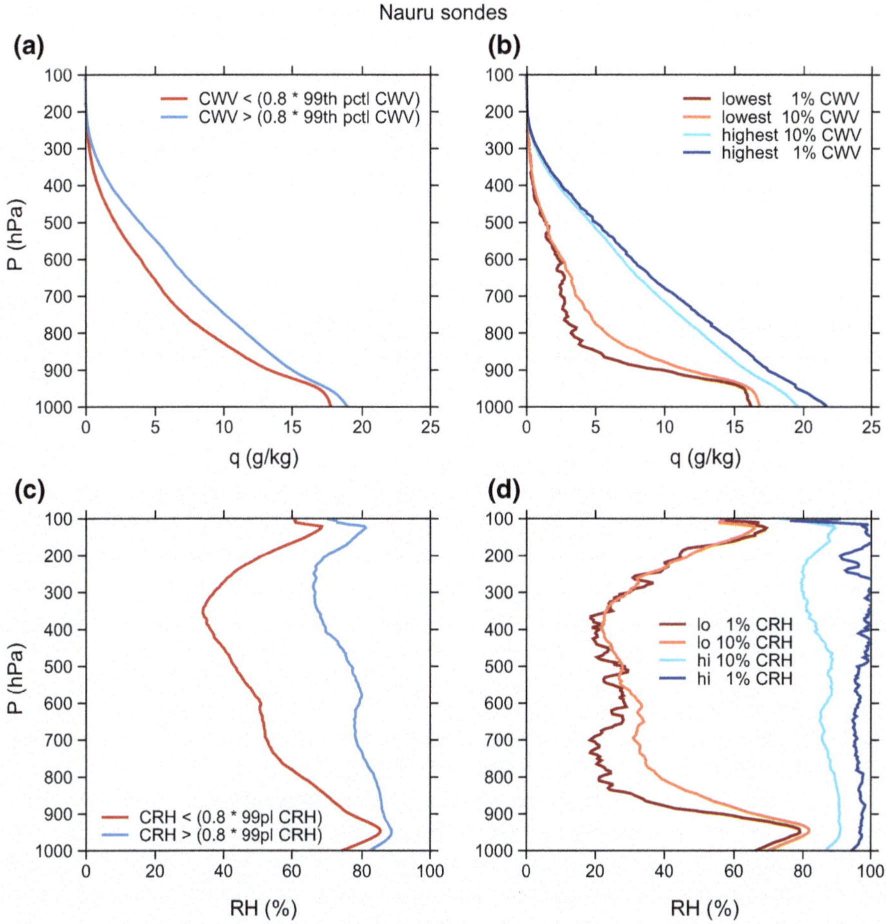

Fig. 6 Nauru radiosondes: **a** mean profiles of water vapor specific humidity (g kg^{-1}) for all sondes with CWV greater than 0.8 times the 99th percentile of CWV and for all sondes less than this threshold, **b** mean profiles for all sondes in the lowest 1%, lowest 10%, highest 10%, and highest 1% as ranked by CWV, **c** mean profiles of relative humidity (%) for all sondes with CRH greater than 0.8 times the 99th percentile of CRH and for all sondes less than this threshold, and **d** mean profiles of relative humidity for all sondes in the lowest 1%, lowest 10%, highest 10%, and highest 1% as ranked by CRH

Table 1 compares surface observations of relative humidity at Nauru with relative humidity at the lowest model level (37 m) in the *ch* simulation from Wing and Cronin (2016). The values for most of the percentiles are comparable, except the simulation has a much lower minimum value than the Nauru observations (32.5% compared to 52.0%).

Figure 7 shows contour plots of all 3491 sondes ranked by CRH and divided into 100 equally populated bins. These show both relative humidity and saturation deficit (saturation specific humidity minus specific humidity). A similar plot for the *ch* simulation from Wing and Cronin (2016) is shown in Fig. 8, and Fig. 7d can be additionally compared with a similar figure from day 90 of the square simulation at 305 K SST in Wing and Emanuel (2014, their Fig. 11). That figure shows that the square simulation has a large spread in relative humidity between 1 and 2 km height of about 100% between moist and dry

Table 1 Values of surface relative humidity (%) at Nauru (averaged from station data over 1 h centered around on each sonde launch time) and lowest model level (37 m) relative humidity (%) in the 305 K Channel simulation from Wing and Cronin (2016)

Percentile	Nauru	Channel simulation
Minimum	52.0	32.5
1st	57.9	58.6
25th	70.8	69.9
50th	76.8	73.5
75th	82.3	77.5
99th	92.8	93.0

The statistics from the channel simulation are computed over the final 25 days of that simulation

regions, while the Nauru sondes show a spread of 60% at most in that layer and a much larger number of bins with small anomalies. The channel simulation (Fig. 8), on the other hand, is much more comparable to the Nauru sondes, indicating that this simulation has realistic humidity variability.

Radiosondes from other tropical locations, such as the Bay of Bengal and the eastern tropical Pacific, also reveal significant variability in mid- and upper-tropospheric relative humidity and little variation in boundary layer moisture (e.g., Zuidema et al. 2006; Zuidema and Mapes 2008), though these locations are subject to large-scale circulations that can bring remote influences from neighboring landmasses. Radiosondes from the equatorial Indian Ocean also demonstrate that most of the relative humidity variability is contained within the middle troposphere (Johnson and Ciesielski 2013), where the signature of self-aggregation may be first detected (Mapes 2016).

There are reasons to expect that the Nauru sondes and tropical sondes from these other locations would not necessarily look exactly like idealized self-aggregation simulations (or indeed, would not be representative of tropical maritime observations more generally). For instance, these sondes are generally launched from islands, which could have local effects on convection, and transport of air from landmasses or higher latitudes could also cause differences compared with idealized conditions. Additionally, we may not necessarily think of idealized aggregated convection as something that would or should be representative of typical tropical conditions anyway.

However, since humidity variability is fundamentally linked to both the contributing processes and large-scale impacts of self-aggregation, it is important to consider possible reasons why the channel simulation appears to have more realistic humidity variability, while the square simulation is too extreme. One possibility is that the channel simulation is "getting the right answer for the wrong reason", for instance because its quasi-2D geometry leads to a spurious strong wind shear similar to that found in 2D simulations in Held et al. (1993). Although the channel simulation does have tropospheric along-channel mean wind and vertical wind shear that are larger than values in the square simulation, the channel values are of order $1\,\mathrm{m\,s^{-1}}$ for both quantities, and this is not overly strong compared with typical tropical mean values. The channel simulation is 192 km wide, which allows for multiple convective systems and associated cold pools to exist and propagate along the shorter dimension. Subsidence in the driest regions is actually stronger in the channel simulation relative to the square simulation, though ascent in the moist regions is weaker. While determining the reasons for the differences between the simulations is beyond the scope of this paper, it is likely that the channel simulation has more

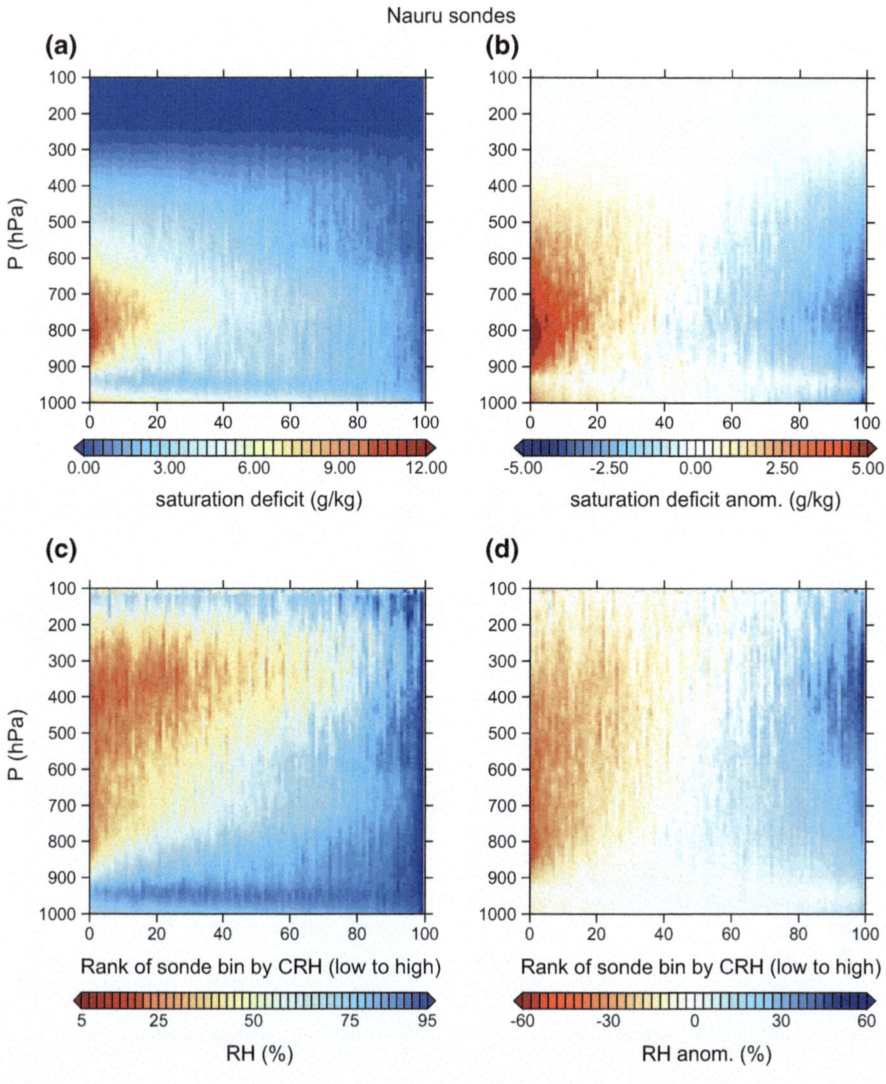

Fig. 7 Nauru radiosondes: **a** saturation deficit $(\mathrm{g\,kg^{-1}})$ for all sondes ranked by CRH and averaged in 100 equally populated bins, **b** anomaly of each bin in **a** from the all-sonde mean saturation deficit at each level, **c** as in **a** but for relative humidity (%), and **d** as in **c** but for relative humidity

mixing and transport between convective and subsidence regions—animations (not shown) reveal that boundaries between convective and subsidence regions are less stationary, and closer to the center of subsidence regions, in the channel simulation.

While the above discussion does not definitively endorse one model domain geometry over another, this type of analysis is informative in starting to address the extent to which idealized aggregated convection is similar to organized convection in the real world, and we hope that it helps frame future comparisons with other data. For instance, analysis tracing air particles back to their time of last condensation within both a modeling

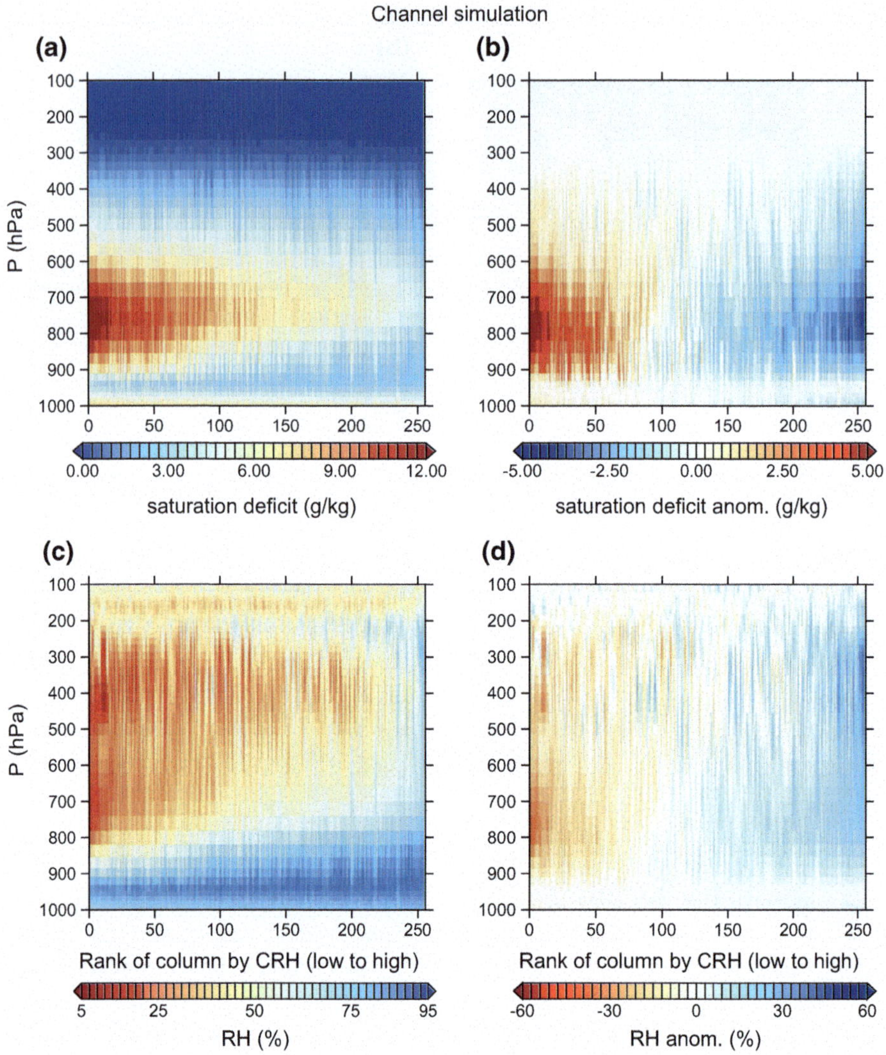

Fig. 8 As in Fig. 7 but for channel simulation at 305 K from Wing and Cronin (2016). **a** Saturation deficit ($g\,kg^{-1}$) for all sondes averaged over 192 km × 48 km blocks and ranked by block-averaged CRH, **b** anomaly of each block in **a** from the domain-mean saturation deficit at each level, **c** as in **a** but for relative humidity (%), and **d** as in **c** but for relative humidity

construct and observations could be helpful (e.g., Pierrehumbert 1998), as well as a spectral analysis to determine if key time scales are matched within both.

3.4 Equatorial Wave Dynamics

Earth's latitudinally varying rotational effects on large-scale horizontal motions result in equatorial wave dynamics which help shape tropical convective organization. For instance, the MJO interacts with equatorially trapped moist Kelvin and Rossby waves, and these dynamics are also important for the development of the Hadley Circulation, the inter-tropical convergence zone (ITCZ), and monsoons. While self-aggregation is generally defined not to include the effects of a latitudinally varying Coriolis parameter, a few studies have looked for the processes that lead to self-aggregation in simulations that do include such effects. Bretherton and Khairoutdinov (2015) found that radiative feedbacks were likely to be important mainly for large-scale convective organization in their near-global RCE channel runs. Arnold and Randall (2015) performed global aquaplanet simulations (using a superparameterization setup in which 2D CRMs are embedded in each large-scale model grid cell) with uniform SST both with and without rotation and found similarities in diabatic feedbacks between the self-aggregation in the non-rotating setup and the MJO in the rotating setup. Holloway (2017) found that simulations of real near-equatorial case studies using a limited-area CRM setup also showed similarities to idealized self-aggregation, although the effects of suppressing interactive radiation were constrained by the imposed lateral boundary conditions. More studies are needed to probe links between self-aggregation and convective organization that interacts with equatorial wave dynamics.

3.5 Ocean Interaction and Feedback

Nearly all studies of self-aggregation have used atmosphere-only simulations. However, there are a few studies that have used coupled models, and they generally find that ocean coupling slows or prevents self-aggregation. For instance, an interactive slab ocean experiment slowed down self-aggregation in Bretherton et al. (2005), possibly because of cloud shading. That experiment had a 60 W m^{-2} imposed ocean cooling to represent large-scale ocean or atmospheric transport, and after aggregation the SST cooled rapidly due to increased longwave cooling. Khairoutdinov and Emanuel (2010) used a 2-m slab ocean but homogenized the SST horizontally at each time step (thus removing effects like cloud shading) and also found that SST dropped after aggregation occurred. They noted hysteresis, since cooler SSTs could still maintain aggregation that had already been present but could not sustain self-aggregation from homogeneous conditions. Popke et al. (2013) performed global-tropics RCE runs using parameterized convection (with no rotation and homogeneous solar forcing) coupled to a slab ocean and found that large convective clusters formed along with transient SST anomalies. Reed et al. (2015) performed similar global-tropics RCE runs and found that, although ocean coupling slows aggregation compared to runs with fixed warm SSTs (302 K) in agreement with other studies, runs with fixed cool SSTs (as low as 295 K) result in much less organization than runs with similar SSTs and an interactive slab ocean, suggesting a possible link between ocean coupling and the sensitivity of aggregation to SST.

Coppin and Bony (2017) also ran global-tropics RCE simulations without rotation and coupled to a slab ocean and found that the coupled RCE system exhibits some internal variability, arising from the interplay between SST, SST gradients and aggregation. The time scale of this variability depends on the depth of the ocean mixed layer, and for a large range of depths, it occurs at the interannual time scale, suggesting a possible link to internal modes of variability in the real tropical ocean atmosphere system such as El Niño Southern Oscillation (ENSO). They also showed that, at this time scale, the relationship between SST and aggregation could be very different from (or even opposite to) that found in prescribed SST simulations or in coupled RCE simulations on long time scales.

Hohenegger and Stevens (2016) ran high-resolution coupled RCE runs (without imposed ocean cooling, but with reduced solar insolation equivalent to that averaged over the full Earth rather than the tropics) and found that aggregation seemed to prevent a runaway greenhouse effect, providing "radiator fins" to the idealized climate in the dry subsiding regions analogous to the role of the subtropics proposed by Pierrehumbert (1995). They also found that slab oceans with small depths can slow or prevent self-aggregation, similar to studies mentioned above. This delay stems from the development of SST gradients which cause a low-level circulation opposing the one that favors self-aggregation. Furthermore, Hohenegger and Stevens (2016) suggest that cloud feedbacks and resulting aggregation and coupled equilibrium states are very different at high resolution using explicit convection versus similar runs using parameterized convection from Popke et al. (2013), showing another example of model disagreement with regards to these processes.

While atmosphere–ocean coupling has been extensively studied for large-scale tropical convective phenomena such as the MJO (cf. DeMott et al. 2015), observational work is needed to explore the interactions between organized tropical convective systems and SST or sea surface salinity across scales. Specifically, this analysis could look at processes important for aggregation in idealized models.

4 Observational Perspectives on Aggregation in a Warming Climate

Several modeling studies suggest that convective aggregation depends on surface temperature, although the exact nature of this dependence remains uncertain. The initiation of aggregation is found to occur more easily at certain temperatures (Coppin and Bony 2015), particularly when considering a given domain size (Wing and Emanuel 2014). Once initiated, the clumping of aggregation in some studies tends to strengthen as the surface temperature rises (Coppin and Bony 2015), though other studies find that the degree of aggregation is relatively insensitive to SST (Wing and Cronin 2016; Holloway and Woolnough 2016; Hohenegger and Stevens 2016). Several interpretations have been proposed for the temperature dependence of the initiation mechanisms (Sect. 2.3). Some of them invoke the nonlinearity of the Clausius–Clapeyron relationship, the sensitivity of the clear-sky longwave radiative cooling of the atmospheric column to lower-tropospheric longwave opacity (Emanuel et al. 2014), or the sensitivity of the low-cloud cover to temperature (e.g., Coppin and Bony 2015; Wing and Cronin 2016; Holloway and Woolnough 2016), and these temperature dependences differ across models. On the other hand, the interaction between temperature, high-cloud radiative effects, and dynamics has been proposed by Bony et al. (2016) as a mechanism for stronger clumping of convection at the aggregated equilibrium state over warmer surfaces. That study argues that, owing to the dependence of static stability on temperature and pressure, as the climate warms anvil clouds not only rise to a higher altitude but also shrink in horizontal area. This behavior, referred to as the "stability-iris" effect in Bony et al. (2016), concentrates the atmospheric cloud radiative effects of anvil clouds and enhances the horizontal gradients in atmospheric radiative cooling (enhancing the cooling in subsiding areas and reducing it in convective areas), which could lead to enhanced convective aggregation.

Given the implications that a dependence of convective aggregation on temperature may have for climate (Sect. 1.1), it is important to verify whether this dependence seen in some models is confirmed by observations. However, very few studies have investigated this issue so far. Long time series of convective aggregation indices have now been produced (e.g., Tobin et al. 2012; Tan et al. 2015), but they have not been analyzed in this perspective yet.

What has been investigated, on the other hand, is the temperature dependence of various large-scale organized convective phenomena that share many characteristics with convective aggregation in idealized models. One of these is the MJO, which likely represents a very large-scale manifestation of convective aggregation in the tropics (Khairoutdinov and Emanuel 2010; Arnold and Randall 2015). There is modeling evidence that MJO activity increases when the climate is warming (e.g., Caballero and Huber 2010; Arnold et al. 2015), though MJO-like behavior has been found even at temperatures as cold as 1 °C (Pritchard and Yang 2016); this finding of increased MJO activity with increased SST in models is qualitatively consistent with observations that suggest linear increases in the intensity and number of MJO events over the last 50 years (Jones and Carvalho 2006).

Tropical cyclones likely constitute another spectacular manifestation of convective aggregation. But unfortunately, no such consistency has yet been reached between their modeled and observed behavior with temperature. Idealized RCE simulations performed in a rotating framework suggest that the number of tropical cyclones decreases as surface temperature rises, while their intensity and precipitation rate increases (Nolan et al. 2007; Held and Zhao 2008; Khairoutdinov and Emanuel 2013). Climate projections made with general circulation models also suggest such a tendency, although the relationship between tropical cyclones and temperature very much depends on the metrics used for warming (Knutson et al. 2013). On the observational side, however, trends in tropical cyclones and their relationship to temperature remain elusive (Stocker et al. 2013). This is partly due to the limited availability and quality of long-term historical records, but also to the large number of global and regional factors that influence the occurrence and intensity of tropical cyclones. In particular, it is difficult to disentangle a trend associated with global warming from records which are either too short or associated with an insufficient geographical sampling. Another source of complication stems from the fact that tropical cyclone activity does not only depend on absolute surface temperature: it is also affected by factors such as the temperature difference between the surface and the tropopause (Emanuel 1987), the local surface temperature relative to the tropical mean (Lin et al. 2015), the wind shear and mid-tropospheric humidity (Tang and Emanuel 2010), and the upper ocean stratification (Emanuel 2015), and these factors are strongly modulated by the decadal to multi-decadal natural climate variability.

To confirm or refute modeling inferences regarding the temperature dependence of convective aggregation, another approach consists of using observations to test the physical processes that contribute to this dependence in models. One such process is the reduction of the anvil cloud amount as the climate warms (Bony et al. 2016). Some observational studies suggest that, on average over the tropics, the anvil cloud amount decreases as the surface temperature increases (Zelinka and Hartmann 2011; Igel et al. 2014) as shown in Fig. 9 reproduced from Igel et al. (2014), but other studies do not find strong evidence for such a relationship (Stein et al. 2017). These differences may result from methodological differences: in contrast with other studies, Stein et al. (2017) consider the dependence of anvil cloud amounts on surface temperature for given precipitation and large-scale forcings, which amounts to comparing situations having a fairly similar convective mass flux, and therefore a weaker change in anvil cloud amount with temperature. Also, Stein et al. (2017) and Igel et al. (2014) are only comparing local SSTs colocated with specific cloud scenes, whereas the reduction of anvil cloud amount with warming may be more sensitive to the tropical mean SST, which is the metric used in Zelinka and Hartmann (2011). And Igel et al. (2014) study anvil cloud per individual cloud object, not

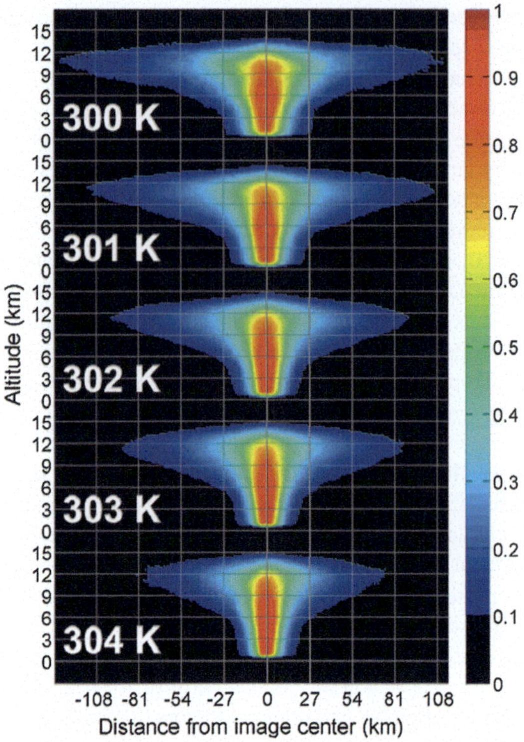

Fig. 9 Composites of anvil cloud objects derived from CloudSat observations: the width of anvil clouds is found to decrease as surface temperature increases. From Igel et al. (2014)

total anvil fraction, but Stein et al. (2017) find that SCAI values increase with SST, meaning that there are more (and smaller) anvil cloud clusters for warmer SSTs, which could cancel out effects of smaller anvil size per cluster. Additional methodological differences may also contribute to these conflicting results, and they will have to be clarified.

More generally, several methodological issues complicate attempts to investigate the relationship between temperature and convective aggregation in observations. First, unlike idealized modeling studies forced by uniform boundary conditions, the Earth's climate is associated with gradients in surface temperature which strongly influence large-scale vertical motions in the tropics. As is widely recognized, at the regional scale clouds and convection are much more influenced by the large-scale atmospheric circulation than by local surface temperature (Hartmann and Michelsen 1993; Bony et al. 1997). For this reason, relationships between convective aggregation and surface temperature derived from regional investigations do not necessarily reflect an intrinsic dependence of aggregation on temperature. Second, there is abundant evidence that the relationship between temperature and water vapor or clouds can differ on short versus long time scales (e.g., Dessler 2010). Relationships inferred from observed climate variations on seasonal or interannual time scales may thus differ from the temperature dependence of convective aggregation on decadal time scales and under long-term climate change. Recent results suggest that this might also be the case for coupled RCE simulations (Coppin and Bony 2017), although the extent to which this result applies to other models remains unknown. Finally, models suggest that convective aggregation can behave very differently in cold and warm climates. For instance, Abbot (2014) predicted stronger convective aggregation on a Snowball Earth than in the modern climate, but observations of clouds and convection are only available for a limited time period and thus for a very narrow range of surface temperatures. To explore possible changes in convective aggregation in more drastically different ranges of temperature, one must consider paleoclimatic changes using proxy data. Techniques of paleotempestology offer opportunities to reconstruct tropical cyclone activity at different periods of the past and for a range of time scales (e.g., Liu and Fearn 2000; Donnelly and Woodruff 2007) and could be very useful for this purpose. The isotopic composition of water is very sensitive to the organization of convection (e.g., Lawrence et al. 2004; Risi et al. 2008), and therefore long-term isotopic records (as well as recent satellite observations of water isotopes) could be used to explore changes in convective organization with climate.

5 Future Observational Aspirations

Much remains to be done in terms of observing convective aggregation. Using satellite observations, the variability of aggregation at different time scales could be investigated, as well as its relationship to local and remote surface and atmospheric conditions. Besides this, it would be nice to investigate whether the physical mechanisms found to play a role in the initiation of aggregation in models can also be observed in nature. For instance, is there evidence for the formation of radiatively driven cold pools in the dry areas of the tropics? Will future space missions such as the ADM-Aeolus wind lidar mission (Reitebuch 2012) help observe the interplay between low clouds and shallow circulations in the vicinity of deep convection? Will they help us observe radiatively driven cold pools? Can we observe signs of convective self-aggregation? To address these questions, one may analyze observations from field experiments such as those collected during AMIE/DYNAMO in the Indian ocean (Feng et al. 2015) or in the tropical Atlantic as part of the NARVAL-EUREC^{4}A campaigns (Stevens et al. 2015; Bony et al. this issue). One may also think of organizing a field campaign specially dedicated to these questions.

The ISSI workshop in February 2016 on "Shallow clouds and water vapor, circulation and climate sensitivity" brought together scientists using numerical simulations to study convective organization with scientists at the forefront of observational work, including experts on remote sensing of clouds and their environment. In this section, we present some perspectives on novel approaches to using satellite data to observe convective aggregation and the processes discussed above. Another paper (Lebsock this issue) also presents some promising new work along these lines using CloudSat and other A-Train satellites, complementing work by Stein et al. (2017). We also propose another possible way forward using a ground-based observational network.

5.1 Evolution of Convective Organization Using Satellite Data

The physical processes of convective self-aggregation involve a range of elements from the dynamics of convective systems to the thermodynamics of their rain-free environment. In this section, we review recent work with unique ideas of exploiting the existing satellite capability to study precipitating cloud systems and the surrounding atmosphere. The potential utility of such satellite observations in addressing different aspects of convective self-aggregation is also discussed.

The variability in the large-scale atmospheric state associated with a life cycle of convective systems has been examined with a suite of satellite measurements by Masunaga

(2012) and several subsequent papers. Since the sporadic nature of low-Earth orbiting (LEO) satellite overpasses with high-inclination orbits makes it difficult to continuously monitor subdaily scale variations, the variability is statistically reproduced by projecting a large number of snapshots obtained from multiple LEO satellites onto a composite time series. For instance, temperature and humidity profiles from the Atmospheric Infrared Sounder (AIRS) aboard the Aqua satellite are combined with the Tropical Rainfall Measuring Mission (TRMM) Precipitation Radar (PR) so that the evolution of the ambient sounding is constructed over the hours before and after convection develops (Masunaga 2012).

Masunaga (2013) applied water and heat budget analysis to this composite time series. The moisture and MSE (denoted by h) budget equations integrated vertically over the troposphere are:

$$\frac{\partial}{\partial t}\langle q\rangle + \langle\overline{\nabla\cdot q\mathbf{v}}\rangle = \overline{E} - \overline{P} \qquad (1)$$

and

$$\frac{\partial}{\partial t}\langle h\rangle + \langle\overline{\nabla\cdot h\mathbf{v}}\rangle = \overline{S} + L_\mathrm{v}\overline{E} + \langle Q_\mathrm{R}\rangle, \qquad (2)$$

where $\langle\cdots\rangle$ designates the vertical integral over the whole troposphere, the overbar denotes horizontal averaging over a large-scale [O(100 km)] domain, q is specific humidity, $\mathbf{v}$ is horizontal wind, E is surface evaporation, P is surface precipitation, S is surface sensible heat flux, L_v is the latent heat of vaporization, and Q_R is the radiative heating rate. Each term on the rhs of (1) and (2) is available from satellite observations and the tendency term on the lhs is evaluated from the composite time series, leaving as the only unknowns the second term of (1) and (2), that is, the horizontal convergence of moisture and MSE convergence. The vertically integrated moisture and MSE convergences, although not directly measurable from satellites, are instead derived as the residual in the budget equations. Note that the quantities calculated for these equations could also be used to

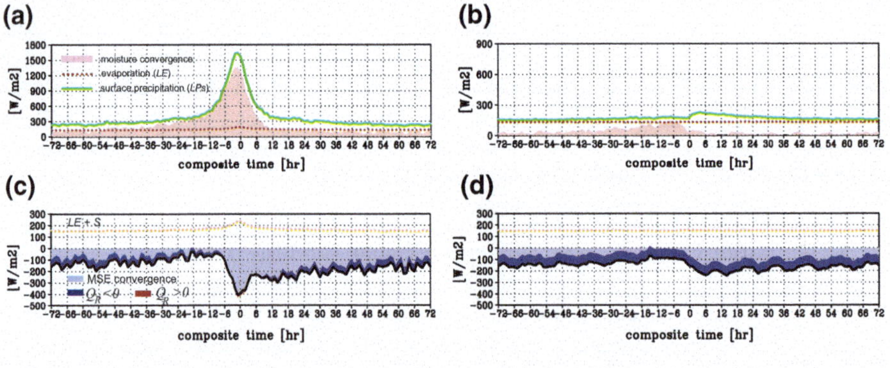

Fig. 10 Satellite-derived moisture and MSE budget parameters in composite time series associated with the development and dissipation of convection. **a** Moisture convergence (*shaded*), surface precipitation (*solid line*), and surface evaporation (*dotted line*) for the organized system regime, **b** as in **a** but for the isolated cumulus regime, **c** MSE convergence (*light-shaded*), radiative heating (*heavy shaded*; *red* where positive and *blue* where negative) on the top of MSE convergence, surface heat flux (*dotted line*) for the organized system regime. **d** As in **c** but for the isolated cumulus regime. All parameters including precipitation and evaporation are plotted in energy flux units ($\mathrm{W\,m^{-2}}$)

calculate the diabatic terms of the MSE spatial variance budget from Wing and Emanuel (2014) as suggested in Sect. 2.4.

Figure 10 shows the composite evolution of different budget terms for both the "organized system" and "isolated cumulus" regimes. These two regimes are separated by applying different thresholds to the areal coverage of TRMM-detected precipitation cells (i.e., <25% for isolated cumuli and >50% for organized systems), aimed at delineating the elements of atmospheric thermodynamics that are favorable or unfavorable for convective organization (Masunaga 2014). The primary moisture source of precipitation is moisture convergence during hours around the peak convection in the organized system regime (Fig. 10a), while precipitation nearly balances out the local moisture supply from surface evaporation in the isolated cumulus regime (Fig. 10b). In both the regimes, evaporation stays almost constant over time at $\sim$100–150 W m^{-2}, which suffices to produce modest rainfall from isolated cumuli but needs to be supplemented by a large dynamically driven import of moisture to feed organized systems.

The dynamics specific to organized convective systems is illustrated in light of the MSE budget (Fig. 10c), where MSE convergence stays overall negative but nearly vanishes to zero as convection intensifies (discussed in detail by Masunaga and L'Ecuyer 2014). The zero MSE convergence, or neutral gross moist stability (GMS), implies that the import of moisture is just large enough to drive the large-scale adiabatic ascent and hence allows a self-sustaining growth of convection (Masunaga 2014). In the isolated cumulus regime (Fig. 10d), MSE convergence vanishes as in the organized system regime but the enhanced radiative cooling, owing to reduced high clouds, appears to work against the further growth of convection that could otherwise occur.

Note that the composite time series above are not to be interpreted as convective self-aggregation itself being in progress. Idealized simulations demonstrate that convective self-aggregation proceeds over a week or two (Tompkins 2001) or a few months (Bretherton et al. 2005; Wing and Emanuel 2014), which is a time scale substantially longer than the life cycle of individual convective systems (a few days at most) as depicted in Fig. 10. The isolated cumulus regime and organized system regime, if put into the context of self-aggregation, may be each a representation of the states before and after the self-aggregation takes place (or outside and inside the area of aggregated convection). From this perspective, the convective self-aggregation could be considered as a "phase transition" from the isolated cumulus regime to the organized system regime. Figure 10c, d suggests that a key role in the transition, if it occurs, would be the magnitude of radiative cooling, which is in line with idealized simulations (Muller and Bony 2015) and theories of convective self-aggregation (Emanuel et al. 2014). This hypothesis may be tested by separating the composite analysis among different degrees of convective aggregation using, for example, SCAI (Tobin et al. 2012). With the other environmental conditions such as SST being equal, a set of composite time series constructed with different SCAI values would provide an observational test bed to examine the self-aggregation processes in the context of moisture and thermal budgets. This would be an interesting line of research to pursue in the future.

5.2 Spaceborne Cloud Radar Approaches

Novel analyses of newer satellite assets that have not traditionally been applied to study convection may offer potential for advancing our understanding of the coupled radiative and hydrological responses to convective aggregation. There is growing acceptance of the utility of spaceborne cloud radars, in particular, for characterizing the distribution, internal structure, spatial organization, updraft intensity, and radiative environments of convection. New methods for discriminating precipitating scenes, isolating convective cores, and profiling radiative fluxes and heating rates both within cloud and in the adjacent cloud-free pixels are becoming sufficiently mature to shed new light on the coupled energy and water cycle impacts of convective aggregation (Haynes et al. 2009; Lebsock and L'Ecuyer 2011; Henderson et al. 2013; Matus and L'Ecuyer 2017).

Igel and Heever (2015), for example, used CloudSat observations to establish a quantitative link between the area of convective anvils and the associated convective cores. Unlike previous studies that relied on coarser or less direct methods for identifying convective updrafts, the high sensitivity and relatively high spatial resolution of the CloudSat Cloud Profiling Radar (CPR) provide an unambiguous means of discriminating precipitating and non-precipitating pixels with associated ice cloud area and vertical structure from a single sensor. Examining nearly 5 years of CloudSat observations over the tropical oceans, Igel et al. (2014) and Igel and Heever (2015) demonstrate that anvil widths systematically decrease while anvil temperatures become colder with increasing SST, as discussed in Sect. 4. In addition, the width of associated cloud object pedestals (the cloud shapes at the base of the anvils) decreases with increasing SST. These findings could be consistent with a trend toward more aggregated convection over warmer oceans, though it should be noted that, as discussed in Sect. 4, these studies look only at anvil size per cloud object, not at total cloud area or how individual cloud objects are spatially distributed. Furthermore, these studies do not explicitly control for precipitation intensity or divide observations into different large-scale circulation regimes, and they look at local SST rather than tropics-wide SST. Stein et al. (2017) use CloudSat-CALIPSO data to link cloud amount to aggregation, showing larger areas of anvil cloud and less low cloud in regions with less large-scale aggregation for a given large-scale rain rate, although they find less dependence of anvil fraction on (local) SST as discussed in Sect. 4.

The greatest potential of cloud radar observations for advancing theories of convective aggregation may, however, reside in recent efforts to infer internal dynamics and related processes (Luo et al. 2010; Nelson et al. 2016). As convection evolves to a more aggregated state, there is reason to anticipate that convective buoyancy and entrainment rates will change owing to the reduced convective area and cloud lateral boundaries. Luo et al. (2010) used the difference between cloud top temperature (CTT) and that of the ambient environment at the radar-defined cloud top height (CTH) to estimate convective buoyancy and entrainment rates in individual convective systems. CloudSat reflectivity observations effectively remove the ambiguity between cloud top temperature and height, allowing buoyancy to be estimated by comparing the observed CTT to the temperature at the CTH in the environmental sounding. Entrainment rates are then estimated through iterative application of an entraining plume model to obtain the best match with observed storm vertical structure. Luo et al. (2010) paint a familiar picture of tri-modal tropical convection made up of shallow, mid-level congestus, and deep convective modes (e.g., Johnson et al. 1999) but further characterize the composite dynamic processes within each mode. Nearly all deep convection has negatively buoyant cloud tops and smaller entrainment rates while congestus can be separated into distinct "transient" and "terminal" modes with positive and negative buoyancy (smaller and larger entrainment rates), respectively.

Given the challenges associated with directly observing the time evolution of convective cloud structures on the scales required to observe convective aggregation, it may be argued that composites of such observation-based estimates of dynamic and thermodynamic processes will be key to testing model-based inferences regarding the driving processes. It is very likely, for example, that transitions from scattered to aggregated states of convection will be accompanied by a shift in the relative frequencies of convective states, leading to corresponding changes in domain-mean buoyancy and entrainment rates that may be measured through a similar approach. While Luo et al. (2010) do not characterize the properties of shallow convection, recent work has demonstrated that evaporation and condensation rates in shallow convection can also be inferred from cloud radar observations, offering the potential to further address the role of shallow convection in the transition from isolated to aggregated convection (Nelson et al. 2016).

Spaceborne cloud radar observations also offer potential for testing hypothesized feedbacks and energy and water cycle impacts of convective aggregation. Luo et al. (2014) use time-differenced infrared brightness temperatures to relate cloud top vertical velocities to convective mass transport and precipitation efficiency, two central physical characteristics linking the causes and effects of convective aggregation. They demonstrate that stronger updrafts correlate with higher precipitation echo-tops, increased convective mass fluxes, and heavier rainfall throughout the tropics. While these studies do not definitively test emerging theories concerning convective aggregation, they attest to the maturity of novel process-related datasets from cloud radar observations and suggest that pursuing new ways of integrating spaceborne cloud radar into future studies of convective aggregation is warranted.

5.3 Feasibility of a Ground-Based Observational Network

Simulations of convective aggregation have shown that there is a marked difference in the water vapor profiles in the dry and moist regions (Fig. 4), and the longwave radiative heating difference between the two regions (Fig. 5) results in an up-gradient flow just above the boundary layer that works to further enhance this moisture gradient (Sect. 2). This characteristic difference between dry and moist regions is an important indicator of aggregation, and therefore something that a field experiment could target. In this section, we discuss the feasibility of such an experiment (for instance as part of a field campaign) using currently available ground-based instruments.

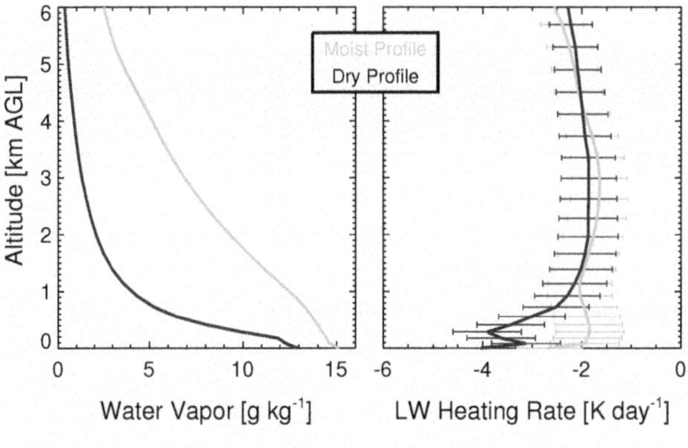

Fig. 11 Water vapor mixing ratio profiles from a dry (*black*) and moist (*gray*) region of an RCE simulation (at day 30) where convective aggregation occurred (*left*), and the corresponding longwave radiative heating rate profiles computed using the RRTM (*right*). The *error bars* on the heating rate profiles were computed by propagating the uncertainties in the Atmospheric Emitted Radiance Interferometer (AERI) retrieved profiles through the RRTM

An example of water vapor profiles from the dry and moist regions in a simulation of aggregation from Muller and Bony (2015), along with longwave radiative heating rates in cloud-free scenes computed using a radiative transfer model (RRTM; Mlawer et al. 1997), are shown in Fig. 11. The differences in the shape of the water vapor profiles result in an extra 2 K day^{-1} clear-sky longwave cooling in the boundary layer. This boundary layer cooling will be enhanced if there are shallow liquid water clouds at the top of the boundary layer; cumulus are often seen in the dry regions of simulations that show aggregation.

The challenge of any field experiment that aims to investigate the results shown by numerical simulations of convective aggregation is the ability to observe water vapor profiles, especially in the boundary layer, with the needed accuracy to yield significant differences in the computed radiative heating rate profiles (cf. Stevens et al. this issue). Many different boundary layer thermodynamic profiling technologies are currently being used; Wulfmeyer et al. (2015) provides a review of these instruments. Satellite sensors have difficulty observing the thermodynamic structure of the boundary layer, especially if there are clouds in the scene; the limitations of satellite observations of water vapor also affect reanalyses (Pincus et al. this issue). Thus, a network of multiple ground-based remote sensors distributed over some area is the best option if a long duration dataset is desired to observe the processes that lead to convective aggregation.

Of ground-based sensors, active remote sensors like water vapor Raman lidar and differential absorption lidar (DIAL) have a special appeal because of their vertical resolution and accuracy. However, there are no commercially available water vapor Raman lidars or DIALs, and thus any network of lidars would consist of systems from multiple research groups where each lidar would have its own sensitivity and uncertainties that may make the analysis of a network of these datasets more challenging. However, there are commercially available microwave radiometers and infrared spectrometers, and thus a network composed only of one of these types of instrument would be homogeneous and potentially easier to analyze.

Passive remote sensors like microwave radiometers and infrared spectrometers observe radiance, and retrieval algorithms are needed to derive thermodynamic profiles from these observations. Several studies have investigated the accuracy and information content of these retrieved profiles. Löhnert et al. (2009) used an instrument system simulation experiment to demonstrate that infrared spectrometers such as the Atmospheric Emitted Radiance Interferometer (AERI, Knuteson et al. 2004a, b) have 2–4 times more information on both the temperature and water vapor profile than microwave radiometers, which leads to improved accuracy in the AERI-retrieved profiles under clear-sky conditions. Blumberg et al. (2015) and Weckwerth et al. (2016) both confirmed that the AERI-retrieved water vapor profile was more accurate than microwave radiometers below cloud base or in cloud-free scenes using real observations.

A natural question is: Do the AERI retrievals have the sensitivity to distinguish between the longwave radiative cooling rate profiles in the dry and moist columns seen in convective aggregation scenarios? If so, then the AERI would be a good choice to include in any ground-based network that is established to study convective aggregation from observations. The AERI retrieval algorithm developed by Turner and Löhnert (2014), which is able to retrieve lower-tropospheric thermodynamic profiles in both clear and cloudy conditions, provides a complete error covariance matrix for each retrieval. Thermodynamic profiles derived from a Monte Carlo sampling of this error covariance matrix were used to derive the RRTM to compute cloud-free longwave radiative heating profiles, and the $1-\sigma$ uncertainties at each level are shown in Fig. 11 (right). This demonstrates that the AERI has the accuracy to determine the radiative heating rate profiles in the two clear-sky scenes. However, in moist convective regions, where clouds are numerous and can dominate radiation fluxes, a combination of AERI and microwave radiometer retrievals may be desirable to ensure sufficiently accurate humidity profiles (Löhnert et al. 2009; Turner and Löhnert 2014). Furthermore, low clouds in the dry region contribute significantly to self-aggregation in numerical simulations, so it would also be desirable to measure vertical profiles of cloud water. This would enable a calculation of the total radiative heating rate profiles. Cloud radars are the only type of instrument capable of this type of measurement, but are likely prohibitively expensive to deploy in a network as proposed here. A first step could be the deployment of ceilometers, which are a standard, relatively inexpensive, autonomous, weak lidar used primarily at airports to determine cloud base height.

A main strength of microwave and AERI measurements is their ability to measure vertical profiles of moisture. If deployed in a network, the profilers in combination also provide spatial context. A further extension integrates the surface-based measurements with a satellite view of the CWV, thereby more fully interrogating the moisture budget expressed in (1) (Hannah et al. 2016). The satellite can also be integrated with satellite-derived perceptions of the precipitation and cloud distribution, while the surface-based network provides further information on low clouds not easily detected from space and fills in measurements in-between satellite overpasses, so that a rich, dense, three-dimensional construction of a moisture field can be constructed that is large enough to encompass both dry and moist regions. A remaining difficulty may be the typically short time spans for a field deployment, muddying an interpretation of self-aggregation from data. Nevertheless, high-resolution large-domain simulations coincident with such field observations, and combined with observed surface fluxes and top-of-atmosphere radiative fluxes, will inspire a deeper confidence in the theory of self-aggregation derived from RCE simulations and help determine the relative importance of contributing processes.

6 Conclusions

Observing convective organization is not a new pursuit, as evidenced by the literature review in this paper. But as we learn more about how convection clumps in idealized models, there are new opportunities to formulate theories of fundamental convective processes and test them (or at least gauge their plausibility) using observations. Models can also be used at more realistic configurations to form a bridge between idealized simulations and observations, and to help us better frame observational studies.

Insights from idealized simulations are already raising many new questions about how the climate interacts with convective organization. But some findings are dependent on model setup or formulation, and the few existing observational studies of aggregation are not completely consistent with each other or with some model findings—we encourage recent efforts to organize an intercomparison of RCE in models over a range of complexities and configurations to help resolve these discrepancies. There is agreement between models and observations that, as convection becomes clumped into fewer moist regions, the subsidence regions become drier, resulting in a drier large-scale mean environment. This drying, and a reduction of upper-tropospheric stratiform cloud, leads to larger OLR and stronger atmospheric cooling. However, how aggregation and its effects interact with local SST on short time scales and with tropics-wide SST on long time scales is still uncertain in both models and observations, as discussed further below.

Initiation processes, such as radiatively driven cold pools and related shallow overturning circulations, are one obvious observational target. Maintenance processes may be even easier to study in observations, since they can be studied in heterogeneous conditions more typical of convection in nature. There are already links between convective self-aggregation processes in models and observed phenomena such as the MJO and tropical cyclones, with feedbacks involving convection, clouds, moisture, radiation, and surface fluxes being important. In fact, the difficulty of global weather and climate models to simulate these phenomena may be related to problems with those models' ability to simulate aggregation processes, as mentioned in Wing et al. (2017).

Time scales are longer in idealized self-aggregation from homogeneous conditions than typical time scales of observed growth of organized mesoscale convection, but we have argued that this does not mean that idealized processes are not relevant for real organized convection. This is because time scales vary a lot in idealized models, and exponential growth implies shorter effective time scales when starting from already existing organization as is often found in nature. Furthermore, although the real world certainly contains additional processes that can organize or disorganize convection (such as those reliant on coastlines and orography), and these may dominate where they are faster or stronger than self-aggregation processes, these are likely to be concentrated in particular regions and at particular (especially smaller) space and time scales. This means that self-aggregation processes may be favored in other regions and, perhaps, on larger spatial scales, and they are still likely to be relevant for many phenomena and for climate. Feedbacks allowing for the maintenance of idealized aggregation may also be important for maintaining organized convection in nature, since disaggregation time scales are relatively short when longwave radiation feedbacks are turned off in idealized simulations.

Some preliminary new findings show that an idealized simulation with elongated channel geometry has a more realistic representation of atmospheric humidity than a simulation with a square domain, which has too broad distribution of humidity and is too dry in the driest regions when compared with radiosonde records from Nauru. This is an example of how observations may be used to discriminate between different model configurations and address concerns that may otherwise cast doubt on the relevance of aggregation studies in general. Determining the reason for the difference in humidity between these two model configurations is beyond the scope of this paper, and some caution should be exercised when interpreting these results until the reasons and their relationship to physical processes are better understood, since it is possible that the channel simulation gets the "right answer for the wrong reason". However, possible reasons for these humidity differences include differences in wind speed caused by stronger overturning circulations in the channel simulations, the development of slightly stronger mean wind and wind shear in the channel simulations, or interactions between large-scale circulations and multiple convective regions causing increased proximity and mixing between moist and dry regions.

Recent work has underlined the potential importance of the sensitivity of aggregation processes to SST and climate change. There are exciting new processes being proposed, such as the "stability-iris" effect (Bony et al. 2016) that predicts smaller anvil fractions in a warmer climate. We are gaining an understanding of how organized convection and climate change may interact, but new questions are being raised about the fidelity of our models. The ability to represent both large-scale circulations and convective processes adequately is still a challenge for idealized models, and the effects of more complex processes such as ocean coupling are only beginning to be explored. There are also challenges in using observations of the recent past to test model behavior in both idealized and more realistic simulations of projected future climate. Specifically, we lack long observational records, and studies of regional snapshots or individual cloud elements over short time periods may not scale up to tropics-wide behavior on long time scales. There are also differences in the SST dependence of initiation processes versus maintenance processes in idealized self-aggregation which need to be further explored in both models and observations. But the problem of aggregation in a warming world is an important one and deserves a sustained research effort.

In addition to encouraging the continuing endeavor of confronting self-aggregation processes with observations in general, we have proposed several specific lines of work which appear promising. These include:

- existence, formation, and structure of radiatively driven cold pools
- MSE spatial variance budget analysis
- particle tracing back to last saturation (model and observations)
- radiosonde time scale analysis
- changes in satellite-observed aggregation state and anvil fraction over last few decades (and further investigation of sensitivity to SST)
- paleotempestology using long-term isotopic records
- satellite analysis of link between water vapor isotopes and aggregation
- multiple-satellite temporal evolution of MSE budget for aggregated convection
- spaceborne radar to infer convective processes
- ground-based observational network

While this list is a good start, there are surely other opportunities to further our knowledge of aggregation processes in idealized models using observations. And as the field advances, there will be more ideas to test and explore. We hope the work reviewed and proposed here is the beginning of an exciting new engagement between modeling and observational studies of organized convection.

Acknowledgements This paper arises from the International Space Science Institute (ISSI) workshop on "Shallow clouds and water vapor, circulation and climate sensitivity". ARM data were used with the cooperation of the U.S. Department of Energy as part of the Atmospheric Radiation Measurement Program Climate Research Facility. We thank Kerry Emanuel, Robert Pincus, Thorwald Stein, James Ruppert, and an anonymous reviewer for helpful comments. AAW acknowledges support from an NSF Postdoctoral Research Fellowship under Award No. AGS-1433251. SB acknowledges support from ERC Grant No. 694768.

References

Abbot DS (2014) Resolved snowball earth clouds. J Clim 27(12):4391–4402. doi:10.1175/JCLI-D-13-00738.1

Arkin PA (1979) The relationship between fractional coverage of high cloud and rainfall accumulations during GATE over the B-Scale array. Mon Weather Rev 107(10):1382–1387. doi:10.1175/1520-0493(1979)107<1382:TRBFCO>2.0.CO;2

Arnold NP, Randall DA (2015) Global-scale convective aggregation: implications for the Madden–Julian oscillation. J Adv Model Earth Syst 7(4):1499–1518. doi:10.1002/2015MS000498

Arnold NP, Branson M, Kuang Z, Randall DA, Tziperman E (2015) MJO intensification with warming in the superparameterized CESM. J Clim 28(7):2706–2724. doi:10.1175/JCLI-D-14-00494.1

Blumberg W, Turner D, Lohnert U, Castleberry S (2015) Ground-based temperature and humidity profiling using spectral infrared and microwave observations. Part 2: Actual retrieval performance in clear sky and cloudy conditions. J Appl Meteorol Clim 54:2305–2319. doi:10.1175/JAMC-D-15-0005.1

Bony S, Lau KM, Sud YC (1997) Sea surface temperature and large-scale circulation influences on tropical greenhouse effect and cloud radiative forcing. J Clim 10(8):2055–2077. doi:10.1175/1520-0442(1997)0102055:SSTALS2.0.CO;2

Bony S, Stevens B, Frierson DMW, Jakob C, Kageyama M, Pincus R, Shepherd TG, Sherwood SC, Siebesma AP, Sobel AH, Watanabe M, Webb MJ (2015) Clouds, circulation and climate sensitivity. Nat Geosci 8(4):261–268

Bony S, Stevens B, Coppin D, Becker T, Reed KA, Voigt A, Medeiros B (2016) Thermodynamic control of anvil cloud amount. Proc Nat Acad Sci 113(32):8927–8932

Bretherton CS, Khairoutdinov MF (2015) Convective self-aggregation feedbacks in near-global cloud-resolving simulations of an aquaplanet. J Adv Model Earth Syst 7(4):1765–1787. doi:10.1002/2015MS000499

Bretherton CS, Blossey PN, Khairoutdinov M (2005) An energy-balance analysis of deep convective self-aggregation above uniform SST. J Atmos Sci 62:4273–4292

Caballero R, Huber M (2010) Spontaneous transition to superrotation in warm climates simulated by CAM3. Geophys Res Lett. doi:10.1029/2010GL043468

Coppin D, Bony S (2015) Physical mechanisms controlling the initiation of convective self-aggregation in a general circulation model. J Adv Model Earth Syst 7(4):2060–2078. doi:10.1002/2015MS000571

Coppin D, Bony S (2017) Internal variability in a coupled general circulation model in radiative–convective equilibrium. Geophys Res Lett. doi:10.1002/2017GL073658

Craig GC, Mack JM (2013) A coarsening model for self-organization of tropical convection. J Geophys Res Atmos 118(16):8761–8769. doi:10.1002/jgrd.50674

Davis CA (2015) The formation of moist vortices and tropical cyclones in idealized simulations. J Atmos Sci 72(9):3499–3516. doi:10.1175/JAS-D-15-0027.1

DeMott CA, Klingaman NP, Woolnough SJ (2015) Atmosphere-ocean coupled processes in the Madden–Julian oscillation. Rev Geophys 53(4):1099–1154. doi:10.1002/2014RG000478

Dessler AE (2010) A determination of the cloud feedback from climate variations over the past decade. Science 330(6010):1523

Donnelly JP, Woodruff JD (2007) Intense hurricane activity over the past 5,000 years controlled by El Nino and the West African monsoon. Nature 447(7143):465–468

Emanuel KA (1987) The dependence of hurricane intensity on climate. Nature 326(6112):483–485

Emanuel K (2015) Effect of upper-ocean evolution on projected trends in tropical cyclone activity. J Clim 28(20):8165–8170. doi:10.1175/JCLI-D-15-0401.1

Emanuel K, Wing AA, Vincent EM (2014) Radiative–convective instability. J Adv Model Earth Syst 6(1):75–90. doi:10.1002/2013MS000270

Feng Z, Hagos S, Rowe AK, Burleyson CD, Martini MN, de Szoeke SP (2015) Mechanisms of convective cloud organization by cold pools over tropical warm ocean during the AMIE/DYNAMO field campaign. J Adv Model Earth Syst 7(2):357–381. doi:10.1002/2014MS000384

Futyan JM, Del Genio AD (2007) Deep convective system evolution over Africa and the tropical Atlantic. J Clim 20(20):5041–5060. doi:10.1175/JCLI4297.1

Hannah W, Mapes B, Elsaesser G (2016) A Lagrangian view of moisture dynamics during DYNAMO. J Atmos Sci 73:1967–1985. doi:10.1175/jas-D-15-0243.1

Hartmann DL, Michelsen ML (1993) Large-scale effects on the regulation of tropical sea surface temperature. J Clim 6(11):2049–2062. doi:10.1175/1520-0442(1993)0062049:LSEOTR2.0.CO;2

Hartmann DL, Hendon HH, Houze RA (1984) Some implications of the mesoscale circulations in tropical cloud clusters for large-scale dynamics and climate. J Atmos Sci 41(1):113–121. doi:10.1175/1520-0469(1984)0410113:SIOTMC;2.0.CO;2

Haynes JM, L'Ecuyer T, Stephens GL, Miller SD, Mitrescu C, Wood NB, Tanelli S (2009) Rainfall retrievals over the ocean with spaceborne high-frequency cloud radar. J Geophys Res. doi:10.1029/2008JD009,973

Haynes JM, Vonder Haar TH, L'Ecuyer T, Henderson D (2013) Radiative heating characteristics of Earth's cloudy atmosphere from vertically resolved active sensors. Geophys Res Lett 40(3):624–630. doi:10.1002/grl.50145

Held IM, Zhao M (2008) Horizontally homogeneous rotating radiative–convective equilibria at GCM resolution. J Atmos Sci 65(6):2003–2013. doi:10.1175/2007JAS2604.1

Held IM, Hemler RS, Ramaswamy V (1993) Radiative–convective equilibrium with explicit two-dimensional moist convection. J Atmos Sci 50(23):3909–3927

Henderson DS, L'Ecuyer T, Stephens G, Partain P, Sekiguchi M (2013) A multi-sensor perspective on the radiative impacts of clouds and aerosols. J Appl Meteorol Climatol 52:853–871

Hennon CC, Papin PP, Zarzar CM, Michael JR, Caudill JA, Douglas CR, Groetsema WC, Lacy JH, Maye ZD, Reid JL, Scales MA, Talley MD, Helms CN (2012) Tropical cloud cluster climatology, variability, and genesis productivity. J Clim 26(10):3046–3066. doi:10.1175/JCLI-D-12-00387.1

Hohenegger C, Stevens B (2016) Coupled radiative convective equilibrium simulations with explicit and parameterized convection. J Adv Model Earth Syst 8(3):1468–1482. doi:10.1002/2016MS000666

Holloway CE (2017) Convective aggregation in realistic convective-scale simulations. J Adv Model Earth Syst. doi:10.1002/2017MS000980

Holloway CE, Neelin JD (2009) Moisture vertical structure, column water vapor, and tropical deep convection. J Atmos Sci 66:1665–1683

Holloway CE, Woolnough SJ (2016) The sensitivity of convective aggregation to diabatic processes in idealized radiative–convective equilibrium simulations. J Adv Model Earth Syst 8(1):166–195. doi:10.1002/2015MS000511

Houze RA Jr (2004) Mesoscale convective systems. Rev Geophys RG4042:3. doi:10.1029/2004RG000150

Igel MR, van den Heever SC (2015) Tropical, oceanic, deep convective cloud morphology as observed by CloudSat. Atmos Chem Phys Discuss 2015:15,977–16,017. doi:10.5194/acpd-15-15977-2015

Igel MR, Drager AJ, van den Heever SC (2014) A CloudSat cloud object partitioning technique and assessment and integration of deep convective anvil sensitivities to sea surface temperature. J Geophys Res Atmos 119(17):10,515–10,535. doi:10.1002/2014JD021717

Jeevanjee N, Romps DM (2013) Convective self-aggregation, cold pools, and domain size. Geophys Res Lett 40(5):994–998. doi:10.1002/grl.50204

Johnson RH, Ciesielski PE (2013) Structure and properties of Madden–Julian oscillations deduced from DYNAMO sounding arrays. J Atmos Sci 70(10):3157–3179. doi:10.1175/JAS-D-13-065.1

Johnson RH, Rickenbach TM, Rutledge S, Ciesielski P, Schubert W (1999) Trimodal characteristics of tropical convection. J Clim 12:2397–2418

Jones C, Carvalho LMV (2006) Changes in the activity of the Madden–Julian oscillation during 1958–2004. J Clim 19(24):6353–6370. doi:10.1175/JCLI3972.1

Khairoutdinov M, Emanuel K (2010) Aggregation of convection and the regulation of climate, preprints. In: 29th conference on hurricanes and tropical meteorology, American Meteorological Society, Tucson, AZ, p P2.69

Khairoutdinov M, Emanuel K (2013) Rotating radiative–convective equilibrium simulated by a cloud-resolving model. J Adv Model Earth Syst 5(4):816–825. doi:10.1002/2013MS000253

Kiladis GN, Wheeler MC, Haertel PT, Straub KH, Roundy PE (2009) Convectively coupled equatorial waves. Rev Geophys. doi:10.1029/2008RG000266

Knuteson RO et al (2004a) The atmospheric emitted radiance interferometer (AERI) part I: instrument design. J Atmos Ocean Technol 21:1763–1776

Knuteson RO et al (2004b) The atmospheric emitted radiance interferometer (AERI) part II: instrument performance. J Atmos Ocean Technol 21:1777–1789

Knutson TR, Sirutis JJ, Vecchi GA, Garner S, Zhao M, Kim HS, Bender M, Tuleya RE, Held IM, Villarini G (2013) Dynamical downscaling projections of twenty-first-century Atlantic hurricane activity: CMIP3 and CMIP5 model-based scenarios. J Clim 26(17):6591–6617. doi:10.1175/JCLI-D-12-00539.1

Laing AG, Fritsch JM (1993a) Mesoscale convective complexes in Africa. Mon Weather Rev 121(8):2254–2263. doi:10.1175/1520-0493(1993)1212254:MCCIA2.0.CO;2

Laing AG, Fritsch JM (1993b) Mesoscale convective complexes over the Indian monsoon region. J Clim 6(5):911–919. doi:10.1175/1520-0442(1993)0060911:MCCOTI2.0.CO;2

Laing AG, Fritsch JM (1997) The global population of mesoscale convective complexes. Q J R Meteorol Soc 123(538):389–405. doi:10.1002/qj.49712353807

Lawrence JR, Gedzelman SD, Dexheimer D, Cho HK, Carrie GD, Gasparini R, Anderson CR, Bowman KP, Biggerstaff MI (2004) Stable isotopic composition of water vapor in the tropics. J Geophys Res Atmos. doi:10.1029/2003JD004046

Lebsock M, L'Ecuyer TS (2011) The retrieval of warm rain from CloudSat. J Geophys Res. doi:10.1029/2011JD016,076

Lin Y, Zhao M, Zhang M (2015) Tropical cyclone rainfall area controlled by relative sea surface temperature. Nat Commun 6:6591

Liu Kb, Fearn ML (2000) Reconstruction of prehistoric landfall frequencies of catastrophic hurricanes in northwestern Florida from lake sediment records. Quat Res 54(2):238–245. doi:10.1006/qres.2000.2166

Löhnert U, Turner D, Crewell S (2009) Ground-based temperature and humidity profiling using spectral infrared and microwave observations. Part 1: Simulated retrieval performance in clear sky conditions. J Appl Meteorol Clim 48:1017–1032. doi:10.1175/2008JAMC2060.1

Long CN, Mather JH, Ackerman TP (2016) The ARM Tropical Western Pacific (TWP) sites. Meteorol Monogr 57:7:1–7:14. doi:10.1175/AMSMONOGRAPHS-D-15-0024.1

Luo Z, Liu GY, Stephens GL (2010) Use of A-Train data to estimate convective buoyancy and entrainment rate. Geophys Res Lett 37:L09,804. doi:10.1029/2010GL042,904

Luo Z, Jeyaratnam J, Iwasaki S, Takahashi H, Anderson R (2014) Convective vertical velocity and cloud internal vertical structure: an A-Train perspective. Geophys Res Lett 41:723–729

Machado LAT, Rossow WB (1993) Structural characteristics and radiative properties of tropical cloud clusters. Mon Weather Rev 121(12):3234–3260. doi:10.1175/1520-0493(1993)121<3234:SCARPO>2.0.CO;2

Mapes BE (1993) Gregarious tropical convection. J Atmos Sci 50:2026–2037

Mapes BE (2001) Water's two height scales: the moist adiabat and the radiative troposphere. Q J R Meteorol Soc 127(577):2353–2366. doi:10.1002/qj.49712757708

Mapes BE (2016) Gregarious convection and radiative feedbacks in idealized worlds. J Adv Model Earth Syst. doi:10.1002/2016MS000651

Mapes BE, Houze RA (1993) Cloud clusters and superclusters over the oceanic warm pool. Mon Weather Rev 121(5):1398–1416. doi:10.1175/1520-0493(1993)1211398:CCASOT2.0.CO;2

Mapes B, Neale R (2011) Parameterizing convective organization to escape the entrainment dilemma. J Adv Model Earth Syst 3(2):M06,004. doi:10.1029/2011MS000042

Mapes B, Tulich S, Lin J, Zuidema P (2006) The mesoscale convection life cycle: building block or prototype for large-scale tropical waves? Dyn Atmos Ocean 42:3–29

Mapes BE, Milliff R, Morzel J (2009) Composite life cycle of maritime tropical mesoscale convective systems in scatterometer and microwave satellite observations. J Atmos Sci 66:199–208

Masunaga H (2012) A satellite study of the atmospheric forcing and response to moist convection over tropical and subtropical oceans. J Atmos Sci 69:150–167

Masunaga H (2013) A satellite study of tropical moist convection and environmental variability: a moisture and thermal budget analysis. J Atmos Sci 70:2443–2466

Masunaga H (2014) Free-tropospheric moisture convergence and tropical convective regimes. Geophys Res Lett 41(23):8611–8618. doi:10.1002/2014GL062301

Masunaga H, L'Ecuyer TS (2014) A mechanism of tropical convection inferred from observed variability in the moist static energy budget. J Atmos Sci 71:3747–3766

Mather JH, Ackerman TP, Clements WE, Barnes FJ, Ivey MD, Hatfield LD, Reynolds RM (1998) An atmospheric radiation and cloud station in the tropical western Pacific. Bull Amer Meteorol Soc 79:627–642

Matus AV, L'Ecuyer TS (2017) The role of cloud phase in Earth's radiation budget. J Geophys Res Atmos 122:2559–2578. doi:10.1002/2016JD025951

Mauritsen T, Stevens B (2015) Missing iris effect as a possible cause of muted hydrological change and high climate sensitivity in models. Nat Geosci 8(5):346–351

Miller D, Fritsch JM (1990) Mesoscale convective complexes in the western Pacific region. Mon Weather Rev 119(12):2978–2992. doi:10.1175/1520-0493(1991)1192978:MCCITW2.0.CO;2

Mlawer EJ, Taubman SJ, Brown PD, Iacono MJ, Clough SA (1997) Radiative transfer for inhomogeneous atmospheres: RRTM, a validated correlated-K model for the longwave. J Geophys Res 102:16663–16682

Muller C (2013) Impact of convective organization on the response of tropical precipitation extremes to warming. J Clim 26(14):5028–5043. doi:10.1175/JCLI-D-12-00655.1

Muller CJ, Held IM (2012) Detailed investigation of the self-aggregation of convection in cloud-resolving simulations. J Atmos Sci 69(8):2551–2565. doi:10.1175/JAS-D-11-0257.1

Muller C, Bony S (2015) What favors convective aggregation and why? Geophys Res Lett 42(13):5626–5634. doi:10.1002/2015GL064260

Neelin JD, Peters O, Hales K (2009) The transition to strong convection. J Atmos Sci 66:2367–2384

Nelson EL, L'Ecuyer TS, Saleeby SM, Berg W, Herbener SR, van den Heever SC (2016) Toward an algorithm for estimating latent heat release in warm rain systems. J Atmos Ocean Technol 33:1309–1329

Nesbitt SW, Zipser EJ, Cecil DJ (2000) A census of precipitation features in the tropics using TRMM: radar, ice scattering, and lightning observations. J Clim 13(23):4087–4106. doi:10.1175/1520-0442(2000)0134087:ACOPFI2.0.CO;2

Nolan DS, Rappin ED, Emanuel KA (2007) Tropical cyclogenesis sensitivity to environmental parameters in radiative–convective equilibrium. Q J R Meteorol Soc 133:2085–2107

Peters O, Neelin JD (2006) Critical phenomena in atmospheric precipitation. Nat Phys 2:393–396

Peters O, Neelin JD, Nesbitt SW (2009) Mesoscale convective systems and critical clusters. J Atmos Sci 66(9):2913–2924

Pierrehumbert RT (1995) Thermostats, radiator fins, and the local runaway greenhouse. J Atmos Sci 52(10):1784–1806. doi:10.1175/1520-0469(1995)0521784:TRFATL2.0.CO;2

Pierrehumbert R (1998) Lateral mixing as a source of subtropical water vapor. Geophys Res Lett 25:151–154

Popke D, Stevens B, Voigt A (2013) Climate and climate change in a radiative–convective equilibrium version of ECHAM6. J Adv Model Earth Syst 5(1):1–14. doi:10.1029/2012MS000191

Pritchard MS, Yang D (2016) Response of the superparameterized Madden–Julian oscillation to extreme climate and basic-state variation challenges a moisture mode view. J Clim 29(13):4995–5008. doi:10.1175/JCLI-D-15-0790.1

Randall DA, Huffman GJ (1980) A stochastic model of cumulus clumping. J Atmos Sci 37:2068–2078

Reed KA, Medeiros B, Bacmeister JT, Lauritzen PH (2015) Global radiative–convective equilibrium in the community atmosphere model, version 5. J Atmos Sci 72(5):2183–2197. doi:10.1175/JAS-D-14-0268.1

Reitebuch O (2012) The spaceborne wind lidar mission ADM-Aeolus. In: Schumann U (ed) Atmospheric physics: background–methods–trends. Springer, Berlin, pp 815–827. doi:10.1007/978-3-642-30183-4_49

Risi C, Bony S, Vimeux F (2008) Influence of convective processes on the isotopic composition (18O and D) of precipitation and water vapor in the tropics: 2. Physical interpretation of the amount effect. J Geophys Res Atmos 100:100. doi:10.1029/2008JD009943

Roca R, Ramanathan V (2000) Scale dependence of monsoonal convective systems over the Indian Ocean. J Clim 13(7):1286–1298. doi:10.1175/1520-0442(2000)0131286:SDOMCS2.0.CO;2

Schumacher C, Houze RA Jr (2003) Stratiform rain in the tropics as seen by the TRMM precipitation radar. J Clim 16:1739–1756

Sessions SL, Sentić S, Herman MJ (2016) The role of radiation in organizing convection in weak temperature gradient simulations. J Adv Model Earth Syst 8(1):244–271. doi:10.1002/2015MS000587

Silvers LG, Stevens B, Mauritsen T, Giorgetta M (2016) Radiative convective equilibrium as a framework for studying the interaction between convection and its large-scale environment. J Adv Model Earth Syst 8(3):1330–1344. doi:10.1002/2016MS000629

Sobel A, Wang S, Kim D (2014) Moist static energy budget of the MJO during DYNAMO. J Atmos Sci 71(11):4276–4291. doi:10.1175/JAS-D-14-0052.1

Stein THM, Holloway CE, Tobin I, Bony S (2017) Observed relationships between cloud vertical structure and convective aggregation over tropical ocean. J Clim 30(6):2187–2207. doi:10.1175/JCLI-D-16-0125.1

Stephens GL, van den Heever S, Pakula L (2008) Radiative–convective feedbacks in idealized states of radiative–convective equilibrium. J Atmos Sci 65(12):3899–3916. doi:10.1175/2008JAS2524.1

Stevens B, Farrell D, Hirsch L, Jansen F, Nuijens L, Serikov I, Brügmann B, Forde M, Linne H, Lonitz K, Prospero JM (2015) The Barbados Cloud Observatory: anchoring investigations of clouds and circulation on the edge of the ITCZ. Bull Am Meteorol Soc 97(5):787–801. doi:10.1175/BAMS-D-14-00247.1

Stocker T, Qin D, Plattner GK, Alexander L, Allen S, Bindoff N, Bréon FM, Church J, Cubasch U, Emori S, Forster P, Friedlingstein P, Gillett N, Gregory J, Hartmann D, Jansen E, Kirtman B, Knutti R, Krishna Kumar K, Lemke P, Marotzke J, Masson-Delmotte V, Meehl G, Mokhov I, Piao S, Ramaswamy V, Randall D, Rhein M, Rojas M, Sabine C, Shindell D, Talley L, Vaughan D, Xie SP (2013) Technical summary. In: Stocker T, Qin D, Plattner GK, Tignor M, Allen S, Boschung J, Nauels A, Xia Y, Bex V, Midgley P (eds) Climate change 2013: the physical science basis. Contribution of working group I to the fifth assessment report of the intergovernmental panel on climate change. Cambridge University Press, Cambridge, pp 33–115. doi:10.1017/CBO9781107415324.005

Su H, Bretherton CS, Chen SS (2000) Self-aggregation and large-scale control of tropical deep convection: a modeling study. J Atmos Sci 57(11):1797–1816. doi:10.1175/1520-0469(2000)0571797:SAALSC2.0.CO;2

Tan J, Jakob C, Lane T (2013) On the identification of the large scale properties of tropical convection using cloud regimes. J Clim 26:6618–6632

Tan J, Jakob C, Rossow WB, Tselioudis G (2015) Increases in tropical rainfall driven by changes in frequency of organized deep convection. Nature 519(7544):451–454

Tang B, Emanuel K (2010) Midlevel ventilation's constraint on tropical cyclone intensity. J Atmos Sci 67:1817–1830

Tobin I, Bony S, Roca R (2012) Observational evidence for relationships between the degree of aggregation of deep convection, water vapor, surface fluxes, and radiation. J Clim 25(20):6885–6904. doi:10.1175/JCLI-D-11-00258.1

Tobin I, Bony S, Holloway CE, Grandpeix JY, Sèze G, Coppin D, Woolnough SJ, Roca R (2013) Does convective aggregation need to be represented in cumulus parameterizations? J Adv Model Earth Syst 5(4):692–703. doi:10.1002/jame.20047

Tompkins AM (2001) Organization of tropical convection in low vertical wind shears: the role of water vapor. J Atmos Sci 58:529–545

Tompkins AM, Craig GC (1998) Radiative–convective equilibrium in a three-dimensional cloud-ensemble model. Q J R Meteorol Soc 124:2073–2097

Tompkins AM, Semie AG (2017) Organization of tropical convection in low vertical wind shears: role of updraft entrainment. J Adv Model Earth Syst. doi:10.1002/2016MS000802

Tselioudis G, Tromeur E, Rossow W, Zerefos C (2010) Decadal changes in tropical convection suggest effects on stratospheric water vapor. Geophys Res Lett 37(L14):806

Turner DD, Löhnert U (2014) Information content and uncertainties in thermodynamic profiles and liquid cloud properties retrieved from the ground-based Atmospheric Emitted Radiance Interferometer (AERI). J Appl Meteorol Clim 53:752–771. doi:10.1175/JAMC-D-13-0126.1

Velasco I, Fritsch JM (1987) Mesoscale convective complexes in the Americas. J Geophys Res Atmos 92(D8):9591–9613. doi:10.1029/JD092iD08p09591

Weckwerth TM, Weber KJ, Turner DD, Spuler SM (2016) Validation of a water vapor micropulse differential absorption lidar (DIAL). J Atmos Ocean Technol 33:2353–2372. doi:10.1175/JTECH-D-16-0119.1

Wing AA (2014) Physical mechanisms controlling self-aggregation of convection in idealized numerical modeling simulations. Ph.D. thesis, MIT

Wing AA, Emanuel KA (2014) Physical mechanisms controlling self-aggregation of convection in idealized numerical modeling simulations. J Adv Model Earth Syst 6(1):59–74. doi:10.1002/2013MS000269

Wing AA, Cronin TW (2016) Self-aggregation of convection in long channel geometry. Q J R Meteorol Soc 142:1–15. doi:10.1002/qj.2628

Wing AA, Camargo SJ, Sobel AH (2016) Role of radiative–convective feedbacks in spontaneous tropical cyclogenesis in idealized numerical simulations. J Atmos Sci 73(7):2633–2642. doi:10.1175/JAS-D-15-0380.1

Wing AA, Emanuel K, Holloway CE, Muller C (2017) Convective self-aggregation in numerical simulations: a review. Surv Geophys. doi:10.1007/s10712-017-9408-4

Wulfmeyer V, Hardesty R, Turner D, Behrendt A, Cadeddu M, Girolamo PD, Schluessel P, van Baelen J, Zus F (2015) A review of the remote sensing of lower-tropospheric thermodynamic profiles and its indispensible role for the understanding and simulation of water and energy cycles. Rev Geophys 53:819–895. doi:10.1002/2014RG000476

Zelinka MD, Hartmann DL (2011) The observed sensitivity of high clouds to mean surface temperature anomalies in the tropics. J Geophys Res Atmos. doi:10.1029/2011JD016459

Zhang C, Mapes BE, Soden BJ (2003) Bimodality in tropical water vapour. Q J R Meteorol Soc 129(594):2847–2866. doi:10.1256/qj.02.166

Zuidema P (2003) Convective clouds over the Bay of Bengal. Mon Weather Rev 131:780–798

Zuidema P, Mapes B (2008) Cloud vertical structure observed from space and ship over the Bay of Bengal and eastern tropical Pacific. J Meteorol Soc Japan 86A:205–218

Zuidema P, Mapes B, Lin J, Fairall C, Wick G (2006) The interaction of clouds and dry air in the eastern tropical Pacific. J Clim 19:4531–4544

Surv Geophys (2017) 38:1237–1254
https://doi.org/10.1007/s10712-017-9443-1

An Observational View of Relationships Between Moisture Aggregation, Cloud, and Radiative Heating Profiles

Matthew D. Lebsock[1] · Tristan S. L'Ecuyer[2] · Robert Pincus[3,4]

Received: 27 March 2017 / Accepted: 19 October 2017 / Published online: 31 October 2017

Abstract Data from several coincident satellite sensors are analyzed to determine the dependence of cloud and precipitation characteristics of tropical regions on the variance in the water vapor field. Increased vapor variance is associated with decreased high cloud fraction and an enhancement of low-level radiative cooling in dry regions of the domain. The result is found across a range of sea surface temperatures and rain rates. This suggests the possibility of an enhanced low-level circulation feeding the moist convecting areas when vapor variance is large. These findings are consistent with idealized models of self-aggregation, in which the aggregation of convection is maintained by a combination of low-level radiative cooling in dry regions and mid-to-upper-level radiative warming in cloudy regions.

Keywords Convective aggregation · Radiation · Water vapor · Satellite · Observations

1 Introduction

Radiative-convective equilibrium, in which heating of the atmosphere by moist convection and precipitation balances radiative cooling, is an idealization of the Earth's tropical atmosphere that neglects advective energy transport. Over a uniform surface, and in domains large enough to contain many convective elements, the null hypothesis would be

The original version of this article was revised due to a retrospective Open Access order.

✉ Matthew D. Lebsock
matthew.d.lebsock@jpl.nasa.gov

1 Jet Propulsion Laboratory, California Institute of Technology, Pasadena, CA 91109, USA

2 University of Wisconsin, Madison, WI 53706, USA

3 Cooperative Institute for Research in Environmental Sciences, University of Colorado, Boulder, CO 80309, USA

4 Physical Sciences Division, NOAA Earth System Research Lab, Boulder, CO 80305, USA

that convection in radiative-convective equilibrium would be distributed roughly uniformly throughout the domain. It has been known for more than 20 years, however, that convection in numerical model simulations of radiative-convective equilibrium frequently gathers itself together, increasing in spatial scale until, under many circumstances, the entire domain contains only a single region of convection (Held et al. 1993; Bretherton et al. 2005; Stephens et al. 2008; Wing and Emanuel 2014; Tompkins and Craig 1998). The phenomenon of convective 'self-aggregation' was originally noted in cloud-resolving models in which deep convection is explicit (Held et al. 1993; Bretherton et al. 2005; Stephens et al. 2008; Wing and Emanuel 2014) but it also appears in global models (Bony et al. 2016; Shi and Bretherton 2014; Reed et al. 2015; Coppin and Bony 2015) in which convection is parameterized.

Self-aggregation has primarily been found to occur in simulations with warm Sea Surface Temperatures (SST) with a possible SST threshold below which aggregations does not occur (e.g., Wing and Emanuel 2014). A precise threshold remains elusive: Self-aggregation has been simulated at SSTs less than 300 K (Wing et al. 2016; Holloway and Woolnough 2016) and there are even simulations with an upper SST bound above which self-aggregation does not occur (Wing and Emanuel 2014). While studies have consistently found a relationship between temperature and self-aggregation, there is not a consensus on the specific nature of this relationship. The remainder of this paper will address aggregation in warm SST environments with the caveat that the phenomena may have broader applicability.

Models show that in a fully aggregated state, the atmosphere consists of a few very moist regions containing strong convection and a much larger dry, subsiding region. The contrast in humidity between dry and moist regions is strong so that the variance of the moisture field increases with the degree of aggregation (Bretherton et al. 2005; Wing et al. 2016). High clouds are less frequent in aggregated states, allowing increased longwave (LW) radiative cooling to space. The precise details of convectively aggregated states are still being explored (Muller and Bony 2015) but the frequent finding that aggregation increases with sea surface temperature suggests a stabilizing feedback on climate reminiscent of the 'Iris hypothesis' (Lindzen et al. 2001), albeit through different mechanisms than originally proposed (Mauritsen and Stevens 2015; Bony et al. 2016).

The degree to which the self-aggregation of convection is relevant to the Earth's atmosphere is not entirely clear. As reviewed carefully by Holloway et al. (2017; this issue), the organization of convection has been thought of for many years as being intimately linked to mesoscale systems organized by gravity and other convectively coupled waves (e.g., Mapes 1993) or large-scale circulations. The Earth's atmosphere does exhibit some characteristics of convective self-aggregation, including the tendency of more organized atmospheres to have lower humidity in clear areas, reduced domain-mean high cloudiness, and increased low cloudiness in non-convective areas, with corresponding impacts on surface and radiative fluxes (Tobin et al. 2012; Stein et al. 2017; Tobin et al. 2013).

As the mechanisms responsible for self-aggregation in radiative-convective equilibrium become more robustly understood, observational tests focusing on those mechanisms become possible (see Holloway et al., this issue; Bony et al., this issue). Here, we expand on the relatively small literature (Tobin et al. 2012, 2013; Stein et al. 2017) examining how the structure of the atmosphere, and the clouds embedded in it, depends on the degree of organization. We exploit a range of colocated observations from the A-Train satellite constellation (Stephens et al. 2002; L'Ecuyer and Jiang 2010) to identify aggregated states in the tropical atmosphere and examine the cloud, precipitation, and radiative structure of these states. We use a new measure of organization based on the spatial variability of the water vapor field. The richness of the observations allows us to identify the circumstances under which aggregation is most frequent and to disentangle the effects of aggregation and mean environmental conditions on the cloud and humidity structure of the atmosphere. We emphasize the important distinction between the observed aggregation that is influenced by external forcing and the idealized concept of self-aggregation, which occurs in the absence of large-scale forcing. The results of this paper must be interpreted with the understanding that these are distinct phenomena.

2 Characterizing Aggregation in Clouds and Their Environment

2.1 Observations from the A-Train

We use data products from the A-Train constellation (Stephens et al. 2002; L'Ecuyer and Jiang 2010). These sensors include the Advanced Microwave Scanning Radiometer for EOS (AMSR-E), the Moderate resolution Imaging Spectroradiometer (MODIS), CloudSat, and CALIPSO. AMSR-E data is available from June 2002 to October 2011 when the instrument spun down. MODIS cloud data is used during this same period. CloudSat/CALIPSO data was used from the period June 2006 to February 2011. All the observations are nearly instantaneous snapshots as opposed to daily average quantities. Results that contain CloudSat/CALIPSO data use the 2006–2011 epoch, whereas results that do not use CloudSat/CALIPSO data use the full A-Train period.

Here, we use Column Water Vapor (CWV) from the AMSR-E sensor (Kawanishi et al. 2003) derived from the version 7 Remote Sensing System algorithms (Wentz and Meissner 2000, 2007). The CWV has an expected precision of 1 kgm^{-2}. The data product is available over ocean surfaces on a 0.25° daily grid with the ascending orbital nodes separated from the descending orbital nodes.

Surface Rain Rate (RR) data are also derived from the AMSR-E sensor using the version 2 Goddard Profiling (GPROF) algorithm (Kummerow et al. 2011). Like the CWV data, the RR is available on a 0.25° daily grid with the ascending orbital nodes separated from the descending orbital nodes.

Cloud data are taken from the collection 5.1, Level 3 Aqua MODIS products. We use the Cloud top pressure (CTP) histograms in the Level 3 products, which are separately stored for ascending and descending nodes (Cloud_Top_Pressure_Day_Histogram_Counts and Cloud_Top_Pressure_Night_Histogram_Counts). The CTP histograms bin the observed cloud counts as a function of 11 bins in 100 hPa increments from the surface to the top of the atmosphere. In addition to the histogram, we use the CTP counts (Cloud_Top_Pressure_Day_Pixel_Counts and Cloud_Top_Pressure_Night_Pixel_Counts) variable in order to calculate cloud fractions from the histograms.

Cloud occurrence profiles are derived from the release-04 2B-Geoprof-Lidar product (Mace et al. 2009) which combines the CloudSat radar cloud mask (Marchand et al. 2008) with the CALIPSO lidar cloud mask (Vaughan et al. 2009). These data are stored on granules that correspond to single orbits; nadir-only sampling means that there is no overlap between the ascending and descending observations.

Precipitation incidence is used from the release-04 CloudSat 2C-Precip-Column product (Haynes et al. 2009; Smalley et al. 2013). Surface rain incidence is defined using the

Precip_flag variable = 3, which corresponds to certain precipitation and a radar reflectivity exceeding 0 dBZ at an altitude of approximately 720 m.

Radiative heating profiles are taken from the CloudSat/CALIPSO 2B-Flxhr-Lidar product (Henderson et al. 2013), which combines meteorological analysis with the cloud and aerosol profile information from CloudSat and CALIPSO and a dynamic land surface as input to a radiative transfer model to compute the profile of radiative fluxes at 240 m vertical resolution. Pixel-level RMS differences between this product and the derived top of the atmosphere (TOA) fluxes from the Clouds and Earth Radiant Energy System (CERES) are 5.7 and 16.5 W m^{-2} for the longwave and shortwave, respectively. However, the biases in pixel-scale retrievals are less than 5 W m^{-2} (Henderson et al. 2013). We use a modified version of the 2B-Flxhr-Lidar product designed to estimate the diurnal mean fluxes. This modified product computes the shortwave fluxes using 12 different solar zenith angles to account for the diurnal precession of the incoming flux. This product does not, however, account for diurnal changes in the cloud or thermodynamic variables, since they are not directly observed by the A-Train constellation. The Flxhr-lidar product includes an estimate of the Cloud Radiative Effect (CRE). The CRE is calculated explicitly by performing the radiative transfer calculation twice: once all sky and once clear sky. The CRE is then calculated as the difference of the clear-sky calculation from the all-sky calculation. All fluxes that follow are defined positive downward.

2.2 Characterizing Aggregation in the Water Vapor Field

Observational studies of convective self-aggregation to date (Tobin et al. 2012; Stein et al. 2017; Tobin et al. 2013) have quantified the degree of aggregation based on the degree to which cold clouds observed with a domain are spatially coherent, and these studies have also noted that clear areas tend to be less humid when convection is more aggregated. We invert this logic using a measure of aggregation defined by the inhomogeneity of the integrated water vapor field. Our motivations are partly practical: Definitions of aggregation based on clouds require processing high volumes of pixel-scale cloud observations, and cloud observations can be sensitive to the details of the observing system including sensor resolution, inherent sensitivity, and algorithmic choices (Pincus et al. 2012) while being subject to much larger high-frequency variability than is vapor. More importantly, it is useful to understand the degree to which variability in water vapor can exist independently of the systematic organization of convection.

We define the degree of aggregation α using the coefficient of variation for water vapor calculated from the 0.25° data on a 5° twice-daily (day/night) grid,

$$\alpha = \frac{\sigma_{\mathrm{CWV}}}{\overline{\mathrm{CWV}}} \tag{1}$$

where the overbar represents the spatial mean and σ is the standard deviation taken over the grid. All other cloud, precipitation, and radiation data are aggregated to a common twice-daily 5° grid. The choice of a 5° grid is arbitrary; however, the conclusions drawn herein do not change when repeating the analysis at 10°. The separation of ascending and descending nodes is important because it keeps each grid box a semi-instantaneous sample in time. Pixels identified as land are filtered out of the analysis with no imposed threshold on the number of ocean pixels that enter each 5° aggregation. The high cloud fraction is calculated from both the MODIS and CloudSat/CALIPSO data, whereas low cloud fraction is only derived from the CloudSat/CALISPO data. From MODIS, the high cloud fraction is calculated using the cloud fraction in the CTP histograms with CTP lower than 400 hPa.

Similarly, from CloudSat the high cloud fraction is calculated as the fraction of cloud cover above 7.5 km.

All 5° regions are filtered for SSTs between 300 and 304 K, which include most of the warmest tropical SSTs. Higher SSTs are infrequent enough that sampling is problematic. The lower SST bound is motivated by modeling studies that suggest that SSTs over 300 K are the relevant regime for self-aggregation. We also note that results are robust down to SSTs of 296 K.

Daily data can occasionally be missing in areas of intense precipitation where high winds affect surface emissivity and large ice water contents cause scattering of the emission signal. These missing data points are disregarded in the calculation of the aggregation. The missing data are most likely the high tail of the CWV distribution and therefore may introduce some systematic bias in the calculation of α; however, this influence is somewhat mitigated by the normalization in Eq. 1 as both the standard deviation and mean will be biased low when data are missing.

Figure 1 shows an example of 1 day's calculation of the aggregation and related data products for the ascending (daytime) orbits.

3 Relationships Among Clouds, Humidity, and Aggregation

Our relatively large data set allows us to examine the geographic distribution of aggregation. Figure 2 shows a map of the degree of aggregation, water vapor, and SST over the period 2002–2011. Aggregation is largest where the gradient of the mean water vapor field is largest, on the edges of the West Pacific warm pool and the Inter-Tropical Convergence Zone (ITCZ), suggesting that aggregation, by this measure, is most common in domains which cover both the ascending and descending regions of either a large-scale or a synoptic circulation. This geographical distribution is inconsistent with some modeling results suggesting that aggregation increases with SST although this may be an artifact of the particular definition of aggregation used here.

Missing CWV retrieval failures might introduce systematic bias in the results that follow. However, Fig. 3 shows that these failures are rare and relatively evenly distributed across the range of α. There is a modest maximum in the failure count for α near 3%. The geographical distribution of failure rate clearly shows that the prevalence of failed CWV retrievals follows the distribution of precipitation with relative maximum in the ITCZ and the warm pool. The fraction of missing pixels only exceeds 2% in a handful of poorly sampled grids. Comparing this distribution to Fig. 2, one cannot find a strong correlation between the occurrences of high, moderate, or low mean aggregation state with the CWV failure rate. For example, while both the ITCZ and the warm pool have relatively elevated precipitation rates and retrieval failure rates, the ITCZ is characterized by large α and the warm pool by low α.

3.1 Cloudiness Depends on Sea Surface Temperature and Aggregation State

One of the most robust features of self-aggregation in idealized simulations is the reduction in high cloud cover with increased aggregation. This feature appears in our data set, echoing results from previous observational studies. Figure 4 shows that the MODIS high cloud cover tends to decrease with increasing aggregation, and results (not shown) using the CloudSat/CALIPSO data confirm this result. The result is consistent with Stein et al.

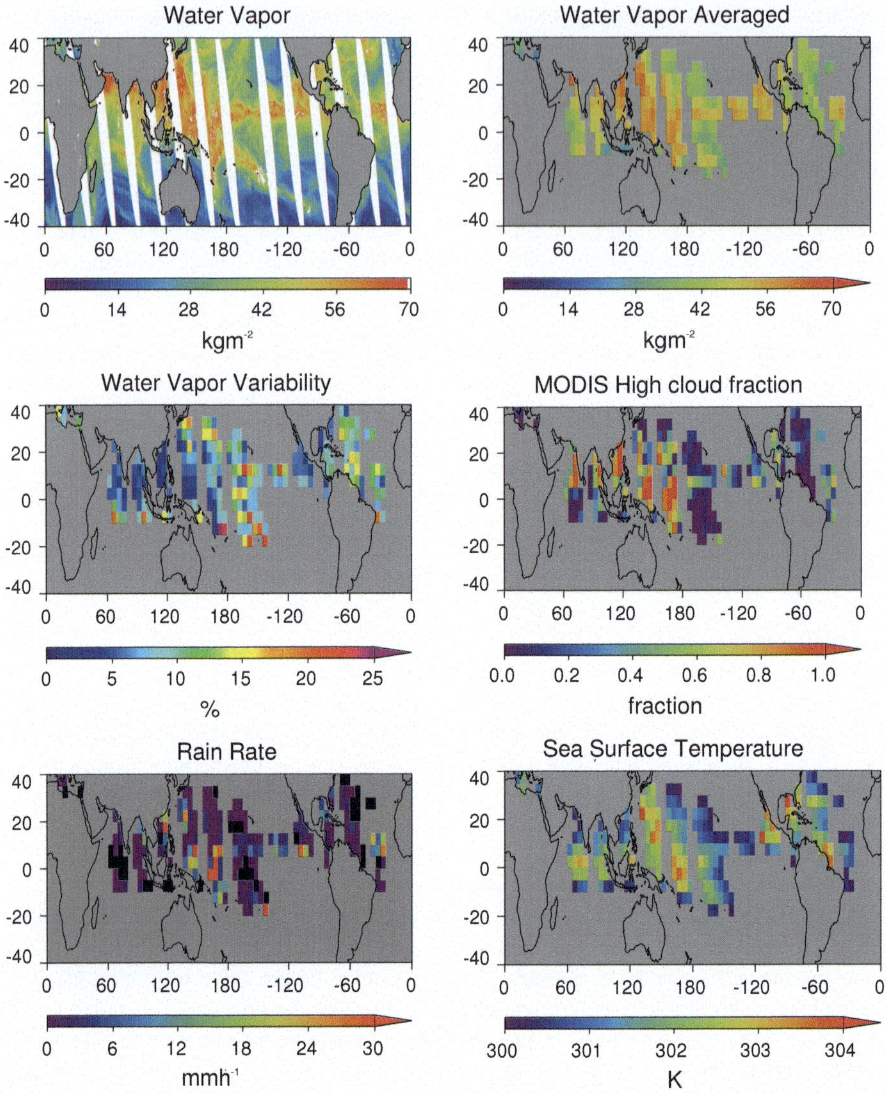

Fig. 1 An example of the data for the ascending orbit of August 04, 2006. The top left panel shows the 0.25° native water vapor fields. All other panels show 5° averaged data for pixels with SST between 300 and 304 K

(2017). That study further finds that the strong dependence of high cloud fraction with aggregation is largely a function of variations in optically thin cirrus cloud. Note that in Fig. 3, panel C shows how the mean water vapor varies with α. It is not surprising to see that $\overline{CWV}$ decreases with α, contributing to changes in α, since $\overline{CWV}$ appears in the denominator of Eq. 1, but these changes in the $\overline{CWV}$ do not explain the majority of the variation in α. Therefore, the degree of aggregation is primarily driven by spatial variation in water vapor, not changes in $\overline{CWV}$. We infer that changes in cloud morphology correlated with α are related to changes in the spatial variability of water vapor as opposed to the

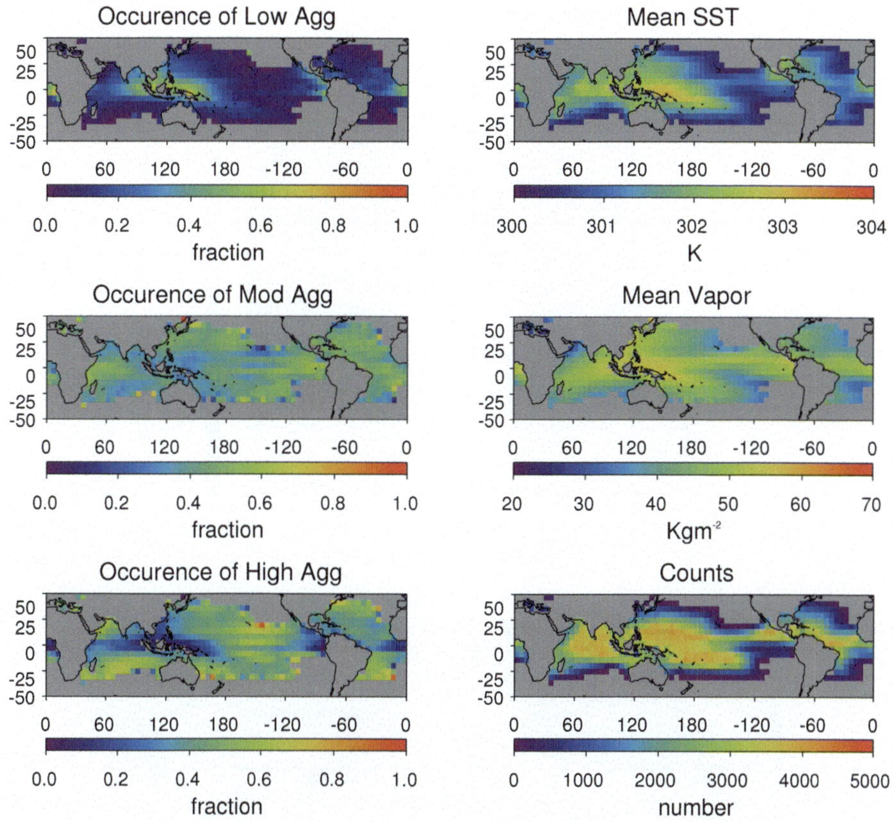

Fig. 2 The left-hand panels show the frequency of occurrence of various aggregation states. Low-aggregation is defined as < 5%, moderate-aggregation is defined as 5–10%, and high-aggregation is defined as > 10%. The right-hand panel shows the mean SST, column water vapor, and sample count. Note the predominance of low-aggregation in the maritime continent with areas of higher aggregation on the boundaries of the ITCZ and the warm pool

mean water vapor. We can also observe from Fig. 4 panel C that dry states tend to display a more aggregated state than do moister states which is consistent with the geographical distributions shown in Fig. 2, which shows a minimum aggregation in the moistest regions.

Figure 5 shows how the vertical profile of cloudiness changes with increases in α (see also Stein et al. 2017, Fig. 3). Profiles are derived from the CloudSat/CALIPSO data which only provide a narrow nadir swath within each sample grid box; however, averaged over a large number of samples, it should provide an unbiased estimate of the mean. This is evident in the fact MODIS and CloudSat/CALIPSO show the same dependence of high clouds on α. We see in Fig. 5 that the total cloud cover is a strong function of mean SST but, for a given SST range, an increase in vapor variability is not only associated with a decrease in high cloud fraction but also a decrease in mid-level clouds indicating a drying out of the mid-troposphere, presumably due to a decrease in the convective area fraction. Smaller changes in the low cloud fraction are observed that depend on the SST. At cooler SST, the low cloud fraction increases slightly with α, whereas it decreases with α at higher SST.

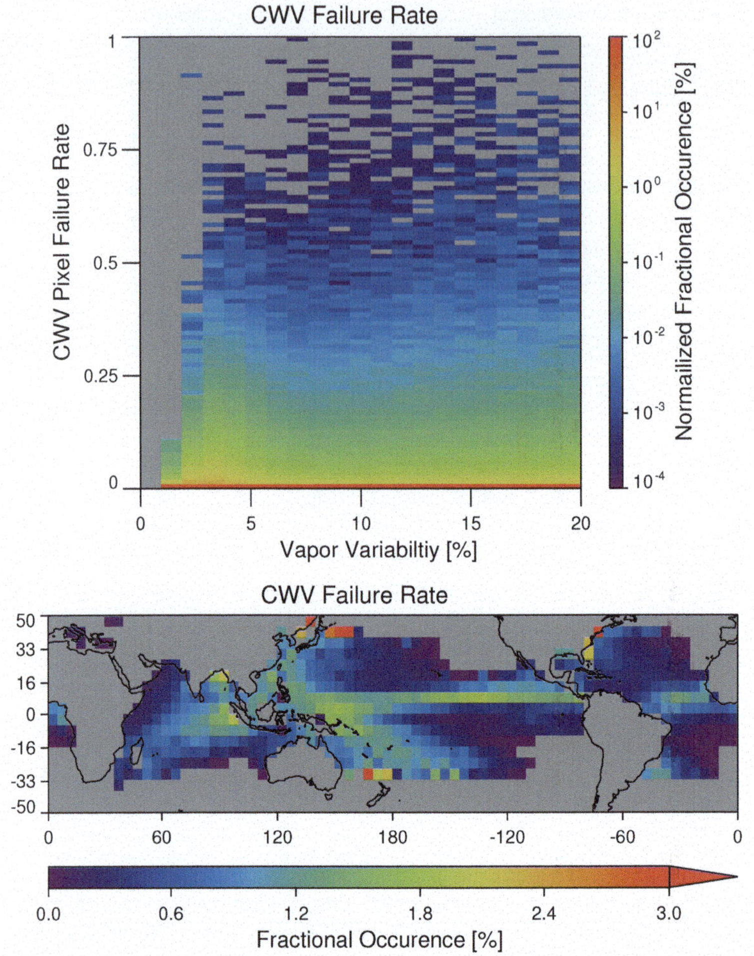

Fig. 3 (top) The distribution of Column Water Vapor (CWV) retrieval failure rates as a function of the vapor variability (α). Results for each α bin are normalized such that the sum for each column adds to 100%. The color scale is logarithmic. Note that for all values of alpha by far the most common occurrence is for a failure rate of 0. (bottom) The geographical distribution of the CWV retrieval failure rate, which clearly shows the imprints of the distribution of precipitation

3.2 Cloudiness, Radiative Heating, and Convective Intensity

Figures 1, 2, 3, 4, and 5 mix many different convective states across the aggregation index α. Might these results be the result of systematic variation in convective activity with the vapor variance, rather than an indication of the aggregation of convection? To address this concern, we further stratify our results by the observed rain rate averaged over each $5° \times 5°$ region following the approach of Stein et al. (2017). This admittedly rough metric for convective intensity is the best available from the A-Train observations. A better measure of convective area fraction might be gleaned from Global Precipitation Measurement (GPM) mission or Tropical Rainfall Measurement Mission radar observations, which can identify convective precipitation using the spatial variance of the radar

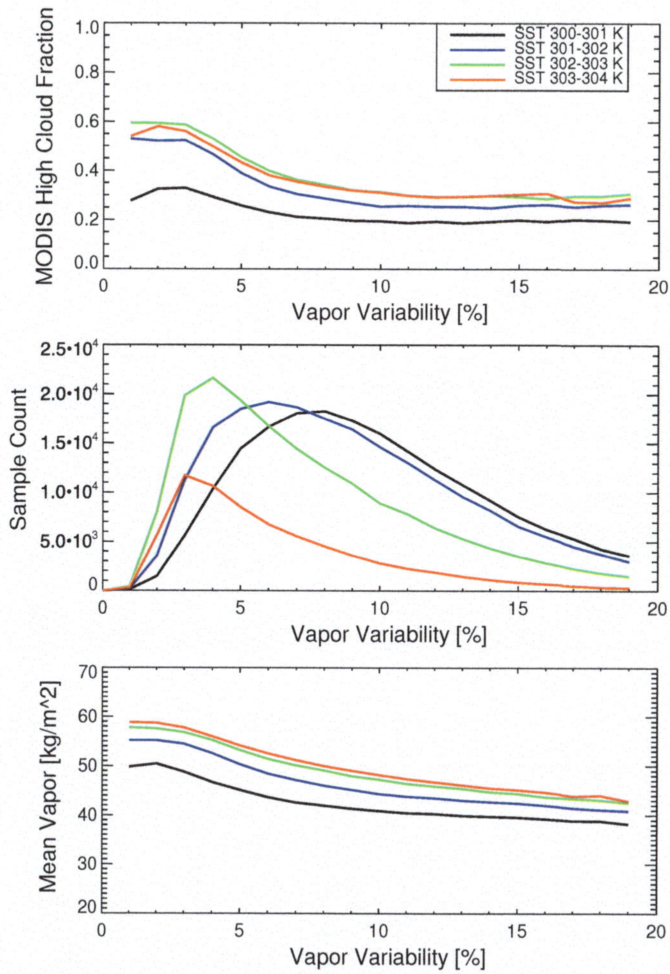

Fig. 4 The MODIS high cloud fraction (< 400 hPa), sample count and mean water vapor as a function of the water vapor aggregation

reflectivity; however, these observations are rarely coincident with the A-Train data used in this study. The GPM mission includes the GPM Microwave Radiometer (GMI), which has similar characteristics to the AMSR-E radiometer, so it would be possible to examine GPM radar observations in terms of the aggregation index defined in this paper.

Figure 6 shows how the mean α depends on both the grid-mean rain rate and SST. There is a tendency toward larger α with decreases in either SST or rain rate. Thus, stratifying results by both SST and rain rate is important to determine whether the dependences of clouds on α are related to variations in α itself or are instead potentially due to correlation of α with precipitation.

High cloud fraction does indeed decrease with increasing α for each rain rate bin and each SST (Fig. 7), with the dependence of high cloud fraction on aggregation similar within each bin. This suggests that the high cloud fraction decreases with α due to

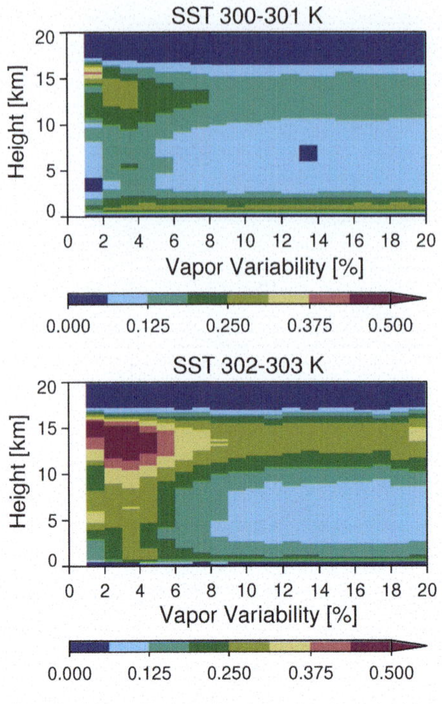

Fig. 5 The height resolved cloud fraction from CloudSat/CALIPSO as a function of the aggregation index

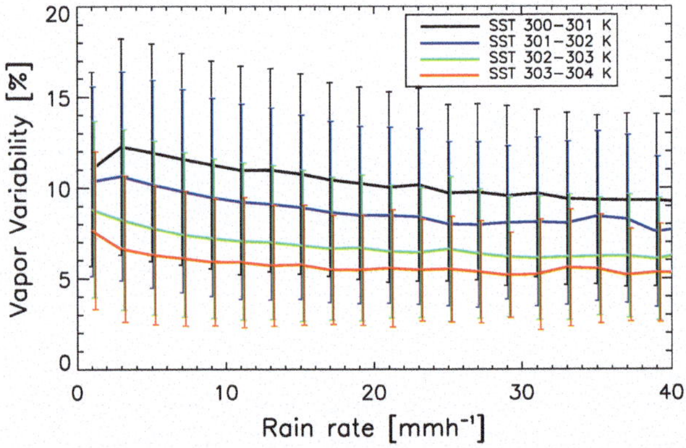

Fig. 6 The mean vapor variability (α) as a function of mean rain rate within a 5° grid box. The error bars show the standard deviation. While virtually any value of α can be observed for any rain rate or SST value, there are clear tendencies for alpha to increase with decreasing rain rate and decreasing SST

aggregation of convection, not via a systematic dependence of convective intensity on the water vapor variance.

Stein et al. (2017) found that low-level cloud fraction increases along with their aggregation metric. Modeling results also suggest that low-level clouds are crucial for the onset of convective aggregation and are one of the several processes that help maintain an established aggregated state (Muller and Bony 2015). Figure 8 shows that the low cloud

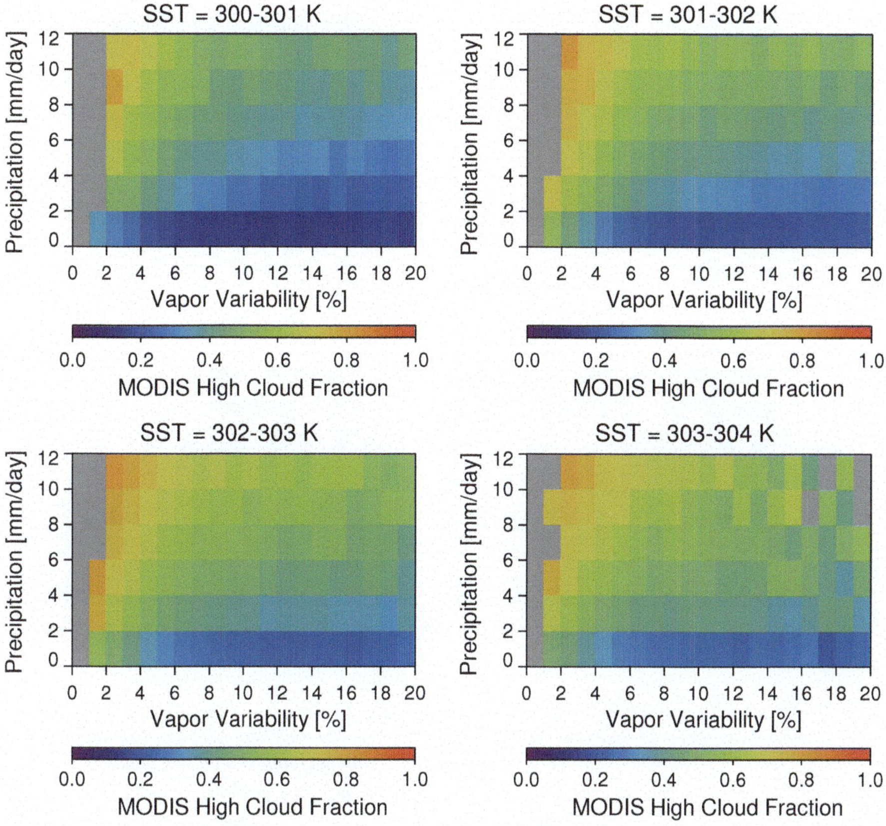

Fig. 7 MODIS high cloud fraction as function of the grid box aggregation, mean precipitation rate, and sea surface temperature (SST)

fraction as deduced from the CloudSat/CALIPSO data shows increases with alpha, in agreement with the Stein et al.'s (2017) result. Low cloud fraction decreases with increasing SST, decreases with increasing rain rate, and low cloud fraction tends to increase with α regardless of the rain rate or SST bin.

The systematic dependence of the cloud cover on α has a substantial influence on the cloud radiative effects. Figure 9 shows the result of the aggregation state on the mean cloud radiative effect at the TOA. As α increases (and high cloud cover decreases), there is increased domain average longwave emission to space compensated by decreased solar reflectance. In general, the shortwave effect is larger than the longwave effect. Results are shown only for the 301–302 K SST bin; qualitatively, similar dependence of the TOA fluxes is found at the other SSTs. For a given rain rate and SST, therefore, net absorption by the earth and atmosphere increases with the degree of aggregation α.

The compensation between longwave and shortwave at the TOA implies a redistribution of heating in the atmospheric column with increased solar heating of the surface and increased longwave cooling of the atmosphere. Longwave radiative cooling is concentrated at the effective emission level, which is governed by cloud top and the water vapor scale height. Indeed, heating rate profiles stratified according to rain rate (Fig. 10) show increasing low-level cooling of the atmosphere with increased aggregation. As α increases,

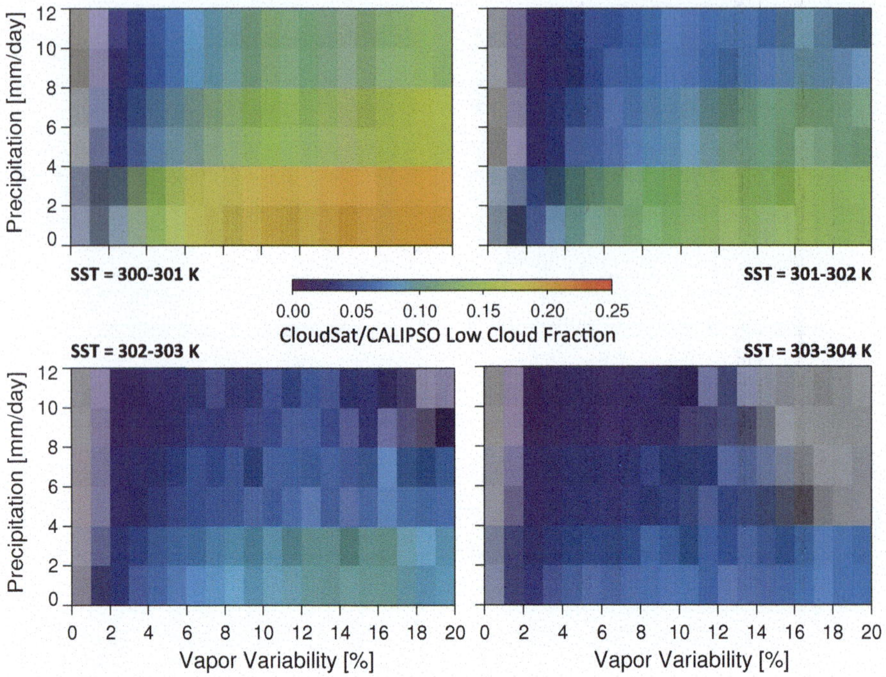

Fig. 8 The low cloud fraction as a function of vapor variability, rain rate, and SST. Low clouds are observed using the combined CloudSat/CALIPSO data set and are defined here as clouds having tops lower than 3 km. Low cloud increases with the vapor variability for the majority of rain rate and SST bins

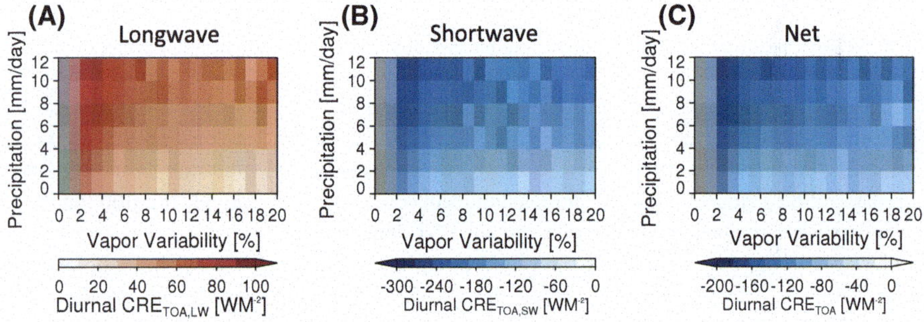

Fig. 9 The top of the atmosphere (TOA) longwave (**a**), shortwave (**b**), and net (**c**) cloud radiative effect as function of the grid box aggregation, mean precipitation rate, and sea surface temperature (SST). The convention is positive downward for all fluxes. Fluxes are derived from the diurnally averaged 2B-Geoprof-Lidar product. These results are for SST between 301 and 302 K. Cloud Radiative Effect is the result for other SST's show a similar dependence on mean precipitation rate and vapor variability

the height of the maximum cooling decreases and the magnitude of the lower tropospheric cooling increases, each of which supports the LW radiative-convective feedback conjecture whereby enhanced low-level atmospheric cooling with aggregation helps sustain the aggregated state through positive feedback on the regional circulation.

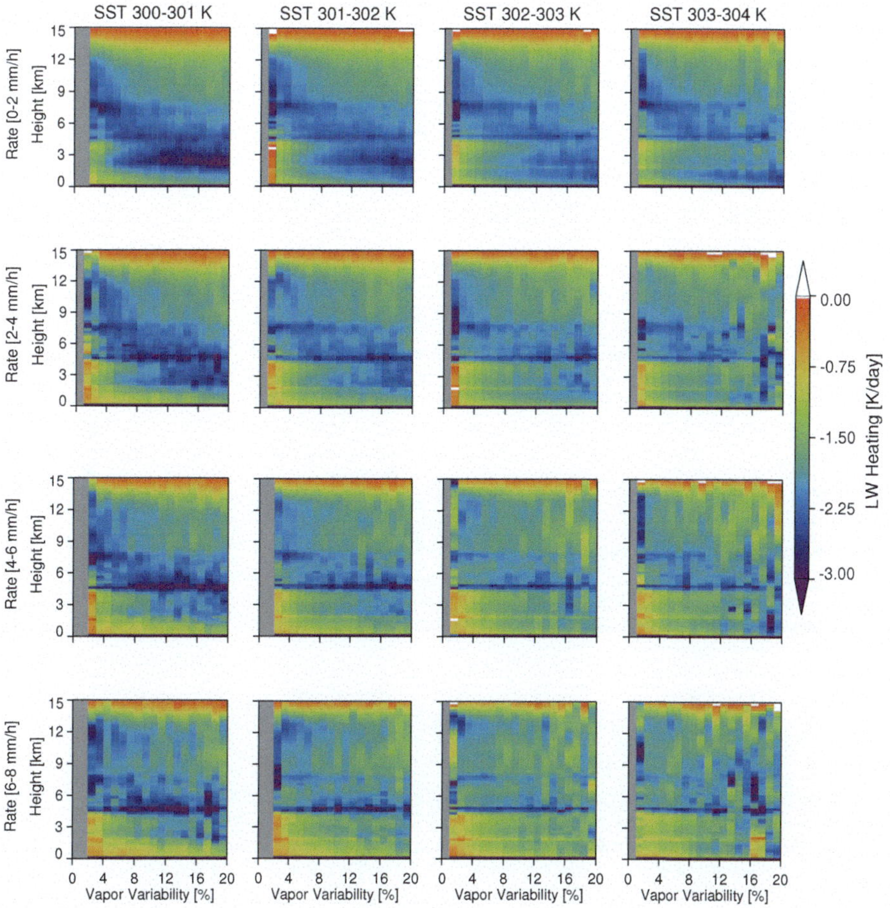

Fig. 10 The vertical profile of the CloudSat/CALIPSO diurnal averaged LW heating rate as a function of the vapor variability

3.3 Does Vapor Aggregation Imply Convective Aggregation?

The results already shown, using a measure of aggregation defined by the water vapor field, show variations of cloudiness consistent with observations stratified by the connectedness of clouds themselves, suggesting that large-scale variance of water vapor and the mesoscale distribution of clouds are tightly linked. In this section, we explore these relationships more carefully.

We examine differences in the structure of the cloud and radiation fields in the dry and moist areas by compositing our observations as a function of mean column-integrated water vapor. Each twice-daily grid box is divided into water vapor octiles, and then cloud and radiation data are composited as a function of each octile (Fig. 11). This analysis averages data across the various aggregation states while retaining information of the covariability of moisture and cloudiness within each domain. Because SST variation across each 5° box is relatively small, this is analogous to energy-budget analyses used to diagnose the mechanisms leading to self-aggregation (Bretherton et al. 2005; Wing and

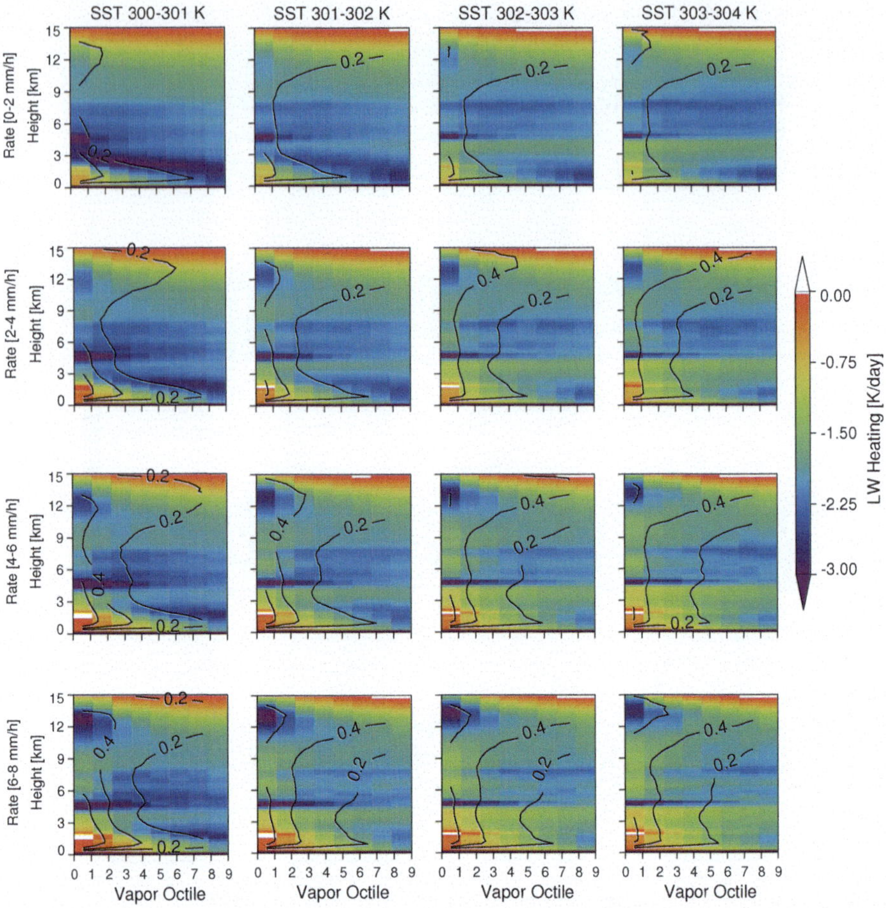

Fig. 11 Composite view of the LW cooling rate (colors) and cloud fraction (contours) as a function of vapor octiles averaged over all aggregation states. By definition, the moistest octiles are on the left and the driest octiles on the right side of the plots

Emanuel 2014; Muller and Bony 2015) although open boundary conditions suggest that inferring circulations from this stratification is unwise.

Within each rain rate/SST regime, the moistest areas have the highest cloud fraction at all levels, while the driest show very little cloud in the middle atmosphere and enhanced low-level cooling. This picture is consistent with modeling results showing preferential convection in the moist regions that can be maintained (or caused) by an enhanced LW radiative cooling in the dry region (c.f. Figure 2 in Muller and Bony (2015)).

Aggregated convection might also be expected to lead to more aggregated precipitation. The hypothesis is tested using a precipitation length scale l_p, defined as the chord length of contiguous areas of precipitation, based on precise precipitation incidence flags from CloudSat (Smalley and L'Ecuyer 2015). On the scale of an individual sample, this chord length may have a great deal of uncertainty due to the nadir sampling of CloudSat and the non-isotropic structure of precipitation. We make the assumption that averaged over a large number of samples, systematic differences in precipitation spatial scale can be inferred from the chord-length measurement. Figure 12 contrasts l_p in regions with very aggregated regions ($\alpha > 10$) with homogenous regions ($\alpha < 5$). Precipitation length scale is generally longer in the moistest octile for the $\alpha > 10$ state than for $\alpha < 5$, whereas it tends to be shorter in the other 7 octiles, regardless of SST or rain rate. This would occur if, for example, rain from convective systems increasingly aggregates in moist regions as water vapor variance increases. Commensurate with these changes in the character of precipitation in the moist region is a decrease in the organization in the dry areas, which may have more isolated shallow convection.

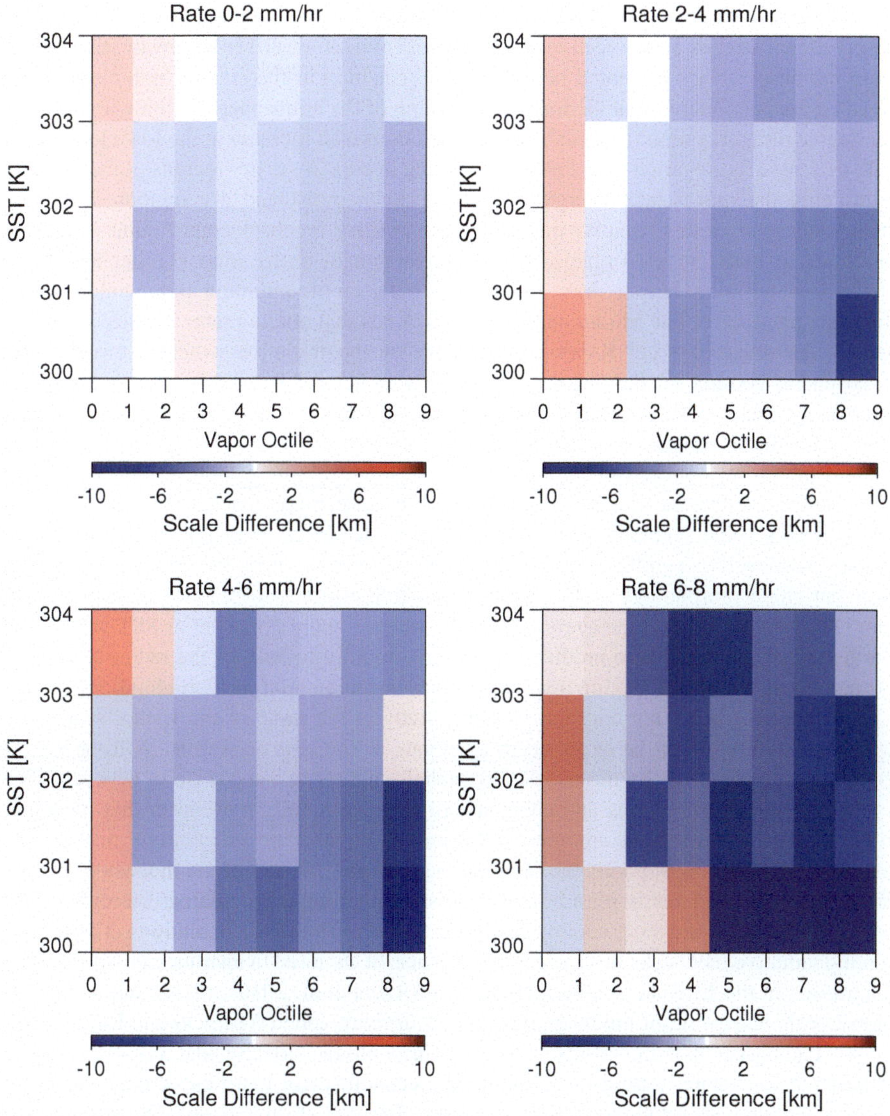

Fig. 12 The difference in the precipitation length scale (l_p) between vapor variance greater than 10% cases and vapor variance less than 5% cases. In the moistest octile, the aggregated cases tend to have a longer precipitation length scale, whereas the precipitation length scale is shorter in drier octiles

4 Summary and Discussion

We have explored the relationships among clouds, precipitation, radiation, and a measure of the convective aggregation given by the variance in the column water vapor field in large domains over the tropical oceans. This study was motivated by a number of modeling studies hypothesizing an increase in convective aggregation with warming SSTs. Cloud modeling studies of this convective self-aggregation robustly find that the water vapor variance increases with aggregation of the convection. Over ocean surfaces, column water vapor is well-measured and relatively continuous. We expect it provides a measure complementary to the infrared cloud observations that have been employed by previous observational studies of convective aggregation (Tobin et al. 2012; Holloway, this issue).

In the observations presented here, we see a reduction in the area of high cloud cover and an associated increase in the longwave cooling of the atmospheric column to space, as the degree of aggregation increases. We further observe an increase in the low cloud cover with increased aggregation. The enhanced cooling occurs in the dry regions of the domain, reinforcing the moist-static energy gradient between moist and dry regions. Modeling studies suggest that this radiative effect acts as a positive feedback contributing with other processes to maintain the organization of convection in moist areas (Muller and Bony 2015). It is important here to draw a distinction between the initiation and maintenance of the aggregated state. The Muller and Bony study finds that cooling rates localized at cloud top on the order of $\sim$ 13 K/day are required for the initiation, whereas broad lower tropospheric cooling on the order of $\sim$ 2 K/day is helpful but not necessary for maintaining aggregation. The observations shown here are on the order of the $\sim$ 2 K/day helpful for the maintenance of the aggregated state.

This picture of convective organization is consistent across a range of SSTs and rain rates, which we take as a loose proxy for convective intensity. Sorting the results by rain rate provides some relevance to the cloud-climate-feedback problem in the context of radiative-convective equilibrium. In particular, the results show that regions with very different cloud morphology and associated radiative effects can produce the same mean rain rate. It follows that if the aggregated state becomes more prevalent as SST warms, the Earth system may be able to produce the required rainfall to balance the radiative cooling of the atmosphere while having a significantly reduced amount of high cloud.

An important point not addressed by this study is the issue of the spatial scale over which convection might be expected to aggregate as the climate warms. Will the aggregation tend to occur on the mesoscale, the global scale, or at some scale in between? The cyclical boundary conditions and constraint of mass continuity may mean that modeling studies of self-aggregation are more relevant to global-scale circulations than to the mesoscale. This study supports the view that aggregated convection on the synoptic scale produces an environment with a bimodal moisture distribution including dry regions that produce a positive radiative cooling feedback on the convective circulations. These reinforcing radiative feedbacks on convection have also been noted in interannual variability as manifested in the El Nino Southern Oscillation (Rädel et al. 2016) and are implicit in the global-scale narrowing of the inter-tropical convergence zone (Wodzicki and Rapp 2016).

We emphasize that this study cannot confirm observationally that convection does indeed self-aggregate; testing this hypothesis mechanistically will require targeted observations and analysis (Holloway et al., this issue; Bony et al., this issue). Our survey does demonstrate that certain features of the cloud morphology present in model simulations of self-aggregation are also present in observations of the routine aggregation found in the tropical atmosphere. This suggests that aggregation of convection in Earth's atmosphere, whatever the mechanism, provides a useful conceptual model through which to view important aspects of cloud feedbacks on climate (Mapes 2016).

Acknowledgments This paper originates from discussions during the International Space Science Institute (ISSI) Workshop on 'Shallow clouds and water vapor, circulation and climate sensitivity.' This research was primarily carried out by ML at the Jet Propulsion Laboratory, California Institute of Technology, under a contract with the National Aeronautics and Space Administration under RTOP/WBS (105357/967701.02.01.02.08). RP was supported by the National Science Foundation under award ATM-1138394. AMSR data are produced by Remote Sensing Systems and were sponsored by the NASA AMSR-E Science Team and the NASA Earth Science MEaSUREs Program. Data are available at www.remss.com. CloudSat data were obtained from the CloudSat Data Processing Center (http://www.cloudsat.cira.colostate.edu/). MODIS data were obtained from the Goddard DAAC.

References

Bony S, Stevens B, Coppin D, Becker T, Reed KA, Voigt A, Medeiros B (2016) Thermodynamic control of anvil cloud amount. Proc Natl Acad Sci 113:8927–8932. doi:10.1073/pnas.1601472113

Bretherton CS, Blossey PN, Khairoutdinov M (2005) An energy-balance analysis of deep convective self-aggregation above uniform SST. J Atmos Sci 62:4273–4292. doi:10.1175/JAS3614.1

Coppin D, Bony S (2015) Physical mechanisms controlling the initiation of convective self-aggregation in a general circulation model: mechanisms of initiation of aggregation. J Adv Model Earth Syst 7:2060–2078. doi:10.1002/2015MS000571

Haynes JM, L'Ecuyer TS, Stephens GL, Miller SD, Mitrescu C, Wood NB, Tanelli S (2009) Rainfall retrieval over the ocean with spaceborne W-band radar. J Geophys Res 114:D00A22. doi:10.1029/2008JD009973

Held IM, Hemler RS, Ramaswamy V (1993) Radiative-convective equilibrium with explicit two-dimensional moist convection. J Atmos Sci 50:3909–3927. doi:10.1175/1520-0469(1993)050<3909:RCEWET>2.0.CO;2

Henderson DS, L'Ecuyer T, Stephens G, Partain P, Sekiguchi M (2013) A multisensor perspective on the radiative impacts of clouds and aerosols. J Appl Meteorol Climatol 52:853–871. doi:10.1175/JAMC-D-12-025.1

Holloway CE, Woolnough SJ (2016) The sensitivity of convective aggregation to diabatic processes in idealized radiative-convective equilibrium simulations: aggregation in idealized simulations. J Adv Model Earth Syst 8:166–195. doi:10.1002/2015MS000511

Holloway CE, Wing AA, Bony S, Muller C, Masunaga H, L'Ecuyer TS, Turner DD, Zuidema P (2017) Observing convective aggregation. Surv Geophys. doi:10.1007/s10712-017-9419-1. http://link.springer.com/10.1007/s10712-017-9419-1. Accessed 30 Aug 2017

Kawanishi T et al (2003) The advanced microwave scanning radiometer for the earth observing system (AMSR-E), NASDA's contribution to the EOS for global energy and water cycle studies. IEEE Trans Geosci Remote Sens 41:184–194. doi:10.1109/TGRS.2002.808331

Kummerow CD, Ringerud S, Crook J, Randel D, Berg W (2011) An observationally generated a priori database for microwave rainfall retrievals. J Atmos Ocean Technol 28:113–130. doi:10.1175/2010JTECHA1468.1

L'Ecuyer TS, Jiang JH (2010) Touring the atmosphere aboard the A-Train. Phys Today 63:36–41. doi:10.1063/1.3463626

Lindzen RS, Chou M-D, Hou AY (2001) Does the earth have an adaptive infrared iris? Bull Am Meteorol Soc 82:417–432. doi:10.1175/1520-0477(2001)082<0417:DTEHAA>2.3.CO;2

Mace GG, Zhang Q, Vaughan M, Marchand R, Stephens G, Trepte C, Winker D (2009) A description of hydrometeor layer occurrence statistics derived from the first year of merged Cloudsat and CALIPSO data. J Geophys Res 114:D00A26. doi:10.1029/2007JD009755

Mapes BE (1993) Gregarious tropical convection. J Atmos Sci 50:2026–2037. doi:10.1175/1520-0469(1993)050<2026:GTC>2.0.CO;2

Mapes BE (2016) Gregarious convection and radiative feedbacks in idealized worlds: aggregation of convection. J Adv Model Earth Syst 8:1029–1033. doi:10.1002/2016MS000651

Marchand R, Mace GG, Ackerman T, Stephens G (2008) Hydrometeor detection using cloudsat—an earth-orbiting 94-GHz cloud radar. J Atmos Ocean Technol 25:519–533. doi:10.1175/2007JTECHA1006.1

Mauritsen T, Stevens B (2015) Missing iris effect as a possible cause of muted hydrological change and high climate sensitivity in models. Nat Geosci 8:346–351. doi:10.1038/ngeo2414

Muller C, Bony S (2015) What favors convective aggregation and why? What favors convective aggregation. Geophys Res Lett 42:5626–5634. doi:10.1002/2015GL064260

Pincus R, Platnick S, Ackerman SA, Hemler RS, Patrick Hofmann RJ (2012) Reconciling simulated and observed views of clouds: MODIS, ISCCP, and the limits of instrument simulators. J Clim 25:4699–4720. doi:10.1175/JCLI-D-11-00267.1

Rädel G, Mauritsen T, Stevens B, Dommenget D, Matei D, Bellomo K, Clement A (2016) Amplification of El Niño by cloud longwave coupling to atmospheric circulation. Nat Geosci 9:106–110. doi:10.1038/ngeo2630

Reed KA, Medeiros B, Bacmeister JT, Lauritzen PH (2015) Global radiative-convective equilibrium in the community atmosphere model, version 5. J Atmos Sci 72:2183–2197. doi:10.1175/JAS-D-14-0268.1

Shi X, Bretherton CS (2014) Large-scale character of an atmosphere in rotating radiative-convective equilibrium. J Adv Model Earth Syst 6:616–629. doi:10.1002/2014MS000342

Smalley M, L'Ecuyer T (2015) A global assessment of the spatial distribution of precipitation occurrence. J Appl Meteorol Climatol 54:2179–2197. doi:10.1175/JAMC-D-15-0019.1

Smalley M, L'Ecuyer T, Lebsock M, Haynes J (2013) A comparison of precipitation occurrence from the NCEP stage IV QPE product and the CloudSat cloud profiling radar. J Hydrometeorol 15:444–458

Stein THM, Holloway CE, Tobin I, Bony S (2017) Observed relationships between cloud vertical structure and convective aggregation over tropical ocean. J Clim 30:2187–2207. https://doi.org/10.1175/JCLI-D-16-0125.1

Stephens GL et al (2002) The cloudsat mission and the a-train. Bull Am Meteorol Soc 83:1771–1790. doi:10.1175/BAMS-83-12-1771

Stephens GL, Van Den Heever S, Pakula L (2008) Radiative–convective feedbacks in idealized states of radiative–convective equilibrium. J Atmos Sci 65:3899–3916

Tobin I, Bony S, Roca R (2012) Observational evidence for relationships between the degree of aggregation of deep convection, water vapor, surface fluxes, and radiation. J Clim 25:6885–6904. doi:10.1175/JCLI-D-11-00258.1

Tobin I, Bony S, Holloway CE, Grandpeix J-Y, Sèze G, Coppin D, Woolnough SJ, Roca R (2013) Does convective aggregation need to be represented in cumulus parameterizations?: Mesoscale convective aggregation. J Adv Model Earth Syst 5:692–703. doi:10.1002/jame.20047

Tompkins AM, Craig GC (1998) Radiative–convective equilibrium in a three-dimensional cloud-ensemble model. Q J R Meteorol Soc 124:2073–2097. doi:10.1002/qj.49712455013

Vaughan MA et al (2009) Fully automated detection of cloud and aerosol layers in the CALIPSO lidar measurements. J Atmos Ocean Technol 26:2034–2050. doi:10.1175/2009JTECHA1228.1

Wentz FJ, Meissner T (2000) AMSR Ocean Algorithm, Version 2, report number 121599A-1, Remote Sensing Systems, Santa Rosa, CA, 66 pp

Wentz FJ, Meissner T (2007) AMSR_E Ocean Algorithms, report number 051707, Remote Sensing Systems, Santa Rosa, CA, 6 pp

Wentz FJ, Meissner T, Gentemann C, Brewer M (2014) Remote sensing systems AQUA AMSR-E daily environmental suite on 0.25 deg grid, version 7.0. Remote Sensing Systems, Santa Rosa, CA. at www.remss.com/missions/amsre

Wing AA, Emanuel KA (2014) Physical mechanisms controlling self-aggregation of convection in idealized numerical modeling simulations: self-aggregation mechanisms. J Adv Model Earth Syst 6:59–74. doi:10.1002/2013MS000269

Wing AA, Emanuel KA, Cronin TW (2016) Self-aggregation of convection in long channel geometry: self-Aggregation in Channel Geometry. Q J R Meteorol Soc 142:1–15. doi:10.1002/qj.2628

Wodzicki KR, Rapp AD (2016) Long-term characterization of the Pacific ITCZ using TRMM, GPCP, and ERA-interim: pacific itcz characterization. J Geophys Res Atmos 121:3153–3170. doi:10.1002/2015JD024458

Surv Geophys (2017) 38:1255
https://doi.org/10.1007/s10712-017-9449-8

CORRECTION

Correction to: An Observational View of Relationships Between Moisture Aggregation, Cloud, and Radiative Heating Profiles

Matthew D. Lebsock[1] · Tristan S. L'Ecuyer[2] · Robert Pincus[3,4]

Published online: 13 November 2017

Correction to: Surv Geophys https://doi.org/10.1007/s10712-017-9443-1

The article "An observational view of relationships between moisture aggregation, cloud and radiative heating profiles", written by Matthew D. Lebsock, Tristan S. L'Ecuyer and Robert Pincus, was originally published electronically on the publisher's internet portal (currently SpringerLink) on 31 October 2017 without open access.

With the author(s)' decision to opt for Open Choice, the copyright of the article was changed on [13 November 2017] to © The Author(s) 2017, and the article is forthwith distributed under the terms of the Creative Commons Attribution 4.0 International License (http://creativecommons.org/licenses/by/4.0/), which permits use, duplication, adaptation, distribution and reproduction in any medium or format, as long as you give appropriate credit to the original author(s) and the source, provide a link to the Creative Commons license and indicate if changes were made. The original article was corrected.

The original article can be found online at https://doi.org/10.1007/s10712-017-9443-1.

✉ Matthew D. Lebsock
matthew.d.lebsock@jpl.nasa.gov

1 Jet Propulsion Laboratory, California Institute of Technology, Pasadena, CA 91109, USA

2 University of Wisconsin, Madison, WI 53706, USA

3 Cooperative Institute for Research in Environmental Sciences, University of Colorado, Boulder, CO 80309, USA

4 Physical Sciences Division, NOAA Earth System Research Lab, Boulder, CO 80305, USA

Surv Geophys (2017) 38:1257–1282
https://doi.org/10.1007/s10712-017-9429-z

Implications of Warm Rain in Shallow Cumulus and Congestus Clouds for Large-Scale Circulations

Louise Nuijens[1] · Kerry Emanuel[2] · Hirohiko Masunaga[3] · Tristan L'Ecuyer[4]

Received: 15 February 2017 / Accepted: 18 September 2017 / Published online: 27 September 2017

Abstract Space-borne observations reveal that 20–40% of marine convective clouds below the freezing level produce rain. In this paper we speculate what the prevalence of warm rain might imply for convection and large-scale circulations over tropical oceans. We present results using a two-column radiative–convective model of hydrostatic, non-linear flow on a non-rotating sphere, with parameterized convection and radiation, and review ongoing efforts in high-resolution modeling and observations of warm rain. The model experiments investigate the response of convection and circulation to sea surface temperature (SST) gradients between the columns and to changes in a parameter that controls the conversion of cloud condensate to rain. Convection over the cold ocean collapses to a shallow mode with tops near 850 hPa, but a congestus mode with tops near 600 hPa can develop at small SST differences when warm rain formation is more efficient. Here, interactive radiation and the response of the circulation are crucial: along with congestus a deeper moist layer develops, which leads to less low-level radiative cooling, a smaller buoyancy gradient between the columns, and therefore a weaker circulation and less subsidence over the cold ocean. The congestus mode is accompanied with more surface precipitation in the subsiding column and less surface precipitation in the deep convecting column. For the shallow mode over colder oceans, circulations also weaken with more efficient warm rain formation, but only marginally. Here, more warm rain reduces convective tops and the boundary layer depth—similar to Large-Eddy Simulation (LES) studies—which reduces the integrated buoyancy gradient. Elucidating the impact of warm rain can benefit from large-domain high-resolution simulations and observations. Parameterizations of warm rain may be constrained through collocated cloud and rain profiling from ground, and concurrent changes in convection and rain in subsiding and

✉ Louise Nuijens
louise.nuijens@tudelft.nl

[1] Delft University of Technology, 2628 CN Delft, Netherlands

[2] Massachusetts Institute of Technology, Cambridge, MA 02139, USA

[3] Nagoya University, Nagoya 464-8601, Japan

[4] University of Wisconsin-Madison, Madison, WI 53706, USA

convecting branches of circulations may be revealed from a collocation of space-borne sensors, including the Global Precipitation Measurement (GPM) and upcoming Aeolus missions.

Keywords Warm rain · Shallow cumulus · Congestus · Circulations · Climate

1 Introduction

Before observations demonstrated that clouds with tops below the freezing level are raining (Byers and Hall 1955; Battan and Braham 1956), scientists believed that ice nuclei are necessary to produce rain. Inspired by those observations, scientists soon discovered the importance of coalescence processes for warm rain formation. Coalescence processes also helped explain why clouds over oceans with a similar depth as clouds over land rain more easily. Namely, oceans are deprived of aerosol, except for sea salt and sulfate. Hence, there are relatively few cloud condensation nuclei over oceans, so that cloud droplets are relatively large, and auto-conversion and accretion processes are very efficient (Kubar et al. 2009).

In the subtropics and tropics, the freezing level is located between 4 and 5 km, and raining clouds include subtropical stratocumulus over cold eastern ocean boundaries; shallow cumulus clouds that are more widespread throughout the subtropics and tropics; and congestus clouds, which tend to be confined to warmer oceans. According to the WMO cloud atlas, congestus is not a cloud type on its own, but a species of cumulus with tops between 2 km and the freezing level (although in the literature congestus is often used to denote cumuli with tops up to 8 km). Although both shallow cumulus and congestus produce drizzle and rain alike, a point we return to below, the distinction is useful, because congestus appears more sensitive to changing large-scale states (Nuijens et al. 2014).

The CloudSat cloud profiling radar (CPR), which is currently the only sensor capable of delineating warm cloud, drizzle and rain, has demonstrated that oceans are covered by 10–50% of warm clouds (Fig. 1a). Of these warm clouds, between 20 and 40% contain

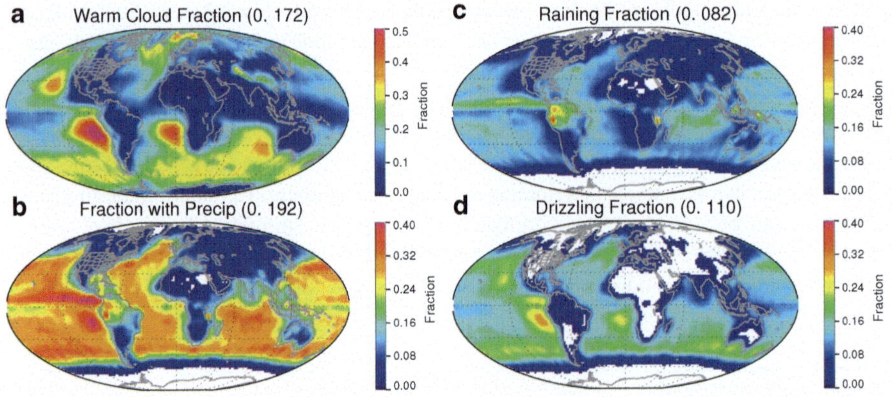

Fig. 1 Global warm rain occurrence observed by CloudSat during 2007–2010. Warm clouds are as those with tops >273 K. Drizzle is defined as having a "near-surface" (the lowest detectable range bin in CloudSat observations) reflectivity Z greater than -7.5 dBZ, while rain is defined as $Z > 0$ dBZ. The rain and drizzle fractions represent the fraction of warm clouds that contain rain or drizzle, respectively. CloudSat data are processed using the 2C-PRECIP-COLUMN and 2C-RAIN-PROFILE algorithms described in Haynes et al. (2009) and Lebsock and L'Ecuyer (2011), respectively

precipitation (Fig. 1b). Somewhat lower fractions are found near ocean boundaries and higher fractions in the Eastern Pacific, but overall most regions have clouds with a precipitating fraction of at least 5–10%. Figure 1c, d further separates precipitation in drizzle and rain, which shows that drizzle is widespread and concentrated over eastern ocean boundaries, whereas rain is concentrated in the downstream trades and near the Intertropical convergence zone (ITCZ), e.g., just north of the Equator in the Pacific and Atlantic oceans, or just south of the Equator in the Indian ocean.

One might not be impressed by those numbers. But those who live on islands have long appreciated the occasional passing of warm rain showers, which on windward sides of hilly islands make for lush vegetation, even during the “dry” season when deep convection abates.

Measurements from the precipitation radar (PR) deployed during the Tropical Rainfall Measuring Mission (TRMM, 1997–2015) first emphasized the prevalence of warm rain over global oceans and their potential impact on large-scale heating rates (Short and Nakamura 2000; Schumacher and Houze 2003; Takayabu et al. 2010). But the TRMM PR has a minimum detection threshold of about 0.4–0.5 $\mathrm{mm\,h^{-1}}$ and therefore misses about 9% of accumulated rain and up to 50% of the occurrence of light rain compared to CloudSat (Berg et al. 2009). Global precipitation estimates from passive microwave sensors also lack a sensitivity to light rain, especially when rain covers only small areas, which is the case for isolated cumulus showers (Burdanowitz et al. 2015). The most widely used Global Precipitation Climatology Project (GPCP) product has long been used to derive surface latent heat fluxes to construct the global mean energy budget, and a lack of warm rain might explain why observed heat fluxes at the ocean surface have not matched observed ocean heating rates. By increasing global precipitation rates by about 16% based on CloudSat data, equivalent to increasing the latent heat flux by 12 $\mathrm{W\,m^{-2}}$, the global mean surface energy budget can be closed (Stephens et al. 2012). But warm rain is probably not the only reason for residuals in the surface energy budget. Largest residuals are found in regions where shallow cumuli dominate but alternate with deeper convection (Kato et al. 2016), but these residuals are larger than warm rain can account for. One hypothesis is that retrieval errors are caused by greater variability in deeper convection along with shallow convection, which warrants a better understanding of the coupling between different types of convection.

Warm rain may also alter the radiation budget by influencing the microphysical and mesoscale structure of clouds, which satellite images, corroborated by in-situ measurements and high-resolution modeling, have demonstrated. Numerous cells of seemingly cloud-free air (pockets of open cells, POCs) surrounded by walls of drizzle can be embedded into an otherwise homogeneous stratocumulus cloud deck (Stevens et al. 2005; Wood et al. 2008). In these cells cloud droplets have been scavenged by drizzle drops, leading to very low cloud droplet number concentrations. Observations have also demonstrated that fields of shallow cumuli accompanied by significant rain are organized into arc-shaped formations (Snodgrass et al. 2009; Zuidema et al. 2012). These are representative for the presence of cold pools, which are produced by the evaporation of rain and convective downdrafts, similar to the cold pools that accompany deep convection (Tompkins 2001).

When rain evaporates, it no longer produces a net (latent) heating, and it redistributes moisture in the atmosphere. The relative fraction of rain that evaporates or reaches the surface is thus important for large-scale heat and moisture budgets. Herbert Riehl derived the first heat budget of the trades and argued that each layer of the lower atmosphere would precipitate (and heat) just as much as would be required to balance the loss of heat from radiation. Hence, from the profile of radiative cooling, one could predict the profile of the rain flux and thus the profile of the moisture flux (Riehl et al. 1951).

But, since Riehl's study, not much research has focused on large-scale controls and impacts of warm rain. More attention has been given to microphysical aspects of warm rain, including the role of aerosols in the onset of warm rain. For instance, aerosol and cloud condensation nuclei (CCN) concentrations control the height at which rain-sized drops first start to form. Figure 2, adopted from Lonitz et al. (2015), shows how radar reflectivity, which measures the sixth moment of the drop-size distribution, increases with height above cloud base for different CCN concentrations. The gray lines illustrate theoretical behavior, assuming a gamma distribution of the drop-size distribution, and the black lines illustrate the behavior of a 1D kinematic bin microphysics model (Seifert and Stevens 2010). According to the CloudSat definition, a reflectivity larger than −7 dBZ corresponds to drizzle, which implies that for a CCN concentration of 100 cm^{-3} drizzle forms when clouds reach 1.7 km. Indeed, the percentage of clouds that develop a maximum reflectivity larger than −7 dBZ (drizzle) or 0 dBZ (rain) increases substantially for cloud tops beyond 1.5 km (Fig. 2b): 20% of clouds contain drizzle, and 20% contain rain. For cloud tops beyond 2 km, most clouds already contain rain (82%).

Warm rain is thus an integral part of shallow convection, which may influence climate in ways that are not yet well measured or modeled on a global scale. Do we understand how the presence of warm rain changes the character of warm clouds and the large-scale circulations in which these clouds are embedded? And what regulates variations in the depth of the convection?

To formulate ideas about the interaction between convection, warm rain and circulations, a conceptual model can be helpful and bypass some of the complexities of global

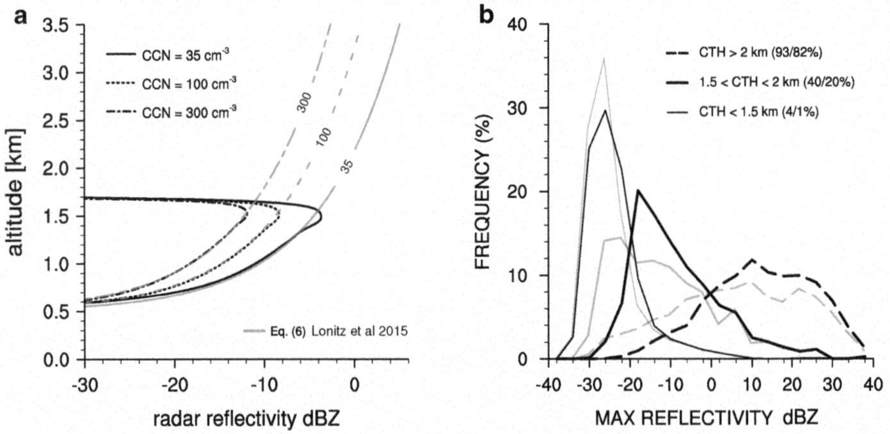

Fig. 2 On the left: the increase in radar reflectivity, a measure of the sixth moment of the rain-drop-size distribution, with height in developing (not-raining) clouds as simulated with a kinematic 1D bin model of microphysics (black lines) and from theory assuming a gamma distribution for the rain-drop-size distribution (gray lines). The figure is adopted from an earlier version in Lonitz et al. (2015). On the right: distributions of the maximum radar reflectivity found anywhere in individual cloud entities with a cloud base < 800 m, as observed at the Barbados Cloud Observatory during 3 years. The distributions in black are for three cloud top height (CTH) categories, and percentages indicate the total number of clouds (in %) which have a maximum reflectivity ≥ −7 dBZ/≥ 0 dBZ. The distributions in light gray are for maximum reflectivities below cloud base only

models. In this paper we use a two-column radiative–convective equilibrium (RCE) model to do so, and Sect. 2 introduces the model and presents first results. Even with such a simplified framework, the interaction between circulation and convection appears intricate, but we will speculate on a few mechanisms that can be tested in further studies or evaluated in models with explicit convection or observations. In Sect. 3 we will discuss ongoing efforts in fine-scale modeling of convection (Large-Eddy Simulations), space-borne and ground-based observations, and we summarize our thoughts in Sect. 4.

2 Warm Rain in Large-Scale Circulations

The prevalence of warm rain demands a better understanding of what warm rain implies for the structure of the lower atmosphere and the energy budget of subtropical and tropical oceans. Because regions with warm clouds are connected to regions with cold clouds through large-scale circulations, the influence of warm rain may also be felt remotely.

A number of studies have hinted that shallow convection can have far-reaching effects. For instance, in an idealized model of tropical climate shallow convection is responsible for the ventilation of boundary layer humidity, which affects the width and the intensity of the intertropical convergence zone (Neggers et al. 2007). Shallow convective mixing is also an attractive tuning factor in numerical weather prediction and global models, because it can substantially influence global distributions of liquid water, cloudiness and the radiation budget (Bechtold et al. 2014; Mauritsen et al. 2012). Even the climate sensitivity of a global model appears to be strongly regulated by local mixing by shallow convection and large-scale shallow overturning circulations (Sherwood et al. 2014). An important process underlying these far-reaching effects is the radiative cooling of the cloud-topped boundary layer. Stronger cloud-radiative effects in the subsiding branch of Walker circulations have been shown to narrow the area of deep convection in the upward branch (Bretherton and Sobel 2002; Peters and Bretherton 2005). Cloud-radiative effects also help aggregate deep convection in cloud-resolving and large eddy simulation (LES) models in radiative-convective equilibrium (RCE) (Muller and Held 2012; Wing and Emanuel 2014; Hohenegger and Stevens 2016).

Observations also show hints that differences in the characteristics of rain are accompanied by differences in the circulation. When shallow overturning circulations in the tropical Eastern Pacific are stronger, the Eastern Pacific is characterized by more large clusters of rain, whereas weaker shallow circulations are accompanied with a larger fraction of smaller isolated raining cells (Chen and Liu 2015). Such findings suggest that important feedbacks between regions of warm and cold cloud may exist, which include not only the vertical structure of moisture and cloud, but also rain.

Ideally global climate models (GCMs) would provide us insight into such feedbacks. But GCMs already disagree on the shallow convection that precedes warm rain (Sherwood et al. 2014) and their complexity makes it challenging to isolate processes. Instead, we aim for a more simplified setting following a series of studies that have used idealized two-box or four-box equilibrium models to study the sensitivity of tropical climate. Notably, Pierrehumbert (1995) demonstrated that the interaction between dry subsiding regions and moist convecting regions is an important determinant of tropical climate. His study and that of others (Miller 1997; Larson et al. 1999) also emphasize the strong sensitivity of tropical circulations to water vapor, cloudiness and radiative cooling in the subsiding region. Other factors that are important for circulation strength are the relative area occupied by subsiding versus convecting areas (Pierrehumbert 1995; Bellon and Le Treut 2003), e.g., the relative size of the two boxes, and the sea surface temperature (SST) gradients as set by surface winds and ocean transport (Sun and Liu 1996; Clement and Seager 1999).

In our first exploration of the sensitivity of circulations to warm rain, we use an extension of a one-dimensional (single-column) RCE model, which numerically solves the hydrostatic equations of motion for non-rotating, nonlinear flow in two side-by-side columns. Rather than assuming convection in the subsiding column is limited to the boundary layer, as in most of the two-column model studies just mentioned, the depth of convection is calculated interactively by the convection scheme. Furthermore, radiation is interactive and a cloud scheme is used. A version of this model for linear flow was first used by Nilsson and Emanuel (1999), which demonstrated that a positive feedback between the circulation, clear-sky water vapor and radiation can destabilize radiative–convective equilibrium and attain a new equilibrium with a thermally direct circulation between the columns. In our simplified setup we prescribe SST gradients, and the two columns are of equal size. The latter is a shortcoming of the model setup that we are aware of. The model also uses parameterized physics and thus carries similar uncertainties as the physics used in GCMs. Nevertheless, the simplified geometry of the model allows us to get a first insight into mechanisms that may be relevant to warm rain in circulations, which we hope are further tested in future studies. The next section describes the model and its setup in more detail.

2.1 A Two-Column RCE Model

The model's columns are oriented in a x, z plane, and the following equations for temperature T, specific humidity q_{v} and the vorticity η are solved by time integration:

$$\frac{\partial T}{\partial t} + u\frac{\partial T}{\partial x} + \omega\frac{\partial T}{\partial p} + \frac{\alpha\omega}{c_p} = \gamma\frac{\partial^2 T}{\partial x^2} + F_{\mathrm{SH}} + F_{\mathrm{R}} + F_{\mathrm{Q1}} \tag{1}$$

$$\frac{\partial q_{\mathrm{v}}}{\partial t} + u\frac{\partial q_{\mathrm{v}}}{\partial x} + \omega\frac{\partial q_{\mathrm{v}}}{\partial p} = \gamma\frac{\partial^2 q_{\mathrm{v}}}{\partial x^2} + F_{\mathrm{LH}} + F_{\mathrm{Q2}} \tag{2}$$

$$\frac{\partial \eta}{\partial t} + u\frac{\partial \eta}{\partial x} + f\frac{\partial v}{\partial p} = \frac{\partial \alpha}{\partial x} + \gamma\frac{\partial^2 \eta}{\partial x^2} + \frac{\partial v(\partial\eta/\partial p)}{\partial p} + \frac{\partial F_{\mathrm{c}}^{u}}{\partial p} \tag{3}$$

whereby the vorticity η of the flow is defined as:

$$\eta = \frac{\partial u}{\partial p} - \frac{\partial \omega}{\partial x} \tag{4}$$

and the specific volume α as:

$$\alpha = \frac{T R_{\mathrm{d}}\,(1 - q_{\mathrm{v}} + q_{\mathrm{v}}/\epsilon)}{p} \tag{5}$$

Here, R_{d} is the gas constant for dry air and ϵ is the ratio of the molecular mass of water vapor and of dry air. Furthermore, u is the zonal wind; ω is the vertical velocity in pressure coordinates; c_{p} is the specific heat capacity of dry air; F_{SH} and F_{LH} are the sensible and latent heat fluxes at the surface; F_{R} is the net radiative heating tendency; F_{Q1} and F_{Q2} are the heat source and moisture source/sink due to convection and condensation; f is the Coriolis parameter; γ represents the inverse of a damping timescale τ; F_{c}^{u} is the tendency of

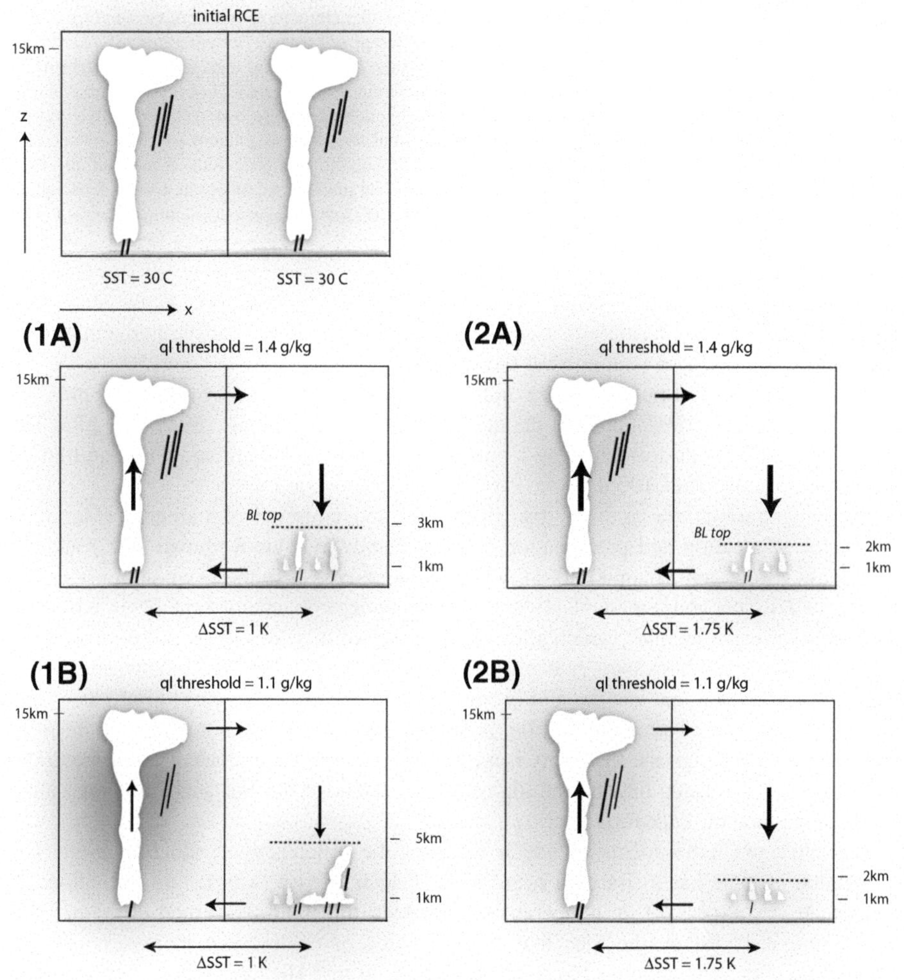

Fig. 3 A sketch of the two-column overturning circulation with deep convection in both columns in the initial RCE state (top). Below two scenarios are sketched whereby deep convection has developed over the warm ocean and shallow convection over the cold ocean. **1A** and **1B** correspond to a SST difference of 1 K between the columns, and **2A** and **2B** to a SST difference of 1.75 K. In experiment (**A**) the condensate threshold for rain formation is $1.4\,\mathrm{g\,kg^{-1}}$ and in experiment (**B**) the threshold is $1.1\,\mathrm{g\,kg^{-1}}$. The horizontal dashed lines denote convective tops over the cold ocean. The arrows denote the large-scale vertical velocity in each column and the zonal flow at the column boundary

the zonal wind due to convective momentum transport; and $\frac{\partial \nu(\partial \eta/\partial p)}{\partial p}$ represents the momentum flux divergence in the boundary layer, whereby ν is a shear viscosity.

The nonlinear flow is thus forced by zonal gradients in α, and α is inversely proportional to the virtual temperature predicted by Eqs. (1) and (2). The Coriolis acceleration is put to zero; hence, the circulation may be considered a mock-Walker circulation. A simple Fickian damping of the flow takes place in the model interior through diffusion at a timescale τ. Furthermore, the momentum flux divergence linearly decreases from a maximum damping near the surface to zero damping above the depth of the boundary layer.

Fig. 4 Sensitivity of convection and precipitation in a Walker-like circulation to the conversion of liquid to rain. The circulation occurs between two columns with different sea surface temperatures (SST), with an increasing SST difference on the *x*-axis (ΔSST). On the *y*-axis are shown: (**a**) convective tops, (**b**) surface precipitation rate, (**c**) column water vapor, (**d**) equivalent potential temperature (θ_e) of the well-mixed layer, (**e**) surface wind speed at the column boundary, (**f**) surface latent heat flux, (**g**) integral of radiative cooling rate from the surface up to 500 hPa, (**h**) integral of the virtual temperature gradient ($\Delta T_v/L$) between the columns from the surface up to the convective top in the cold column (*L* is the width of one column). Both black lines are for the column over the colder ocean, with dashed lines for a condensate-to-rain threshold of $1.4\,g\,kg^{-1}$, and solid lines for a threshold of $1.1\,g\,kg^{-1}$ (more efficient warm rain formation). The gray lines are for the column over the warm ocean instead

Momentum is also damped through convective momentum transport. At the surface, a linearized surface drag formula is used as the boundary condition for the horizontal flow, and a free-slip condition is applied at the model top.

The two columns are of equal size and 1500 km wide. A vertical grid of 100 pressure levels is used, with a resolution of about 125 m that becomes finer above 100 hPa. The model integration is performed using a time step of 1 min and continued until equilibrium is reached, usually after 100 days.

Parameterizations are used for the absorption and emission of radiation (Morcrette 1991), for convection and precipitation (Emanuel and Zivkovic-Rothman 1999) and for cloudiness (Bony and Emanuel 2001), which is an extension to the clear-sky-only calculations in Nilsson and Emanuel (1999). The Emanuel convection scheme is based on buoyancy sorting principles, which allows a spectrum of mixtures to ascend or descend to their level of neutral buoyancy. Notably, the scheme does not distinguish between shallow and deep convection and has the ability (in the past deemed a disability) to produce light rain in the absence of deep convection. A simple buoyancy closure determines the mass flux at cloud base, e.g., the mass flux is adjusted to maintain sub-cloud layer air neutrally buoyant when displaced beyond the top of the sub-cloud layer. Surface fluxes are calculated using standard bulk formulae.

We carry out experiments in which we alter the efficiency of rain formation. The convection scheme has a straightforward way of dealing with (warm) rain formation: all condensate in excess of a temperature-dependent threshold is turned to rain. This assumption is based on the idea that the efficiency of coalescence increases with the amount of condensate and thus the presence of large drops. The threshold is constant up to the freezing level and decreases beyond this level in light of the Bergeron–Findeisen process. Rain and its associated heat are added to a single hydrostatic unsaturated downdraft, and rain evaporates as a function of the temperature and humidity of the environment and the downdraft (Emanuel 1991).

Given the models' physics and lack of a separate boundary layer scheme, our experiments exclude stratocumulus clouds, but we acknowledge that also these shallow clouds produce a significant amount of warm rain that may be relevant to circulating equilibria in the tropical atmosphere (Sect. 1).

2.2 Circulating Equilibria in the Two-Column System

A sketch of the two-column system is shown in Fig. 3. The columns are first run into RCE at a uniform prescribed sea surface temperature (SST) of 30 °C (top panel). No circulation exists when both columns have an identical RCE state (top panel). Consequently, the SST in one of the columns is lowered with increments of 0.25 K up to an SST difference (ΔSST) of 2 K. Alternatively, we can increase the SST in one of the columns from a colder

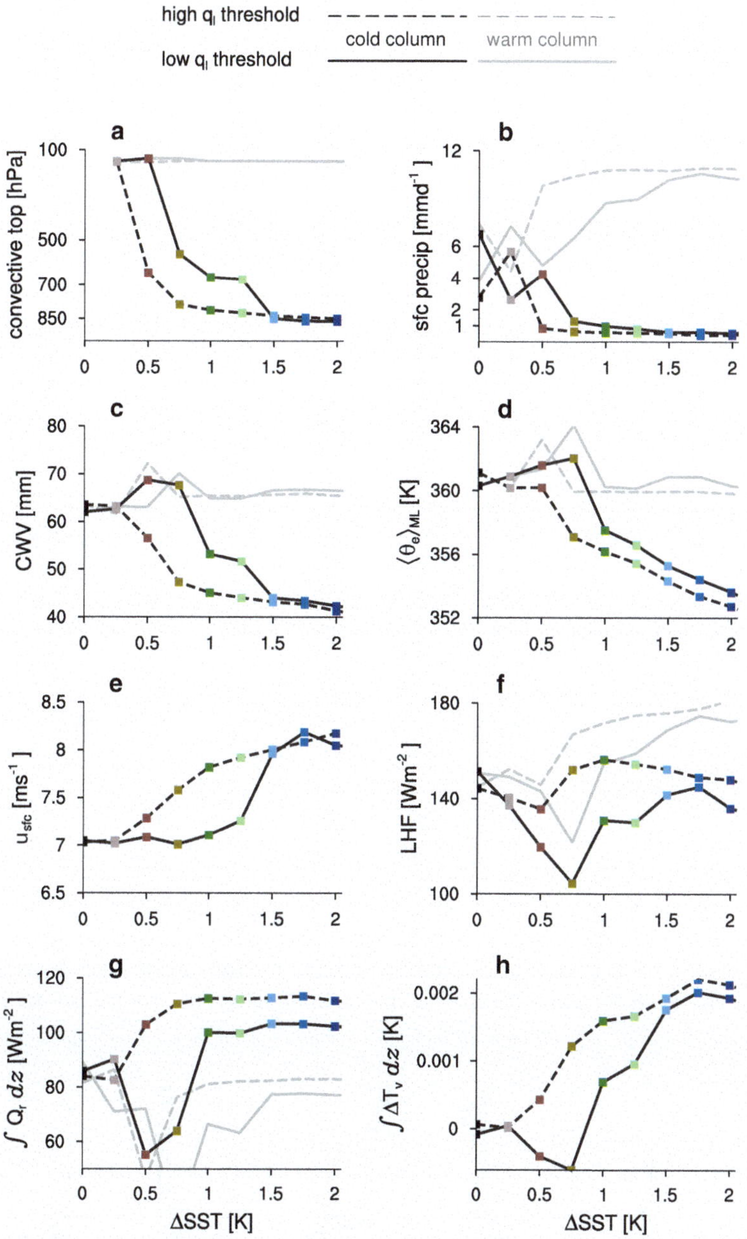

RCE state. This gives qualitatively the same results, but some hysteresis is present, because of differences in the initial moisture structure.

The black dashed lines in Fig. 4a show how the tops of convection over the cold ocean collapse with ΔSST for a condensate-to-rain threshold of $1.4\,\mathrm{g\,kg^{-1}}$, with blue-hued markers for the largest ΔSST. Here, convective tops are defined as the maximum level of positive convective mass flux. For all ΔSST convection over the warm ocean remains deep with tops up to 150 hPa (the gray dashed lines). Evidently, a ΔSST = 0.5 K is enough to collapse convective tops to roughly 600 hPa (about 4 km), and a ΔSST = 0.75 K collapses convection to roughly 800 hPa (about 2 km). Even the shallowest modes produce precipitation with surface rates just below 1 mm day^{-1}.

Because convection collapses, the atmosphere above the tops of convection experiences a net (radiative) cooling. This results in a temperature difference with the other column, which triggers a circulation: subsidence develops over the cold ocean ($\omega_{\text{cold}} > 0$) and rising motion over the warm ocean ($\omega_{\text{warm}} < 0$). Near the tropopause the flow is directed from the warm to the cold column ($u > 0$) and near the surface from the cold to the warm column ($u < 0$), as illustrated in Fig. 3-1A, 2A (middle panels) for a SST difference (ΔSST) of 1, respectively, 1.75 K. A new circulating equilibrium develops in which radiative cooling above the cloud layer over the cold ocean is now balanced by subsidence warming instead of convective heating. Over the warm ocean the mean rising motion introduces extra cooling next to radiative cooling, both of which are balanced by deep convective heating.

Nonlinear behavior at low ΔSST, such as in the surface precipitation rates, are caused by a reverse in the circulation. For ΔSST $\leq$ 0.5 K convection over the cold ocean has not yet collapsed, and a weak oscillatory circulation can develop. Furthermore, the convection scheme favors detrainment near mid-levels, which can produce large cloud fractions that significantly lower radiative cooling in the lower troposphere, and also reverse the circulation.

The profiles of ω and u over the cold ocean show how divergence and near-surface wind speeds increase with ΔSST (Fig. 5), but saturate as ΔSST > 1.5 K, and even weaken again at very large ΔSST = 5 K. The strength of the circulation may be understood through a simplified form of the equation for the flow's vorticity (Eq. 3). For non-rotating flow and ignoring horizontal advection, damping, surface friction and convective momentum transport, we may write:

$$\frac{\partial \eta}{\partial t} = \frac{\partial \alpha}{\partial x} + \frac{\partial}{\partial p}\left(\nu \frac{\partial \eta}{\partial p}\right) \tag{6}$$

The first term on the right-hand side measures the buoyancy gradient between the columns and tends to increase the vorticity. The second term is the momentum flux divergence in the boundary layer. This shear stress tends to decrease the vorticity, but may be assumed to vanish at the top of the convective layer. Hence, if we integrate from the surface p_s up to

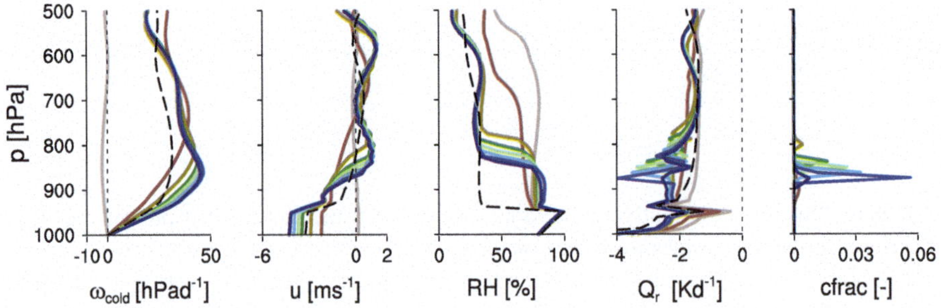

Fig. 5 Vertical profiles of the vertical velocity ω, the horizontal wind speed u, the relative humidity RH, the heating tendency due to radiation Q_r and the cloud fraction, for circulations at ΔSST between 0.25 and 2 K. Colors correspond to those used in Fig. 4. All profiles are for the column over the colder ocean surface and for the higher threshold of condensate-to-rain conversion (1.4 g kg^{-1})

the top of the boundary layer p_h and rearrange the stress term to the left-hand side, we can write:

$$v_s \left(\frac{\partial \eta}{\partial p} \right)_s = \int_{p_s}^{p_h} \frac{\partial \alpha}{\partial x} dp \tag{7}$$

The term $\left(\frac{\partial \eta}{\partial p}\right)_s$ on the left-hand side represents a measure of the strength of the circulation.

Equation (7) thus shows how the circulation is not only a function of ΔSST, but of buoyancy differences over the entire boundary layer h. Those buoyancy differences are strongly regulated by the radiative cooling (Q_r) over the cold ocean, which changes as the free troposphere dries with increasing ΔSST, which is best seen from the RH profiles in Fig. 5. Q_r tends to peak where temperature and moisture gradients and cloudiness are large, for instance at the inversion and at the mixed layer top (950 hPa). The maximum in Q_r increases as the inversion lowers. Larger cloudiness below the inversion can do so, but alone the interaction of long wave radiation with the clear-sky humidity profile would be sufficient (for example, see Stevens et al. 2017 in this same book collection). Large increases in ΔSST further dry the convective layer, leading to less liquid water and cloudiness, and therefore less radiative cooling and a weaker circulation (black dashed lines, Fig. 5).

In these experiments condensate is turned into rain at a threshold of $1.4\,\mathrm{g\,kg^{-1}}$. For liquid water lapse rates at these temperatures, a threshold of $1.4\,\mathrm{g\,kg^{-1}}$ is exceeded after about 600 m. A lower threshold of $1.1\,\mathrm{g\,kg^{-1}}$ is exceeded after about 450 m, which is a small difference, but one which has a relatively large impact on the character of convection and the circulation at intermediate ΔSSTs (the black solid lines in Fig. 4). Convection still collapses, but is more stepwise, with convective tops preferably located near 600 hPa at ΔSSTs = 1 K, indicative of congestus clouds. Although convective tops are higher and column water vapor increases (Fig. 4c), surface precipitation rates remain near or below $1\,\mathrm{mm\,day^{-1}}$ (Fig. 4b).

Also in nature surface precipitation rates slowly increase with column water vapor values between 20 and 60 mm and only pick up beyond a critical value (Holloway and Neelin 2009). One idea that could explain the slow increase in surface precipitation is that in order for precipitation and latent heating to increase, the atmosphere would need to cool more or warm less. Less warming could result from weaker subsidence as congestus develops. However, subsidence warming is still largely balanced by Q_r, and Q_r itself decreases as congestus deepens the moist layer.

In the next two sections, some more details of the changes in circulation for the shallow cumulus mode at ΔSST = 1.75 K (Fig. 3-2A, B) and the congestus mode at ΔSST = 1 K (Fig. 3-1A, B) are discussed.

2.3 Shallow Cumulus

At large ΔSST ($\geq$ 1.5 K) the cold ocean column develops shallow cumulus with tops up to 850 hPa and surface precipitation rates $<1\,\mathrm{mm\,day^{-1}}$, suggestive of light rain or drizzle. For this shallow mode, experiments with a condensate threshold of 1.4 and $1.1\,\mathrm{g\,kg^{-1}}$ are notably similar. However, when looking closely a few interesting differences can be seen, further illustrated with vertical profiles of the difference in potential temperature between low and high condensate thresholds ($\Delta\theta$), and profiles of specific humidity, radiative cooling, convective heating and the vertical velocity (Fig. 6). The dashed and solid lines refer to experiments with q_l thresholds of $1.4\,\mathrm{g\,kg^{-1}}$, respectively, $1.1\,\mathrm{g\,kg^{-1}}$.

Convection is slightly shallower when rain formation is more efficient ($q_l = 1.1\,\mathrm{g\,kg^{-1}}$), and the surface precipitation rate is marginally smaller (Fig. 4b). The vertical humidity profile shows that the boundary layer is also a little shallower with a moister sub-cloud and cloud layer ($\langle\theta_e\rangle_{ML}$ in Fig. 4d, q in Fig. 6). Accordingly, the surface evaporation is lower (Fig. 4f). The evaporation of cloud condensate takes place in a somewhat thinner and stronger inversion layer, with a stronger peak in cooling. As we will discuss later, in Sect. 3, these results are consistent with LES studies, which show that rain regulates inversion height by removing condensate that would otherwise deepen the boundary layer.

Because of shallower convection, the integrated buoyancy difference between the two columns is smaller (Fig. 4h). Without any change in the viscosity, this implies a weaker circulation (Eq. 7). The maximum ω_{cold} at h is smaller, along with smaller ω_{cold} throughout the rest of the troposphere. Less subsidence drying leads to a slightly moister free troposphere, consistent with somewhat lower radiative cooling rates.

For the congestus mode the change in the circulation with warm rain efficiency can be explained in a similar manner, as we will see next. However, in that regime convective tops

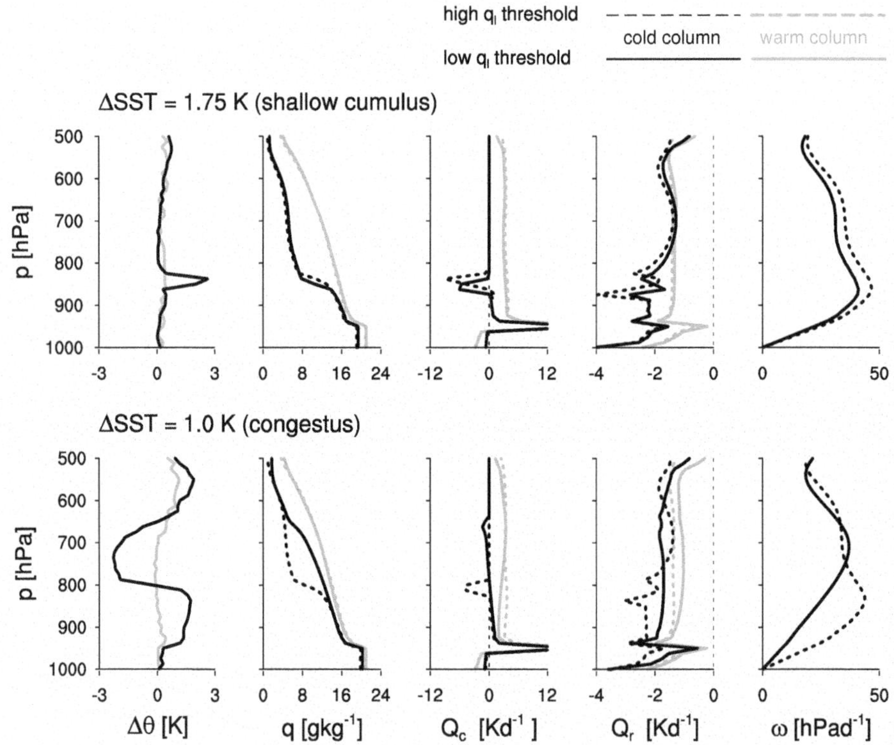

Fig. 6 For the circulations in Fig. 4 at ΔSST = 1 and 1.75 K we show vertical profiles of the difference in potential temperature between the two experiments $\Delta\theta = \theta_{ql=1.1} - \theta_{ql=1.4}$, specific humidity q, the heating tendency due to convection Q_c, the heating tendency due to radiation Q_r and the vertical velocity ω. Black and gray lines are for the cold, respectively, and warm column, and solid and dashed lines are for the low, respectively, and high threshold on liquid water for conversion to rain

increase with more efficient rain formation and are accompanied by larger changes in circulation and convection over the warm ocean.

2.4 Congestus

For ΔSST between 0.75 and 1.5 K convection is more sensitive to the change in condensate-to-rain threshold. More efficient rain formation ($q_l = 1.1\,\mathrm{g\,kg^{-1}}$) gives rise to deeper convection and larger surface precipitation rates (Fig. 4). When more condensate is turned to rain, at a lower altitude, updraft buoyancy increases. For ΔSST = 1 K a marginal increase in convective heating Q_c can be seen between cloud base (950 hPa) and 850 hPa with more efficient rain formation (Fig. 6). The level where convective heating turns into cooling has shifted upward, and most of the detrainment and evaporative cooling takes place at cloud tops near 650 hPa. At this level, where cooling is pronounced, subsidence peaks.

Again, the change in condensate-to-rain threshold is small, yet important in this scheme where convection is modeled based on buoyancy sorting principles. Each mixture of air ascends or descends to its level of neutral buoyancy, which might involve several episodes of mixing, especially when precipitation changes the amount of condensate during ascent/descent. Dealing with multiple mixing episodes is bypassed in the current scheme by insisting that mixed air detrains at levels where its liquid water potential temperature is equal to that of the environment (Emanuel 1991). This is illustrated in Fig. 7, which shows the liquid water potential temperature (or liquid water static energy h_w) of a lifted parcel ($h_{w,p}$, dashed lines) and that of the environment ($h_w = h$, solid lines) before mixing. $h_{w,p}$ is conserved and equal to the dry static energy h of the sub-cloud layer when there is no precipitation and no mixing. But upon precipitation h_w increases. In this case, the lower condensate-to-rain threshold of $1.1\,\mathrm{g\,kg^{-1}}$ has almost shifted the liquid water static energy of lifted parcels at ΔSST = 1 K to that of lifted parcels at ΔSST = 0.5 K. Indeed, these two experiments have almost similar congestus tops (Fig. 4a). In other words, the SST is crucial at setting convective tops, but the precipitation efficiency may allow convection over colder SSTs to reach a similar depth as convection over warmer SSTs.

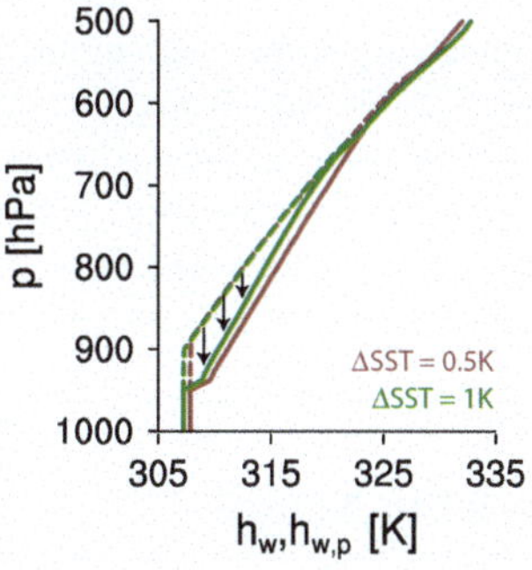

Fig. 7 The liquid water static energy of the environment h_w (solid lines), and of a lifted parcel $h_{w,p}$ before mixing (dashed lines), as calculated in the convection scheme. Two sets of lines are shown, which both represent the column over the colder ocean: one for the experiment with ΔSST = 1 K and a threshold of $1.1\,\mathrm{g\,kg^{-1}}$ (green) and one for the experiment with ΔSST = 0.5 K and a threshold of $1.4\,\mathrm{g\,kg^{-1}}$ (brown). The arrows indicate the levels at which the parcel would detrain upon mixing in this scheme. Increasing $h_{w,p}$ implies that air will not descend as far before detraining

Along with these changes, the thermodynamic structure over the cold ocean, the circulation, and the character of convection over the warm ocean change. For instance, at a threshold of $1.1\,\mathrm{g\,kg^{-1}}$ the mixed layer has a larger equivalent potential temperature (θ_e, in Fig. 4d). The larger mixed layer θ_e can be explained by less low-level radiative cooling, a decrease in subsidence drying and more precipitation. For the congestus modes, the larger mixed layer θ_e is accompanied by larger column water vapor (Fig. 4c), which is also evident from the deeper moist layer in the vertical profiles in Fig. 6. Similar to the shallow cumulus mode, the circulation decreases in strength at a lower condensate-to-rain threshold. Whereas for the shallow mode the reduction in h helps explain the weaker circulation (Eq. 7), in the congestus regime the reduced humidity gradient (followed by the temperature gradient) is responsible for a smaller buoyancy gradient and a weaker circulation (Fig. 4h).

One may question if the response to warm rain efficiency is the same if there were no interaction between the two columns, i.e., when the circulation is fixed and there is no feedback of subsidence to latent heating. LES studies of shallow convection generally impose a fixed subsidence rate and therefore constrain the depth of convection a priori. If we fix the profile of ω_{cold} (using the profile of the experiment with a threshold of $1.4\,\mathrm{g\,kg^{-1}}$) and run a single-column experiment with a lower threshold of $1.1\,\mathrm{g\,kg^{-1}}$ convection is shallower instead of deeper (green lines in Fig. 8). To balance the fixed moderate subsidence warming while having larger latent heating in the cloud layer, the model has to find a new equilibrium with larger radiative cooling in the cloud layer. This layer will be shallower, with a stronger inversion and a drier overlying free troposphere. In this case, the response is similar to that of the shallow mode and similar to LES studies discussed in Sect. 3.1.

The circulation is thus critical to the impact of warm rain in the subsiding column. It leads to small changes in deep convection over the warm ocean. As convection over the cold ocean precipitates more, convection over the warm ocean precipitates less (Fig. 4b). Less precipitation is consistent with a weaker circulation and less cooling from mean vertical ascent over the warm ocean. There is also less cloud-base mass flux, more detrainment and larger cloud fractions near 400 hPa (Fig. 8j, k), which leads to less radiative cooling at low and mid-levels (Fig. 8n).

An intricate interaction between convection, rain microphysics, radiation and the circulation is thus responsible for congestus modes in the two-column model. In the next section, we turn our focus to reviewing some ongoing efforts in fine-scale modeling of convection and space-borne observations, which may be used to address ideas suggested by the two-column model. What limits the congestus modes at mid-levels specifically is discussed in more detail in an upcoming manuscript (Nuijens and Emanuel, in preparation), along with a longer discussion on the influence of model resolution, domain size and the momentum of the flow.

3 From a Conceptual Model to Nature

The two-column model suggests that changes in the depth of convection and warm rain in subsiding regions may be accompanied by changes in convection in ascending regions. The model also suggests that small-scale processes such as mixing and warm rain formation, followed by feedbacks through the circulation, have a noticeable impact on the character of convection and the circulation. But in our setup we have made a number of simplifications,

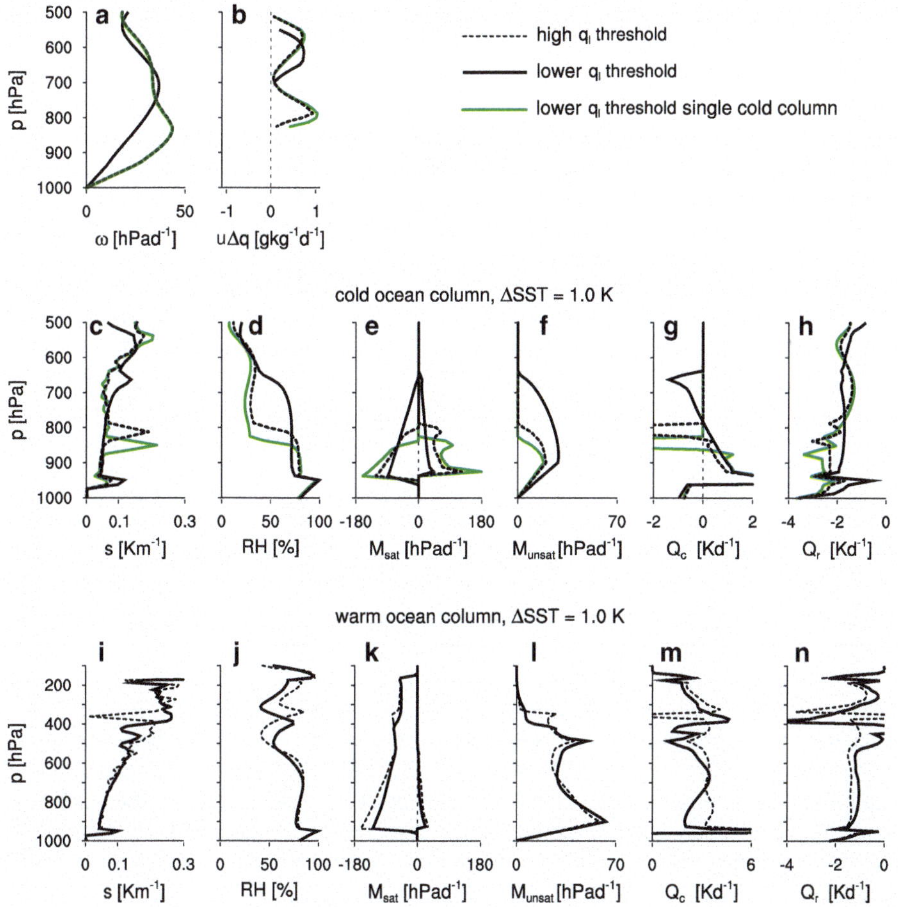

Fig. 8 Vertical profiles of circulation, convection and thermodynamic structure for the experiment with ΔSST = 1 K, which develops congestus with more efficient rain formation. Solid and dashed black lines correspond to condensate-to-rain thresholds of 1.1 and 1.4 g kg^{-1} (same as in the bottom panel of Fig. 6). Green lines show an experiment with a threshold of 1.1 g kg^{-1} but fixed ω_{cold} and near-surface u and fixed moisture convergence. The latter are taken from the experiment with a 1.4 g kg^{-1} threshold. Shown are: (top panel) the vertical velocity and moisture convergence, and (middle and bottom panels) the potential temperature lapse rate s, relative humidity, the saturated mass flux M_{sat}, both upward and downward components, the unsaturated mass flux driven by precipitation falling outside of the cloud M_{unsat}, convective heating Q_{c} and radiative cooling Q_{r}

most importantly that the columns are of equal size. As suggested by Pierrehumbert (1995) and Bellon and Le Treut (2003), an important follow-up to this study would be to change the relative size of the columns, or even adding columns. Our findings also crucially depend on the parameterized convection and cloudiness. Hence, our findings should be interpreted as ideas, which need further testing with models that explicitly simulate convection.

For instance, cloud-resolving models (CRMs) that are run on very large domains, even spanning ocean basins, may be forced with a surface temperature gradient to study the sensitivity of circulations to convection (Bretherton et al. 2006). In another approach,

CRMs on two domains may be used to simulate convection explicitly, and the two domains can be coupled through a circulation derived from the weak-temperature gradient approximation (Daleu et al. 2012).

Because shallow convection and cloudiness occur on scales smaller than conventional CRM grids, Large-Eddy simulations (LESs) may be preferred to investigate warm rain implications. So far LESs have been run on domains too small for circulations to develop, and too small for convection to organize itself into moist clusters surrounded by dry regions—a process that may crucially impact climate (Pierrehumbert 1995; Mauritsen et al. 2012). Nevertheless, LES studies have given some insight into the influence of warm rain on the thermodynamic structure of the lower atmosphere. We shall discuss these in the next section and draw out similarities with the two-column model results. Furthermore, we discuss how ground-based or airborne observations may help constrain warm rain formation, which remains uncertain even in LES.

3.1 Large-Eddy Simulations

Large-Eddy simulation has long been a tool to study turbulent flows including the cloudy boundary layer on limited horizontal domains ($20 \times 20 \times 4\,\mathrm{km}^3$) at fine grids ($100 \times 100 \times 40\,\mathrm{m}^3$). LES is also increasingly used to simulate deeper convection (up to 10–12 km) on horizontal domains of $50 \times 50\,\mathrm{km}^2$ and larger and with global-scale simulations underway. Microphysics in LES are typically parameterized using either a bulk scheme, which prognoses one or two moments of the drop-size distribution (only the total mass of rain, or the mass and number of rain drops), or a bin scheme, which uses a discretized version of the drop-size distribution and attempts to model the full evolution of the droplet spectrum. For the bulk schemes, the total cloud mass is inferred from an equilibrium assumption and the cloud droplet number is specified. Hence, these schemes implicitly assume an aerosol or cloud condensation nuclei concentration. By varying the cloud droplet number concentration, the sensitivity to warm rain formation has been studied.

Such sensitivity studies have demonstrated that at larger cloud droplet number concentrations shallow cumuli get deeper before they rain (Stevens and Seifert 2008; Seifert et al. 2015). Therefore, larger cloud droplet number concentrations produce less rainfall initially. But the response of clouds and the boundary layer will mitigate this initial effect. Namely, the removal of liquid water via rain reduces evaporative cooling and mixing near cloud tops and thus the entrainment of warm free tropospheric air into the boundary layer and the deepening of convection and the boundary layer. Therefore, after a long enough (>30 h) simulation time, differences in cloud and rain statistics for different cloud droplet number (aerosol) concentrations are small (Xue and Feingold 2006).

The regulation of inversion height by warm rain has been noted in early bulk theories of shallow cumulus convection and also matters for the sensitivity of shallow cumuli to global warming scenarios. Larger SSTs and less large-scale subsidence under global warming lead to a deepening of shallow cumuli. This increases the entrainment of dry air into the boundary layer, which dries the cumulus layer drier and reduces cloud fraction. But rainfall puts a notable limit to such deepening, so that changes in cloudiness with global warming are overall small (Blossey et al. 2013; Bretherton et al. 2013; Vogel et al. 2016).

In the two-column model, the response of the shallow mode to more efficient warm rain (Fig. 6, top panels), as well as the response of the congestus mode under fixed subsidence (Fig. 8, green lines), is similar to these LES studies. But in LES the subsidence rate is generally fixed. Our two-column model experiments suggest that changes in subsidence with (latent) heating can change the response to rain at SSTs that are favorable for congestus.

A recent study using LES demonstrates processes that are important in the transition from a shallow to a congestus regime, and we may use these to loosely evaluate changes in atmospheric structure produced by the two-column model. Vogel et al. (2016) and Vogel and Nuijens (in preparation) use a constant exponential profile of subsidence, prescribed SSTs and interactive radiation, for a model domain of $51.2 \times 51.2 \times 10\,\mathrm{km}^3$ with a resolution of 50 m in the horizontal and 10 m in the vertical. To reproduce many of the differences in cloud and boundary layer structure that are observed during months with predominantly shallow or deeper convection (congestus), an increase in SST and decrease in subsidence (at a reference scale height) are sufficient.

A 2K increase in SST and $1.5\,\mathrm{mm\,s^{-1}}$ decrease in ω lead to a deepening of convection from 2 to 7 km, a quadrupling of surface precipitation and a 15% decrease in cloud cover. The cloud and boundary layer structure of these control (CTL) and CTL.2K.ω6 simulations is shown in Fig. 9 and reveals that as congestus develops, moisture and temperature are much better mixed in the vertical, and the inversion is weaker. Because the inversion is weaker, less cloudiness develops near the inversion.

Critical to reproducing the character of convection and cloudiness in the two regimes is the role of interactive radiation, which can both stabilize and destabilize convection (Vogel and Nuijens, in preparation). For instance, interactive radiation is crucial for stabilizing convection and developing the stratiform outflow layers near the inversion for the CTL simulation. Sometimes interactive radiation also leads to a response one might not expect. For instance, when the free troposphere is drier, convection gets deeper and rains more. This is illustrated with the DRY simulation in Fig. 9, which has $1\,\mathrm{g\,kg^{-1}}$ less water vapor in the free troposphere. This enhances the radiative cooling in the moist convective layer and inversion layer, which boosts the buoyancy of cloudy updrafts that reach those layers. As congestus develops and moisture is mixed over a deeper layer, radiative cooling within the cloud and sub-cloud layer and its maximum value decrease (Fig. 9e). Because of the deeper moist layer, the surface latent heat flux also decreases (Fig. 10). Both factors—less destabilization from radiative cooling and a lower surface latent heat flux—could imply that convection self-limits itself. However, in these simulations there is no feedback

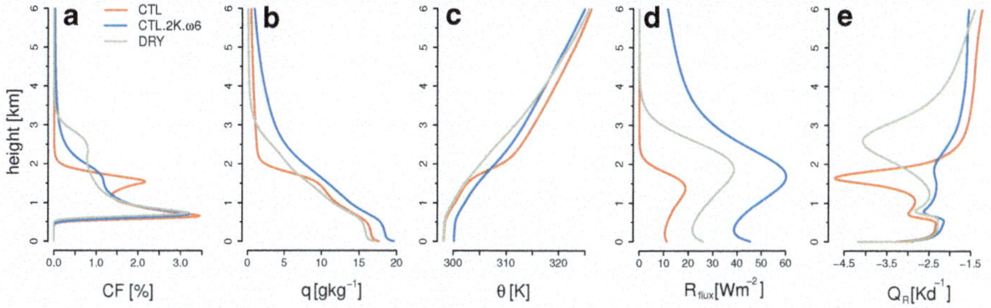

Fig. 9 Domain-averaged profiles of cloud fraction, specific humidity, potential temperature, precipitation flux and radiative cooling from LES. The CTL (orange) and DRY (gray) simulations differ only in their initial profile of specific humidity in the free troposphere, whereby the CTL run is about $1\,\mathrm{g\,kg^{-1}}$ more humid. The CTL.2K.ω6 (blue) simulations has a 2K larger SST and a $1.5\,\mathrm{mm\,s^{-1}}$ reduction in the prescribed ω profile compared to the CTL simulation

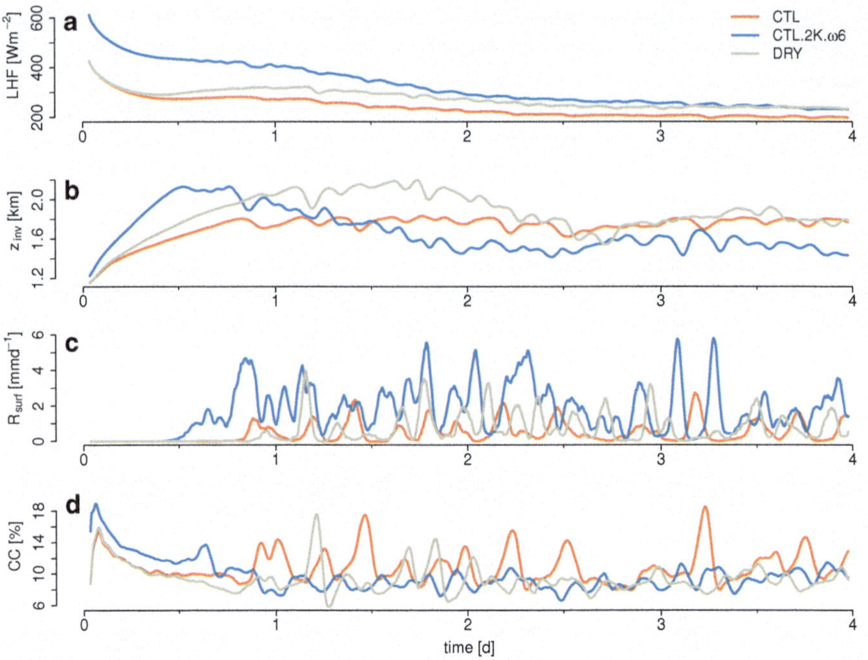

Fig. 10 Time series of the surface latent heat flux, the inversion height, the surface precipitation rate and total cloud cover for the CTL (orange), DRY (gray) and CTL.2K.ω6 (blue) simulations shown in Fig. 9

through the circulation such as in the two-column model, where the reduced cooling leads to a weakening of subsidence.

Average surface rain rates for the congestus clouds in LES are between 20 and 40 W m^{-2} (Fig. 10). But the simulated rainfall is very intermittent, which is caused by the limited number of deeper clusters and cold pools that this domain size can support. The profile of the rain flux also shows that a fraction of rain evaporates in the lower cloud and sub-cloud layer (Fig. 9d). The evaporation of rain triggers downdrafts which pull down cooler air that can spread out in the sub-cloud layer like a density current (cold pools). Such cold pools have been long known to exist for deep convection (see also Zuidema et al. in this collection). Within cold pools convection is suppressed, but at the downwind (colliding) boundaries of cold pools new convection can be triggered. This leads to arc-shaped cloud formations with clear skies in between as seen from satellite imagery (Snodgrass et al. 2009; Zuidema et al. 2012), and which the LES reproduces. The cold pools from shallow cumuli and congestus are mostly dry in their center, similar to deep convection, because rain rates are sufficiently strong to bring down relatively dry air from higher altitudes. This is different from the cold pools observed in open cell stratocumulus decks, which tend to be moist instead. At what rain rates either dry or moist cold pools develops and for which fraction of rain falling through unsaturated air are still open questions. And even in LES rain microphysics still carry a considerable uncertainty (VanZanten et al. 2011; Seifert and Heus 2013; Li et al. 2015). LES development would thus greatly benefit from progress made in deriving vertical profiles of rain and cloud from ground-based remote sensing networks, which we discuss next.

3.2 Observations

Relationships Between Cloud and Rain Estimates of how much condensate is turned into rain would help validate microphysical models used in LES, which in turn could help inform models of rain for large-scale models, such as the simple condensate-to-rain threshold used in the convection scheme in the two-column model. Useful first steps would be estimating how much water is removed via rain. Surface measurements of rainfall over a larger area (such as from a weather radar), alongside measurements of surface evaporation (from buoys or ships), and measurements of advection (from sounding arrays) could provide such estimates. For instance, during the Rain in Cumulus over the Ocean (RICO) field campaign surface precipitation rates comprised about a fifth of the surface evaporation rate (Nuijens et al. 2009). The upcoming EUREC^{4}A field study (see Bony et al. 2017 in this collection) will provide the measurements needed for such a study.

But surface precipitation rates alone do not reveal at what levels rain has evaporated, or how much rain has fallen through clouds or clear sky. Those differences are important for understanding how moisture is distributed vertically or how convective downdrafts form. This requires better estimates of the vertical profile of the rain flux alongside that of cloud condensate. But an inherent problem is that the profile of cloud condensate and rain cannot be measured simultaneously with a single radar wavelength. Existing methods use a vertically pointing cloud radar (typically 36 GHz or K_a band) to measure the sixth moment of the drop-size distribution, which may be turned into a profile of cloud liquid water by using the liquid water path obtained from a microwave radiometer, and by making assumptions on the drop-size distribution. Through synergy of instruments at dedicated field sites (such as the CloudNet network in Europe or the US Department of Energy's Atmospheric Radiation Measurement (ARM) sites), refined algorithms have been developed to do so. Such data sets seem ripe for further exploration, but are not without challenges. As drizzle or rain-sized drops develop the drop-size distribution changes, and assumptions underlying these algorithms need to be adjusted. Furthermore, when rain rates are sufficiently high the (cloud) radar beam will become attenuated, and the microwave radiometer has to be shut down. Hence, to estimate the profile of rain that develops in statistically similar clouds, a radar with a smaller frequency (larger wavelength), which suffers less from attenuation, has to be employed at a nearby location, where it has a similar radar footprint. To accurately estimate rain evaporation, the radar should be almost as sensitive as the cloud radar, and hence, a K or K_u band radar would be ideal. Unfortunately, this would exclude deep convection whose rain rates can be so high that even a K or K_u band radar signal will be attenuated.

Globally, the CloudSat Cloud Profiling Radar (CPR) can be used to relate measured cloud top heights to surface precipitation rates. But also this is not straightforward. A 94 GHz W band radar suffers from significant attenuation as rain intensity increases, and surface backscatter remains an issue. A recently Bayesian Monte Carlo algorithm uses a cloud-resolving model database to link observed vertical and integrated measurements of liquid clouds to latent heating structures and precipitation rates at the surface (Nelson et al. 2016). A histogram of surface rain rates of each CloudSat CPR profile, stratified by cloud top heights (Fig. 11), shows that clouds with tops beyond 1.6 km produce rain that reaches the surface (similar to what we infer from Fig. 2). Clouds with tops beyond 2 km clearly produce even higher rain rates, but beyond 3 km, the increase in rain rates is overall small, with the exception of very high rain rates, yet these are very rare.

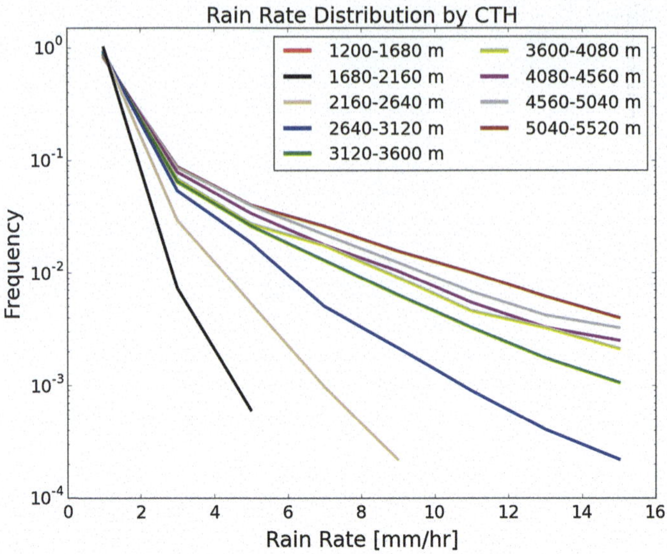

Fig. 11 Histogram of near-surface rain rates stratified by cloud top heights over tropical oceans as generated with the algorithm of Nelson et al. (2016) using CloudSat CPR data

The successor of the TRMM mission, the Global Precipitation Measurement (GPM) mission, has launched a dual-frequency precipitation radar (DPR) in 2014. This radar has a K_a and a K_u band, which suffer less from attenuation than CloudSat's W band. The DPR flies in a non-Sun synchronous orbit with an inclination angle of 65°, allowing it to measure extratropical clouds and the Arctic and Antarctic circles, which TRMM did not sample. Along with a microwave imager, the DPR will measure the vertical structure of precipitation intensities. By combining these measurements with traditional radiometers that already onboard satellites, the GPM mission promises to provide three-hourly rain estimates almost globally (Hou et al. 2013). This will allow studies on relationships between rain and the large-scale flow that can be done on much shorter timescales than has been possible so far.

Temporal Relationships One idea suggested by the two-column model experiments is that periods with shallow cumuli and drizzle occur in a different circulation than periods with congestus and rain. Data from the Barbados Cloud Observatory, currently the only remote sensing platform in the tropics, have already demonstrated that congestus and rain vary predominantly on timescales of days to weeks (Nuijens et al. 2014), which suggests that their occurrence is favored during certain large-scale states.

Linking clouds to circulations has been mostly done by long-term averaging of satellite observations over geographical regions or dynamical regimes. But to study variability on daily to synoptic timescales a different approach is needed, because most (polar orbiting) satellites sample a given location too infrequently (one exception being the new GPM mission with its three-hourly precipitation rates globally). One approach is to project different polar orbiting satellites onto a composite time axis centered on a rain event, a strategy that has been used in Masunaga (2013) and Masunaga and L'Ecuyer (2014). The result is a statistical time series on hourly and daily timescales, which precedes and follows those rain events. To illustrate this approach, we can use the areal rain coverage data from TRMM over the subtropical Pacific as a proxy for congestus. This is an area with considerable warm rain (Fig. 1). Because TRMM's minimum detection threshold is 17 dBZ, much larger than values observed for cloud tops up to 2 km (see also Fig. 2), TRMM likely does not observe drizzle, but only the more intense warm rain showers from congestus. We assume that congestus has an areal rain coverage less than 25%, and use all those events to create the time series in Fig. 12, which shows the evolution of profiles of cloud cover (measured by CloudSat's CPR), dry static energy and humidity, and the moisture budget.

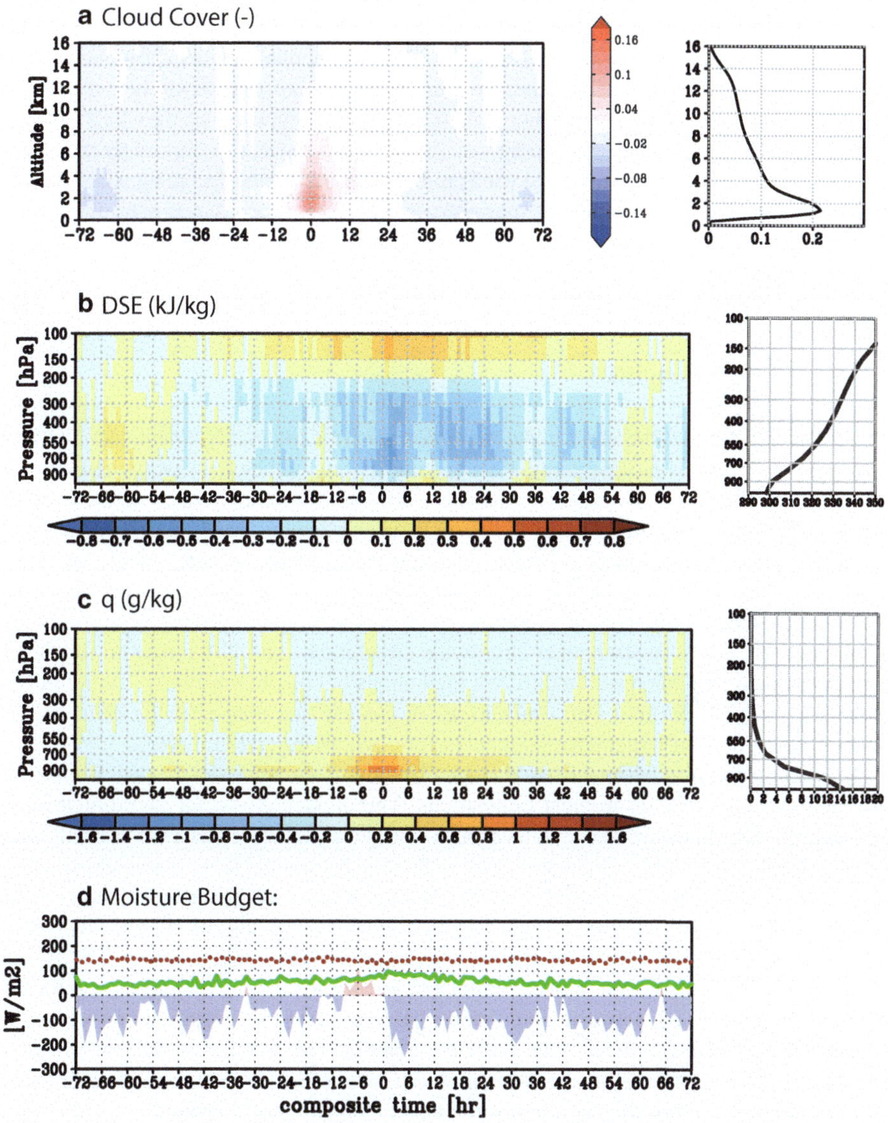

Fig. 12 Top panel (**a**) shows a composite time series of CloudSat cloud fraction anomalies from the background state for congestus rain events over the subtropical Pacific at 0 h. Similar composites for the profile of water vapor mixing ratio (g kg^{-1}) and dry static energy DSE (kJ kg^{-1} are shown in panels (**b**) and (**c**). The bottom panel (**d**) shows the composite moisture budget, including the surface evaporation rate (solid green), the surface evaporation (red dotted) and the vertically integrated moisture convergence (in shading, with convergence in light red and divergence in blue)

Figure 12a shows that after the rain event, cloudiness in the lower troposphere is larger for up to 12 h. The surface precipitation rate increases only little at the time of the rain event, which suggests that other precipitating clouds are always nearby (Fig. 12d). We also observe that the surface evaporation is rather invariant, which suggests it does not play a major role in controlling rain, unless its heterogeneity is not captured by the measurements.

Between two days before and after the rain event a cool anomaly in the lower and mid-troposphere is present (Fig. 12b). This cooling can destabilize the atmosphere and promote convection. Furthermore, during and following rain events the lower troposphere moistens, especially up to about 3 km (700 hPa) (Fig. 12c). Radiative cooling might explain the cooling anomalies, because these anomalies (even those in the upper atmosphere 72–36 h preceding the rain event) roughly correspond to the moist anomalies. But without knowing the winds and thus advection, we cannot draw any conclusions yet. This approach should therefore be repeated with additional satellite data, where the zonal wind profiles measured by the upcoming ADM Aeolus mission are particularly interesting. We may also use GPM data, which is more sensitive to light rain, and use cloud top heights instead of rain coverage as proxies for convection.

4 Concluding Thoughts

During the last decade, the cloud radar deployed aboard CloudSat has demonstrated that warm rain over oceans is ubiquitous. Recognizing its importance for shallow convection and low-level cloudiness, parameterizations of warm rain formation have been included in high-resolution models, such as LES. LES studies have demonstrated that warm rain can significantly alter the character of shallow convection, such as the depth of clouds, their organization and low-level cloudiness. In turn, shallow convection has been shown to impact large-scale circulations and climate. At least, the radiative driving from low-level cloud has been shown to strengthen large-scale circulations, and differences in low-level cloudiness among GCMs result in different predictions of climate sensitivity. Therefore, in this paper we question how warm rain itself—a process that may alter the character of shallow convection on larger scales—matters for circulations and climate.

We presented new experiments with an idealized two-column model to speculate on the influence of warm rain on tropical circulations. This model solves for two-dimensional non-rotating flow between two columns on a fairly fine vertical grid (125 m) and uses parameterized convection, cloudiness and radiation. Naturally, these parameterizations carry uncertainties, alike those used in GCMs. But the simpler dynamics in the two-column framework allow us to gain some insight into mechanisms involving warm rain, which may serve as a starting point for future studies using models with explicit convection. A circulation with deep convection in one column and subsidence in the other column is forced by prescribing the ΔSST between the columns. The circulation and depth of convection in the subsiding column as a function of ΔSST are found to depend on an intricate interaction between convection, warm rain, the circulation and radiation. The most interesting findings with respect to the sensitivity to warm rain can be summarized as follows:

1. At large ΔSST efficient warm rain formation lowers shallow cumulus tops and the inversion height over the cold ocean. This leads to a small reduction in the integrated buoyancy gradient between subsiding and ascending columns (Fig. 3-2A, B).
2. At smaller ΔSST efficient warm rain formation raises shallow cumulus tops, leading to congestus clouds. These congestus clouds are accompanied with deeper moist

boundary layers and a reduction in the integrated buoyancy gradient between the columns (Fig. 3-1A, B).

3. Efficient warm rain formation can thus weaken the circulation across a range of ΔSST (but especially at smaller ΔSST). Surface precipitation rates over the cold ocean consequently increase, while surface precipitation and cloud-base mass fluxes over the warm ocean decrease.
4. Congestus clouds can develop because of the extra latent heating with more efficient warm rain, which raises parcel buoyancy. But the weakening of the circulation (weaker subsidence) is important for maintaining the congestus mode. Here, the strong reduction of low-level radiative cooling in response to low-level moistening by congestus is crucial.

We may thus postulate that warm rain formation has a negative feedback on the strength of circulations by regulating the depth and thermal structure of the lower troposphere. In other words, raining or deeper shallow cumuli in regions with subsidence may also weaken circulations, besides strengthening them through radiative cooling. An interesting observation in that regard is that periods with stronger near-surface winds are accompanied with deeper cloud layers and significant rain showers (Nuijens et al. 2009, 2015). This could suggest that, as circulations strengthen, convection in the subsiding branches responds by deepening and raining more, which can slow down the circulation.

Because the convection scheme we use is based on the premise that microphysical processes are important for the humidity structure of the atmosphere, a sensitivity to warm rain formation may not be a surprise. Given the uncertainty associated with any convection scheme, the above results should be considered speculative and merely a basis for further testing with models and observations.

An important step forward in improving warm rain processes in models is to use existing ground-based remote sensing. For instance, collocated vertical profiles of cloud and rain from ground-based radar can be exploited to constrain how much cloud condensate detrains and moistens the atmosphere, compared to how much condensate reaches the surface via precipitation. Even in LES such processes remain uncertain. Furthermore, deriving large-scale winds through the use of sounding arrays, combined with intensive vertical profiling of cloud, rain and the thermal structure of the atmosphere, can shed light on relationships between convection and the large-scale flow. These measurements are planned for the EUREC^{4}A field campaign (see Bony et al. 2017 in this collection). Finally, to test ideas suggested by the two-column model framework, satellite remote sensing, in particular the new GPM and Aeolus missions, can be used. Namely, these may identify whether long-term variations in circulation strength are linked to variations in rain in subsiding and ascending branches.

Acknowledgements This paper arises from the International Space Science Institute (ISSI) workshop on Shallow clouds and water vapor, circulation and climate sensitivity. The first author wishes to express a special thanks to Reimar Luest and the Max Kade Foundation for the opportunity to spend a year at MIT. Thanks go to Ethan Nelson, Raphaela Vogel, Katrin Lonitz and Vivek Sant for their help in creating figures, and to Cathy Hohenegger, Chris Bretherton and Robert Pincus for insightful discussions. Lastly, we thank three reviewers for providing insightful feedback.

References

Battan LJ, Braham RR (1956) A study of convective precipitation based on cloud and radar observations. J Meteorol 13(6):587–591

Bechtold P, Sandu I, Klocke D, Semane N, Ahlgrimm M, Beljaars A, Forbes R, Rodwell M (2014) The role of shallow convection in ECMWFs Integrated Forecasting System. Ecmwf technicalmemoranda, ECMWF

Bellon G, Le Treut HL (2003) Large-scale and evaporation-wind feedbacks in a box model of the tropical climate. Geophys Res Lett 30(22):2145. doi:10.1029/2003GL017895

Berg W, L'Ecuyer T, Haynes JM (2009) The distribution of rainfall over oceans from spaceborne radars. J Appl Meteorol Climatol 49(3):535–543

Blossey PN, Bretherton CS, Zhang M, Cheng A, Endo S, Heus T, Liu Y, Lock AP, de Roode SR, Xu KM (2013) Marine low cloud sensitivity to an idealized climate change: the CGILS LES intercomparison. J Adv Model Earth Syst 5(2):234–258

Bony S, Emanuel KA (2001) A parameterization of the cloudiness associated with cumulus convection; evaluation using TOGA COARE data. J Atmos Sci 58:3158–3183

Bony S, Stevens B, Ament F, Bigorre S, Chazette P, Crewell S, Delanoë J, Emanuel K, Farrell D, Flamant C, Gross S, Hirsch L, Karstensen J, Mayer B, Nuijens L, Ruppert JH Jr., Sandu I, Siebesma P, Speich S, Szczap F, Totems J, Vogel R, Wendisch M, Wirth M (2017) EUREC^{4}A: A field campaign to elucidate the couplings between clouds, convection and circulation. Surv Geophys. doi:10.1007/s10712-017-9428-0

Bretherton CS, Sobel AH (2002) A simple model of a convectively coupled walker circulation using the weak temperature gradient approximation. J Clim 15(20):2907–2920

Bretherton CS, Blossey PN, Peters ME (2006) Interpretation of simple and cloud-resolving simulations of moist convection radiation interaction with a Mock–Walker circulation. Theor Comput Fluid Dyn 20(5):421–442

Bretherton CS, Blossey PN, Jones CR (2013) Mechanisms of marine low cloud sensitivity to idealized climate perturbations: a single-LES exploration extending the CGILS cases. J Adv Model Earth Syst 5(2):316–337

Burdanowitz J, Nuijens L, Stevens B, Klepp C (2015) Evaluating light rain from satellite- and ground-based remote sensing data over the subtropical North Atlantic. J Appl Meteorol Climatol 54(3):556–572

Byers HR, Hall RK (1955) A census of cumulus height versus precipitation in the vicinity of Puerto Rico during the Winter and Spring of 1953–1954. J Meteorol 12(2):176–178

Chen B, Liu C (2015) Warm organized rain systems over the tropical Eastern Pacific. J Clim 29(9):3403–3422

Clement A, Seager R (1999) Climate and the tropical oceans. J Clim 12(12):3383–3401. doi:10.1175/1520-0442(1999)012<3383:CATTO>2.0.CO;2

Daleu CL, Woolnough SJ, Plant RS (2012) Cloud-resolving model simulations with one- and two-way couplings via the weak temperature gradient approximation. J Atmos Sci 69(12):3683–3699. doi:10.1175/JAS-D-12-058.1

Emanuel KA (1991) A scheme for representing cumulus convection in large-scale models. J Atmos Sci 48(21):2313–2329

Emanuel KA, Zivkovic-Rothman M (1999) Development and evaluation of a convection scheme for use in climate models. J Atmos Sci 56:1766–1782

Haynes JM, L'Ecuyer TS, Stephens GL, Miller SD, Mitrescu C, Wood NB, Tanelli S (2009) Rainfall retrieval over the ocean with spaceborne W-band radar. J Geophys Res Atmos 114(D8):n/a–n/a

Hohenegger C, Stevens B (2016) Coupled radiative convective equilibrium simulations with explicit and parameterized convection. J Adv Model Earth Syst 8(3):1468–1482

Holloway CE, Neelin JD (2009) Moisture vertical structure, column water vapor, and tropical deep convection. J Atmos Sci 66(6):1665–1683

Hou AY, Kakar RK, Neeck S, Azarbarzin AA, Kummerow CD, Kojima M, Oki R, Nakamura K, Iguchi T (2013) The global precipitation measurement mission. Bull Am Meteorol Soc 95(5):701–722

Kato S, Xu KM, Wong T, Loeb NG, Rose FG, Trenberth KE, Thorsen TJ (2016) Investigation of the residual in column-integrated atmospheric energy balance using cloud objects. J Clim 29(20):7435–7452

Kubar TL, Hartmann DL, Wood R (2009) Understanding the importance of microphysics and macrophysics for warm rain in marine low clouds. Part I: satellite observations. J Atmos Sci 66(10):2953–2972

Larson K, Hartmann DL, Klein SA (1999) The role of clouds, water vapor, circulation, and boundary layer structure in the sensitivity of the tropical climate. J Clim 12(8 PART 1):2359–2374. doi:10.1175/1520-0442(1999)012<2359:TROCWV>2.0.CO;2

Lebsock MD, L'Ecuyer TS (2011) The retrieval of warm rain from CloudSat. J Geophys Res Atmos. doi:10.1029/2010JD015415

Li Z, Zuidema P, Zhu P, Morrison H (2015) The sensitivity of simulated shallow cumulus convection and cold pools to microphysics. J Atmos Sci 72(9):3340–3355

Lonitz K, Stevens B, Nuijens L, Sant V, Hirsch L, Seifert A (2015) The signature of aerosols and meteorology in long-term cloud radar observations of trade wind cumuli. J Atmos Sci 72(12):4643–4659

Masunaga H (2013) A satellite study of tropical moist convection and environmental variability: a moisture and thermal budget analysis. J Atmos Sci 70(8):2443–2466

Masunaga H, L'Ecuyer TS (2014) A mechanism of tropical convection inferred from observed variability in the moist static energy budget. J Atmos Sci 71(10):3747–3766

Mauritsen T, Stevens B, Roeckner E, Crueger T, Esch M, Giorgetta M, Haak H, Jungclaus J, Klocke D, Matei D, Mikolajewicz U, Notz D, Pincus R, Schmidt H, Tomassini L (2012) Tuning the climate of a global model. J Adv Model Earth Syst. doi:10.1029/2012MS000154

Miller RL (1997) Tropical thermostats and low cloud cover. J Clim 10(3):409–440. doi:10.1175/1520-0442(1997)010<0409:TTALCC>2.0.CO;2

Morcrette JJ (1991) Radiation and cloud radiative properties in the European Centre for medium range weather forecasts forecasting system. J Geophys Res Atmos 96(D5):9121–9132

Muller CJ, Held IM (2012) Detailed investigation of the self-aggregation of convection in cloud-resolving simulations. J Atmos Sci 69(8):2551–2565

Neggers RAJ, Neelin JD, Stevens B (2007) Impact mechanisms of shallow cumulus convection on tropical climate dynamics. J Clim 20(11):2623–2642

Nelson EL, L'Ecuyer TS, Saleeby SM, Berg W, Herbener SR, van den Heever SC (2016) Toward an algorithm for estimating latent heat release in warm rain systems. J Atmos Ocean Technol 33(6):1309–1329

Nilsson J, Emanuel K (1999) Equilibrium atmospheres of a two-column radiative–convective model. Q J R Meteorol Soc 125:2239–2264

Nuijens L, Stevens B, Siebesma aP (2009) The environment of precipitating shallow cumulus convection. J Atmos Sci 66(7):1962–1979

Nuijens L, Serikov I, Hirsch L, Lonitz K, Stevens B (2014) The distribution and variability of low-level cloud in the North-Atlantic trades. QJRMS 140:2364–2374

Nuijens L, Medeiros B, Sandu I, Ahlgrimm M (2015) Observed and modeled patterns of covariability between low-level cloudiness and the structure of the trade-wind layer. J Adv Model Earth Syst 7(4):1741–1764

Peters ME, Bretherton CS (2005) A simplified model of the Walker circulation with an interactive ocean mixed layer and cloud-radiative feedbacks. J Clim 18(20):4216–4234

Pierrehumbert RT (1995) Thermostats, radiator fins, and the local runaway greenhouse. J Atmos Sci 52(10):1784–1806. doi:10.1175/1520-0469(1995)052<1784:TRFATL>2.0.CO;2

Riehl H, Yeh TC, Malkus JS, la Seur NE (1951) The north-east trade of the Pacific Ocean. Q J R Meteorol Soc 77(334):598–626

Schumacher C, Houze Ra (2003) Stratiform rain in the tropics as seen by the TRMM precipitation radar. J Clim 16(11):1739–1756

Seifert A, Heus T (2013) Large-eddy simulation of organized precipitating trade wind cumulus clouds. Atmos Chem Phys 13(11):5631–5645

Seifert A, Stevens B (2010) Microphysical scaling relations in a kinematic model of isolated shallow cumulus clouds. J Atmos Sci 67(5):1575–1590

Seifert A, Heus T, Pincus R, Stevens B (2015) Large-eddy simulation of the transient and near-equilibrium behavior of precipitating shallow convection. JAdv Model Earth Syst 7(4):1918–1937

Sherwood SC, Bony S, Dufresne JL (2014) Spread in model climate sensitivity traced to atmospheric convective mixing. Nature 505(7481):37–42

Short DA, Nakamura K (2000) TRMM radar observations of shallow precipitation over the Tropical Oceans. J Clim 13(23):4107–4124

Snodgrass ER, Di Girolamo L, Rauber RM (2009) Precipitation characteristics of trade wind clouds during rico derived from radar, satellite, and aircraft measurements. J Appl Meteorol Climatol 48(3):464–483

Stephens GL, Li J, Wild M, Clayson CA, Loeb N, Kato S, L'Ecuyer T, Stackhouse PW, Lebsock M, Andrews T (2012) An update on Earth's energy balance in light of the latest global observations. Nature Geosci 5(10):691–696

Stevens B, Seifert A (2008) Understanding macrophysical outcomes of microphysical choices in simulations of shallow cumulus convection. J Meteorol Soc Jpn Ser II 86A:143–162

Stevens B, Vali G, Comstock K, Wood R, Zanten MCV, Austin PH, Bretherton CS, Lenschow DH (2005) Pockets of open cells and drizzle in marine stratocumulus. Bull Am Meteorol Soc 86(1):51–57

Stevens B, Brogniez H, Kiemle C, Lacour J-L, Crevoisier C, Kiliani J (2017) Structure and dynamical influence of water vapor in the lower tropical troposphere. Surv Geophys. doi:10.1007/s10712-017-9420-8

Sun DZ, Liu Z (1996) Dynamic ocean–atmosphere coupling: a thermostat for the tropics. Science 272(5265):1148–1150

Takayabu YN, Shige S, Tao WK, Hirota N (2010) Shallow and deep latent heating modes over Tropical Oceans observed with TRMM PR spectral latent heating data. J Clim 23(8):2030–2046

Tompkins AM (2001) Organization of tropical convection in low vertical wind shears: the role of cold pools. J Atmos Sci 58(13):1650–1672

VanZanten MC, Stevens B, Nuijens L, Siebesma AP, Ackerman AS, Burnet F, Cheng A, Couvreux F, Jiang H, Khairoutdinov M, Kogan Y, Lewellen DC, Mechem D, Nakamura K, Noda A, Shipway BJ, Slawinska J, Wang S, Wyszogrodzki A (2011) Controls on precipitation and cloudiness in simulations of trade-wind cumulus as observed during RICO. J Adv Model Earth Syst 3(2):M06,001

Vogel R, Nuijens L, Stevens B (2016) The role of precipitation and spatial organization in the response of trade-wind clouds to warming. J Adv Model Earth Syst 8(2):843–862

Wing AA, Emanuel KA (2014) Physical mechanisms controlling self-aggregation of convection in idealized numerical modeling simulations. J Adv Model Earth Syst 6(1):59–74

Wood R, Comstock KK, Bretherton CS, Cornish C, Tomlinson J, Collins DR, Fairall C (2008) Open cellular structure in marine stratocumulus sheets. J Geophys Res Atmos. doi:10.1029/2007JD009371

Xue H, Feingold G (2006) Large-eddy simulations of trade wind cumuli: investigation of aerosol indirect effects. J Atmos Sci 63(6):1605–1622

Zuidema P, Li Z, Hill RJ, Bariteau L, Rilling B, Fairall C, Brewer WA, Albrecht B, Hare J (2012) On trade wind cumulus cold pools. J Atmos Sci 69(1):258–280

Surv Geophys (2017) 38:1283–1305
https://doi.org/10.1007/s10712-017-9447-x

A Survey of Precipitation-Induced Atmospheric Cold Pools over Oceans and Their Interactions with the Larger-Scale Environment

Paquita Zuidema[1] · Giuseppe Torri[2] · Caroline Muller[3] · Arunchandra Chandra[1]

Received: 2 April 2017 / Accepted: 29 October 2017 / Published online: 14 November 2017

Abstract Pools of air cooled by partial rain evaporation span up to several hundreds of kilometers in nature and typically last less than 1 day, ultimately losing their identity to the large-scale flow. These fundamentally differ in character from the radiatively-driven dry pools defining convective aggregation. Advancement in remote sensing and in computer capabilities has promoted exploration of how precipitation-induced cold pool processes modify the convective spectrum and life cycle. This contribution surveys current understanding of such cold pools over the tropical and subtropical oceans. In shallow convection with low rain rates, the cold pools moisten, preserving the near-surface equivalent potential temperature or increasing it if the surface moisture fluxes cannot ventilate beyond the new surface layer; both conditions indicate downdraft origin air from within the boundary layer. When rain rates exceed $\sim$ 2 mm h^{-1}, convective-scale downdrafts can bring down drier air of lower equivalent potential temperature from above the boundary layer. The resulting density currents facilitate the lifting of locally thermodynamically favorable air and can impose an arc-shaped mesoscale cloud organization. This organization allows clouds capable of reaching 4–5 km within otherwise dry environments. These are more commonly observed in the northern hemisphere trade wind regime, where the flow to the intertropical

✉ Paquita Zuidema
pzuidema@miami.edu

Giuseppe Torri
torri@fas.harvard.edu

Caroline Muller
carolinemuller123@gmail.com

Arunchandra Chandra
achandra@miami.edu

1 Rosenstiel School of Marine and Atmospheric Science, University of Miami, 4600 Rickenbacker Causeway, Miami, FL 33149, USA

2 Department of Earth and Planetary Sciences, Harvard University, 20 Oxford St., Cambridge, MA 02138, USA

3 CNRS Laboratoire d'Hydrodynamique de l'École Polytechnique, Palaiseau, France

convergence zone is unimpeded by the equator. Their near-surface air properties share much with those shown from cold pools sampled in the equatorial Indian Ocean. Cold pools are most effective at influencing the mesoscale organization when the atmosphere is moist in the lower free troposphere and dry above, suggesting an optimal range of water vapor paths. Outstanding questions on the relationship between cold pools, their accompanying moisture distribution and cloud cover are detailed further. Near-surface water vapor rings are documented in one model inside but near the cold pool edge; these are not consistent with observations, but do improve with smaller horizontal grid spacings.

Keywords Convective cold pools · Tropical convection · Shallow cumulus convection

1 Introduction

Cold pools are defined by the American Meteorological Society (AMS) as "a region or pool of relatively cold air surrounded by warmer air," or, "any large-scale mass of cold air" (Glickman 2000). In one example, nighttime radiative cooling of a land surface can create a near-surface cold pool of air. Over the open ocean, far from land, the cooling of near-surface air is arguably exclusively caused by the partial evaporation of precipitation within subsaturated air. This precipitation can be liquid or solid, though in subtropical and tropical regions the precipitation is entirely liquid near the surface. The precipitation-induced downdrafts introduce denser air underneath warmer, lighter environmental air, giving rise to a density current. The increased surface pressure establishes a horizontal pressure gradient force that drives the cold pool air outward, establishing an outflow boundary known as a gust front. The air inside the cold pool is less able to support surface-based buoyancy-driven convection. The gust front itself behaves as a material surface, and air moving over the gust front can encourage further secondary convection through mechanical lifting and anomalous buoyancy. Such convectively produced atmospheric cold pools are typically observed to span 10–200 km in diameter, and to last for less than a day, after which they lose their individual identity to the large-scale or synoptic flow. This range of sizes and lifetimes identifies them as mesoscale features, although cold pools are often embedded in and can help define the edges of larger synoptic systems (e.g., squall lines or hurricane rainbands).

Although the latent cooling from evaporation of precipitation is necessary for creating a cold pool, not all precipitation leads to well-defined cold pools at the bottom of the atmosphere. Precipitative downdrafts are driven by both condensate loading and evaporation, and, if the resulting downdraft is not strong enough to reach the surface, elevated patches of moist, cool air remain. These most likely will not fit the AMS cold pool definition, in that they are unlikely to be completely surrounded by warmer air. One example, common in stratocumulus clouds because of their relatively smaller precipitation drop sizes (e.g., Wood 2005b), is light precipitation (drizzle) evaporating below cloud base (e.g., Wood 2005a). The local cooling can serve to destabilize the subcloud layer and reinforce the updrafts in stratocumulus clouds (e.g., Feingold et al. 1996) and can also be visualized in lower-level scud clouds. In the shallow cumulus regime, such an elevated enhanced moisture layer may aid future convection (e.g., Li et al. 2014; Schlemmer and Hohenegger 2014). As drops reach larger sizes through collision-coalescence (e.g., Baker et al. 2009) and their corresponding size-dependent fall velocities increase, drops can reach the ground with only partial evaporation.

The convective organization established by surface-based precipitation-induced cold pools differs fundamentally from the larger-scale convective self-aggregation discussed elsewhere in this volume (e.g., Holloway et al. 2017; Wing et al. 2017). In convective self-aggregation, radiative subsidence produces an expansive (> 1000 km), convectively suppressed, dry region (a "dry pool") that is growing in time, neighboring a moist, deep convective region. Although an analogy between the radiative dry pool and evaporation-driven cold pools is often made, there are important differences. In the simulation shown in Fig. 1a and b, the driest precipitable water (PW) percentile corresponds to the "dry pool," and the moistest PW percentile to the convective regime capable of supporting cold pools. A dry region is seen to form and expand, confining the deep convection to the remaining moist area. Convective self-aggregation is most clearly exhibited within radiative-convective equilibrium models of tropical convection. Observations in nature of convective self-aggregation remain elusive (Holloway et al. 2017) and difficult to attribute to specific feedbacks, but one study examining observed aggregation depicts properties similar to modeled self-aggregation, including large-scale drying (Tobin et al. 2012).

Convective organization leads to an enhancement of moisture gradients, as moist regions become moister and dry regions become drier, evident in Fig. 1. The mesoscale convective cold pools within the convectively self-aggregated regions can act to oppose the aggregation, by transporting water vapor from moist to dry areas, reflecting the divergence resulting from strong downdrafts in the subcloud layer in the model moist cold pools (Fig. 1e). In nature, the most dramatic visual example of this is arguably cold-pool-containing squall lines emanating from the African monsoon into the Sahara desert (e.g., Flamant et al. 2009; Trzeciak et al. 2017). Over the tropical ocean, cold pools within

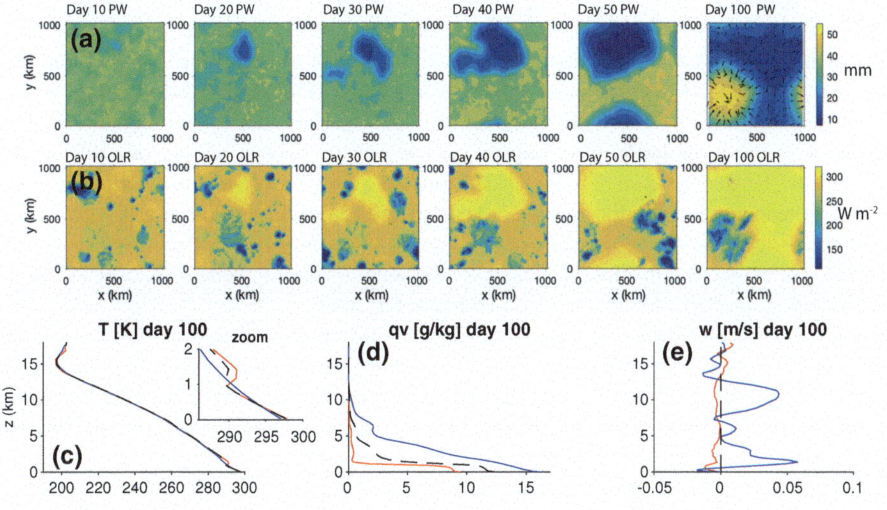

Fig. 1 Time evolution showing days 10, 20, 30, 40, 50 and 100 of **a** precipitable water PW(t) and **b** outgoing longwave radiation OLR(*t*) in a cloud-resolving model called System for Atmospheric Model (SAM) with doubly periodic boundary conditions and without large-scale forcing. The last panel of PW (day 100) also shows the low-level winds, which are seen to converge into the moist aggregate. Vertical profiles of **c** temperature, **d** water vapor and **e** vertical velocities in the radiatively subsiding dry pool, identified by PW $\leq$ 20th percentile, and in the precipitation-induced cold pools, identified by the near-surface (first atmospheric level $z = 37$ m) temperature $T_{air} \leq$ 20th percentile, and precipitation > 0 at the end of the simulation. An inset in the first 2 km added to **c** to indicate the inversion in the dry pool at the top of the subcloud layer

simulated squall lines help broaden the precipitating intertropical convergence zone poleward (Nolan et al. 2016). Precipitatively-generated cold pools also counteract convective self-aggregation through suppression of local convection by divergence of the near-surface air (Fig. 1e) (Jeevanjee and Romps 2013; Muller and Bony 2015). In theory, such mesoscale subsidence within a precipitation cold pool could morph into a new, larger radiatively induced "dry pool," although this model behavior has not yet been witnessed, to our knowledge (see also Held et al. 1993; Muller and Held 2012).

Cold pools act to disperse convection. Atmospheric cold pools over oceans are receiving attention for their ability to reorganize the mesoscale cloud distributions and potentially facilitate the transitions from high- to low-albedo shallow cloud cover. In the Tropics, the expansion of the spatial and height distribution of convection by cold pools may facilitate transitions from shallow to deep convective regimes, and therefore the eastward propagation of the Madden–Julian oscillation (MJO) into moistening environments (Rowe and Houze 2015; Feng et al. 2015; Ruppert and Johnson 2015; Schlemmer and Hohenegger 2016; Hannah et al. 2016; Ciesielski et al. 2017).

These foci on cold pool impacts justify the timeliness of a survey of cold pool characteristics and processes in different convective regimes. New observational capabilities and strategies were employed during the Rain in Cumulus over Ocean campaign (RICO; Rauber et al. 2007) and Dynamics of the MJO campaign (DYNAMO; Yoneyama et al. 2013) and will be during the upcoming Elucidating the role of clouds-circulation coupling in climate campaign (EUREC^{4}A; Bony et al. 2017). These advance early observational studies stymied by imprecise data collocation and measurement misunderstanding (e.g., Warner et al. 1979; LeMone 1980) that had difficulty perceiving the larger-scale mesoscale organization. Measurements either came from one point, e.g., a ship (Addis et al. 1984; Young et al. 1995; Saxen and Rutledge 1998) or from aircraft (e.g., Zipser 1977; Kingsmill and Houze 1999), convoluting space and time. Modeling capabilities are improving also, either in their spatial grid spacing (Romps and Jeevanjee 2016; Grant and van den Heever 2016) or domain size (Schlemmer and Hohenegger 2014, 2016), and at times both (Seifert and Heus 2013; Vogel et al. 2016). The recent years have also seen advances in the formulation of cold pool parameterizations (e.g., Qian et al. 1998; Rozbicki et al. 1999; Rio et al. 2009, 2013; Grandpeix and Lafore 2010; Grandpeix et al. 2010; Hohenegger and Bretherton 2011; Del Genio et al. 2015; Pantillon et al. 2015) and their coupling with convective schemes. These provide an avenue for rectifying a model diurnal cycle that is too closely linked to the solar cycle in models (Rio et al. 2009; Grandpeix et al. 2010; Schlemmer and Hohenegger 2014).

The focus in this survey is over ocean in the subtropical and tropical latitudes. Cold pools are extremely important over land for initiating convection, where their depth and gust fronts are often substantial (e.g., Bryan and Parker 2010), surface sensible fluxes are significant and aerosol effects are more pronounced (e.g., Grant and van den Heever 2015, 2016; Schlemmer and Hohenegger 2016). Cold pools are also important for squall lines and tornadogenesis (e.g., Markowski and Richardson 2014). The omission of land-based cold pools is merely to keep the scope of this particular survey tractable. The focus on the oceanic regions equatorward of $\sim$ 30°N also emphasizes near-surface latent cooling induced by the partial evaporation of the liquid phase, as opposed to cooling by ice melting or sublimation. Cold pool behavior and significance vary with the depth of the originating convection, its degree of organization and relationship to the large-scale environment (trade wind vs. equatorial). This is reflected in the structure of the survey, which begins with shallow convection and moves toward deeper convection. The subtropical and trade wind regions are defined by climatologically steady winds imposing a background wind shear, extending to Barbados in the examples provided and cited literature. The tropical ocean examples and cited literature are primarily near-equatorial, where wind shear can be ignored at times, and convection is readily able to span the full free troposphere.

2 Cold Pools from Boundary Layers not Exceeding 2 km Altitude

The strong inversion capping subtropical stratocumulus clouds maintains cloud top heights at $\sim$1.5 km or less (Zuidema et al. 2009), yet drizzle is ubiquitous (Leon et al. 2008). Some of the precipitating clouds occupying slightly deeper boundary layers (e.g., Mechem et al. 2012) are capable of downdrafting air that can reach the surface and develop cold pools (e.g., Fig. 2), also documented in Savic-Jovcic and Stevens (2008) and Wood et al. (2011). Aircraft measurements within stratocumulus cold pools indicate that near the surface the equivalent potential temperature (θ_e) tends to increase, rather than maintain a constant value or decrease (Zanten and Stevens 2005; Savic-Jovcic and Stevens 2008; Terai and Wood 2013).[1] The increase in θ_e is attributed to an accumulation of the surface fluxes underneath a cold pool capping stratification. Cold pool depth is difficult to observe. Inferences made from pressure increases suggest altitudes of $\sim$ 300 m (Terai and Wood 2013, and references therein). Further aloft, evaporation at constant moist static energy still brings air closer to saturation, increasing its susceptibility to convection. Wind speed convergence of the convectively favorable air near the cold pool edges helps perpetuate open-celled organization (Feingold et al. 2010), ventilating the accumulated surface fluxes and mixing air through the boundary layer. Thus, alterations to both the thermodynamic vertical structure and surface dynamics provide mechanisms by which cold pools can contribute to the longevity of precipitation within stratocumulus regions, in both closed- and open-celled organizations. Precipitation is most pronounced pre-dawn (e.g., Burleyson et al. 2013), suggesting that cold pools are also most effective at night.

Radar observations indicate that the speed of advection of cold pools is similar to that of the cloud layer (Wilbanks et al. 2015). This suggests that cold pools are better thought of as tracers or artifacts of stratocumulus precipitation as opposed to drivers. This is also concluded from simulations relying on fixed cloud droplet numbers (Zhou et al. 2017) and is consistent with simulations showing little influence on cloud organization from the inhomogeneization of surface fluxes by cold pools (Kazil et al. 2014). Precipitation is necessary for transitions from closed- to open-cell stratocumulus cloud organizations (e.g., Savic-Jovcic and Stevens 2008; Xue et al. 2008; Wang and Feingold 2009; Feingold et al. 2010; Wood et al. 2011), thereby implicating cold pools in mesoscale organization transitions, but only indirectly. Aerosol number concentrations must also deplete sufficiently for open-celled or cumuli organization to form (Wood et al. 2011; Zhou et al. 2017; Yamaguchi et al. 2015). That said, aerosol budgets are unlikely to be influenced by cold pools per se, as the wind increases at cold pool fronts are shortlived (Terai and Wood 2013), except perhaps in strongly aerosol-depleted conditions (Kazil et al. 2014). It is also noteworthy that, while an important motivation for studying shallow mesoscale transitions is their influence on cloud fraction and the planetary albedo, a clear relationship between

[1] A notable contradictory observation, of stratocumulus clouds reaching only 1.5 km yet able to transport drier air downward from aloft, is documented in Jensen et al. (2000); the conditions explaining the difference remain underexplored.

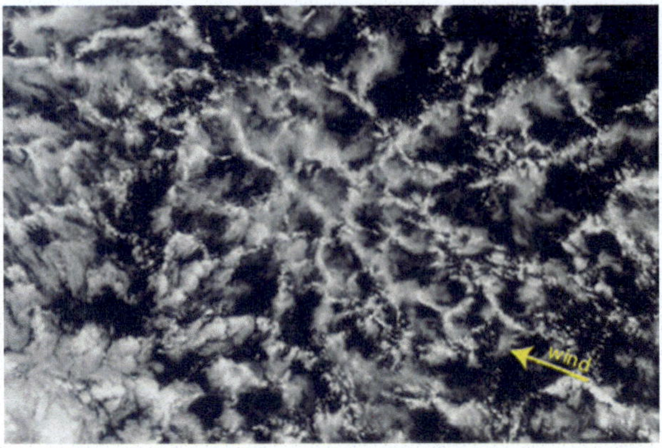

Fig. 2 September 14, 2016 MODIS *Terra* 10:30 am LT visible image near Ascension Island (14°W, 8°S), located in the upper-left-hand corner of the image. The highest cloud tops reach 1.7 km, as detected by a cloud radar located on Ascension as part of the DOE Layered Atlantic Smoke Interactions with Clouds campaign (Zuidema et al. 2016). Radiosonde winds indicate west-northwestward boundary layer flow (yellow vector)

precipitation and cloud cover is not apparent in Fig. 2, as the precipitation feeds an upper stratiform layer below the trade wind inversion. Nuijens et al. (2015) clarify that it is this cloud layer, rather than a lower cloud layer at the lifting condensation level, that is most variable.

As air advects from the stratocumulus regions to warmer sea surface temperatures, the altitude of the trade wind capping inversion rises only slowly (Schubert 1995), but the weakening temperature inversion does begin to permit deeper convection. Early experiments studying the suppressed trade wind/tropical convective environment, namely the Atlantic Trade wind Experiment (ATEX; Augstein et al. 1973) and the Barbados Oceanographic and Meteorological Experiment (BOMEX; Nitta and Esbensen 1974), estimated the fractional area occupied by actively ascending cumuli to be a negligible 0.02 (Augstein et al. 1973). The prevailing cloud organization was isolated cumuli or cloud lines aligned with the mean wind that precipitated very little if at all (LeMone and Pennell 1976; Nair et al. 1998), suggesting that cold pools can likely be ignored for this environment (e.g., Albrecht 1993; Siebesma et al. 2003). This supported a paradigm begun with Riehl et al. (1951) in which precipitation within the boundary layer does not change the moist static energy, but rather the cooling introduced by evaporating rain is energetically balanced by an increase in moisture.

3 Cold Pools from Convection Reaching the Mid-Troposphere

The diffusing trade wind inversion strength does allow some deeper clouds to develop, however, and along with them, cold pools. Larger drop sizes, encouraged by a stronger collision-coalescence process, allow precipitation to return approximately one-third of the surface evaporation to the ocean in the trade wind cumulus region (Snodgrass et al. 2009). Visible in Figs. 3 and 4 are examples of isolated convection in the northeast Atlantic trade winds able to reach 4 km coexisting with cloud lines lacking precipitation. These images are representative of the northeast Atlantic (Zuidema et al. 2012; Nuijens et al. 2017). The

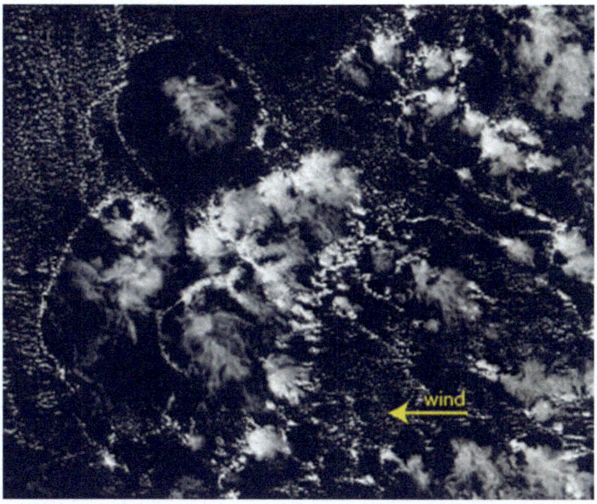

Fig. 3 December 19, 2013, MODIS *Aqua* 13:30 pm LT visible image, east of Barbados (46.5–50°W, 17.5–20°N) coincident with the Next-generation Airborne Remote Sensing for Validation (NARVAL) aircraft campaign (Stevens et al. 2016, 2017). The wind is flowing from right to left. The two largest cold pools span approximately 100 km and are better defined on the downwind side, particularly the left side of the image. The higher cloud tops within the cold pool centers reach approximately 4 km

Fig. 4 Aircraft view of a cold pool taken on August 25, 2016, southeast of Barbados. Noteworthy are the cloud lines to the left of the image, with the nearby cold pool convection organized in a circle, a portion of which is oriented perpendicular to the cloud lines in the left-hand side. Isolated convection reaching a higher altitude is detraining, most likely into a layer of increased stability

deeper clouds spawn downdrafts capable of bringing down air that is drier than the near-surface air, contributing to a lowering of the equivalent potential temperature (Zuidema et al. 2012). The presence of less convectively favorable air near the surface, spreading out as a density current, explains an organization of mesoscale cloud arcs circumscribing mostly cloud-free regions. The small areal coverage of the deeper clouds can thus alter a much larger area of near-surface air. More quantitatively, the satellite-derived cloud fraction producing rain rates exceeding 1 mm h^{-1} is a mere 0.02 for wintertime Caribbean cumuli (Snodgrass et al. 2009), over an area of approximately $\sim 10^4$ km^2 scanned by a precipitation radar, also consistent with estimates from large-eddy-scale simulations (Neggers et al. 2002). The diurnal cycle is weak (Snodgrass et al. 2009).

A further notable feature of Figs. 3 and 4 is their siting at approximately 10°N of the intertropical convergence zone (ITCZ), after a multi-day equatorward advection of boundary layer air away from the stratocumulus regions. At what point along the journey shallow convection can deepen substantially remains underexplored. Space-based lidar suggests the northern hemisphere ocean basins are more conducive of 'deeper' shallow convection (Fig. 1 of Medeiros et al. 2010), perhaps because northern hemisphere surface parcels do not advect over cooler equatorial waters in their journey to the ITCZ. The environmental conditions supporting shallow convection that can become deep enough to support coherent downdrafts bringing down air of lower θ_e from above the cloudy boundary layer is also not known and may depend on the history of the air parcel as well.

Liquid-only clouds reaching 4 km produce cold pools that are very similar to those produced by tropical deep convection extending throughout the depth of the troposphere, shown next.

4 Cold Pools from Deep Tropical Convection

The mean properties of ~ 300 cold pools composited from conventional surface meteorological datasets from Gan Island (0.6°S, 73.1°E) and the Research Vessel *Roger Revelle* located ~ 700 km to the east, at 80.5°E are shown in Fig. 5. A cold pool is identified through a temperature drop of 0.5 K, applied to a 5-min time series that has been smoothed using a Haar wavelet filter, combined with the requirement of no rain within the hour prior. The temperature drops are normalized to their mean time span of 20 min. The cold pool

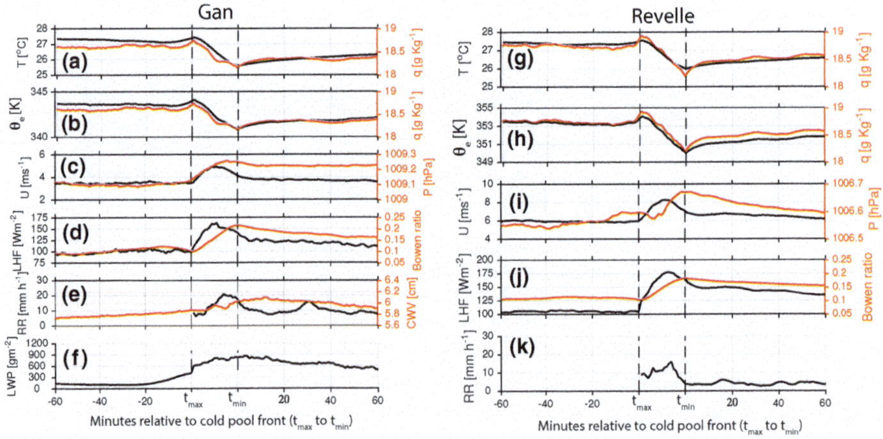

Fig. 5 Cold pool composite based on 193 cold pools at Gan island (**a–f**) and 103 at the *Revelle* (**g–k**) between October 6, 2011 to December 31, 2011, from one prior to 1 h past the cold pool frontal passage. The cold pool front is defined through a temperature drop of 0.5 K applied to the 5-min time series smoothed using a Haar wavelet filter. The front is normalized to a 20-min time interval, corresponding to the mean frontal passage time. **a**, **g** Air temperature (black) and water vapor mixing ratio (red); **b**, **h** water vapor mixing ratio (red) and equivalent potential temperature (black); **c**, **i** surface wind speed (black) and surface pressure (red); **d**, **j** latent heat fluxes (LHF, black) and the Bowen ratio (surface heat fluxes/latent heat fluxes; red); **e**, **k** rain rate (black) and column water vapor (red, **e**) only); **f** liquid water path

sample is drawn from the full tropical convective spectrum. The composite-mean includes a near-surface drying of ~ 1 g kg^{-1} and decrease in the equivalent potential temperature of 2 K. The mean wind speed increases by ~ 2 m s^{-1}, but only for 15–30 min. The pressure increase corresponds to a mean cold pool depth of approximately 300 m, is that stronger temperature drops are accompanied by stronger decreases in the water vapor mixing ratio (and θ_e), indicating either higher origin heights for the downdrafts or more coherent structures less modified by environmental mixing (de Szoeke et al. 2017). The mean near-surface relative humidity of the cold pools indicate subsaturation (not shown), suggesting the downdrafts rarely if ever maintain saturation, despite being initially saturated. All of these traits are shared with cold pools in the northeast Atlantic trade wind regime (Zuidema et al. 2012).

Cold pools are much more likely to occur underneath obscuring cirrus shields (e.g., de Szoeke et al. 2017), but one example of a tropical cold pool visible from space is shown in Figs. 6 and 7, coinciding with the DYNAMO field campaign (Yoneyama et al. 2013). A vertically pointing cloud radar and meteorological station operated within the range of a scanning precipitation radar, with a range-height indicator scan dedicated to the precipitation radar azimuth encompassing the cloud radar (Feng et al. 2014), allowing a precise collocation of cold pool convective features with its surface features. The cold pool spanning ~ 120 km is outlined on the 0615 UTC image, reaching the surface site at approximately 0915 UTC, where the vertically pointing radar reveals a cloud depth of 12 km in places lacking wind shear. The visible image shows other cold pools that are less obvious in the radar image. The surface meteorological time series shows a water vapor path increasing from 5.1 to 5.6 cm prior to the cold pool, whose edge has a temperature drop of ~ 5 K, a more quickly recovering drop in the vapor mixing ratio, and a short, almost unidentifiable, wind speed increase. The relative humidity at altitudes above 5 km was $\leq$ 50% (not shown).

Stratiform precipitation occurring later in the day prevented recovery of the near-surface temperature and maintained a near-surface relative humidity of 85–90%. Earlier observational studies focused on longer-lasting cold pools also selected from mesoscale systems with stratiform precipitation (e.g., Young et al. 1995), and longer-lasting surface fluxes changes are documented (Saxen and Rutledge 1998). For these, the altered surface fluxes, with their higher Bowen ratio, may be enough to influence the mean. This is one difference from cold pools in the trade wind regime. Deep tropical convection with stratiform precipitation can also sustain mesoscale downdrafts of warmer air. Early aircraft measurements concluded mesoscale downdrafts are too warm to reach the surface (Zipser 1977), but this is contradicted by Kilpatrick and Xie (2015), who relied on satellite scatterometer data combined with surface buoys.

Other influences on tropical cold pool characteristics, besides the level of mesoscale organization, include the atmospheric moisture distribution and the amount of mixing with environmental air during the downdraft. Since tropical deep convection easily attains cloud top heights of 8 km and upwards (Fig. 8), the downdraft air can in theory originate from a higher altitude. The modification of near-surface air properties is indeed more pronounced when the cloud top heights of the parent convection are higher (Fig. 8). Most studies point to an origin altitude for the downdraft air of 2 km or less, however (Betts 1976; Betts and Dias 1979; Torri and Kuang 2016a; de Szoeke et al. 2017; Schiro and Neelin 2017). That higher clouds are associated with stronger cold pools may reflect a correlation between downdraft width and clouds that are wider as a result of organization, discouraging environmental dilution of the downdraft air (Schlemmer and Hohenegger 2014; Schiro and

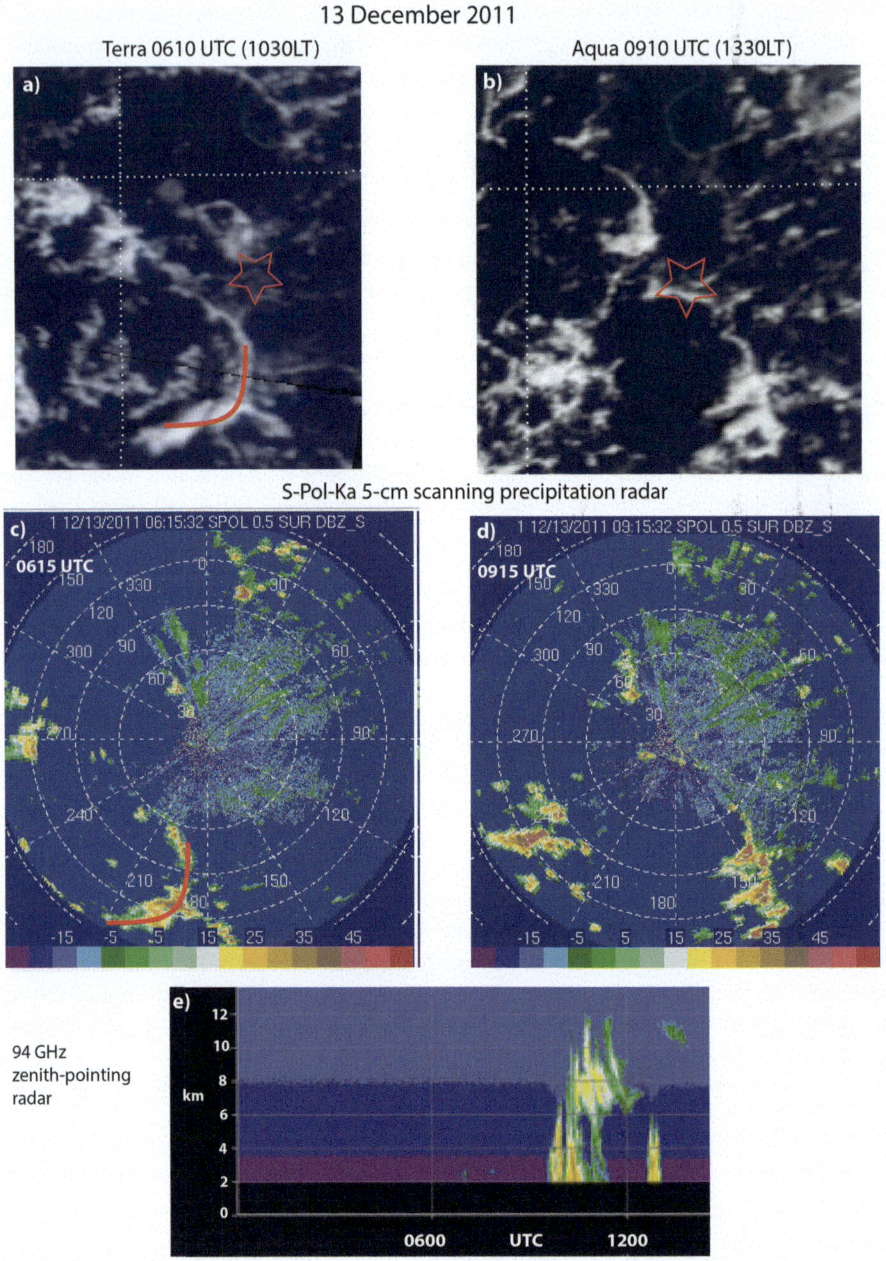

Fig. 6 **a** *Terra* and **b** *Aqua* December 13, 2011 MODIS visible satellite imagery of cold pool mesoscale organization in the equatorial Indian Ocean with **c**, **d** near-in-time 5-cm scanning precipitation radar surveillance scan imagery corresponding to **a**, **b**. **e** A 94-GHz vertically pointing radar located on Gan Island indicates the height of the corresponding convection. Red stars indicate Gan island (0.6°S, 73.1°E). The red half-circles in panels **a**, **c** correspond to the cold pool arriving at Gan island ~ 3 h later

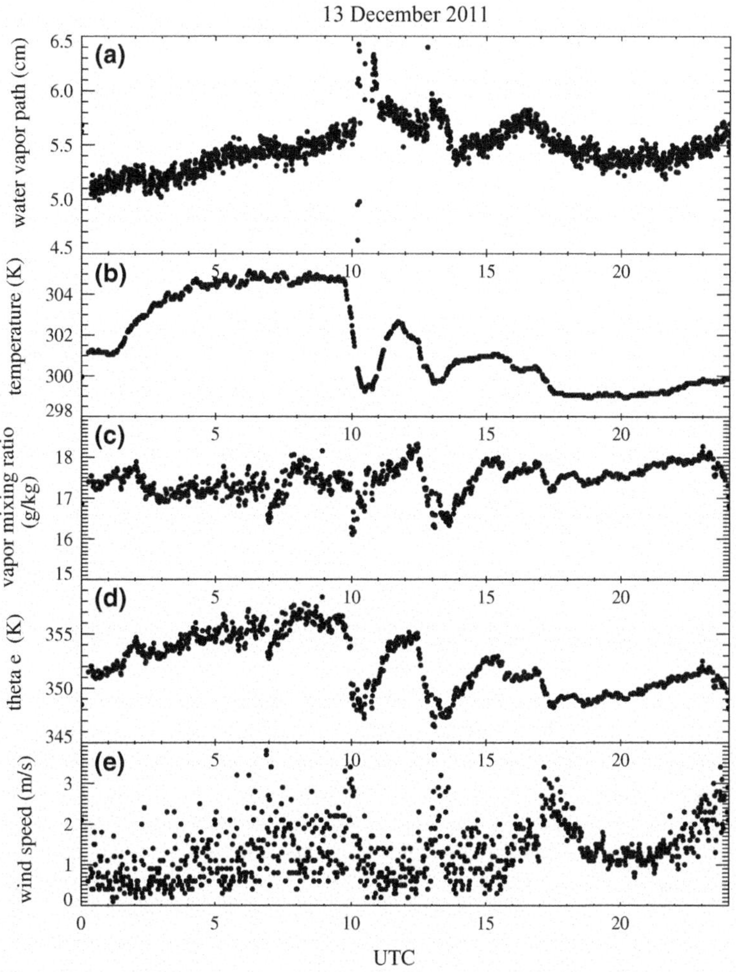

Fig. 7 December 13, 2011 time series from a surface meteorological station and microwave radiometer on Gan Island of **a** water vapor path, **b** temperature, **c** water vapor mixing ratio, **d** equivalent potential temperature and **e** wind speed

Neelin 2017; Mapes et al. 2017). The low-altitude origin discounts contributions to the convective downdrafts (treated separately from stratiform precipitation) by the melting or sublimation of ice particles (Srivastava 1987).

Convection able to impinge upon relatively dry mid-tropospheres should in theory also produce stronger cold pools, by allowing more evaporation into the downdraft while simultaneously transporting drier air of lower θ_e to the surface (e.g., Chen et al. 2016). Many examples of drier atmospheres coexisting with more isolated convection are documented within shallow-to-deep transition studies (Feng et al. 2014; Ruppert and Johnson 2015), as well as with linearly organized convection such as squall lines (Takemi and Satomura 2000; Mapes et al. 2017; Schiro and Neelin 2017). A correspondence between stronger cold pools and drier mid-tropospheres has not yet been robustly identified in observations, but is consistent with published emphases (see also Takemi et al. 2004). One

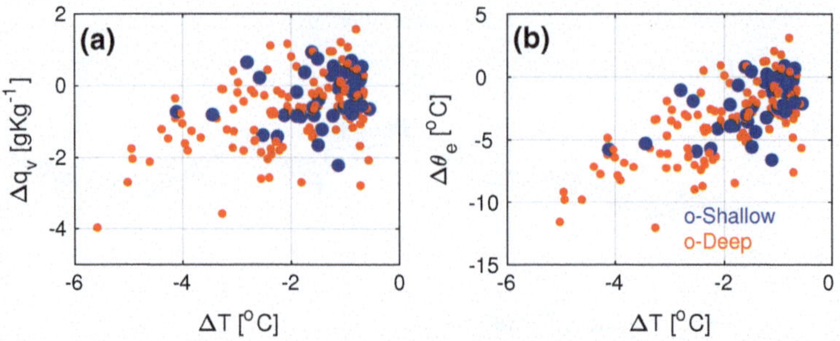

Fig. 8 Changes in **a** water vapor mixing ratio and **b** equivalent potential temperature low as a function of cloud top height (< 4 km in blue, > 4 km in red), from a surface meteorological station at Gan island (0.6°S and 73.2°W), in the equatorial Indian ocean. Cloud top heights inferred from a 35-GHz-band zenith-pointing radar. The initial temperature and moisture values correspond to the 1-min smoothed value before the cold pool temperature drop, and the minimum temperature thereafter represents the end of the cold pool frontal passage, with a criterion of a temperature drop of at least 0.5 K applied

important compensation is the condensate loading within the downdrafts, as more condensate will decrease the surface parcel buoyancy (James and Markowski 2010; Torri and Kuang 2016a).

Tropical cold pools are more frequent during times of lower outgoing longwave radiation, reflecting more organized deep convection with larger cloud covers, cooler sea surface temperatures (SSTs), and cooler near-surface layers that are closer to saturation (de Szoeke et al. 2017), and more fully saturated atmospheres. The number of observed cold pools only varies weakly with time of day (de Szoeke et al. 2017), though satellite measurements suggest mesoscale downdrafts able to reach the surface occur 8–12 h after the peak rainfall (Kilpatrick and Xie 2015). The boundary layer is cooled and moistened over a large area by the evaporation of stratiform precipitation, reducing the buoyancy of individual surface parcels. Nevertheless, cold pools and the surface-originating convection they reflect clearly do occur (de Szoeke et al. 2017), perhaps more than originally thought at the time of Houze and Betts (1981). This is also evident in space-based cloud radar observations of cumulus congestus/cumulonimbus occurring underneath upper-level stratiform cloud (e.g., Riley et al. 2011). The moisture distribution is influenced more by layer-lifting and less by surface-based buoyancy (see, e.g., in this issue, Mapes et al. (2017), dating back to at least Houze and Betts (1981), so that it cannot be argued that cold pools are important for redistributing moisture.

A connection between deep convective cold pools and those in the trade wind regions may then be their relationship to the column water vapor path. In the trade wind regions, column water vapor paths of 4.5–5.0 cm correspond to moister free tropospheres capable of supporting deeper convection and more cold pools (Zuidema et al. 2012). The DYNAMO water vapor path frequency distribution contains a plateau at 5.8–5.9 cm, and a mean of 5.1 cm (Zhang et al. 2017), suggesting the slightly drier conditions in which cold pools can be more influential are close in value to those for the trade wind regions. This is consistent with the subcloud moisture field remaining critical for convective initiation (Kingsmill and Houze 1999; Seifert and Heus 2013), particularly at cloud base level (Takemi and Satomura 2000), as well as anomalous moisture in the lower free troposphere (Sherwood et al. 2010), since column water vapor paths between 4.5 and 5.5 cm allow drier mid-tropospheres to coexist with deep, moist lower free tropospheres (Zhang et al. 2017).

One long-standing question in the Tropics, to which cold pools are relevant, is what maintains its boundary layer. The powerful Arakawa and Schubert (1974) parameterization assumes a quasi-equilibrium moisture closure in which surface moisture fluxes are balanced by the clear-sky entrainment of air drier than the saturation surface vapor mixing ratio. This is questioned within Raymond (1995), which concluded that surface fluxes must be balanced by convective downdrafts rather than clear-sky entrainment, because downdrafts provide a larger reduction in near surface θ_e. In the more comprehensive observational assessment of de Szoeke et al. (2017) based on DYNAMO soundings, downdrafts contribute to 20–30% of the boundary layer moisture and temperature budget, on days with diminished sea surface temperature (SST), and less otherwise. This suggests the effects of downdrafts are secondary, if not negligible, to those from clear-air entrainment through the boundary layer top, on the boundary layer θ_e. Thayer-Calder and Randall (2015) in an updated simulation also conclude clear-air entrainment predominantly balances surface evaporation. Torri and Kuang (2016a) also find a small contribution from convective downdrafts to the net flux of moist static energy into the boundary layer, and a much larger contribution from turbulent mixing across the boundary layer top, using a Lagrangian particle tracking analysis. These recent, independent studies all indicate a secondary role for downdrafts on the large-scale boundary layer moist static energy budget.

5 Remaining Questions

This survey motivates two remaining, intertwined questions.

5.1 The Relationship of Trade Wind Cold Pools to Cloud Cover

An outstanding question with shallow convection remains the still poorly known relationship between convection and cloud cover in the trade wind region, where the cloud cover is important for the planetary albedo. Modeling simulations, despite their many advances (e.g., Seifert and Heus 2013), remain inconclusive (van Zanten et al. 2011), to a substantial degree because many microphysical parameterizations are not optimal for trade wind cumuli representations (Li et al. 2015). While the cold pool itself discourages further surface-based convection, detrainment elsewhere in the troposphere increases cloud cover. In Fig. 4, the deeper convection is detraining at a higher altitude, affecting the overall cloud cover, arguably also visible in Fig. 3. Such detrainment can occur into layers of increased stability, and although the altitude of the higher stable layer is not clear in Fig. 2, the 0 °C level, which occurs at $\sim$ 500 hPa, is enriched with such layers (Zuidema 1998; Stevens et al. 2017).

This undertaking also means critically assessing microphysical parameterizations and their interaction with representations of environmental mixing. Experimentation with two popular microphysical schemes has revealed significant differences in cloud fraction ensuing from cold pools that are similarly simulated (Li et al. 2015), and relevant observations should also be acquired to assess results from arguably better-suited schemes (Seifert and Heus 2013). The sensitivity of the convective downdraft to condensate loading and mixing above the boundary layer is also important. In addition, this means further evaluating the effects of low- and high-wind shear, as wind shear can interact dynamically with the convective structure (Li et al. 2014). Related to all these questions is also the timescale of the boundary layer recovery within a cold pool, and how often cold pools are able to fully recover before being impinged upon by a subsequent cold pool; observations suggest cold pools occur in clumps (e.g., Fig. 2). If so, and embedded in larger-scale moisture envelopes, those in tandem can alter the atmospheric longwave radiative cooling on longer time and larger spatial scales than of just an individual cold pool.

As computational capabilities continue to improve, the trade-off between spatial grid spacing and domain size in the modeling of cold pools can be expected to diminish. The enabled sophisticated examinations of cold pools with their larger-scale environment include examining the relative importance of colliding cold pools as opposed to single pools. The further development of tracking algorithms (Torri et al. 2015; Feng et al. 2015; Böing 2016; Drager and van den Heever 2017) within such more complex simulations will allow for a better understanding of collisions and their impact of the cold pool life cycle. Such efforts will also further parameterization development.

5.2 Thermodynamic Secondary Initiation Processes

Models, through their ability to capture the full four-dimensional fields (Fig. 9), highlight an important distinction between dynamical and thermodynamical influences on secondary convection. Early ideas for triggering new convection focused on dynamics, in which the cold density currents interacted with the prevailing near-surface wind shear to encourage upright updrafts favorable for convection (Rotunno et al. 1988). The trade wind regions in particular experience steady winds, with a maximum wind speed at cloud base decreasing both above and below. Dynamical forcing of boundary layer air is stronger when the cold pool gust front is aligned with the mean wind at the downwind side of a trade wind region cold pool, and the vorticity interactions help explain the more pronounced convection on the downwind side of a cold pool (e.g., in Figs. 3 and 4, further explored in Li et al. 2014). Stronger dynamical forcing is also related to higher downdraft heights in Jeevanjee and Romps (2015).

In tropical oceanic regions lacking rotation and strong wind shear (Fig. 5), a separate paradigm emphasizing the role of water vapor in initiating new convection has also been influential (Tompkins 2001). In the study of Tompkins (2001), cloud-resolving simulations using a doubly periodic domain of 90 km, at a horizontal grid spacing of 350 m, and with

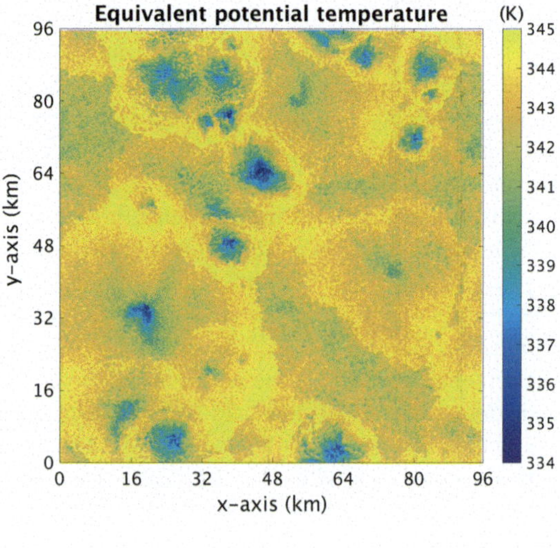

Fig. 9 A snapshot of the near-surface equivalent potential temperature distribution from Torri and Kuang (2016b)

no imposed mean wind vertical structure, accumulate water vapor within the full outer one-third of the cold pool, by 0.25 g kg $^{-1}$ in the mean. The moistening remained inside the gust front edge. The anomalous moisture was attributed to the evaporation of precipitation into temperature-recovered air before the downdraft of air from a higher-altitude source could reach the surface, and a straightforward calculation of how much rain could evaporate into the subcloud layer is consistent with the 0.25 g kg^{-1} increase evident in the composite mean (Langhans and Romps 2015). In this way, rain evaporation not only moistens the air, but it also raises the air's θ_e (which is otherwise conserved with phase changes), creating a positive feedback in triggering further convection. Surface fluxes (and winds) were discounted because these would be diminishing at the cold pool edge, as also evident in the observational composite (Fig. 5).

More recent modeling experimentation with 'single-bubble' convection further emphasizes the existence and importance of enhanced near-surface moisture resulting from the parent convection, but attributes the source of the moisture primarily to surface fluxes (Langhans and Romps 2015; Romps and Jeevanjee 2016). Rain evaporation nevertheless remains important by contributing the anomalous moisture capable of reaching the cloud base (Torri and Kuang 2016b). Similar to Tompkins (2001), the moisture in both these studies remains inside the cold pool edge.

The observational composites in Fig. 5 do not show widespread enhancements of water vapor mixing ratio q_v interior to the cold pool. A slight observed enhancement in q_v, of 0.25 g kg^{-1}, instead, appears at the cold pool edge, and prior to the wind increase. The increase in temperature of $\sim$ 0.1 K could conceivably reflect how the temperature drop is identified, but that would not explain the q_v increase. Such prior moisture increases were also noted in the tethered balloon measurements analyzed within Addis et al. (1984) and associated with upward vertical velocities. Their presence prior to the wind speed increases points to a convergence of the surface wind, and with it, moisture. The slight temperature increase could also be explained in this way.

The question then arises why models produce water vapor rings (e.g., Langhans and Romps 2015). Model cold pools can be composited similarly to observations, using time series at model grid points in the lowest model layer. Three examples, all based on the System for Atmospheric Modeling (SAM; Khairoutdinov and Randall 2006), are shown in Figs. 10 and 11. The composites shown in Fig. 10 incorporate large-scale forcings driven DYNAMO observations (Wang et al. 2015) and utilize a one-km horizontal grid spacing. Figure 11 show composites based on the Torri and Kuang (2016b) simulations, with an additional simulation performed using a smaller grid spacing (80 vs. 250 m), but otherwise identical. All three simulations lack wind shear.

Similarly to the Tompkins (2001) composites, a water vapor mixing ratio increase is seen to occur within the cold pool frontal edge, before a decrease after the frontal passage. This occurs within all three simulations. The simulations in Fig. 11 do place the water vapor increase closer to the front, but are themselves not sensitive to the horizontal grid spacing. The model q_v increases exceed those observed, also evident in Fig. 3 of Feng et al. (2015). These comparisons suggest models may overproduce water vapor rings. Postulated explanations can include an under-entrainment of drier air into the gust front edge or excessive evaporation, reflecting turbulent mixing and microphysical processes that are difficult to represent accurately, or a surface flux-wind feedback that is too strong. In Fig. 11, the simulation with the smaller grid spacing produces surface fluxes that are slightly lower than in the other simulation (102.7 vs. 100.6 W m^{-2} for the latent heat fluxes, and 9.58 vs. 9.48 W m^{-2} for the sensible heat fluxes), and may be due to different

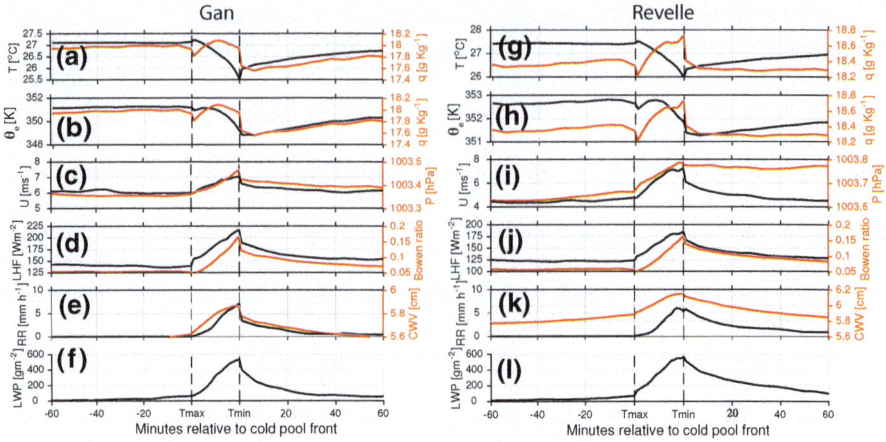

Fig. 10 Composites of **a**, **g**: air temperature (black) and water vapor mixing ratio (red); **b**, **h** water vapor mixing ratio (red) and equivalent potential temperature (black); **c**, **i** surface wind speed (black) and surface pressure (red); **d**, **j** latent heat fluxes (LHF, black) and the Bowen ratio (surface to latent heat fluxes; red); **e**, **k** rain rate (black) and column water vapor (red, **e**) only); **f**, **l** liquid water path, based on SAM simulations. The left **a–f** panel applies to Gan island from two time periods combined, October 8–16 and December 18–25, 2011. The right **g–l** panel applies to *R/V Revelle* simulations spanning October 2 to November 1, 2011. The doubly periodic simulations at a one-km horizontal grid spacing incorporate daily large-scale forcings developed for the DYNAMO time period from the campaign observations (Wang et al. 2015). 5-min model output was evaluated at 16 grid points within a 256 km by 256 km spatial domain

gust front velocities, while evaporation rates are slightly higher when the grid spacing is smaller. More work is required to confidently explain the differences beyond the scope of this current contribution, with this example primarily introduced to highlight the need for a benchmark dataset of both observations and model simulations.

In addition, model simulations of cold pools generally make trade-offs between domain size and grid spacing.[2] Studies that use larger domain sizes to represent cold pool mesoscale organization are more likely to attribute anomalous moisture sources for secondary convection outside of the cold pool (Li et al. 2014; Schlemmer and Hohenegger 2016). This is also consistent with the observed inhomogeneity of the secondary convection around the mesoscale arc (Figs. 2, 4) locations, where a dynamical lifting of a surface parcel is most able to access moisture, will be the most likely to see convection thrive (see also Torri et al. 2015). Further advantages of larger domains are the ability to capture colliding cold pools (e.g., Fig. 9), and, when incorporating the large-scale forcing at the boundary as opposed to using doubly periodic domains, the ability to capture cold pool asymmetries (Li et al. 2014, 2015). Wind shear aloft can modify the cloud's geometry, thus limiting cloud deepening (Zuidema et al. 2012; Li et al. 2014) but can also increase evaporation (Schlemmer and Hohenegger 2014), as well as allow the precipitation to fall outside of the main updraft core, and has bearing on the overall cloud fraction.

[2] For example, Romps and Jeevanjee (2016) use a horizontal grid spacing of 50 m and a vertical grid spacing as high as 10 m in the bottom 600 m of the computational domain. Grant and van den Heever (2016) imposed a 50-m grid spacing in the horizontal and 25 m in the vertical, within a two-dimensional cold pool. In contrast, Schlemmer and Hohenegger (2014) and Schlemmer and Hohenegger (2016) use a computational domain of 256 × 256 km^2 with a horizontal grid spacing of 250 m, and Li et al. (2014) and Li et al. (2015) use a large outer domain of 972×972 km^2 nesting an inner domain with a 100-m grid spacing. Seifert and Heus (2013) use a 50 × 50 km^2 domain with grid spacings of 25–100 m.

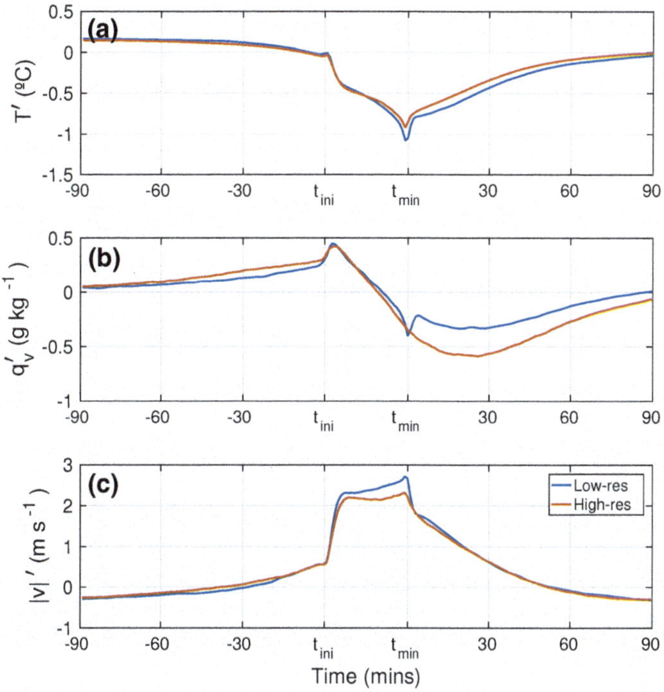

Fig. 11 Composites of **a** temperature, **b** water vapor and **c** wind speed based on the Torri and Kuang (2016b) model simulations, at 250- and 80-m horizontal grid spacings (blue and red lines, respectively), shown as anomalies from a 2-h mean to control for model drift. Cold pool selection is based on a virtual temperature drop of 0.5 K within a 10-min time period at 10,000 random points within the 30-meter bottom layer (t_{ini}). The history between t_{ini} and t_{min} interpolated to a fixed length, similar to Fig. 5. Domain size is 64 km by 64 km, with the model output saved every 30 min over 2.5 days

The dominating remaining observational challenge arguably remains the need to better resolve the four-dimensional humidity field. The idea that the updraft air feeding convection tends to have a high θ_e (or equivalently, moist static energy) is not controversial, either observationally (Kingsmill and Houze 1999) or theoretically (Emanuel et al. 1994). Consensus about the source of the anomalous moisture and equivalent potential energy (θ_e) for the secondary convection that is triggered by cold pools has not yet been reached. The distribution of moisture near cloud base remains poorly known, along with the vertical structure of cold pools. This is unlikely to be solved soon via remote sensing (see, e.g., Mapes et al. 2017; Zhang et al. 2017), with what is possibly the best passive remote sensing technique, using spectral infrared (Blumberg et al. 2015) still primarily limited to clear skies. Differential absorption and Raman lidars can profile moisture (Kiemle et al. 2017) but are obstructed by cloud, although extrapolation from points nearby could provide insight, and in combination with Doppler wind lidar might be able to resolve moist updrafts. The measurement of the most important state variables can also be readily done as the standard package on a small research aircraft and now also on drones and unmanned aerial vehicles (e.g., http://vandenheever.atmos.colostate.edu/vdhpage/c3loud-ex/index.php), or using small, light balloons as pseudo-Lagrangian drifters (https://sites.psu.edu/pmarkowski/2017/06/07/markowski-and-richardson-fly-dozens-of-airborne-probes-into-three-severe-storms-in-oklahoma-and-kansas/). This would shed further light on the relationship of the thermodynamics to the cold pool dynamics as well as conceivably interaction with surface fluxes (Ross et al. 2004; Gentine et al. 2016; Grant and van den Heever 2016).

Future challenges also remain on the modeling front. For example, the influence of entrainment at the gust front remains an open question. Work has been done to determine the entrainment in gravity currents (see e.g., Hacker et al. 1996; Hallworth et al. 1996; Fragoso et al. 2013), but mostly in idealized scenarios, which raises questions on the applicability of these results to real-world cold pools. To address this question using numerical models is a challenging task as the simulations required for this purpose would have to be conducted in a large enough domain, and, at the same time, with a spatial grid spacing capable of properly representing turbulent mixing at the gust front (see Grant and van den Heever (2016) for such an attempt). Such efforts will also improve parameterizations of the boundary layer turbulent transports towards representing clear-air entrainment and up/downdrafts correctly (see also Tompkins and Semie 2017).

Acknowledgements PZ and AC gratefully acknowledge support from NOAA Climate Variability Program Grant NA13OAR4310157 and PZ from National Science Foundation grant AGS-132189. GT acknowledges support from DOE Awarrd DE-SC0018120 and NSF Grant AGS-1649819. CM gratefully acknowledges CINES/GENCI and PRACE/TGCC in France for providing access and support to their computing platforms OCCIGEN and CURIE. The satellite imagery in Fig. 1 was acquired through the NASA Worldview website. We acknowledge use of DOE ARM Mobile Facility data within Fig. 5. We are thankful to Steve Krueger and Adam Kochanski for the SAM simulation data diagnosed in Fig. 10. We thank two anonymous reviewers for input that greatly improved the manuscript. This contribution is a result of the International Space Science Institute (ISSI) Workshop on "Shallow clouds, water vapor, circulation and climate sensitivity."

References

Addis RP, Garstang M, Emmitt GD (1984) Downdrafts from tropical oceanic cumuli. Bound Layer Meteorol 28:23–49

Albrecht BA (1993) Effects of precipitation on the thermodynamic structure of the trade-wind boundary layer. J Geophys Res 98:7327–7337

Arakawa A, Schubert W (1974) Interaction of a cumulus cloud ensemble with the large-scale environment, Part 1. J Atmos Sci 45:3188–3211

Augstein E, Riehl H, Ostapoff F, Wagner V (1973) Mass and energy transports in an undisturbed Atlantic trade-wind flow. Mon Weather Rev 101:101–111

Baker B, Mo Q, Lawson P, O'Connor D, Korolev A (2009) Drop size distributions and the lack of small drops in RICO rain shafts. J Atmos Sci 48:616–623

Betts AK (1976) The thermodynamic transformation of the tropical subcloud layer by precipitation and downdrafts. J Atmos Sci 33:1008–1020

Betts AK, Dias MFS (1979) Unsaturated downdraft thermodynamics in cumulonimbus. J Atmos Sci 36:1060–1067

Blumberg W, Turner D, Lohnert U, Castleberry S (2015) Ground-based temperature and humidity profiling using spectral infrared and microwave observations. Part 2: actual retrieval performance in clear sky and cloudy conditions. J Appl Meteorol Clim 54:2305–2319. https://doi.org/10.1175/JAMC-D-15-0005.1

Böing SJ (2016) An object-based model for convective cold pool dynamics. Math Clim Weather Forecast. https://doi.org/10.1515/mcwf-2016-0003.

Bony S, Stevens B, Ament F, Crewell S, Delanoe J, D Farrell et al (2017) Eurec4a: a field campaign to elucidate the couplings between clouds, convection and circulation. Surv Geophys https://doi.org/10.1007/s10712-017-9428-0

Bryan GH, Parker MD (2010) Observations of a squall line and its near environment using high-frequency rawinsonde launches during VORTEX2. Mon Weaather Rev 138(11):4076–4097. https://doi.org/10.1175/2010MWR3359.1

Burleyson CD, de Szoeke SP, Yuter SE, Wilbanks M, Brewer WA (2013) Ship-based observations of the diurnal cycle of southeast pacific marine stratocumulus clouds and precipitation. J Atmos Sci 70(12):3876–3894. https://doi.org/10.1175/JAS-D-13-01.1

Chen S, Kerns B, Guy N, Jorgensen D, Delanoe J, Viltard N, Zappa C, Judt F, Lee C, Savarin A (2016) Aircraft observations of dry air, ITCZ, convective cloud systems and cold pools in MJO during DYNAMO. Bull Am Meteorol Soc. https://doi.org/10.1175/bams-d-13-00196.1

Ciesielski P, Johnson RH, Jiang X, Zhang Y, Xie S (2017) Relationships between radiation, clouds, and convection during DYNAMO. J Geophys Res. https://doi.org/10.1002/2016JD025965

de Szoeke S, Skyllingstad ED, Zuidema P, Chandra A (2017) Cold pools and their influence on the tropical marine boundary layer. J Atmos Sci. https://doi.org/10.1175/jas-d-16-0264.1

Del Genio AD, Wu J, Wolf AB, Chen Y, Yao MS, Kim D (2015) Constraints on cumulus parameterization from simulations of observed MJO events. J Clim 28(16):6419–6442. https://doi.org/10.1175/JCLI-D-14-00832.1

Drager AJ, van den Heever SC (2017) Characterizing convective cold pools. J. Adv. Model Earth Syst. https://doi.org/10.1002/2016MS000788

Emanuel KA, Neelin JD, Bretherton CS (1994) On large-scale circulations in convecting atmospheres. Q J R Meteorol Soc 120:1111–1143

Feingold G, Stevens B, Cotton WR, Frisch AS (1996) The relationship between drop in-cloud residence time and drizzle production in numerically simulated stratocumulus clouds. J Atmos Sci 53:1108–1122

Feingold G, Koren I, Wang H, Xue H, Brewer WA (2010) Precipitation-generated oscillations in open cellular cloud fields. Nature 466:849–852. https://doi.org/10.1038/nature09314

Feng Z, McFarlane SA, Schumacher C, Ellis S, Comstock J, Bharadwaj N (2014) Constructing a merged cloud-precipitation radar dataset for tropical convective clouds during the DYNAMO/AMIE experiment at Addu Atoll. J Atmos Ocean Technol 31:1021–1042

Feng Z, Hagos S, Rowe AK, Burleyson CD, Martini MN, de Szoeke SP (2015) Mechanisms of convective cloud organization by cold pools over tropical warm ocean during the amie/dynamo field campaign. J Adv Model Earth Syst. https://doi.org/10.1002/ 2014MS000384

Flamant C, Knippertz P, Parker DJ, Chaboureau JP, Lavaysse C, Agusti-Panareda A, Kergoat L (2009) The impact of a mesoscale convective system cold pool on the northward propagation of the intertropical discontinuity over west africa. Q J R Meteorol Soc 135:139–159. https://doi.org/10.1002/qj.357

Fragoso AT, Patterson MD, Wettlaufer JS (2013) Mixing in gravity currents. J Fluid Mech. https://doi.org/10.1017/jfm.2013.475

Gentine P, Garelli A, Park SB, Nie J, Torri G, Kuang Z (2016) Role of surface heat fluxes underneath cold pools. Geophys Res Lett 43(2):874–883. https://doi.org/10.1002/2015GL067262

Glickman T (2000) Glossary of meteorology, 2nd edn. AMS, Flensburg, p 855

Grandpeix JY, Lafore JP (2010) A density current parameterization coupled with Emanuel's convection scheme. Part I: the models. J Atmos Sci. https://doi.org/10.1175/2009jas3044.1

Grandpeix JY, Lafore JP, Cheruy F (2010) A density current parameterization coupled with Emanuel's convection scheme. Part II: 1D simulations. J Atmos Sci 67(4):898–922. https://doi.org/10.1175/2009JAS3045.1

Grant L, van den Heever S (2016) Cold pool dissipation. J Geophys Res 121(3):1138–1155. https://doi.org/10.1002/2015JD023813

Grant LD, van den Heever S (2015) Cold pool and precipitation responses to aerosol loading: modulation by dry layers. J Atmos Sci 72:1398–1408

Hacker J, Linden P, Dalziel S (1996) Mixing in lock-release gravity currents. Dyn Atmos Oceans 24(1):183–195. https://doi.org/10.1016/0377-0265(95)00443-2

Hallworth MA, Huppert HE, Phillips JC, Sparks RSJ (1996) Entrainment into two-dimensional and axisymmetric turbulent gravity currents. J Fluid Mech 308:289–311. https://doi.org/10.1017/S0022112096001486

Hannah WM, Mapes B, Elsaesser G (2016) A lagrangian view of moisture dynamics during DYNAMO. J Atmos Sci 73:1967–1985. https://doi.org/10.1175/jas-d-15-0243.1

Held IM, Hemler RS, Ramaswamy V (1993) Radiative-convective equilibrium with explicit two-dimensional moist convection. J Atmos Sci 50(23):3909–3927

Hohenegger C, Bretherton CS (2011) Simulating deep convection with a shallow convection scheme. Atmos Chem Phys 11(20):10389–10406. https://doi.org/10.5194/acp-11-10389-2011
Holloway CE, Wing AA, Bony S, Muller C, Masunaga H, L'Ecuyer T, Turner D, Zuidema P (2017) Observing convective aggregation. Rev Geophys 131:780
Houze RA, Betts A (1981) Convection in GATE. Rev Geophys Space Phys 19:541–576
James RP, Markowski PM (2010) A numerical investigation of the effects of dry air aloft on deep convection. Mon Weather Rev 138:140–161. https://doi.org/10.1175/2009MWR3018.1
Jeevanjee N, Romps DM (2013) Convective self-aggregation, cold pools, and domain size. Geophys Res Lett 40:994–998. https://doi.org/10.1002/grl.50204
Jeevanjee N, Romps DM (2015) Effective buoyancy, inertial pressure, and the mechanical generation of boundary layer mass flux by cold pools. J Atmos Sci 72(8):3199–3213. https://doi.org/10.1175/JAS-D-14-0349.1
Jensen JB, Lee S, Krummel PB, Katzfey J, Gogoasa D (2000) Precipitation in marine cumulus and stratocumulus. Part I: thermodynamic and dynamic observations of closed cell circulations and cumulus bands. Atmos Res 54:117–155
Kazil J, Feingold G, Wang H, Yamaguchi T (2014) On the interaction between marine boundary layer cellular cloudiness and surface heat fluxes. Atmos Chem Phys 14(1):61–79. https://doi.org/10.5194/acp-14-61-2014
Khairoutdinov M, Randall D (2006) High-resolution simulations of shallow- to-deep convection transition over land. J Atmos Sci 63:3421–3436
Kiemle C, Gro S, Wirth M, Bugliaro L (2017) Airborne lidar observations of water vapor variability in tropical shallow convective environment. Surv Geophys. https://doi.org/10.1007/s10712-017-9431-5
Kilpatrick TJ, Xie SP (2015) ASCAT observations of downdrafts from mesoscale convective systems. Geophys Res Lett 42:1951–1958. https://doi.org/10.1002/2015GL063025
Kingsmill DE, Houze RA (1999) Thermodynamic characteristics of air flowing into and out of precipitating convection over the west Pacific warm pool. Q J R Meteorol Soc 125:1209–1229
Langhans W, Romps DM (2015) The origin of water vapor rings in tropical oceanic cold pools. Geophys Res Lett 42:7824–7834. https://doi.org/10.1002/2015GL065623
LeMone MA (1980) On the difficulty of measuring temperature and humidity in cloud: comments on "Shallow convection on day 261 of GATE: Mesoscale arcs". Mon Weather Rev 108:1702–1705
LeMone MA, Pennell WT (1976) The relationship of trade wind cumulus distribution to subcloud layer fluxes and structure. Mon Weather Rev 104:524–539
Leon D, Wang Z, Liu D (2008) Climatology of drizzle in marine boundary layer clouds based on one year of data from CloudSat and CALIPSO. J Geophys Res. https://doi.org/10.1029/2008JD009835
Li Z, Zuidema P, Zhu P (2014) Simulated convective invigoration processes at trade-wind cumulus cold pool boundaries. J Atmos Sci 71:2823–2841. https://doi.org/10.1175/JAS-D-13-0184.1
Li Z, Zuidema P, Zhu P, Morrison H (2015) The sensitivity of simulated shallow cumulus convection and cold pools to microphysics. J Atmos Sci 72:3340–3355. https://doi.org/10.1175/JAS-D-14-0099.1
Mapes B, Chandra A, Kuang Z, Zuidema P (2017) Importance profiles for water vapor. Surv Geophys. https://doi.org/10.1007/s10712-017-9427-1
Markowski PM, Richardson YP (2014) The influence of environmental low-level shear and cold pools on tornadogenesis: insights from idealized simulations. J Atmos Sci 71(1):243–275. https://doi.org/10.1175/JAS-D-13-0159.1
Mechem D, Yuter SE, de Szoeke SP (2012) Thermodynamic and aerosol controls in southeast pacific stratocumulus. J Atmos Sci 69:1250–1266. https://doi.org/10.1175/JAS-D-11-0165.1
Medeiros B, Nuijens L, Antoniazzi C, Stevens B (2010) Lowlatitude boundary layer clouds as seen by CALIPSO. J Geophys Res. https://doi.org/10.1029/2010JD014437
Muller C, Bony S (2015) What favors convective aggregation and why? Geophys Res Lett 42:5626–5634. https://doi.org/10.1002/2015gl064260
Muller CJ, Held IM (2012) Detailed investigation of the self-aggregation of convection in cloud-resolving simulations. J Atmos Sci 69:2551–2565. https://doi.org/10.1175/JAS-D-11-0257.1
Nair US, Weger RC, Kuo KS, Welch RM (1998) Clustering, randomness, and regularity in cloud fields. 5. The nature of regular cumulus cloud fields. J Geophys Res 103:11,363–11,380
Neggers RA, Siebesma AP, Jonker HJJ (2002) A multiparcel model for shallow cumulus convection. J Atmos Sci 59:1655–1667
Nitta T, Esbensen S (1974) Heat and moisture budget analyses using BOMEX data. Mon Weather Rev 102:17–28
Nolan DS, Tulich SN, Blanco JE (2016) ITCZ structure as determined by parameterized versus explicit convection in aquachannel and aquapatch simulations. J Adv Model Earth Syst 8:425–452. https://doi.org/10.1002/2015MS000560

Nuijens L, Medeiros B, Sandu I, Ahlgrimm M (2015) The behavior of trade-wind cloudiness in observations and models: the major cloud components and their variability. J Adv Model Earth Syst 7(2):600–616. https://doi.org/10.1002/2014MS000390
Nuijens L, Emanuel K, Masunaga H, L'Ecuyer T (2017) Implications for warm rain in shallow cumulus and congestus clouds for large-scale circulations. Surv Geophys. https://doi.org/10.1007/s10712-017-9429-z
Pantillon F, Knippertz P, Marsham JH, Birch CE (2015) A parameterization of convective dust storms for models with mass-flux convection schemes. J Atmos Sci 72(6):2545–2561. https://doi.org/10.1175/JAS-D-14-0341.1
Qian L, Young GS, Frank WM (1998) A convective wake parameterization scheme for use in general circulation models. Mon Weather Rev 126(2):456–469. https://doi.org/10.1175/1520-0493(1998)126%3c0456:ACWPSF%3e2.0.CO;2
Rauber RM et al. (2007) Rain in shallow cumulus over the ocean: the RICO campaign. Bull Am Meteorol Soc 88:1912–1928
Raymond D (1995) Regulation of moist convection over the west pacific warm pool. J Atmos Sci 52:3945–3959
Riehl H, Yeh C, Malkus J, Seur NL (1951) The northeast trade of the Pacific ocean. Q J R Meteorol Soc 77:598–626
Riley EM, Mapes BE, Tulich SN (2011) Clouds associated with the MaddenJulian oscillation: a new perspective from CloudSat. J Atmos Sci 68(12):3032–3051. https://doi.org/10.1175/JAS-D-11-030.1
Rio C, Hourdin F, Grandpeix JY, Lafore JP (2009) Shifting the diurnal cycle of parameterized deep convection over land. Geophys Res Lett. https://doi.org/10.1029/ 2008GL036779
Rio C, Grandpeix JY, Hourdin F, Guichard F, Couvreux F, Lafore JP, Fridlind A, Mrowiec A, Roehrig R, Rochetin N, Lefebvre MP, Idelkadi A (2013) Control of deep convection by sub-cloud lifting processes: the alp closure in the lmdz5b general circulation model. Clim Dyn 40(9):2271–2292. https://doi.org/10.1007/s00382-012-1506-x
Romps DM, Jeevanjee N (2016) On the sizes and lifetimes of cold pools. Q J R Meteorol Soc 142(696):1517–1527
Ross AN, Tompkins AM, Parker DJ (2004) Simple models of the role of surface fluxes in convective cold pool evolution. J Atmos Sci 61(13):1582–1595. https://doi.org/10.1175/1520-0469(2004)061%3c1582:SMOTRO%3e2.0.CO;2
Rotunno R, Klemp JB, Weisman ML (1988) A theory for strong, long-lived squall lines. J Atmos Sci 45:463–485. https://doi.org/10.1175/1520-0469(1988)045%3c0463:ATFSLL%3e2.0.CO;2
Rowe AK, Houze RA (2015) Cloud organization and growth during the transition from suppressed to active MJO convection. J Geophys Res 120:10,324–10,350. https://doi.org/10.1002/2014JD022948
Rozbicki JJ, Young GS, Qian L (1999) Test of a convective wake parameterization in the single-column version of ccm3. Mon Weather Rev 127(6):1347–1361. https://doi.org/10.1175/1520-0493(1999)127%3c1347:TOACWP%3e2.0.CO;2
Ruppert JH, Johnson RH (2015) Diurnally modulated cumulus moistening in the preonset stage of the Madden–Julian oscillation during dynamo. J Atmos Sci 72:1622–1647. https://doi.org/10.1175/JAS-D-14-0218.1
Savic-Jovcic V, Stevens B (2008) The structure and mesoscale organization of precipitating stratocumulus. J Atmos Sci 65:1587–1605. https://doi.org/10.1175/2007JAS2456.1
Saxen TR, Rutledge SA (1998) Surface fluxes and boundary layer recovery in TOGA COARE: sensitivity to convective organization. J Atmos Sci 55:2763–2781
Schiro KA, Neelin JD (2017) Tropical continental downdraft characteristics: mesoscale systems versus unorganized convection. Atmos Chem Phys Discuss 2017:1–29. https://doi.org/10.5194/acp-2017-684
Schlemmer L, Hohenegger C (2014) The formation of wider and deeper clouds as a result of cold-pool dynamics. J Atmos Sci 71:2842–2858. https://doi.org/10.1175/JAS-D-13-0170.1
Schlemmer L, Hohenegger C (2016) Modifications of the atmospheric moisture field as a result of cold-pool dynamics. Q J R Meteorol Soc 142:30–42. https://doi.org/10.1002/qj2625
Schubert W (1995) Dynamical adjustment of the trade-wind inversion layer. J Atmos Sci 52:2941–2952
Seifert A, Heus T (2013) Large-eddy simulation of organized precipitating trade-wind cumulus clouds. Atmos Chem Phys 13:5631–5645. https://doi.org/10.5194/acp-13-5631-2013
Sherwood SC, Roca R, Weckwerth TM, Andronova NG (2010) Tropospheric water vapor, convection, and climate. Rev Geophys. https://doi.org/10.1029/2009RG000301
Siebesma AP, Bretherton CS, Brown BA, Chlond A, Cuxart J, Duynkerke P, Jiang H, Khairoutdinov M, Lewellen D, Moeng CH, Sanchez E, Stevens B, Stevens DE (2003) A large-eddy simulation study of shallow cumulus convection. J Atmos Sci 60:1201–1219
Snodgrass ER, Girolamo LD, Rauber RM (2009) Precipitation characteristics of trade wind clouds during RICO derived from radar, satellite, and aircraft measurements. J Appl Meteorol Clim 48:464–483

Srivastava RC (1987) A model of intense downdrafts driven by the melting and evaporation of precipitation. J Atmos Sci 44:1752–1774

Stevens B, Farrell D, Hirsch L, Jansen F, Nuijens L, Serikov I, ugmann BB, Forde M, Linne H, Lonitz K, Prospero J (2016) The Barbados cloud observatory: anchoring investigations of clouds and circulation on the edge of the ITCZ. Bull Am Meteorol Soc 97:787–801. https://doi.org/10.1175/BAMS-D-14-00247.1

Stevens B, Brogniez H, Kiemle C, Lacour JL, Crevoisier C, Kiliani J (2017) Structure and dynamical influence of water vapor in the lower tropical troposphere. Surv Geophys. https://doi.org/10.1007/s10712-017-9420-8

Takemi T, Satomura T (2000) Numerical experiments on the mechanisms for the development and maintenance of long-lived squall lines in dry environments. J Atmos Sci 57:1718–1740

Takemi T, Hirayama O, Liu C (2004) Factors responsible for the vertical development of tropical oceanic cumulus convection. Geophys Res Lett. https://doi.org/10.1029/2004GL020225

Terai C, Wood R (2013) Aircraft observations of cold pools under marine stratocumulus. Atmos Chem Phys 13:9899–9914. https://doi.org/10.5194/acp-13-9899-2013

Thayer-Calder K, Randall D (2015) A numerical investigation of boundary layer quasi-equilibrium. Geophys Res Lett 42:550–556. https://doi.org/10.1002/2014GL062649

Tobin I, Bony S, Roca R (2012) Observational evidence for relationships between the degree of aggregation of deep convection, water vapor, surface fluxes, and radiation. J Clim 25:6885–6904

Tompkins AM (2001) Organization of tropical convection in low vertical wind shears: the role of cold pools. J Atmos Sci 58:1650–1672

Tompkins AM, Semie AG (2017) Organization of tropical convection in low vertical wind shears: role of updraft entrainment. J Adv Model Earth Syst 9:1046–1068. https://doi.org/10.1002/2016MS000802

Torri G, Kuang Z (2016a) A lagrangian study of precipitation-driven downdrafts. J Atmos Sci. https://doi.org/10.1175/JAS-D-15-0222s1

Torri G, Kuang Z (2016b) Rain evaporation and moist patches in tropical boundary layers. Geophys Res Lett 43:9895–9902. https://doi.org/10.1002/2016GL070893

Torri G, Kuang Z, Tian Y (2015) Mechanisms for convection triggering by cold pools. Geophys Res Lett 42(6):1943–1950. https://doi.org/10.1002/2015GL063227

Trzeciak TM, Garcia-Carreras L, Marsham JH (2017) Cross-saharan transport of water vapor via recycled cold pool outflows from moist convection. Geophys Res Lett 44:1554–1563. https://doi.org/10.1002/2016GL072108

van Zanten MC, Stevens B (2005) Observations of the structure of heavily precipitating marine stratocumulus. J Atmos Sci 62:4327–4342

van Zanten MC, Stevens B, Nuijens L, Siebesma AP, Ackerman A, Burnet F, Cheng A, Couvreux F, Jiang H, Khairoutdinov M, Kogan Y, Lewellen DC, Mechem D, Nakamura K, Noda A, Shipway BJ, Slawinska J, Wang S, Wyszogrodzki A (2011) Controls on precipitation and cloudiness in simulations of trade-wind cumulus as observed during RICO. J Adv Model Earth Syst. https://doi.org/10.1029/2011MS000056

Vogel R, Nuijens L, Stevens B (2016) The role of precipitation and spatial organization in the response of trade-wind clouds to warming. J Adv Model Earth Syst 8:843–862. https://doi.org/10.1002/2015MS000568

Wang H, Feingold G (2009) Modeling mesoscale cellular structure and drizzle in marine stratocumulus. Part I: impact of drizzle on the formation and evolution of open cells. J Atmos Sci 66:3237–3256. https://doi.org/10.1175/2009JAS3022.1

Wang S, Sobel A, Fridlind A, Feng Z, Comstock J, Minnis P, Nordeen M (2015) Simulations of cloud-radiation interaction using large-scale forcing derived from the CINDY/DYNAMO northern sounding array. J Adv Model Earth Syst 7(3):1472–1498. https://doi.org/10.1002/2015MS000461

Warner C, Simpson J, Martin DW, Suchman D, Mosher FR, Reinking RF (1979) Shallow convection on day 261 of GATE: mesoscale arcs. Mon Weather Rev 107:1617–1635

Wilbanks M, Yuter SE, deSzoeke SP, Brewer WA, Miller MA, Hall AM, Burleyson CD (2015) Near-surface density currents observed in the southeast pacific stratocumulus- topped marine boundary layer. Mon Weather Rev 143:3532–3555

Wing AA, Emanuel K, Holloway C, Muller C (2017) Convective self-aggregation in numerical simulations: a review. Surv Geophys. https://doi.org/10.1007/s10712-017-9408-4

Wood R (2005a) Drizzle in stratiform boundary layer clouds. Part I: vertical and horizontal structure. J Atmos Sci 62:3012–3033

Wood R (2005b) Drizzle in stratiform boundary layer clouds. Part II: microphysical aspects. J Atmos Sci 62:3034–3050

Wood R, Bretherton CS, Leon D, Clarke AD, Zuidema P, Allen G, Coe H (2011) An aircraft case study of the spatial transition from closed to open mesoscale cellular convection over the Southeast Pacific. Atmos Chem Phys 11:2341–2370. https://doi.org/10.5194/acp-11-2341-2011

Xue H, Feingold G, Stevens B (2008) Aerosol effects on clouds, precipitation, and the organization of shallow cumulus convection. J Atmos Sci 65:392–406. https://doi.org/10.1175/2007JAS2428.1

Yamaguchi T, Feingold G, Kazil J, McComiskey A (2015) Stratocumulus to cumulus transition in the presence of elevated smoke layers. Geophys Res Lett. https://doi.org/10.1002/2015GL066544

Yoneyama K, Zhang C, Long CN (2013) Tracking pulses of the Madden–Julian oscillation. Bull Am Meteorol Soc 94:1871–1891. https://doi.org/10.1175/bams-d-12-00157.1

Young GS, Perugini SM, Fairall CW (1995) Convective wakes in the equatorial western Pacific during TOGA. Mon Weather Rev 123:110–123

Zhang J, Zuidema P, Turner DD, Cadeddu MP (2017) Tropical humidity vertical structure inferred from a microwave radiometer during dynamo. J Appl Meteor Clim (Under review)

Zhou X, Heus T, Kollias P (2017) Influences of drizzle on stratocumulus cloudiness and organization. J Geophys Res 122(13):6989–7003. https://doi.org/10.1002/2017JD026641

Zipser E (1977) Mesoscale and convective-scale downdrafts as distinct components of squall line structure. Mon Weather Rev 105:1568–1589

Zuidema P (1998) The 600–800-mb minimum in tropical cloudiness observed during TOGA COARE. J Atmos Sci 55:2220–2228

Zuidema P, Painemal D, deSzoeke S, Fairall C (2009) Stratocumulus cloud top height estimates and their climatic implications. J Clim 22:4652–4666

Zuidema P, Li Z, Hill R, Bariteau L, Rilling B, Fairall C, Brewer WA, Albrecht B, Hare J (2012) On trade-wind cumulus cold pools. J Atmos Sci 69:258–277. https://doi.org/10.1175/jas-d-11-0143.1

Zuidema P, Redemann J, Haywood J, Wood R, Piketh S, Hipondoka M, Formenti P (2016) Smoke and clouds above the southeast atlantic: upcoming field campaigns probe absorbing aerosol's impact on climate. Bull Am Meteorol Soc 97:1131–1135. https://doi.org/10.1175/bams-d-15-00082.1

Surv Geophys (2017) 38:1307–1329
https://doi.org/10.1007/s10712-017-9433-3

Low-Cloud Feedbacks from Cloud-Controlling Factors: A Review

Stephen A. Klein[1] · Alex Hall[2] · Joel R. Norris[3] · Robert Pincus[4,5]

Received: 26 April 2017 / Accepted: 21 September 2017 / Published online: 24 October 2017

Abstract The response to warming of tropical low-level clouds including both marine stratocumulus and trade cumulus is a major source of uncertainty in projections of future climate. Climate model simulations of the response vary widely, reflecting the difficulty the models have in simulating these clouds. These inadequacies have led to alternative approaches to predict low-cloud feedbacks. Here, we review an observational approach that relies on the assumption that observed relationships between low clouds and the "cloud-controlling factors" of the large-scale environment are invariant across time-scales. With this assumption, and given predictions of how the cloud-controlling factors change with climate warming, one can predict low-cloud feedbacks without using any model simulation of low clouds. We discuss both fundamental and implementation issues with this approach and suggest steps that could reduce uncertainty in the predicted low-cloud feedback. Recent studies using this approach predict that the tropical low-cloud feedback is positive mainly due to the observation that reflection of solar radiation by low clouds decreases as temperature increases, holding all other cloud-controlling factors fixed. The positive feedback from temperature is partially offset by a negative feedback from the tendency for the inversion strength to increase in a warming world, with other cloud-controlling factors playing a smaller role. A consensus estimate from these studies for the contribution of tropical low clouds to the global mean cloud feedback is

✉ Stephen A. Klein
klein21@llnl.gov

[1] Cloud Processes Research Group, Lawrence Livermore National Laboratory, 7000 East Avenue, L-103, Livermore, CA 94551, USA

[2] Department of Atmospheric and Oceanic Sciences, University of California, Los Angeles, CA 90095, USA

[3] Scripps Institution of Oceanography, University of California, San Diego, La Jolla, CA 92093, USA

[4] Cooperative Institute for Research in Environmental Sciences, University of Colorado, Boulder, CO 80309, USA

[5] Physical Sciences Division, NOAA Earth System Research Laboratory, Boulder, CO 80305, USA

0.25 ± 0.18 W m^{-2} K^{-1} (90% confidence interval), suggesting it is very unlikely that tropical low clouds reduce total global cloud feedback. Because the prediction of positive tropical low-cloud feedback with this approach is consistent with independent evidence from low-cloud feedback studies using high-resolution cloud models, progress is being made in reducing this key climate uncertainty.

Keywords Climate change · Cloud feedbacks · Low clouds

1 Seeking Observational Constraints on Low-Cloud Feedbacks

How clouds respond to the climate warming is a major uncertainty in climate change science that hinders prediction of the temperature sensitivity to radiative perturbations (Boucher et al. 2013). At the center of this uncertainty is the response of tropical oceanic low clouds, which is the single cloud type that explains the most spread of climate model predictions of cloud feedbacks (Bony and Dufresne 2005). A recent study estimates that low clouds globally explain around 50% of the inter-model variance of the global mean cloud feedback (Zelinka et al. 2016).

The widely varying responses of low clouds are perhaps unsurprising because climate models struggle to simulate these clouds. Tropical low clouds involve highly interactive processes of radiative transfer, turbulent and convective mixing and cloud physics that are imperfectly represented by climate model parameterizations. The parameterizations are necessary because the space and time-scales that climate models resolve are coarse relative to the space and time-scales of tropical low clouds.

The problems simulating low clouds motivate approaches to determine tropical low-cloud feedbacks that do not directly rely upon climate model simulations. One approach is to use large-eddy simulations that resolve low-cloud processes to predict the low-cloud changes forced by the climate changes in the environment (Rieck et al. 2012; Zhang et al. 2012; Blossey et al. 2013; Bretherton 2015). A second approach relies on observations of clouds to predict how they will respond to changes in the large-scale environment typical of climate warming. This observational approach is the subject of this paper.

At the heart of the observational approach is the fact that tropical low clouds are not randomly distributed but instead tend to vary with characteristics of the large-scale

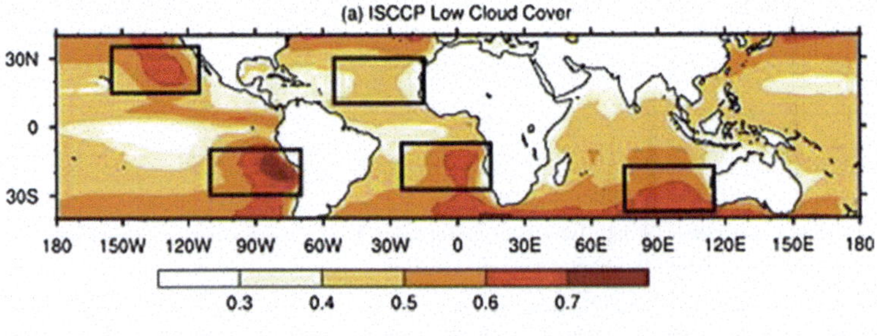

Fig. 1 Low-cloud cover from the International Satellite Cloud Climatology Project. Rectangles indicate the preferred regions of tropical low clouds of the stratocumulus type. These regions were studied in Q15, but they were also studied by M16, B16, and M17. Another common low-cloud type is trade cumulus which typically occur in the regions to the west of the rectangles in the figure. These clouds were studied directly by M17 and to some extent by B16. Low clouds are also common in the subsidence portions of the tropical-extra-tropical transition zone between 20° and 40° latitude in each hemisphere, and these clouds were studied in Z15 and M17. (Figure from Q15)

environment (Fig. 1). The conditions that favor extensive sheets of low clouds such as stratocumulus include a relatively cold sea-surface temperature (*SST*) and a strong capping temperature inversion, among others. Elsewhere in the tropics, *SST* is warmer and the inversion weaker even as the air is still subsiding, favoring the smaller cloud fractions typical of trade cumulus clouds. Assuming low clouds are a response to their environment, environmental conditions influencing low clouds may be called "cloud-controlling factors" (Stevens and Brenguier 2009).

The basis of the observational approach for predicting low-cloud feedbacks from their controlling factors is the following: suppose we know how sensitive the clouds are to each cloud-controlling factor, as derived from observations of cloud variability in the present climate, and we have an idea of how each of the factors will change with climate warming, as derived from climate models and confirmed by physical reasoning. Then we can predict how the low clouds will change with climate warming under the assumption that the sensitivities of clouds to their controlling factors are time-scale invariant. This approach has been taken in five recent studies (Qu et al. 2015b; Zhai et al. 2015; Myers and Norris 2016; Brient and Schneider 2016; McCoy et al. 2017, in chronological order; hereafter these studies will be named "Q15," "Z15," "M16," "B16" and "M17," respectively). In this paper, we review these studies. From them we form a consensus estimate of the average tropical low-cloud feedback for marine subsiding regions (including both stratocumulus and trade cumulus) that can be used in an estimate of Earth's climate sensitivity. We also examine issues with this approach and how uncertainties in its predictions might be reduced.

2 Cloud-Controlling Factors

The studies considered make the assumption that anomalies in some measure of tropical low clouds ΔC relevant to radiative fluxes (such as low-cloud fraction or shortwave cloud-radiative effect) can be represented by a first-order Taylor expansion in cloud-controlling factors x_i:

$$\Delta C = \sum_i \left.\frac{\partial C}{\partial x_i}\right|_{x_j \neq x_i} \Delta x_i \quad (1)$$

In (1), the partial derivative $\frac{\partial C}{\partial x_i}$ represents the sensitivity of low clouds to a cloud-controlling factor and is assumed to be the same regardless of the time-scale over which anomalies are calculated ("time-scale invariant"). This time-scale is often inter-annual, but it could also be weekly or over decades and centuries. Any time-scale is valid provided it is greater than about 2–3 days, the longest time-scale over which the boundary layer and its clouds respond to changes in the cloud-controlling factors (Schubert et al. 1979; Bretherton 1993). Temporal averaging reduces but does not eliminate any disequilibrium between the clouds and their controlling factors.

Table 1 lists the most important controlling factors for tropical low clouds. The table also explains why tropical low clouds depend on each controlling factor, and cites a supporting observational and/or large-eddy simulation modeling study. While a wide body of research supports each factor, they do not all have the same level of theoretical understanding or observational and modeling support. The relationship of increased low cloud to increased inversion strength is the most robust relationship. Covariance between factors, such as free-tropospheric relative humidity with downward longwave radiative

Table 1 Most prominent cloud-controlling factors affecting tropical low clouds, their physical explanation, and their support from observational and large-eddy simulation modeling studies

Cloud-controlling factor	Physical explanation	Observational support	Modeling support
Strengthened inversion stability	Reduced mixing across inversion keeps boundary layer shallower, more humid and more cloudy	Wood and Bretherton (2006)	Bretherton et al. (2013)
Reduced subsidence	Deeper boundary layer increases cloud	Myers and Norris (2013)	Blossey et al. (2013)
Increased horizontal cold advection	Greater destabilization of the surface–atmosphere interface increases upward buoyancy flux promoting more clouds	Norris and Iacobellis (2005)	N/A
Increased free-tropospheric humidity	Entrainment drying is reduced, thus moistening the boundary layer and increasing cloud	M16	van der Dussen et al. (2015)
Decreased downward longwave radiation	Reduced downward longwave radiation increases cloud-top radiative cooling, driving more turbulence supporting cloud	Christensen et al. (2013)	Bretherton et al. (2013)
Colder Sea-surface temperature (*SST*)	Colder temperature reduces the efficiency of entrainment necessitating more cloud to produce a given entrainment rate	Q15	Bretherton and Blossey (2014)
Increased surface wind speed	Increased surface driven shear mixing increases latent heat flux and cloud	Brueck et al. (2015)	Bretherton et al. (2013)

In the first column, the direction of the cloud-controlling factor corresponds to that that would increase low clouds. Only the single most prominent study supporting the cloud-controlling factor is listed in the third and fourth columns

N/A indicates the absence of a study demonstrating the role of the factor in tropical low clouds

flux, makes it difficult to conclusively identify the individual role of some factors from observations, even if these are easily distinguished in modeling studies. Reliable large-scale observations of some controlling factors are sometimes unavailable. It is unlikely that Table 1 is missing any important cloud-controlling factors since it includes the majority of the external large-scale variables in the energy and moisture budget equations for the boundary layer (Stevens and Brenguier 2009). Nonetheless, the list may be missing some known (e.g., aerosol) and unknown factors that likely only play a minor role in tropical low-cloud feedbacks to climate warming.

3 Low-Cloud Feedbacks

In the forcing-adjustment-feedback framework (Sherwood et al. 2015), changes in global mean top-of-atmosphere radiative flux (R) due to individual feedbacks such as clouds are assumed to be linearly related to changes in global mean surface air temperature (T_g). The contribution of tropical low clouds to the global cloud feedback can be thought of as the product of the fraction of the planet dominated by tropical low clouds (a) with the sensitivity to changes in T_g of the local cloud-induced changes in top-of-atmosphere radiation (e.g., using shortwave cloud-radiative effect):

$$\frac{dR}{dT_g} = a\frac{dC}{dT_g} \tag{2}$$

If we view the local cloud response as resulting from changes in the local cloud-controlling factors, we can use (1) to expand the local cloud feedback $\frac{dC}{dT_g}$ as:

$$\frac{dC}{dT_g} = \sum_i \frac{\partial C}{\partial x_i}\frac{dx_i}{dT_g} \tag{3}$$

In (3), the partial derivatives $\frac{\partial C}{\partial x_i}$ are the radiative sensitivities of cloud to the controlling factors and $\frac{dx_i}{dT_g}$ measures how each cloud-controlling factor x_i varies with increases in T_g on climate change time-scales. Equation (3) expresses the concept that clouds respond to the local values of the cloud-controlling factors while cloud-controlling factors may depend on non-local factors (such as the large-scale circulation of the atmosphere), which can be (imperfectly) parameterized as a function of T_g.

Multi-linear regression analysis of observations provides the sensitivities of clouds to their controlling factors on inter-annual or shorter time-scales (but no shorter than 8 days in the studies reviewed here), whereas analysis of climate model simulations reveals how the factors vary with long-term climate change. Cloud sensitivities can also be calculated from model simulations and compared to those calculated from observations on the time-scales for which they are available.

Figure 2 shows the end result from one of the studies covered by this review (M16). In particular, panel (a) compares the local cloud feedback predicted by Eq. (3) (called "constrained") with that simulated by climate models (called "actual"); panel (b) shows each individual term from the right-hand side of Eq. (3). As Fig. 2 shows, M16 deduce a positive cloud feedback primarily because $\frac{\partial C}{\partial SST}$ is positive in the satellite cloud observational datasets they use. However, the positive contribution from *SST* increases is offset by a negative contribution from changes in the Estimated Inversion Strength (*EIS*). This contribution results from the facts that (1) climate models universally predict, with robust physical justification, that *EIS* will increase with warming (Webb et al. 2013; Qu et al. 2015a), and (2) cloud amount and the associated reflection of solar radiation increases strongly with increases in *EIS* in observations. *EIS* increases in warming simulations are driven by increased *SST* gradients between tropical low cloud and deep convection regions as well as increased land–ocean surface temperature contrast (Qu et al. 2015a). The other factors examined in M16, namely horizontal temperature advection, free-tropospheric humidity and subsidence, make smaller but collectively non-negligible negative contributions to the predicted cloud feedback.

The five studies in our review make different choices with respect to the observational datasets, cloud-controlling factors and spatiotemporal variability examined (Table 2). Despite these differences, the following commonalities emerge: (1) *SST* is the most important cloud-controlling factor for climate change cloud feedbacks; (2) tropical low clouds are observed to decrease in extent or radiative impact with increasing *SST*, leading to the prediction of positive tropical low-cloud feedbacks to climate change; (3) the four studies that consider *EIS* agree that although *EIS* contributes a negative feedback, it only partially offsets the positive feedback from *SST*; and (4) the three studies that consider additional factors beyond *EIS* and *SST* agree that these additional factors collectively make only a minor contribution to tropical low-cloud feedback.

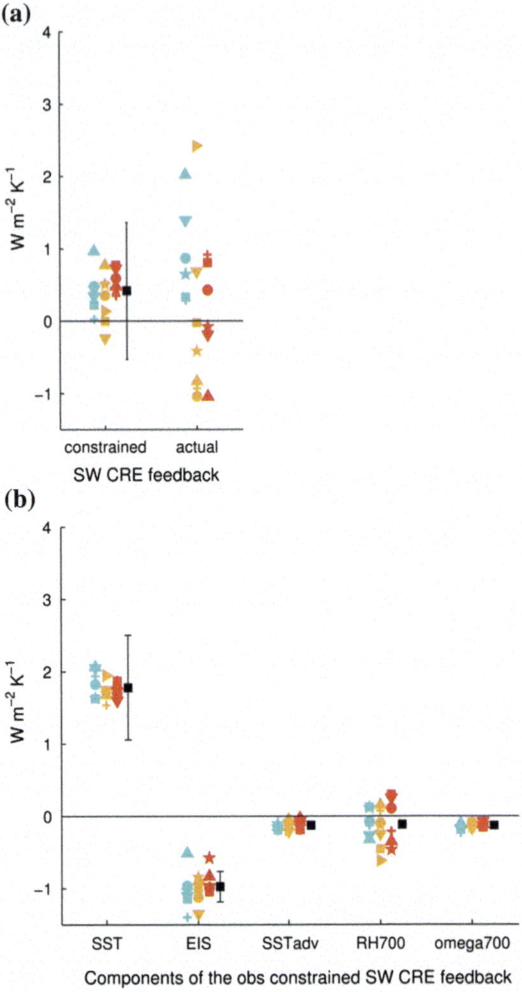

Fig. 2 **a** Local tropical low-cloud feedback predicted from the observed sensitivity of clouds to their controlling factors (called "constrained") and that actually simulated by climate models; and **b** the components of the predicted cloud feedback from each controlling factor according to Eq. (3). The estimate in black is computing using the model-mean changes in factors and shows a 95% confidence interval calculated from the uncertainty in the cloud sensitivities calculated from observations. In panel **b** and for the constrained predictions in panel **a**, the spread in model predictions is due solely to inter-model differences in how cloud-controlling factors change with rises in global mean surface air temperature. Symbol color classifies climate models according to how well they reproduce the observed cloud sensitivities (cyan = above average, orange = average, red = below average). Acronym definitions in the figure are: "EIS"—Estimated Inversion Strength, "SSTadv"—horizontal temperature advection, "RH700"—relative humidity at 700 hPa, "omega700"—subsidence velocity at 700 hPa, and "SW CRE"—Shortwave Cloud-Radiative Effect. (Figure from M16)

Figure 3 displays the quantitative predictions of the local tropical low-cloud feedback from these observationally based studies, along with values predicted from large-eddy simulations and global climate models; the "Appendix" explains how these predictions were derived. Some observational studies have more than one estimate because they consider multiple satellite cloud datasets (Q15 and M16), geographical areas (M17) or temporal scales of variability (B16). Nearly all observational estimates of the local tropical low-cloud feedback are positive and many values cluster near 1 W m^{-2} K^{-1}.

4 Implications for Climate Sensitivity

Do the cloud feedback estimates from the observational studies reviewed here help narrow the uncertainty in the climate change response of tropical low clouds? Local cloud feedback values from the cloud-controlling factor studies range from − 1.0 to

Table 2 List of observational cloud data, cloud-controlling factors, and the spatial–temporal variability examined in each study employing observations of clouds and their controlling factors to predict tropical low-cloud feedbacks

Study	Satellite cloud observations used	Cloud-controlling factors included	Variability used to define cloud sensitivities
Qu et al. (2015b)—Q15	Cloud fraction observations from ISCCP (1984–2009), MISR (2000–2013), MODIS (2002–2014), and PATMOS-x (1982–2009)	*EIS* and *SST* (model 1); *EIS*, latent heat flux, specific humidity lapse rate, free-tropospheric humidity, subsidence rate, surface wind speed, and horizontal temperature advection (model 2)	Inter-annual temporal variability in the annual means of the tropical low-cloud regions containing stratocumulus and cumulus with stratocumulus (Fig. 1)
Zhai et al. (2015)—Z15	Cloud fraction from merged CloudSat and CALIPSO data (2006–2010)	*SST*	Seasonal cycle temporal variability in the monthly means averaged over subsidence portions of the 20°–40° latitude band
Myers and Norris (2016)—M16	Shortwave cloud-radiative effect from CERES-EBAF (2000–2012) and ISCCP-FD (1984–1999)	*EIS*, *SST*, free-tropospheric humidity, subsidence rate, and horizontal temperature advection	Combined spatial and temporal variability in the inter-annual anomalies of monthly means of tropical low-cloud regions containing stratocumulus and cumulus with stratocumulus
Brient and Schneider (2016)—B16	Shortwave cloud-radiative effect from CERES-EBAF (2000–2015) and low-cloud fraction from CALIPSO-GOCCP (2006–2014)	*EIS* and *SST*	Temporal variability at 3 time-scales (intra-annual, seasonal cycle, and inter-annual) and using inter-annual anomalies in the monthly means averaged over geographically varying tropical regions in the lowest quartile of 500 hPa relative humidity
McCoy et al. (2017)—M17	Cloud fraction from MODIS (2002–2014)	*EIS*, *SST*, free-tropospheric humidity, subsidence rate, and surface wind speed	Combined spatial and temporal variability in 8-day mean data within 3 oceanic regions: 40°N–40°S, trade cumulus, and mixed stratocumulus and trade cumulus regions

Acronym definitions and references for the satellite cloud observations are: *ISCCP* International Satellite Cloud Climatology Project (Rossow and Schiffer 1999), *MISR* Multiangle Imaging SpectroRadiometer (Marchand and Ackerman 2010), *MODIS* Moderate Resolution Imaging Spectroradiometer (Platnick et al. 2003), *PATMOS-x* Pathfinder Atmospheres Extended (Foster and Heidinger 2013), merged CloudSat/Calipso (Mace et al. 2009), *CERES-EBAF* Clouds and the Earth's Radiant Energy System Energy Balanced and Filled (Loeb et al. 2009), *ISCCP-FD* International Satellite Cloud Climatology Project Radiative Flux Dataset (Zhang et al. 2004), and *CALIPSO-GOCCP* Cloud–Aerosol Lidar and Infrared Pathfinder Satellite Observations GCM-Oriented CALIPSO Cloud Product (Chepfer et al. 2010)

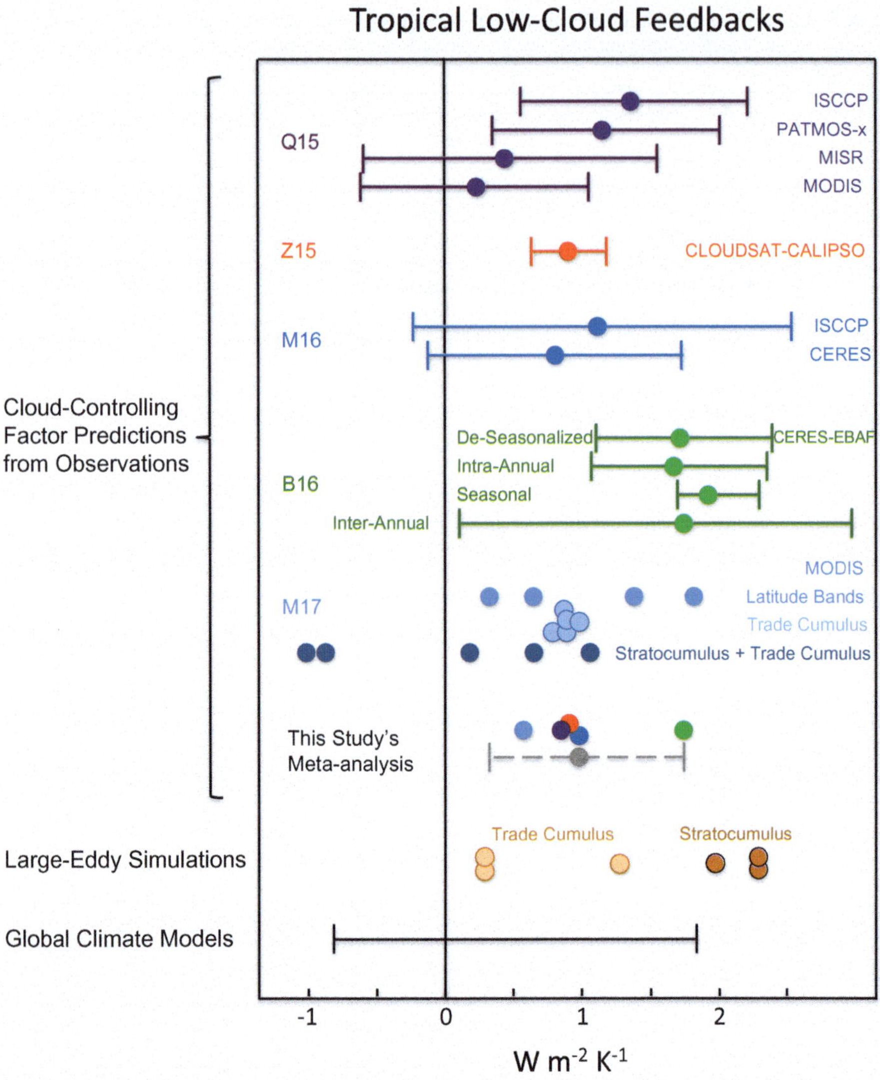

Fig. 3 Values of local tropical low-cloud feedbacks predicted from recent observational studies, large-eddy simulations and global climate models. Local feedbacks are defined as the local change in top-of-atmosphere radiation from tropical low clouds per degree increase in global mean surface air temperature. Bar widths for observational studies (unavailable for M17) and this study's meta-analysis represent 90% confidence intervals. Values from individual large-eddy simulation studies are shown. The bar width for global climate models indicates the range of model results. See the "Appendix" for details

$+ 1.9$ W m^{-2} K^{-1}. This range would appear to offer no constraint on the climate model range, $- 0.8$ to $+ 1.8$ W m^{-2} K^{-1}, as seen in Fig. 3. Still it is worth recognizing that many observational estimates are concentrated in a narrower range. We synthesize these results to form a consensus estimate through a meta-analysis of these studies. A formal approach would consider the uncertainty of each study and account for their degree of independence, but measures of uncertainty are not supplied uniformly for these studies, and no confidence intervals are supplied by M17 at all. Instead we proceed approximately by assuming that each study represents a partially independent result, which we justify by noting the diversity of observational satellite cloud datasets, geographic domains, and time periods employed. We also assume that each study gives a representative estimate of the cloud feedback averaged over tropical low-cloud regions; this is further discussed below (issue *I3*). Our consensus estimate is made by averaging all central estimates to form a single value of cloud feedback for each study (shown near our meta-estimate in Fig. 3). Then we compute the five-study mean and 90% confidence interval as 1.645 times the sample standard deviation, consistent with a normal distribution. The meta-analysis uncertainty describes the uncertainty across the ansätze employed by each of the five studies but not the uncertainty within each study. Nonetheless, this uncertainty estimate seems appropriate as it produces a 90% confidence interval whose width is within 10% of the average interval width in individual studies shown in Fig. 3.

The meta-analysis produces a local tropical low-cloud feedback of 1.0 ± 0.7 W m^{-2} K^{-1}. Our estimate suggests that climate models with negative tropical low-cloud feedback are unrealistic, but still leaves an uncertainty range of ~ 50% (= 1.4/2.6) of that of current climate models.

To determine how the local response of tropical low clouds contributes to climate sensitivity, we first calculate the tropical low-cloud contribution to global cloud feedback by multiplying the local cloud feedback by the fraction of the planet covered by tropical low-cloud regions, following (2). Under the assumptions that (a) subsidence regions cover 2/3 of the tropical oceans, (b) oceans cover 3/4 of the tropics, and (c) the tropics cover ½ of the planet, we estimate that tropical oceanic subsidence regions cover approximately 1/4 of the planet ($a = 1/4$). Thus, we arrive at a contribution of tropical low clouds to the global mean cloud feedback of 0.25 ± 0.18 W m^{-2} K^{-1}.

We then calculate an approximate equilibrium climate sensitivity *ECS* according to $ECS = F_{2\text{co}_2}/(-\lambda)$, where $F_{2\text{co}_2}$ is the effective radiative forcing for a doubling of carbon dioxide (CO_2) and λ is the climate feedback parameter (Dufresne and Bony 2008). The climate feedback parameter is equal to the sum of the Planck response and feedbacks from water vapor, lapse, surface albedo and clouds. We use average climate model values for the forcing and non-cloud feedbacks as reported in Caldwell et al. (2016) (Table 3). Further assuming a high-cloud altitude feedback (Zelinka et al. 2016) of + 0.2 W m^{-2} K^{-1}, but no other cloud feedbacks, we compute an *ECS* of 2.4 K. Adding the central estimate of + 0.25 W m^{-2} K^{-1} for the tropical low-cloud feedback from our meta-analysis to the high-cloud altitude feedback, we arrive at a central estimate for *ECS* of 3.0 K. Thus, if the tropical low-cloud feedback is positive with the magnitude suggested by these observational studies, *ECS* would be in the middle of its canonical range of 1.5–4.5 K (Stocker et al. 2013).

5 Sources of Uncertainty

In interpreting cloud feedbacks derived from observations of clouds and their controlling factors, a number of issues merit discussion. We roughly divide these into two categories: those of a fundamental nature (*F1–F4*) that may limit the validity of this observational approach, and those related to implementation (*I1–I5*) that may limit the accuracy of the feedback estimated with a presumed valid approach. The latter issues, if addressed, might allow for tighter constraints on tropical low-cloud feedback.

Table 3 Values of the CO_2 radiative forcing and various feedbacks used in the calculation of equilibrium climate sensitivity

Term	Value
Radiative forcing for doubling of CO_2 concentration (F_{2CO_2})	3.43 W m^{-2}
Planck feedback	– 3.15 W m^{-2} K^{-1}
Water vapor feedback	1.69 W m^{-2} K^{-1}
Lapse rate feedback	– 0.53 W m^{-2} K^{-1}
Surface albedo feedback	0.38 W m^{-2} K^{-1}
High-cloud altitude feedback	0.20 W m^{-2} K^{-1}

These values are the multi-model-mean values from Caldwell et al. (2016) computed by linear regression of the first 150 years of the abrupt quadrupling of CO_2 climate model experiments. As such, the forcing is an effective radiative forcing that includes the rapid cloud adjustments. The values are averaged only for models passing the clear-sky linearity test used to test the accuracy of the radiative kernel approach to quantify feedbacks. See Caldwell et al. (2016) for details

5.1 Fundamental Issues

5.1.1 F1. Are Cloud Sensitivities Time-scale Invariant?

The approach used in these five studies relies heavily on the assumption that the sensitivity of clouds to their controlling factors remains constant across any time-scale longer than a few days—the longest time-scale over which the boundary layer is still in a state of transient adjustment to changes in the cloud-controlling factors. We can test this proposition by examining results from studies that consider multiple time-scales. B16 calculates sensitivities at 3 time-scales: intra-annual, seasonal cycle, and inter-annual (from monthly mean data). Table 5 indicates consistency (within their uncertainty estimates) across these time-scales for the *SST* sensitivity. Although this is less true for the *EIS* sensitivity, the final estimates of their cloud feedback are still consistent across time-scale (Fig. 3). M17 calculates sensitivities at 2 time-scales: using 8-day means and annual means. They find that the sensitivities vary by less than a factor of two between those two time-scales, with one exception, namely for $\frac{\partial C}{\partial SST}$ from the regions of mixed stratocumulus and trade cumulus. Because M17 do not supply uncertainty coefficients, one cannot judge if this difference is significant. Separately, deSzoeke et al. (2016) make a thorough analysis of the time-scale dependence of the relationship between low cloud and *EIS*, finding that while the amount of low-cloud variance explained by *EIS* varies with time-scale, the sensitivity of low cloud to *EIS* varies by less than a factor of two between daily, monthly, and inter-annual time-scales examined. Klein (1997) examined low-cloud variability at a single point in the Northeast Pacific and found that the signs of the correlation coefficients between low-cloud fraction and several controlling factors remain fixed across time-scales from daily to monthly. Some variations in the cloud sensitivities are expected due to statistical uncertainty in sensitivity coefficients, and from this available evidence one cannot disprove the notion that cloud sensitivities are time-scale invariant. Certainly the sensitivities agree qualitatively and in sign across time-scales, if not in exact magnitude. To fully address this question, more observational studies calculating cloud sensitivities with error estimates at multiple time-scales are needed.

5.1.2 F2. Are Clouds Responding to the Controlling Factors?

When regression analysis is applied to observations to derive sensitivity coefficients, it is assumed that these reflect the influence of the factors on the clouds, rather than the influence of the clouds on the factors. But how confident are we that this is the case? This concern is most obviously relevant for variables internal to the boundary layer. For example, relative humidity in the boundary layer or the state of cloud organization would be questionable candidates for a controlling factor and is not listed in Table 1 for this reason. For the cloud-controlling factors listed in Table 1, substantial observational evidence exists that cloud properties are best correlated to upwind (Klein et al. 1995; Klein 1997; Mauger and Norris 2010) or earlier (deSzoeke et al. 2016) sampling of the factors. These lines of evidence reinforce the notion that these quantities are external and large-scale characteristics of the atmosphere or ocean which influence the boundary layer and its clouds, rather than the other way around.

The relationship between clouds and *SST* deserves extra discussion in this connection, given the major role for $\frac{\partial C}{\partial SST}$ in determining tropical low-cloud feedback. Modeling studies demonstrate that a positive radiative feedback from tropical low clouds can amplify low frequency (multi-year and decadal) *SST* variability (Bellomo et al. 2014, 2015), so there is no doubt that clouds affect *SST*. Nonetheless, it is also clear from large-eddy simulations (Blossey et al. 2013) and observational evidence (Klein et al. 1995; Klein 1997; Mauger and Norris 2010) that clouds respond to *SST* over just a few days. For an ocean mixed-layer depth of 50 m, it takes about 300 days to produce an *SST* anomaly in response to cloud-radiative anomalies that is consistent with the observed value of $\frac{\partial C}{\partial SST}$ (deSzoeke et al. 2016); covariations of cloud with *SST* at time-scales shorter than 300 days would therefore reflect the influence of *SST* on cloud, and not the other way around. The fact that M17 find similar values of $\frac{\partial C}{\partial SST}$ at 8-day time-scales as at inter-annual time-scales (with the exception of the region with mixed stratocumulus and trade cumulus) suggests that two-way interactions between cloud and *SST* do not cause $\frac{\partial C}{\partial SST}$ to be different at the longer time-scales. Furthermore, climate model simulations with prescribed *SST*, which by definition do not have the two-way interactions of clouds and *SST*, produce a value of $\frac{\partial C}{\partial SST}$ reasonably close to those derived from simulations with fully coupled ocean–atmosphere models (X. Qu personal communication).

To understand the reason for this similarity across time-scales, we appeal to our understanding of the water vapor feedback. One expects water vapor anomalies to adjust to changes in the underlying *SST* so that relative humidity is approximately conserved. One expects this to be true even if water vapor did not produce the longwave radiative anomalies that fed back on the *SST* changes. Thus, the diagnosed sensitivity of water vapor to surface temperature is the same, whether the interaction is one-way or two-way. In a similar way, we may also think of cloud anomalies as being in a state of mutual adjustment with underlying *SST*, so as to maintain a boundary layer that is thermodynamically consistent with its environment. This would be the case whether or not *SST* has enough time to be affected by the top-of-atmosphere energy budget perturbation that the cloud anomaly produces.

5.1.3 F3. Uncertainty in the Climate Change Prediction of Cloud-Controlling Factors

The cloud-controlling factors are among the more trustworthy variables of climate models because they are aspects of the resolved large-scale state. While climate models generally agree on their predicted climate changes, any inter-model spread contributes to spread in the predicted low-cloud feedback with this observational approach. This can be seen in Fig. 2, where the spread among climate models (which are displayed as colored symbols) for the "constrained" column in panel a and for each factor in panel b arises solely from inter-model spread in the climate changes in cloud-controlling factors $\frac{dx_i}{dT_g}$. For the total feedback ("constrained" column in panel a), inter-model spread is comparable to, but slightly smaller than the spread due to the uncertainty in the observed cloud sensitivities (shown by the black uncertainty bar). Comparison to panel b indicates that much of the total feedback spread is due to inter-model differences in the predicted changes in *EIS* and free-tropospheric humidity. For *SST*—the factor with the largest average cloud feedback contribution—inter-model spread in $\frac{dSST}{dT_g}$ in the period examined (121–140 years after CO_2 quadrupling) is smaller than the uncertainty in the observed value of $\frac{\partial C}{\partial SST}$. We conclude from this figure, as well as the analysis of Q15, that the uncertainty in the predicted climate change of the cloud-controlling factors is a significant component of the cloud feedback uncertainty but not quite as large as the uncertainty in $\frac{\partial C}{\partial SST}$. Reducing this uncertainty would include diagnosing the influences on the cloud-controlling factors and identifying constraints on the climate-model-simulated changes. A step in this direction for *EIS* was taken by Qu et al. (2015a). Also, a preliminary investigation finds that the normalized changes in *SST* and associated cold advection are positively correlated across models with the low-cloud feedback itself (Tim Myers, personal communication). This correlation is consistent with more positive low-cloud feedbacks locally warming the ocean more (relative to the global mean temperature increase). Normalized changes in other cloud-controlling factors including free-tropospheric humidity, subsidence and *EIS* do not have an apparent relationship to the low-cloud feedback itself, consistent with the expectation that their large-scale nature makes them additionally sensitive to remote influences. Until this uncertainty is reduced, the "constrained" column in Fig. 2a suggests that the uncertainty in the local cloud feedback will not be smaller than $\pm$ 0.5 W m^{-2} K^{-1}. While this uncertainty is considerably smaller than that of the individual observational estimates in Fig. 3, it is not very much smaller than the $\pm$ 0.7 W m^{-2} K^{-1} uncertainty in our meta-estimate.

5.1.4 F4. Time-Dependency of Cloud-Controlling Factors During a Climate Change

Our cloud feedbacks estimates have been made under the assumption that changes in cloud-controlling factors $\frac{dx_i}{dT_g}$ are constant in time. A growing body of evidence (Andrews et al. 2015; Rugenstein et al. 2016) suggests that cloud feedbacks to climate change are sensitive to the spatial pattern of *SST* warming, which evolves during simulated time-dependent climate change. Of particular importance to tropical low-cloud feedbacks is the differential rate of warming between tropical ascent and subsidence regions: if tropical ascent regions initially warm more rapidly than tropical subsidence regions, *EIS* in tropical subsidence regions will increase through the influence of the large-scale circulation (Caldwell and Bretherton 2009; Qu et al. 2015a). This will contribute to low cloud increases and hence smaller low-cloud feedbacks (Zhou et al. 2016). When more warming later appears in the subsidence regions, tropical low-cloud feedbacks will become more positive. This behavior is most apparent in the simulations with abrupt quadrupling of CO_2 (Andrews et al. 2015; Rugenstein et al. 2016), but it also occurs in decadal feedbacks inferred for the last century (Gregory and Andrews 2016; Zhou et al. 2016). This does not negate the framework of Eqs. (1–3). Rather it suggests that there would be value in allowing that the $\frac{dx_i}{dT_g}$, especially $\frac{dEIS}{dT_g}$ and $\frac{dSST}{dT_g}$, might vary with time, even as the cloud sensitivities $\frac{\partial C}{\partial x_i}$ to local conditions remain constant.

As interesting as *SST* pattern effects are, they are unlikely to have a first-order impact on the century time-scale tropical low-cloud feedback. With typical values of the cloud sensitivities, a negative tropical local low-cloud feedback would not occur unless the ratio of *EIS* to *SST* change is $\sim$ 1, several times larger than the typical ratio of 0.2 exhibited by climate models. Such a large value might happen for decadal variability (Zhou et al. 2016), but is extremely unlikely to happen for century time-scale warming. Oceanic heat transport on the century time-scale prevents warming in tropical subsidence regions from differing much from warming in tropical ascent regions. For century time-scale forced climate change such as 100 + years after abrupt quadrupling of CO_2 or by the end of the twenty-first century in a scenario simulation, the values of cloud-controlling factors simulated by climate models are such that the tropical low-cloud feedbacks are decidedly positive, given the observed cloud sensitivities.

5.2 Implementation Issues

5.2.1 I1. Imperfect Observations of Clouds and Their Controlling Factors

Figure 3 shows that the central estimate spread among the four satellite cloud fraction datasets in Q15 is 1.1 W m^{-2} K^{-1}, but the spread among two satellite cloud-radiative effect datasets in M16 is only 0.4 W m^{-2} K^{-1}. Such differences could arise from uncertainties in cloud observations. Indeed, the five studies in this review employed a wide range of satellite-derived cloud metrics, including estimates of cloud fraction and shortwave cloud-radiative effect (Table 2). However, the effect of these choices on cloud feedback estimates is difficult to determine. Q15 and M16 use multiple cloud datasets, but the datasets cover different years, and thus differences are not solely due to measurement or algorithmic changes. The larger spread among Q15 estimates might be consistent with the fact that cloud fraction is more difficult to measure. As a result, there may be greater differences between cloud fraction datasets than those describing cloud-radiative effect (Maddux et al. 2010; Pincus et al. 2012).

The differences in feedback estimates could also come from observational uncertainty in cloud-controlling factors. Unfortunately, no study has quantified this effect. *SST* is extremely well-observed from satellite, but observational uncertainty might be not negligible for the other factors that often rely on reanalysis data (Pincus et al. this issue). For example, M17's estimate of $\frac{\partial C}{\partial EIS}$ using satellite *EIS* observations appears consistent with M16's estimate using *EIS* from reanalysis data. But this is not a clean comparison because the satellite observations are used in data assimilation, among other reasons.

Clearly, more research into the impact on predicted low-cloud feedbacks of observational uncertainty in clouds and their controlling factors would be helpful. At the same time, because the error bars on low-cloud feedback estimates derived from diverse cloud and cloud-controlling factors overlap substantially, we judge it unlikely that estimates of cloud feedback would change significantly if observational uncertainties in cloud or cloud-controlling factors were better quantified and reduced.

5.2.2 I2. Limited Duration of the Observational Record

The majority of studies use inter-annual variability to determine $\frac{\partial C}{\partial SST}$ and $\frac{\partial C}{\partial EIS}$, and the typical length (15–25 years) of the more reliable satellite records offers very limited numbers of independent samples. This suggests the limited duration of the observational record is a major contributor to uncertainty in the estimates shown in Fig. 3. A compounding problem arises from the covariance of *EIS* and *SST* for current climate variability on monthly and longer time-scales. However, the uncertainty in $\frac{\partial C}{\partial EIS}$ is probably smaller than the uncertainty in $\frac{\partial C}{\partial SST}$ because sub-monthly variations of *EIS*, which typically do not co-occur with large *SST* fluctuations, confirm the value of the *EIS* sensitivity (M17, deSzoeke et al. 2016). As time goes by, longer satellite records will gradually reduce uncertainty in $\frac{\partial C}{\partial SST}$ due to limited observational duration.

5.2.3 I3. Limited Spatial Sampling of the Observations

The results of large-eddy simulation suggest a systematic difference in the cloud feedback between regions dominated by trade cumulus and regions dominated by stratocumulus (Fig. 3). Three of the observational studies used here, however, (Q15, M16, B16) primarily analyze variations in the stratocumulus regions. These studies' estimates of low-cloud feedback may be biased because they do not sample trade cumulus that might have a smaller feedback. However, M17's feedback for trade cumulus is close to our meta-estimate and is in fact larger than their feedback for stratocumulus regions. The observational analysis for latitude bands in M17 and Z15 also produces feedbacks that do not depart significantly from our meta-estimate. For individual cloud sensitivities, M17 found general agreement between regions for most factors, with the exception of subsidence (Myers and Norris 2013; deSzoeke et al. 2016). Observational studies focused specifically on trade cumulus exhibit relationships of low clouds to cloud-controlling factors with the same sign as in M17 (Brueck et al. 2015; Nuijens et al. 2015). In conclusion, there is not enough evidence at this time to demonstrate that differing spatial sampling in the observational studies leads to a biased estimate of the mean feedback for tropical low-cloud regions. Further observational studies, particularly for trade cumulus regions, are needed.

The unusual spatial sampling in B16 bears further examination. In B16, the particular locations analyzed vary in time, unlike those in the other studies. Moreover, they obtain a single data point for each month by averaging data across all points they select, no matter how wide their geographical separation. This means the cloud sensitivities they calculate may not necessarily represent a local relationship between cloudiness and *SST* or *EIS*.

5.2.4 I4. Imprecise Statistical Modeling

A key question is whether clouds vary linearly with their controlling factors. Although low clouds result from the interactions of inherently nonlinear processes, there is ample evidence that a linear approach can explain cloud variations at spatial scales greater than

100 km and time-scales longer than a few days. For example, observations show that a linear relationship with inversion strength can explain over 80% of the variance in the seasonal cycle of tropical and extra-tropical marine low clouds (Klein and Hartmann 1993; Wood and Bretherton 2006). Over decadal time-scales, Seethala et al. (2015) find that observed tropical low-cloud changes can be well explained with a linear model using *SST*, *EIS*, and horizontal temperature advection as cloud-controlling factors. In large-eddy simulations, changes in shortwave radiation reflected by low clouds in response to the simultaneous changes in many cloud-controlling factors are within 10% of the linear sum of changes in simulations forced by individual cloud-controlling factors (Bretherton et al. 2013).

There is a hazard in applying any statistical model "out of sample," an inherent risk whenever sensitivities inferred from a system's variability are used to infer information about the system's response to a perturbation. But this does not appear to be an important concern for low-cloud-controlling factors. In tropical subsidence regions, both inter-annual variability in *SST* (1–2 K $\sim$ two standard deviations, Deser et al. 2010) and the amplitude of its seasonal cycle (2–4 K, Shea et al. 1992) are generally comparable in magnitude to the 2–3 K increases typical of a response to CO_2 doubling. If cloud changes are indeed linear within the ranges of variability and climate change, the cloud sensitivities derived from variability ought to approximately agree with those associated with climate change. The reviewed studies find approximate agreement when we compare the actual cloud feedback simulated by the climate model to that predicted by (3) when the sensitivities to each factor are derived from each model's simulation of current climate variability. Figure 2a of M16 shows that the linear model of (3) gives a very good prediction for the actual cloud feedback for those climate models whose cloud sensitivities are closer to observations, but less so for the climate models with more erroneous cloud sensitivities. Clouds in the latter models are likely sensitive to cloud-controlling factors not found in nature and also excluded from the linear prediction model. This good agreement (sometime regardless of model fidelity) was also found by Z15, Q15, and B16, although in some instances the feedback from (3) overestimates the actual feedback. The across-model agreement between cloud variability in the current climate and the cloud feedback to climate change illustrates a type of "emergent constraint" relationship (Klein and Hall 2015). This range of evidence provides support for a linear model of tropical low-cloud changes, although residuals between the actual cloud feedback and that predicted by (3) should be expected.

5.2.5 I5. Incomplete Set of Cloud-Controlling Factors

The number of factors used varies across the reviewed studies (Table 2); in Q15 and B16, one can directly examine the sensitivity to this issue. Q15 arrive at similar predictions whether they use two or seven factors, although the climate-model-predicted feedback with seven factors would be 15–30% smaller than the feedback predicted with two factors (*SST* and *EIS*). This is consistent with M16's result that the factors other than *SST* and *EIS* produce small negative feedbacks (Fig. 2b) whose collective sum is $-0.4\ \mathrm{W\ m^{-2}\ K^{-1}}$. B16 find that $\frac{\partial C}{\partial SST}$ is $\sim$ 30% smaller when a two-factor (*SST* and *EIS*) regression model is used instead of a single factor (*SST*), consistent with the general anti-correlation of *SST* and *EIS* within natural climate variability. This suggests that studies (Z15, B16, Q15 two-factor model shown in Fig. 3, M17) calculating feedbacks with a reduced set of cloud-controlling factors may have a small positive bias in their predicted low-cloud feedback.

6 Summary and Final Remarks

Tropical low-cloud feedback is a key uncertainty for climate change. In this paper, we reviewed recent studies that predict the tropical low-cloud feedback using the observed sensitivities of clouds to controlling factors of the large-scale environment. The strength of this approach is that it relies primarily on observations of the cloud response to controlling factors and does not depend on the simulation of clouds by climate models. (It does rely on model predictions of how the controlling factors change with climate, however.) Although we only discuss studies of tropical low clouds, there is also evidence that this approach would also be useful for predicting and understanding low cloud amount and reflectivity feedbacks over the middle-latitude oceans (Gordon and Klein 2014; Ceppi et al. 2016; Terai et al. 2016; Grise and Medeiros 2017).

Studies taking this approach agree that the tropical low-cloud feedback is positive. Our synthesis of the results from these studies is that the contribution of tropical low clouds to the global mean cloud feedback is 0.25 ± 0.18 W m^{-2} K^{-1}, indicating that climate models with negative tropical low-cloud feedbacks are implausible. Our synthesis suggests a central estimate for climate sensitivity of 3.0 K. Longer observational records offer perhaps the best near-term prospects for reducing uncertainty, but ultimately smaller uncertainties would also require greater certainty in the prediction of climate changes in cloud-controlling factors. More observational studies targeting trade cumulus regions would also be desirable (Brueck et al. 2015; Bony et al. 2017).

Our observational estimate of tropical low-cloud feedback is consistent with independent estimates from large-eddy simulation models forced by climate-model-simulated changes in cloud-controlling factors. The range of local cloud feedbacks from large-eddy simulations is 0.3–2.3 W m^{-2} K^{-1} (Fig. 3). This overlaps reasonably well with our observational estimate of the local cloud feedback of 0.3–1.7 W m^{-2} K^{-1}.

Even if we know what the tropical low-cloud feedback should be based upon observations and large-eddy simulations, getting climate models to reproduce a feedback of this magnitude is not straightforward. Although some climate models are in agreement with our estimate of the tropical low-cloud feedback, it remains to be seen if they are in agreement for the right reasons. This motivates additional research to understand the physical basis for the cloud sensitivities (particularly for $\frac{\partial C}{\partial SST}$) through both observations (Brient et al. 2016) and large-eddy simulations (Bretherton and Blossey 2014), and whether the physics is correctly modeled in global climate models (Zhang et al. 2013; Sherwood et al. 2014; Vial et al. 2017

Acknowledgements This paper arises from the International Space Science Institute workshop on "Clouds and water vapor, circulation and climate sensitivity." The first author thanks Daniel McCoy, Tim Myers, Xin Qu, and Mark Zelinka for discussions and calculations related to the details of their papers. The first author thanks Louise Nuijens, Bjorn Stevens, and Mark Webb for their review of this paper. The efforts of SAK and AH are supported by the Regional and Global Climate Modeling program of the United States Department of Energy's Office of Science under a project entitled under the project "Identifying Robust Cloud Feedbacks in Observations and Models." RP is grateful for financial support from the US National Science Foundation under award AGS-1138394. The efforts of SAK were performed under the auspices of the United States Department of Energy by Lawrence Livermore National Laboratory under contract DE-AC52-07NA27344.

Appendix: Details Used in Synthesizing Studies

In order for Fig. 3 to provide a meaningful comparison of feedbacks between studies, the original estimates must be converted into a common measure. The common measure is the local cloud feedback: namely by how much the absorbed local net radiation at the top-of-atmosphere in tropical low-cloud regions changes per degree increase in the global mean surface air temperature, as given by (3). We also aim to synchronize error bars so that they each represent 90% confidence intervals, the typical confidence interval used in Intergovernmental Panel on Climate Change reports.

In this synthesis, there are two common issues affecting multiple studies. First, all observational studies except M16 only provide estimates of the cloud sensitivities $\frac{\partial C}{\partial x_i}$, so we must supply values of the cloud-controlling factor changes $\frac{\mathrm{d}x_i}{\mathrm{d}T_g}$ in (3). We specify that $\frac{\mathrm{d}SST}{\mathrm{d}T_g}$, the ratio of changes in local *SST* to changes in global mean surface air temperature, T_g, is 0.7 which is a typical value for climate model simulations of climate warming (Andrews et al. 2015). A value less than unity reflects model predictions of greater warming over land relative to oceans, high latitudes relative to low latitudes, and (least important) tropical ascent regions relative to tropical subsidence regions. We also specify that $\frac{\mathrm{d}EIS}{\mathrm{d}T_g} = 0.14$, matching climate model results that the ratio of temperature-mediated *EIS* changes to local *SST* increases in tropical subsidence regions is around 20% (Webb et al. 2013; Qu et al. 2015a).

Second, questions arise whether the cloud sensitivities measured with observations of either "cloud fraction" (in Q15, Z15, and M17) or "cloud-radiative effect" (in M16 and B16) are a direct measure of the impact of tropical low clouds on the top-of-atmosphere radiation budget. Tropical low clouds have only a small impact on the top-of-atmosphere longwave radiation budget, so we focus on determining the impacts of tropical low clouds on the shortwave radiation budget. The use by M16 and B16 of the shortwave cloud-radiative effect (which is defined as clear-sky fluxes minus all-sky fluxes) as a surrogate for these cloud impacts is known to be a good approximation since clear-sky shortwave radiation undergoes relatively smaller changes over the ice-free oceans (Hakuba et al. 2016). However, the results are less clear when using observations of cloud fraction since cloud feedbacks may also result from changes in other cloud properties, especially the distribution of optical thickness (Zelinka et al. 2012). Cloud fraction itself is also relatively sensitive to details of the observing system (Maddux et al. 2010; Pincus et al. 2012) and how this system changes over time (Norris and Evan 2015). Nonetheless, these concerns are somewhat mitigated by the fact that the observed variability in cloud reflectance in tropical low-cloud regions is primarily driven by changes in low-cloud fraction (Klein and Hartmann 1993; George and Wood 2010). To that end, we convert the cloud fraction sensitivities from Q15, Z15, and M17 into cloud feedbacks by multiplying by the sensitivity of the top-of-atmosphere radiation budget to a unit increase in low-cloud fraction. In particular, we use a 1 W m^{-2} decrease per % increase in low-cloud fraction based upon Klein and Hartmann (1993) who analyzed the relationship between top-of-atmosphere net radiation and ISCCP cloud fraction in tropical low-cloud regions. We note this factor is within 10% of an average factor derived by comparing cloud-radiative effect sensitivities to CALIPSO-GOCCP cloud fraction sensitivities (Tables 2, 3 of B16).

We now present the study-specific details used to derive the estimates shown in Fig. 3.

Q15

The cloud feedback for Q15 is calculated using the reported values from different satellite cloud observations of the cloud sensitivities $\frac{\partial C}{\partial SST}$ and $\frac{\partial C}{\partial EIS}$ (Table S4 of Q15), together with our specified values of $\frac{dSST}{dT_g}$ and $\frac{dEIS}{dT_g}$. The 5–95% confidence intervals for total sensitivity are calculated assuming that Q15's reported values of 90% confidence intervals for the sensitivities $\frac{\partial C}{\partial SST}$ and $\frac{\partial C}{\partial EIS}$ (scaled by 0.2) add in quadrature. This assumes that *EIS* and *SST* are normally distributed and uncorrelated. Finally, we convert Q15 measures of low-cloud fraction into cloud feedbacks by multiplying by the -1 W m^{-2} per % cloud fraction factor.

Z15

Z15 determine from seasonal cycle satellite observations that low-cloud fraction decreases at a rate of 1.28% cloud fraction per degree *SST* increase with a 3-sigma (standard deviation) uncertainty of 0.56% cloud fraction per degree. For a 90% confidence interval, the uncertainty would be equal to 1.81 times the standard deviation of the slope estimate, assuming that the slopes are governed by a Student's *t* distribution with 10 degrees of freedom (= 2 less than the 12 months used in a regression). Thus, we estimate that the 90% uncertainty in this slope is 0.56*(1.81/3) = 0.34% cloud fraction per degree. As *SST* is the only cloud-controlling factor in Z15, this slope is equal to the climate change time-scale cloud fraction sensitivity and a cloud feedback can be computed by multiplying by $\frac{dSST}{dT_g} = 0.7$ and the -1 W m^{-2} per % cloud fraction factor. This yields a value of $+0.90 \pm 0.24$ W m^{-2} K^{-1} to the 90% confidence interval of the local low-cloud feedback from Z15.

M16

M16 report a local tropical cloud feedback of $+0.4 \pm 0.9$ W m^{-2} K^{-1}. This estimate combines separate estimates from two independent observational datasets in two different time periods. In order to illustrate the level of agreement, we show the results for each dataset separately. Tim Myers kindly provided these estimates which are $+0.7 \pm 1.7$ W m^{-2} K^{-1} for ISCCP-FD and $+0.3 \pm 1.1$ W m^{-2} K^{-1} for CERES-EBAF. We make two modifications to convert these estimates into our desired quantity. First, the confidence intervals in M16 are 95% confidence intervals calculated assuming perfect knowledge of the changes in cloud-controlling factors, and 95% uncertainty in the sensitivities of clouds to controlling factors. We convert the uncertainty estimates to 90% confidence intervals by multiplying by (1.645/1.96), the ratio of the *t*-distribution standard variables corresponding to 90 and 95% confidence intervals for large number samples. Second, M16 calculate cloud feedbacks from the difference between year 121–140 of the abrupt quadrupling of CO_2 climate model experiment and a control integration, so that differences in cloud-controlling factors result not only from increases in temperature but also from adjustments to the CO_2 radiative forcing (Gregory and Webb 2008; Sherwood et al. 2015). Thus, the M16 "feedback" includes cloud changes from rapid adjustments in cloud-controlling factors that needs to be removed in order to have an improved estimate of the temperature-mediated changes in the cloud feedback as represented by (3). The most significant of the cloud adjustments to remove from M16 are those in response to the rapid adjustment of *EIS*. From the difference of the two climate states, M16 estimate an *EIS* change of $\frac{dEIS}{dT_g} = 0.28$. As this is twice of our desired value of that $\frac{dEIS}{dT_g} = 0.14$, the *EIS* component of the temperature-mediated cloud feedback in M16 is overestimated by a factor of two. Using M16's reported sensitivity of top-of-atmosphere shortwave cloud-radiative effect to *EIS*, we calculate that the M16 feedback should further be adjusted upward by 0.5 W m^{-2} K^{-1}. With this second change, we arrive at our estimates of $+ 1.2 \pm 1.4$ W m^{-2} K^{-1} for ISCCP-FD and $+ 0.8 \pm 0.9$ W m^{-2} K^{-1} for CERES-EBAF for the 90% confidence intervals of the local low-cloud feedback from M16.

B16

Table 5 of B16 reports the sensitivity of the albedo cloud-radiative effect to two cloud-controlling factors, *SST* and *EIS*, using de-seasonalized variations and variations band-passed filtered to 3 time-scales. To recover the cloud feedback of (3), we use our specified values of $\frac{dSST}{dT_g}$ and $\frac{dEIS}{dT_g}$, and then multiply the sum by the average insolation of 387.9 W m^{-2} that B16 calculates for their examined regions. The central estimate of the calculated cloud feedback shown in Fig. 3 is produced using their central estimate of the cloud sensitivities. B16 also report 90% confidence intervals for these sensitivities. But because B16 use a boot-strap procedure, their confidence intervals are not symmetric about the central estimate. In order to produce 90% confidence intervals for the cloud feedback which are also asymmetric about their central estimate, the following approximate procedure was used. First a provisional lower bound is calculated using the lower bounds for the *EIS* and *SST* sensitivities. Likewise, a provisional upper bound is calculated using the upper bounds for the *EIS* and *SST* sensitivities. At the same time, we calculate our target value of the difference between the 5th and 95th percentiles of the cloud feedback distribution by assuming that the 5th–95th percentile difference in the *SST* and *EIS* (scaled by 0.2) sensitivities add in quadrature. We then modify our provisional upper and lower bounds such that the difference between upper and lower bounds equals our target value of the difference between the 5th and 95th percentiles without changing the mean value of the upper and lower bounds. By this procedure, we recover approximate 90% confidence intervals for the cloud feedback that are asymmetric about the central estimate.

M17

M17 use their observed estimates of the sensitivity of cloud to *EIS* and *SST* to calculate a cloud fraction change for a 1 degree rise in local *SST* and 0.2 degree rise in *EIS*. As the ratio of *EIS* to *SST* changes is the same as our desired value, we only need to multiply their estimates by $\frac{dSST}{dT_g} = 0.7$ and the –1 W m^{-2} per % cloud fraction factor to yield the local cloud feedback according to (3). Their cloud sensitivities are calculated from observations in three types of regions: (a) 4 latitude bands between 40°N and 40°S, (b) 5 regions with predominately trade cumulus clouds, and (c) 5 regions that contain a mix of stratocumulus and trade cumulus clouds. M17 do not calculate any confidence intervals, and thus in Fig. 3 we report all their estimates without 90% confidence intervals.

Large-Eddy Simulation Cloud Feedbacks

For large-eddy simulations, we use the estimates of local cloud feedback from Table 1 of Bretherton (2015) along with his characterization of cloud regime simplified to either stratocumulus or trade cumulus. For simplicity, we characterize the LES for his transition regime as stratocumulus. We also include the estimate from the large-eddy simulation of precipitating trade cumulus in Vogel et al. (2016). Per degree of local *SST*, their simulations have a radiation change that spans the range of 0.3–0.55 W m^{-2} K^{-1}. We assign the feedback from this study to the midpoint of this range and multiply by $\frac{dSST}{dT_g} = 0.7$ to arrive at a value of the local cloud feedback of 0.3 W m^{-2} K^{-1} for this study. We note that the exact environmental changes forcing the large-eddy simulations vary in these studies. For example, some simulations omit changes to *EIS* (Vogel et al. 2016), while others may include changes in additional environmental parameters such as wind speed, subsidence and CO_2 concentration (Bretherton 2015). Comparison is justified based upon the expectation that the temperature response is the dominant factor contributing to the cloud feedback.

Global Climate Model Cloud Feedbacks

Figure 3 shows a range of climate model feedbacks for tropical low-cloud regions. This estimate was derived from Q15, Z15, M16 (second column of panel a in Fig. 2), and B16, each of whom examined the cloud responses to climate change simulated by climate models in each of their studied regions. Because the regions studied differ, the estimates of cloud feedback from climate models will differ. The cloud feedback estimates may also differ because these studies examined different model ensembles (Coupled Model Intercomparison Project Version 3 vs. Version 5), model experiments (scenarios such as A1B or the Representative Concentrations Pathway 4.5 or 8.5 versus idealized experiments such as the abrupt quadrupling or 1% per year increase of CO_2), and model variables (cloud fraction versus shortwave cloud-radiative effect). Cloud fraction sensitivities are converted to cloud feedbacks by multiplying by the previously mentioned factor of − 1 W m^{-2} per % cloud fraction. This is appropriate because Fig. 2 of Qu et al. (2014) showed that in climate models the sensitivity of the top-of-atmosphere shortwave cloud-radiative effect to cloud fraction in tropical low-cloud regions is close to this factor. The differing model experiments means that the rapid cloud adjustments to CO_2 are included in some studies. In addition to these four estimates, we also consider the average cloud feedbacks in tropical subsidence regions calculated from the abrupt CO_2 quadrupling simulations analyzed in Caldwell et al. (2016). These feedbacks primarily reflect the shortwave feedbacks from low clouds due to the absence of upper-level clouds. Despite these differences, the upper and lower bounds of the climate model values across the five studies differ by no more than 0.7 W m^{-2} K^{-1} from the multi-study mean values of − 0.8 and + 1.8 W m^{-2} K^{-1} shown in Fig. 3.

References

Andrews T, Gregory JM, Webb MJ (2015) The dependence of radiative forcing and feedback on evolving patterns of sea surface temperature change in climate models. J Clim 28:1630–1648. doi:10.1175/JCLI-D-14-00545.1

Bellomo K, Clement A, Mauritsen T, Radel G, Stevens B (2014) Simulating the role of subtropical stratocumulus clouds in driving Pacific climate variability. J Clim 27:5119–5131. doi:10.1175/JCLI-D-13-00548.1

Bellomo K, Clement A, Mauritsen T, Radel G, Stevens B (2015) The influence of cloud feedbacks on Equatorial Atlantic variability. J Clim 28:2725–2744. doi:10.1175/JCLI-D-14-00495.1

Blossey PN et al (2013) Marine low cloud sensitivity to an idealized climate change: the CGILS LES intercomparison. J Adv Model Earth Syst. doi:10.1002/jame.20025

Bony S, Dufresne J-L (2005) Marine boundary layer clouds at the heart of tropical cloud feedback uncertainties in climate models. Geophys Res Lett 32:L20806. doi:10.1029/2005GL023851

Bony S, Stevens B, Ament F, Bigorre S, Chazette P, Crewell S, Delanoe J, Emanuel K, Farrell D, Flamant C, Gross S, Hirsch L, Karstensen J, Mayer B, Nuijens L, Ruppert Jr. JH, Sandu I, Siebesma P, Speich S, Szczap F, Totems J, Vogel R, Wendisch M, Wirth M (2017) EUREC^{4}A: a field campaign to elucidate the couplings between clouds, convection and circulation. Surv Geophys. doi:10.1007/s10712-017-9428-0

Boucher O et al (2013) Clouds and aerosols. In: Stocker TF, Qin D, Plattner G-K, Tignor M, Allen SK, Boschung J, Nauels A, Xia Y, Bex V, Midgley PM (eds) Climate change 2013: the physical science basis. Contribution of working group I to the fifth assessment report of the intergovernmental panel on climate change. Cambridge University, Cambridge, pp 571–657

Bretherton CS (1993) Understanding Albrecht's model of trade cumulus cloud fields. J Atmos Sci 50:2264–2283

Bretherton CS (2015) Insights into low-latitude cloud feedbacks from high-resolution models. Philos Trans R Soc A. doi:10.1098/rsta.2014.0415

Bretherton CS, Blossey PN (2014) Low cloud reduction in a greenhouse-warmed climate: results from Lagrangian LES of a subtropical marine cloudiness transition. J Adv Model Earth Syst. doi:10.1002/2013MS000250

Bretherton CS, Blossey PN, Jones CR (2013) Mechanisms of marine low cloud sensitivity to idealized climate perturbations: a single-LES exploration extending the CGILS cases. J Adv Model Earth Syst 5:316–337

Brient F, Schneider T (2016) Constraints on climate sensitivity from space-based measurements of low-cloud reflection. J Clim 29:5821–5834. doi:10.1175/JCLI-D-15-00897.1

Brient F, Schneider T, Tan Z, Bony S, Qu X, Hall A (2016) Shallowness of tropical low clouds as a predictor of climate models' response to warming. Clim Dyn 47:433–449. doi:10.1007/s00382-015-2846-0

Brueck M, Nuijens L, Stevens B (2015) On the seasonal and synoptic time-scale variability of the North Atlantic trade wind region and its low-level clouds. J Atmos Sci 72:1428–1446

Caldwell PM, Bretherton CS (2009) Response of a subtropical stratocumulus-capped mixed layer to climate and aerosol changes. J Clim 22:20–38. doi:10.1175/2008JCLI1967.1

Caldwell PM, Zelinka MD, Taylor KE, Marvel K (2016) Quantifying the sources of intermodel spread in equilibrium climate sensitivity. J Clim 29:513–524. doi:10.1175/JCLI-D-15-00352.1

Ceppi P, McCoy DT, Hartmann DL (2016) Observational evidence for a negative shortwave cloud feedback in middle to high latitudes. Geophys Res Lett 43:1331–1339. doi:10.1002/2015GL067499

Chepfer H, Bony S, Winker D, Cesana G, Dufresne J, Minnis P, Stubenrauch C, Zeng S (2010) The GCM-oriented CALIPSO cloud product (CALIPSO-GOCCP). J Geophys 115:D00H16. doi:10.1029/2009JD012251

Christensen MW, Carri GG, Stephens GL, Cotton WR (2013) Radiative impacts of free-tropospheric clouds on the properties of marine stratocumulus. J Atmos Sci 70:3102–3118. doi:10.1175/JAS-D-12-0287.1

Deser C, Alexander MA, Xie S-P, Phillips AS (2010) 2010: sea surface temperature variability: patterns and mechanisms. Ann Rev Mar Sci 2:115–143. doi:10.1146/annurev-marine-120408-151453

De Szoeke SP, Verlinden KL, Yuter SE, Mechem DB (2016) The time scales of variability of marine low clouds. J Clim 29:6463–6481. doi:10.1175/JCLI-D-15-0460.1

Dufresne J-L, Bony S (2008) An assessment of the primary sources of spread of global warming estimates from coupled atmosphere–ocean models. J Clim 21:5135–5144. doi:10.1175/2008JCLI2239.1

Foster MJ, Heidinger A (2013) PATMOS-x: results from a diurnally corrected 30-yr satellite cloud climatology. J Clim 26:414–425. doi:10.1175/JCLI-D-11-00666.1

George RC, Wood R (2010) Subseasonal variability of low cloud properties over the southeast Pacific Ocean. Atmos Chem Phys 10:4047–4063. doi:10.5194/acp-10-4047-2010

Gordon ND, Klein SA (2014) Low-cloud optical depth feedback in climate models. J Geophys Res Atmos 119:6052–6065

Gregory JM, Andrews T (2016) Variation in climate sensitivity and feedback parameters during the historical period. Geophys Res Lett 43:3911–3920. doi:10.1002/2016GL068406

Gregory JM, Webb M (2008) Tropospheric adjustment induces a cloud component in CO_2 forcing. J Clim 21:58–71. doi:10.1175/2007JCLI1834.1

Grise KM, Medeiros B (2017) Understanding the varied influence of mid-latitude jet position on clouds and cloud radiative effects in observations and global climate models. J Clim. doi:10.1175/JCLI-D-16-00295.1 **(in press)**

Hakuba MZ, Folini D, Wild M (2016) On the zonal near-constancy of fractional solar absorption in the atmosphere. J Clim 29:3423–3440. doi:10.1175/JCLI-D-15-0277.1

Klein SA, Hall A (2015) Emergent constraints for cloud feedbacks. Curr Clim Change Rep 1:276–287. doi:10.1007/s40641-015-0027-1

Klein SA, Hartmann DL (1993) The seasonal cycle of low stratiform clouds. J Clim 6:1587–1606. doi:10.1175/1520-0442(1993)006,1587:TSCOLS.2.0.CO;2

Klein SA, Hartmann DL, Norris JR (1995) On the relationships among low-cloud structure, sea surface temperature, and atmospheric circulation in the summertime Northeast Pacific. J Clim 8:1140–1155

Klein SA (1997) Synoptic variability of low-cloud properties and meteorological parameters in the subtropical trade wind boundary layer. J Clim 10:2018–2039. doi:10.1175/1520-0442(1997)010<2018:SVOLCP>2.0.CO;2

Loeb N, Wielicki B, Doelling D, Smith G, Keyes D, Kato S, Manalo-Smith N, Wong T (2009) Toward optimal closure of the Earth's top-of-atmosphere radiation budget. J Clim 22:748–766. doi:10.1175/2008JCLI2637.1

Mace GG, Zhang Q, Vaughan M, Marchand R, Stephens G, Trepte C, Winker D (2009) A description of hydrometeor layer occurrence statistics derived from the first year of merged Cloudsat and CALIPSO data. J Geophys Res 114:D00A26. doi:10.1029/2007JD009755

Maddux BC, Ackerman SA, Platnick S (2010) Viewing geometry dependencies in MODIS cloud products. J Atmos Ocean Technol 27:1519–1528. doi:10.1175/2010JTECHA1432.1

Marchand R, Ackerman TP (2010) An analysis of cloud cover in multiscale modeling framework global climate model simulations using 4 and 1 km horizontal grids. J Geophys Res 115:D16207. doi:10.1029/2009JD013423

Mauger GS, Norris JR (2010) Assessing the impact of meteorological history on subtropical cloud fraction. J Clim 23:2926–2940. doi:10.1175/2010JCLI3272.1

McCoy DT, Eastman R, Hartmann DL, Wood R (2017) The change in low-cloud cover in a warmed climate inferred from AIRS, MODIS, and ECMWF-interim analyses. J Clim. doi:10.1175/JCLI-D-15-0734.1 **(in press)**

Myers TA, Norris JR (2013) Observational evidence that enhanced subsidence reduces subtropical marine boundary layer cloudiness. J Clim 26:7507–7524. doi:10.1175/JCLI-D-12-00736.1

Myers TA, Norris JR (2016) Reducing the uncertainty in subtropical cloud feedback. Geophys Res Lett 43:2144–2148. doi:10.1002/2015GL067416

Norris JR, Evan AT (2015) Empirical removal of artifacts from the ISCCP and PATMOS-x satellite cloud records. J Atmos Ocean Technol 32:691–702. doi:10.1175/JTECH-D-14-00058.1

Norris JR, Iacobellis SF (2005) North Pacific cloud feedbacks inferred from synoptic-scale dynamic and thermodynamic relationships. J Clim 18:4862–4878. doi:10.1175/JCLI3558.1

Nuijens L, Medeiros B, Sandu I, Ahlgrimm M (2015) Observed and modeled patterns of covariability between low-level cloudiness and the structure of the trade-wind layer. J Adv Model Earth Syst 7:1741–1764. doi:10.1002/2015MS000483

Pincus R, Platnick S, Ackerman SA, Hemler RS, Hofmann RJP (2012) Reconciling simulated and observed views of clouds: MODIS, ISCCP, and the limits of instrument simulators. J Clim 25:4699–4720. doi:10.1175/JCLI-D-11-00267.1

Pincus et al (this issue) The distribution of water vapor over low-latitude oceans: current best estimates, errors, and impacts. Surv Geophy (in press)

Platnick S, King MD, Ackerman SA, Menzel WP, Baum BA, Riedi JC, Frey RA (2003) The MODIS cloud products: algorithms and examples from Terra. IEEE Trans Geosci Remote Sens 41:459–473. doi:10.1109/TGRS.2002.808301

Qu X, Hall A, Klein SA, Caldwell PM (2014) On the spread of changes in marine low cloud cover in climate model simulations of the 21st century. Clim Dyn 42:2603–2626. doi:10.1007/s00382-013-1945-z

Qu X, Hall A, Klein SA, Caldwell PM (2015a) The strength of the tropical inversion and its response to climate change in 18 CMIP5 models. Clim Dyn 45:375–396. doi:10.1007/s00382-014-2441-9

Qu X, Hall A, Klein SA, DeAngelis AM (2015b) Positive tropical marine low-cloud cover feedback inferred from cloud-controlling factors. Geophys Res Lett 42:7767–7775. doi:10.1002/2015GL065627

Rieck M, Nuijens L, Stevens B (2012) Marine boundary layer cloud feedbacks in a constant relative humidity atmosphere. J Atmos Sci 69:2538–2550

Rossow WB, Schiffer RA (1999) Advances in understanding clouds from ISCCP. Bull Am Meteorol Soc 80:2261–2287

Rugenstein MAA et al (2016) Multiannual ocean-atmosphere adjustments to radiative forcing. J Clim 29:5643–5659. doi:10.1175/JCLI-D-16-0312.1

Schubert WH, Wakefield JS, Steiner EJ, Cox SK (1979) Marine stratocumulus convection. Part II: horizontally inhomogeneous solutions. J Atmos Sci 36:1308–1324

Seethala C, Norris JR, Myers TA (2015) How has subtropical stratocumulus and associated meteorology changed since the 1980s? J Clim 28:8396–8410. doi:10.1175/JCLI-D-15-0120.1

Shea DJ, Trenberth KE, Reynolds RW (1992) A global monthly sea surface temperature climatology. J Clim 5:987–1001

Sherwood SC, Bony S, DuFresne J-L (2014) Spread in model climate sensitivity traced to atmospheric convective mixing. Nature 505:37–42. doi:10.1038/nature12829

Sherwood SC et al (2015) Adjustments in the forcing-feedback framework for understanding climate change. Bull Am Meteorol Soc 96:228–277. doi:10.1175/BAMS-D-13-00167.1

Stevens B, Brenguier J-L (2009) Cloud-controlling factors: low clouds. In: Heintzenberg J, Charlson R (eds) Clouds in the perturbed climate system. MIT Press, Cambridge, pp 173–196

Stocker TF et al (2013) Technical summary. In: Stocker TF, Qin D, Plattner G-K, Tignor M, Allen SK, Boschung J, Nauels A, Xia Y, Bex V, Midgley PM (eds) Climate change 2013: the physical science basis. Contribution of working group I to the fifth assessment report of the intergovernmental panel on climate change. Cambridge University, Cambridge, pp 33–115

Terai CR, Klein SA, Zelinka MD (2016) Constraining the low-cloud optical depth feedback at middle and high latitudes using satellite observations. J Geophys Res Atmos 121:9696–9716. doi:10.1002/2016JD025233

van der Dussen JJ, de Roode SR, Gesso SD, Siebesma AP (2015) An LES model study of the influence of the free tropospheric thermodynamic conditions on the stratocumulus response to a climate perturbation. J Adv Model Earth Syst 7:670–691. doi:10.1002/2014MS000380

Vial J, Bony S, Stevens B, Vogel R (2017) Mechanisms and model diversity of trade-wind shallow cumulus cloud feedbacks: a review. Surv Geophys. doi:10.1007/s10712-017-9418-2

Vogel R, Nuijens L, Stevens B (2016) The role of precipitation and spatial organization in the response to trade-wind clouds to warming. J Adv Model Earth Syst 8:843–862. doi:10.1002/2015MS000568

Webb M, Lambert FH, Gregory JM (2013) Origins of difference in climate sensitivity, forcing, and feedback in climate models. Clim Dyn 40:677–707. doi:10.1007/s00382-012-1336-x

Wood R, Bretherton CS (2006) On the relationship between stratiform low cloud cover and lower-tropospheric stability. J Clim 19:6425–6432. doi:10.1175/JCLI3988.1

Zelinka MD, Klein SA, Hartmann DL (2012) Computing and partitioning cloud feedbacks using cloud property histograms. Part II: attribution to the nature of cloud changes. J Clim 25:3736–3754

Zelinka MD, Zhou C, Klein SA (2016) Insights from a refined decomposition of cloud feedbacks. Geophys Res Lett 43:9249–9259. doi:10.1002/2016GL069917

Zhai C, Jiang JH, Su H (2015) Long-term cloud change imprinted in seasonal cloud variation: more evidence of high climate sensitivity. Geophys Res Lett 42:8729–8737. doi:10.1002/2015GL065911

Zhang Y, Rossow WB, Lacis AA, Oinas V, Mishchenko MI (2004) Calculation of radiative fluxes from the surface to top of atmosphere based on ISCCP and other global data sets: refinements of the radiative transfer model and the input data. J Geophys Res 109:D19105. doi:10.1029/2003JD004457

Zhang M, Bretherton CS, Blossey PN, Bony S, Brient F, Golaz J-C (2012) The CGILS experimental design to investigate low cloud feedbacks in general circulation models by using single-column and large-eddy simulation models. J Adv Model Earth Syst 4:M12001. doi:10.1029/2012MS000182

Zhang M et al (2013) CGILS: results from the first phase of an international project to understand the physical mechanisms of low cloud feedbacks in single column models. J Adv Model Earth Syst. doi:10.1002/2013MS000246

Zhou C, Zelinka MD, Klein SA (2016) Impact of decadal cloud variations on the Earth's energy budget. Nat Geosci 9:871–874. doi:10.1038/ngeo2828

Surv Geophys (2017) 38:1331–1353
https://doi.org/10.1007/s10712-017-9418-2

Mechanisms and Model Diversity of Trade-Wind Shallow Cumulus Cloud Feedbacks: A Review

Jessica Vial[1] · Sandrine Bony[2] · Bjorn Stevens[3] · Raphaela Vogel[3]

Received: 13 January 2017 / Accepted: 3 June 2017 / Published online: 14 July 2017

Abstract Shallow cumulus clouds in the trade-wind regions are at the heart of the long standing uncertainty in climate sensitivity estimates. In current climate models, cloud feedbacks are strongly influenced by cloud-base cloud amount in the trades. Therefore, understanding the key factors controlling cloudiness near cloud-base in shallow convective regimes has emerged as an important topic of investigation. We review physical understanding of these key controlling factors and discuss the value of the different approaches that have been developed so far, based on global and high-resolution model experimentations and process-oriented analyses across a range of models and for observations. The trade-wind cloud feedbacks appear to depend on two important aspects: (1) how cloudiness near cloud-base is controlled by the local interplay between turbulent, convective and radiative processes; (2) how these processes interact with their surrounding environment and are influenced by mesoscale organization. Our synthesis of studies that have explored these aspects suggests that the large diversity of model responses is related to fundamental differences in how the processes controlling trade cumulus operate in models, notably, whether they are parameterized or resolved. In models with parameterized convection, cloudiness near cloud-base is very sensitive to the vigor of convective mixing in response to changes in environmental conditions. This is in contrast with results from high-resolution models, which suggest that cloudiness near cloud-base is nearly invariant with warming and independent of large-scale environmental changes. Uncertainties are difficult to narrow using current observations, as the trade cumulus variability and its relation to large-scale environmental factors strongly depend on the time and/or spatial scales at

✉ Jessica Vial
jessica.vial@lmd.jussieu.fr

[1] Laboratoire d'Océanographie et du Climat: Expérimentations et Approches Numériques (LOCEAN), Université Pierre et Marie Curie, Boîte 100 - 4, place Jussieu, 75252 Paris Cedex 05, France

[2] Laboratoire de Météorologie Dynamique (LMD), CNRS, Université Pierre et Marie Curie, Boîte 99 - 4, place Jussieu, 75252 Paris Cedex 05, France

[3] Max Planck Institute for Meteorology (MPI), Bundesstr 53, 20146 Hamburg, Germany

which the mechanisms are evaluated. New opportunities for testing physical understanding of the factors controlling shallow cumulus cloud responses using observations and high-resolution modeling on large domains are discussed.

Keywords Climate sensitivity · Global climate models · High-resolution models · Low-cloud feedbacks · Observations · Single-column models · Trade-wind shallow cumulus clouds

1 Introduction

Over the past decades, marine boundary-layer clouds have emerged as a central issue for the projection and understanding of anthropogenic climate change. Because shallow cumulus and stratocumulus clouds cover large areas of the tropical and subtropical oceans, their response to global warming substantially impacts the Earth's radiative budget. Climate models predict different low-level cloud responses to a warming climate, which results in a large dispersion in model-based estimates of climate sensitivity (Bony and Dufresne 2005; Webb et al. 2006). In the fifth Intergovernmental Panel on Climate Change (IPCC) assessment report (Boucher et al. 2013), global climate models (GCM) generally produce a positive low-level cloud feedback ranging between -0.09 and $0.63\ \mathrm{W\ m^{-2}\ K^{-1}}$ (Boucher et al. 2013; Zelinka et al. 2016), which is primarily associated with a reduction in low-level cloud cover (Rieck et al. 2012; Bretherton et al. 2013; Brient and Bony 2013; Webb and Lock 2013; Zhang et al. 2013; Qu et al. 2014; Zelinka et al. 2016). Despite the apparent robustness in the sign of the low-cloud feedback among GCM (Zelinka et al. 2016), climate models suffer from important systematic biases in the present-day representation of marine boundary-layer clouds (e.g., Nuijens et al. 2015b) and physical mechanisms underlying cloud changes sometimes operate differently depending on whether they are parameterized (as in GCM) or largely resolved (as in high-resolution models). As a result, the confidence in the sign of the low-cloud feedback and therefore in the magnitude of climate sensitivity remains fairly low (Vial et al. 2016; Sherwood et al. 2014; Brient et al. 2015).

Although boundary-layer clouds are an integral part of a tightly coupled system, the structure and dynamics of these clouds appear to depend primarily on local processes acting at timescales that are much shorter than the large-scale dynamics (Neggers 2015a). These processes, which include turbulent and convective mixing, cloud radiative forcing and microphysics, remain unresolved at the typical grid size of standard GCM and thus have to be represented through parameterizations. Unfortunately parameterizations remain limited and model-based estimates of low-level cloud feedback and climate sensitivity depend on how cloud-related processes are parameterized (Zhang et al. 2013; Qu et al. 2014; Vial et al. 2016).

The confidence attributed to low-level cloud changes in a warming climate can only be improved by advancing the comprehension of the key processes that influence these clouds, ideally to the point where our understanding of factors controlling the cloud response can be tested against data (Klein and Hall 2015). Moreover, better process understanding of low-cloud changes contributes to the development and/or improvement in physical parameterizations and thus to the reduction in systematic model biases. Important contributions arose from the analysis of low-cloud feedbacks across a hierarchy of numerical models (Wyant et al. 2009; Brient and Bony 2012; Rieck et al. 2012; Blossey et al. 2013; Bretherton et al. 2013; Webb and Lock 2013; Zhang et al. 2013; Medeiros et al. 2015; Tan et al. 2017), through perturbed-physics model experimentations (Watanabe et al. 2012; Brient and Bony 2013; Tomassini et al. 2014; Zhao 2014; Webb et al. 2015; Vial et al. 2016) and by the use of process-oriented diagnostics in models and observations (e.g., Brient et al. 2015; Nuijens et al. 2015b).

This review aims to synthesize what is known about marine boundary-layer cloud feedbacks from observation- and model-based studies, focusing on the physical understanding of processes underlying the cloud response of fair-weather cumulus. As these clouds are most frequently observed in the trade-wind regions, they are often referred to as trade cumulus. Because, in climate models, trade cumulus cloud feedbacks are governed to a large extent by changes in cloud fraction near cloud-base in a warming climate (Brient and Bony 2013; Brient et al. 2015; Vial et al. 2016), a better understanding of the mechanisms that control cloudiness at lowest levels deserves particular attention. A number of studies have addressed this question over the past decades, including global and high-resolution modeling, and observational studies. But it appears that the cloud controlling factors on present-day timescales and the cloud feedback mechanisms in response to climate perturbations remain uncertain in this specific cloud regime.

Whereas inconsistencies in the response of stratocumulus to warming are thought to arise from differences in the balance of opposing feedback processes that are increasingly well understood (Bretherton 2015), the diversity of model responses of fair-weather cumulus appears to be more related to fundamental differences in how processes operate in models with parameterized, as opposed to resolved convection. Accordingly, we structured this review paper so as to emphasize two divergent interpretations of trade cumulus cloud feedbacks and mechanisms, as they emerged across the past decades, from the perspective of large-scale model parameterizations or from the perspective of Large-Eddy Simulations (LES). In Sect. 2, we discuss the first perspective, derived from the analysis of GCM. It considers changes in cloud-base cloud fraction as the main driver of trade cumulus cloud feedbacks and brings out the important role of parameterized convective mass fluxes in the diversity of model responses. In contrast, the interpretation of shallow cumulus cloud feedbacks at the process scale, based on theoretical considerations (Sect. 3) and LES (Sect. 4), suggests that cloud-base cloud fraction remains nearly invariant in response to climate change perturbations and that uncertainty in cumulus cloud feedbacks among LES is primarily driven by cloud changes near the trade inversion. In Sect. 5, we attempt to use a unified framework for GCM and LES results, to better interpret these contrasting views of trade cumulus cloud feedbacks and help consider the issue from a broader perspective. Finally, in Sect. 6 we discuss observational support for model-based trade-wind cumulus cloud mechanisms and consider opportunities for more discriminating observational tests.

2 Interpreting Model Differences in Trade-Wind Cloud Responses to Warming in General Circulation Models

Because GCM are designed to simulate the evolution of the climate system at the global scale for hundreds of years, computational constraints limit the spatial resolution with which they can represent circulation systems. The effect of small-scale physical processes (such as turbulent and convective transports) on the resolved large-scale circulation must be parameterized. These parameterizations involve a large number of assumptions and numerical approximations that can affect the balance of the physical processes responsible for cloud formation and variability. This therefore causes large differences in cloud-topped boundary-layer structures among models (Brient et al. 2015; Nuijens et al. 2015b). Furthermore, at the time when parameterizations were developed for numerical weather prediction, the processes controlling low-level cloudiness were probably less of an interest as those clouds only represent a small contribution to the total cloud cover in many circulation regimes. Therefore, for the purpose of getting the total cloud cover right, parameterizations were tuned and harmonized to give a good representation of the present climate (e.g., Tiedtke 1989), which only indirectly constrains how cloud might respond to a changing climate.

2.1 Boundary-Layer Moisture Budget

To better understand the behavior of the parameterized physics within GCM, we consider the budget equation of moisture, which in its simplest form (Eq. 1) describes the time rate of change of water vapor (q) as a function of source and sink terms, namely condensation (c) and evaporation (e), respectively:

$$\frac{Dq}{Dt} = c - e \tag{1}$$

To solve this equation in a numerical model, we use its Eulerian form (Eq. 2), which then includes a local rate of change in q ($\partial q/\partial t$) and its evolution resulting from transport ($\mathbf{U} \cdot \nabla q$):

$$\frac{\partial q}{\partial t} + \mathbf{U} \cdot \nabla q = c - e \tag{2}$$

To solve Eq. (2) in a large-scale model, the transport term is separated into two different types of transport: one by resolved fluid motions ($\overline{\mathbf{U}} \cdot \nabla \overline{q}$) and the other by unresolved fluid motions ($\partial(\overline{\omega' q'})/\partial p$, assuming horizontal homogeneity). In a GCM, the unresolved fluid motions are further broken down into two terms (convection and turbulence), so that to get the evolution of q requires different parameterized processes to interact with one another in a consistent way. Thus, the budget equation of moisture in a GCM can be written as:

$$\frac{\partial \overline{q}}{\partial t} = -\left[(\overline{\mathbf{v}} \cdot \nabla \overline{q}) + \overline{\omega} \frac{\partial \overline{q}}{\partial p} \right]_{\mathrm{LS}} - \left. \frac{\partial(\overline{\omega' q'})}{\partial p} \right|_{\mathrm{turb}} - \left. \frac{\partial(\overline{\omega' q'})}{\partial p} \right|_{\mathrm{conv}} - (\overline{c} - \overline{e}) \tag{3}$$

where physical parameterized processes affecting specific humidity and thus low-level clouds in subsidence regimes usually arise from separate schemes for turbulent diffusion in the boundary layer (turb), convection (conv) and net grid-scale condensation ($c - e$, which includes cloud formation, precipitation and evaporation and thus determines to a large extent the conversion to cloud water).

Large-scale low-level divergent winds in subsidence regimes act to export mass out of the boundary layer, which lowers the boundary layer. This is compensated by turbulent mixing that deepens and then dries the boundary layer as dry free tropospheric air is entrained into the boundary layer. In steady-state climates, this drying effect is compensated by moistening from the sum of the physical processes: the turbulence scheme is a source of moisture at lowest tropospheric levels, the convection scheme (when it is active) vertically transports moisture over the depth of the trade-wind layer from cloud-base up to overlying layers below the inversion or in the lower free troposphere and thus dries at levels near cloud-base (this transport is now commonly called *lower-tropospheric convective mixing* or *shallow convective mixing*), and the condensation scheme, which is the direct source of cloud water, is usually a sink term for the boundary-layer moisture budget.

Coordinated multi-model intercomparison studies such as those conducted by CFMIP (the Cloud Feedback Model Intercomparison Project; Webb et al. 2016) offer a way to sample model structural uncertainties for a given idealized framework and perturbation. The single-column model (SCM) intercomparison carried out as part of the CGILS (CFMIP-GASS Intercomparison of LES and SCM; Zhang et al. 2012, 2013; Blossey et al. 2013; Bretherton et al. 2013) project focused on marine boundary-layer clouds under idealized large-scale forcings representative of three different cloud regimes. This review focuses on those cases where cumulus convection plays a role in the coupling.

Different models balance their moisture budgets in regions of shallow cumulus in very different ways. This is illustrated in Fig. 1 (taken from a regime of mixed cumulus and stratocumulus convection), where differences in the convective mixing terms (tendencies) stand out when comparing how models maintain the present state and its response to warming. The ways in which these different balances influence the response to warming can be seen by considering what happens in a warmer climate. Because surface latent heat fluxes are expected to increase with warming (by about 2%/K—cf. Qu et al. 2015; Tan et al. 2017), we expect a larger turbulent moisture flux convergence in the cloud layer. In addition, the large-scale subsidence is reduced owing to the weakening of the tropical circulation. These two effects lead to increased cloud water (thicker and/or more abundant clouds). However, when convection plays a role, the enhanced moistening via turbulence and large-scale vertical advection is to a large extent compensated by enhanced drying from the export of condensate and the shallow convection (in a warming climate). If the rate of drying from the shallow convection is greater than the rate of moistening from turbulence and large-scale vertical advection, then we expect less condensation and less cloudiness, which would constitute a positive cloud feedback on the radiative forcing (as in Fig. 1c). Zhang et al. (2013)'s findings suggest that cloud feedbacks tend to be negative in

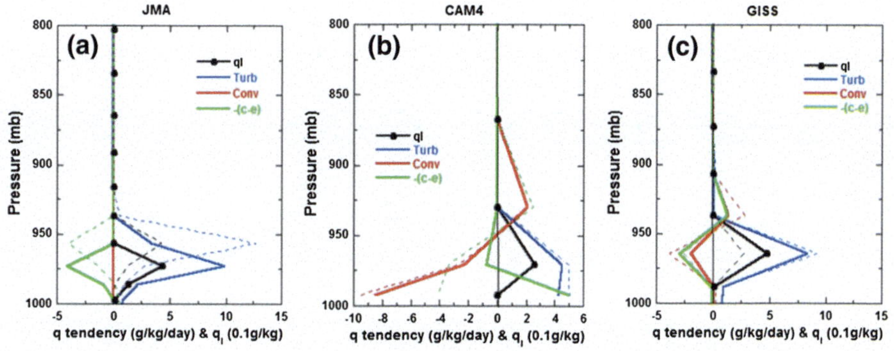

Fig. 1 Physical tendencies of moisture (in g kg^{-1} day^{-1}) for decoupled stratocumulus (s11) in the present-day climate (*solid lines*) and in a warmer climate (*dash lines*): *turb* for the turbulence scheme, *conv* for the convection scheme, *c–e* for the net condensation scheme. *ql* represents the grid-averaged cloud liquid water (0.1 g kg^{-1}, *black dotted line*). A sample of three SCM, having very different behaviors, is shown: **a** JMA (Japan Meteorological Agency), **b** CAM4 (Community Atmospheric Model Version 4), and **c** GISS (Goddard Institute for Space Studies). Note that although these profiles apply to decoupled stratocumulus, the sampled model diversity presented here remains relevant for shallow cumulus clouds. From Zhang et al. (2013)

models where parameterized convection is not playing an important role in balancing the moisture budget. The inter-model spread in this cloud regime for this SCM intercomparison is presented in Fig. 2 (in yellow). This large model diversity in shallow cumulus cloud feedbacks is primarily due to differences in cloud fraction changes at lowest atmospheric levels, where the effect of convective drying is the most important.

2.2 The Role of Shallow Convective Mixing

In a warmer climate, the enhanced rate of drying by the shallow convection is similar to the *thermodynamic response* described by Rieck et al. (2012), Blossey et al. (2013) and Bretherton et al. (2013) on the basis of their analysis of LES results. More specifically, it was found that when just a surface (and/or atmospheric) warming is applied (while keeping the subsidence unchanged), the moisture gradient between the saturated air at surface and the drier free tropospheric air increases, yielding more efficient drying of the boundary layer by cloud top entrainment and/or vertical mixing by shallow convection (for a given entrainment/mixing rate). It is noteworthy as well that, in both LES and GCM, the presence of a stronger humidity gradient can also be interpreted as an enhanced subsidence drying (from an Eulerian point of view, which takes the equilibrium depth of the boundary layer fort granted); this provides an additional drying on top of the convective drying.

To better understand how convective mixing influences cloud amount, Vial et al. (2016) developed an analysis framework which allowed them to explore how changes in the convective mixing influence cloudiness in conditions reminiscent of trade cumulus convection. Using a single-column configuration of the Institut Pierre Simon Laplace (IPSL) model, they performed experiments using two different convective parameterization schemes. Their framework starts from the well-recognized result that the boundary-layer cloud fraction is mainly influenced by two antagonistic mechanisms: (1) the shallow

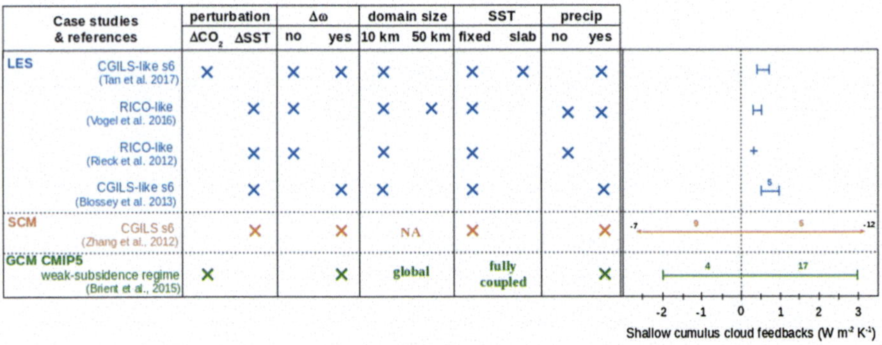

Fig. 2 Synthesis of trade-wind shallow cumulus cloud feedback strength (in W m^{-2} K^{-1}) as simulated by different types of numerical models: LES (*blue*), SCM (*orange*) and CMIP5 GCM (*green*). For LES/SCM, different case studies are considered: CGILS-like s6 (Zhang et al. 2012; Blossey et al. 2013; Tan et al. 2017) and RICO-like (Rieck et al. 2012; Vogel et al. 2016). For each study, we provide, where applicable, information on the perturbed experiment used as surrogate for climate change (ΔSST or ΔCO_2), if the large-scale subsidence (ω) is perturbed or not, the domain size (small domain of $\sim$10 km or large domain of $\sim$50 km), if SST is prescribed (fixed SST) or interactive (the atmosphere is coupled to a slab ocean), and if precipitation is allowed or not. For multi-model studies, we indicate the number of models that simulate a positive or negative feedback (*colored numbers* on the *top* of the *arrow bars*). The *black numbers* at the extremities of the SCM *arrow bar* correspond to the minimum and maximum values of the simulated feedbacks

convective mixing that dries the lower atmosphere and reduces the cloud fraction (Stevens 2007; Rieck et al. 2012; Zhang et al. 2013; Brient et al. 2015) and (2) the boundary-layer turbulent moistening (or latent heat flux) that enhances the cloud amount at low levels (Rieck et al. 2012; Webb and Lock 2013; Zhang et al. 2013; Brient et al. 2015). They thus expressed the sensitivity of the boundary-layer cloud fraction ($\mathrm{d}f$) to a change in convective mixing ($\mathrm{d}\mu$) and latent heat flux (E) as:

$$\mathrm{d}f = \mathcal{C}\mathrm{d}\mu + \mathcal{T}\mathrm{d}E \tag{4}$$

where the first term on the right-hand side describes the sensitivity of cloud fraction to convective ($\mathcal{C}$) mixing, the second to turbulent ($\mathcal{T}$) mixing. The model thus attempts to encapsulate the interplay between the two parameterizations used to model the transport of eddies as in Eq. (3). More specifically:

- $\mathcal{C}$ is the reduced cloud fraction when lower-tropospheric convective drying is enhanced under the effect of increased mixing ($\mathcal{C} \equiv \left.\frac{\partial f}{\partial \mu}\right|_E < 0$)
- $\mathcal{T}$ is the increased cloud fraction when lower-tropospheric turbulent moistening is enhanced through increased latent heat flux ($\mathcal{T} = \left.\frac{\partial f}{\partial E}\right|_\mu > 0$)

Using a series of sensitivity experiments, they showed that it was possible to linearly relate the surface latent heat fluxes to changes in the convective mixing ($\mathrm{d}\mu$) and changes in the net boundary-layer cloud radiative effect ($\mathrm{d}R$) as:

$$\begin{aligned} \mathrm{d}E &= \lambda\mathrm{d}\mu + \lambda_r\mathrm{d}R \\ \mathrm{d}E &= (\lambda + \alpha\mathcal{C}\lambda_r)\mathrm{d}\mu \end{aligned} \tag{5}$$

where the variations in the net cloud radiative effect are essentially driven by the longwave cloud radiative cooling ($R > 0$ by convention) and linearly related to $\mathrm{d}f$, such as $\mathrm{d}R = \alpha\mathrm{d}f = \alpha\mathcal{C}\mathrm{d}\mu + \alpha\mathcal{T}\mathrm{d}E$ [see Vial et al. (2016) for more details on the simplifications that lead to the final form of Eq. (5)].

In Eq. (5), λ and λ_r describe the two additional mechanisms that influence the latent heat flux, which can then modulate the sensitivity in boundary-layer cloud fraction to a change in convective mixing [see Vial et al. (2016) for more details on how λ and λ_r are defined; here we just provide their physical description]:

- λ is the increased latent heat flux through lower-tropospheric drying induced by the convective mixing ($\lambda > 0$), which damps the reduction in cloudiness.
- λ_r is the reduced latent heat flux as the lower troposphere stabilizes under the effect of reduced low-cloud radiative cooling ($\lambda_r > 0$), which enhances the reduction in cloudiness.

By replacing $\mathrm{d}E$ into Eq. (4), the sensitivity of the boundary-layer cloud fraction to a change in convective mixing can be expressed as:

$$\mathrm{d}f = [\mathcal{C} + \mathcal{T}(\lambda + \alpha\mathcal{C}\lambda_r)]\mathrm{d}\mu \tag{6}$$

Using Eq. (6), the relative importance that the model assigns to the two processes (i.e., convective mixing and radiative cooling) can thus be measured by the magnitude of λ and λ_r. In the IPSL model, this depends to some extent on the closure of the convective parameterization. When this model uses a closure in stability (e.g., the convective available potential energy—CAPE), it exhibits a stronger sensitivity of low-level clouds to convective mixing in the present-day climate and a stronger low-level cloud feedback in response to surface warming, due to the prevailing coupling between latent heat flux and cloud radiative cooling (λ_r). In contrast, when the IPSL model is run using a closure in subcloud moisture convergence, the coupling between latent heat flux and convective mixing (λ) dominates, which results in a lower sensitivity of cloudiness to convective mixing in the present-day climate and a weaker low-cloud feedback in a warming climate (Vial et al. 2016).

However, the closure of the convective parameterization is not the only assumption that can affect boundary-layer cloud feedbacks. In the CGILS SCM intercomparison (Zhang et al. 2013), two models having the same closure of the convective parameterization (CAPE) exhibit cloud feedbacks of opposite signs (the models differ also by entrainment/detrainment assumptions: one model includes lateral entrainment into the convective plumes, while the other does not). It is very challenging to determine how the different parameterizations fix the behavior of boundary-layer clouds, because they all are tightly connected to each other and with other parameterized and/or resolved processes (e.g., Vial et al. 2016). That said, this illustrates how different parameterization assumptions can affect the balance of the physical processes and boundary-layer cloud feedbacks, often in ways that were not considered when the schemes were designed. Following the Zhang et al. (2013) study, other process-oriented studies have then suggested that shallow convective mixing (and also more generally parameterized convection) appears as a leading source of inter-model spread in cloud feedbacks (Sherwood et al. 2014; Brient et al. 2015; Kamae et al. 2016; Vial et al. 2016).

Although convection is likely an important source of model diversity in the response of clouds in some regimes, the importance of other processes can also be important. This is shown for instance in experiments wherein convective cloud parameterizations are eliminated (Webb et al. 2015) and support the idea that the treatment of turbulence and cloud radiative effects also influences the evaporation and cloud amount (Vial et al. 2016).

Brient et al. (2015) have proposed another mechanism that could influence the change in convective mixing in a warmer climate, and thus the low-cloud feedback. Based on their analysis of the Coupled Model Intercomparison Project (CMIP5, Taylor et al. 2012) ensemble, they argue that increased near-surface stability in a warming climate weakens the sensible heat flux and limits the increase in latent heat flux. This in turn reduces the buoyancy flux and yields a shallowing of moisture mixing (due to weaker turbulent mixing) within the boundary layer and thus a shallowing of low-level clouds (with only subtle changes in cloud fraction). In their study, about half of the models favor this mechanism with respect to enhanced lower-tropospheric convective mixing as a result of increased surface evaporation. For these models, the low-cloud feedback is weaker (less positive). In contrast, in models where the changes in surface fluxes are more strongly related to changes in the trade-wind vertical humidity gradient (rather than near-surface stability), the moisture mixing deepens, yielding deeper clouds with a reduced cloud fraction at lowest levels and a more positive cloud feedback. In all models, the convective mixing is enhanced in a warmer climate, but models that simulate a low-cloud shallowing, with warming, are more influenced by the weakening of turbulent mixing (due to reduced surface sensible heat flux) and models that simulate a low-cloud deepening with warming are more influenced by the strengthening of convective mixing (due to increased surface evaporation).

A number of recent studies have used observations to evaluate which of the hypothesized mechanisms better describe the cloud response to changes in large-scale environmental conditions (e.g., Clement et al. 2009; Qu et al. 2014, 2015; Brient and Schneider

2016). These studies generally indicate that it might be the lower-troposphere mixing, although a complete demonstration of this mechanism using current observations remains difficult (this is a point we return to in Sect. 6).

The above discussion reflects our understanding of shallow cumulus cloud feedbacks and mechanisms from the perspective of large-scale model parameterizations of the trade-wind boundary layers (in GCM and SCM). In those models, cloudiness near cloud-base is the main driver of shallow cumulus cloud feedbacks and is strongly controlled by local interplays between turbulent, convective and radiative processes as a response to changes in large-scale environmental factors (e.g., surface/atmospheric temperature, vertical humidity gradient, subsidence). This is in contrast to what one finds in high-resolution modeling (e.g., LES), in which cloud fraction near cloud-base is nearly invariant with warming and independent of large-scale environmental factors that vary on long time-scales. As a result, trade cumulus cloud feedbacks as simulated by LES are much smaller than usually simulated in GCM or SCM (Fig. 2). As discussed in the following sections, this contrasting behavior between GCM and LES appears to be related to the fact that large-scale climate models might lack cloud-base regulation processes between the cloud and subcloud layer, which in nature act to couple the turbulent fluxes in the subcloud layer with the convective fluxes within the cloud layer. In the following section, we provide the theoretical background used to rationalize the apparent constancy in trade-wind cloud fraction near cloud-base. Shallow cumulus cloud changes and mechanisms as simulated by LES are then reviewed in Sect. 4.

3 A Mass Budget Perspective on Cloud-Base Cloud Fraction

Unlike what happens in most large-scale models, conceptual models of the layers of shallow convection [e.g., single-bulk layer models for the entirety of the trade-wind layer in Betts and Ridgway (1989) or subcloud layer models in Betts (1976)] emphasize how exchanges between the cloud and subcloud (well mixed) layers adjust the amount of mass in the subcloud layer so that its height remains close to the lifting condensation level (LCL). Such a process would imply that the humidity at cloud-base remains roughly constant. A closure of this form was used in early models of trade-wind cumulus (Albrecht et al. 1979; Betts and Ridgway 1989; Stevens 2006). By immediately adjusting the sub-cloud layer height to the LCL, these models essentially fix the humidity at cloud-base and by implication allow little room for cloudiness at cloud-base to vary with the cloud-base convective mass flux, M.

The mass budget of the subcloud layer (illustrated in Fig. 3) provides the theoretical backdrop for this idea. Neglecting variations of density, ρ, within the shallow subcloud layer, the total mass (per unit area) of the layer can be written as ρh, where h is the depth of the layer, and

$$\rho \frac{\mathrm{d}h}{\mathrm{d}t} = \frac{1}{g}[\mathcal{E} - \omega - M] \tag{7}$$

This equation recognizes three source or sink terms: (1) the entrainment ($\mathcal{E} > 0$) of air from the cloud layer into the subcloud layer, a mass source (of relatively dry and warm air); (2) the large-scale divergence of mass out of the layer, which by continuity is equal to the large-scale subsidence velocity (ω) at h, for $\omega > 0$ a mass sink; and (3) a convective mass flux ($M > 0$), whereby cumulus convection evacuates mass out of the subcloud layer, a

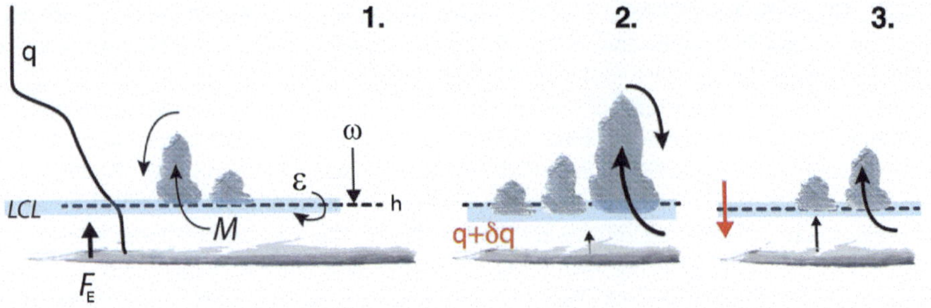

Fig. 3 Illustration of the cumulus-valve mechanism. In (*1*) the system is at equilibrium, with a surface evaporation flux (F_E), a typical trade-wind humidity profile (q) roughly constant up to the top of the well-mixed layer at h (*dotted line*), clouds starting to form at the lifting condensation levels (LCL, represented by the *blue layer*) and the processes controlling the mass budget of the well-mixed layer (Eq. 7): the entrainment rate at h ($\mathcal{E}$), the large-scale subsidence at h (w) and the convective mass flux (M). In (*2*) the humidity profile within the well-mixed layer is increased by dq. This reduces the surface evaporation, lowers the LCL, enhances the fraction of air parcels (including the cloud core fraction, a_c) and through the mass flux closure ($M = a_c w_c$) increases M. Larger M transports more moisture upward, which deepens trade-wind clouds and also yields more downward mixing of dry free tropospheric air to the mixed layer. In (*3*) a new equilibrium is reached whereby increased mixing has lowered the mixed-layer top close to the LCL again. Therefore, the fraction of saturated air parcels is reduced again (including a_c) and thus M is weakened. From Nuijens et al. (2015b)

further sink. Assuming that the subcloud layer is well mixed, and neglecting downdrafts, only the entrainment term changes the properties of the subcloud layer air. The other source terms in Eq. 7 export mass with the same properties as the subcloud layer.

Neggers et al. (2006) adjust the subcloud layer height, h, to the LCL through a closure on M. This cloud-base mass flux can be interpreted as being composed of the product of an effective area of convective active mass export out of the subcloud layer, a_c, and the mean velocity of this export, w_c, such that

$$M = \rho g(a_c w_c). \tag{8}$$

The Neggers et al. (2006) closure for M follows by parameterizing w_c as being proportional to the convective scale velocity, $w_c \propto (h\mathcal{B})^{1/3}$ where $\mathcal{B}$ is the surface buoyancy flux and a_c as being proportional to the disequilibrium between the LCL and h, or the humidity at the top of the subcloud layer. This means that, for a given w_c; the larger the difference between h and the LCL, the larger is a_c, and hence the larger is M.

To understand how this closure maintains h near the LCL, consider the perturbed scenario whereby the humidity of the subcloud layer is increased. As a result, the LCL will lower and the surface fluxes will decrease. The reduction in $\mathcal{B}$ has a small effect on w_c but this is more than offset by the increase in a_c arising from the larger difference between the LCL relative to h. As a result M is increased, thereby exporting more mass out of the subcloud layer and lowering h, bringing it closer to the LCL. This process is also illustrated schematically in panels 2 and 3 in Fig. 3. Note that the moistening of the subcloud layer also affects the entrainment term, both by changing the surface fluxes and slightly affecting the stability at the top of the subcloud layer, but for the purpose of our discussion these can be considered to be negligible. In practice, this mechanism can be thought of as a moisture convergence closure on M. It is sometimes called the *cumulus-valve mechanism* because the clouds act as a valve which helps maintain the top of the subcloud layer, h, close to the

LCL and thus acts as a negative feedback of convection on the humidity, and presumably cloudiness, at the base of the cumulus layer.

The cumulus-valve mechanism has been evaluated at a specific shallow cumulus location but also in an tropical climate model with full interaction with the large-scale flow. Neggers et al. (2006) argue that the cloud fractions that result from the implementation of this closure are consistent with what is known about the climatology of shallow cumulus clouds from observations. A close inspection of their results shows that a_c indeed increases with M, which implies a relatively moister subcloud layer, as air-moves though the trades over warmer waters. This in contrast to what one finds in parameterizations used in many climate models, whereby increasing M, without adjusting h (which given the coarseness of the Eulerian coordinate in most parameterizations is hardly possible) dries, instead of shoals, the subcloud layer. This happens because, by not resolving variations in h, any convective mass out of the layer has to (by definition) be compensated by a flux of mass in (i.e., an implicit entrainment). Therefore, the GCM parameterizations effectively are increasing $\mathcal{E}$ to compensate for an increase in M; increased entrainment dries and warms the subcloud layer. A more careful accounting for the terms influencing the boundary-layer mass budget would (in the absence of a downdraft mass flux) not imply a subcloud layer drying, but rather a shoaling.

The above discussion illustrates how, when it comes to the humidity of the subcloud layer, old debates regarding closures for the convective mass flux have, it seems, unintended implications. In particular, the idea of the cumulus valve raises the question as to whether the strongly negative coupling between low-level cloudiness and convective mixing in many climate models (as shown in Sherwood et al. 2014; Brient et al. 2015; Vial et al. 2016; Kamae et al. 2016) may be a consequence of parameterizing the convective mass flux in a manner that does not sufficiently account for its link to the mass budget of the subcloud layer. Based on these ideas, and (as discussed in the following sections) the support they receive from measurements and large-eddy simulations, it is tempting to argue that many climate models generate cloud-base cloud fractions that are overly, or even wrongly, sensitive to the magnitude of the cumulus mass flux. In the case of the measurements, the lack of observations of key terms, such as the mass flux, hinders a conclusive interpretation using this framework (Bony et al. in revision). Evidence from LES presumes that the relative humidity at the top of the subcloud layer is the best determinant of cloud amount at cloud-base, and that LES—whose predictions of cloud-base cloud amount have not been critically evaluated against data (see Bony et al. in revision)—is a good surrogate for nature.

4 High-Resolution Simulation of Shallow Cumulus Cloud Changes and Mechanisms

Unlike in climate models, where cloudiness near cloud-base is strongly controlled by convective and turbulent parameterizations as a response to changes in the large-scale environment (such as subsidence, surface temperature and vertical gradient of humidity—see Sect. 2), large-eddy simulation aims to explicitly resolve these convective and turbulent processes. Until quite recently computational restrictions only permitted LES over relatively small domains, which then required the parameterization of larger scale processes, usually by assuming that they can be specified independently of how turbulent and convective processes themselves develop. LES over larger domains are increasingly relaxing this assumption. Here we review what we know about shallow cumulus from LES, and whether LES is indeed doing a good job at capturing the observed vertical distribution and variability of shallow cumulus cloudiness. In so doing, we evaluate to what extent we can reject the strong cloud-base response to warming seen in many climate models, or at least what observations would be required to improve confidence in one or the other hypothesis.

4.1 Trade-Wind Shallow Cumulus Cloud Response to Warming in LES

Overall, LES studies exhibit very small changes in cloudiness near cloud-base in response to surface and/or atmospheric warming. This suggests that the cumulus-valve mechanism (Sect. 3) may robustly constrain cloudiness at cloud-base in response to strong climate change perturbations (up to 8 K surface and atmospheric warming in Rieck et al. 2012). On the other hand, and unlike current climate models, LES models show that cloud changes near the inversion are the primary contributor to the total change in cloud cover (Rieck et al. 2012; Blossey et al. 2013; Vogel et al. 2016). The corresponding changes in cloud radiative effects appear robustly positive among LES studies, but much smaller than changes routinely simulated in global or single-column models (Fig. 2).

Cloud changes in a warming climate along with moistening tendencies in LES are illustrated in Figs. 4 and 5 (taken from Vogel et al. 2016).

In the absence of mesoscale organization and precipitation, the response of trade-wind cumulus to warming, as represented by LES, can be understood through simple bulk arguments (Rieck et al. 2012; Vogel et al. 2016). In a warmer climate, larger absolute humidity gradients imply that for a boundary layer of the same depth, which thus has the same rate of deepening to balance an assumed constant subsidence, the entrainment drying

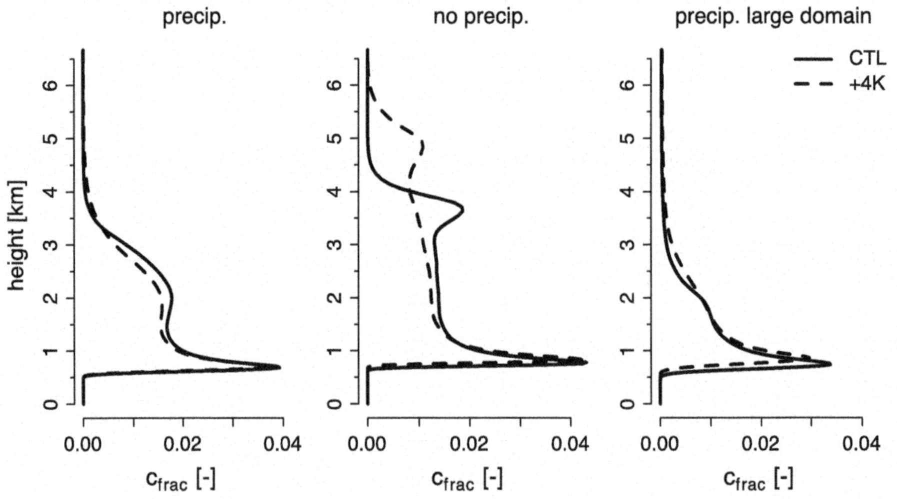

Fig. 4 Domain-averaged vertical profiles of trade-wind shallow cumulus cloud fraction in LES (using the University of California Los Angeles—UCLA model) of precipitating clouds (*left*) and non-precipitating clouds (*middle*) over a small domain (~ 13 km), and precipitating clouds over a larger domain (~ 50 km; *right panel*). The experimental setup is similar to that in Bellon and Stevens (2012) and consists of prescribed initial conditions and large-scale forcings (sea surface temperature, subsidence, radiative cooling and geostrophic wind) representative of the trade-wind regions. Results are shown for the present-day climate (*solid*) and as a response to a uniform warming of +4K at constant relative humidity (*dashed*). Figure adapted from Vogel et al. (2016)

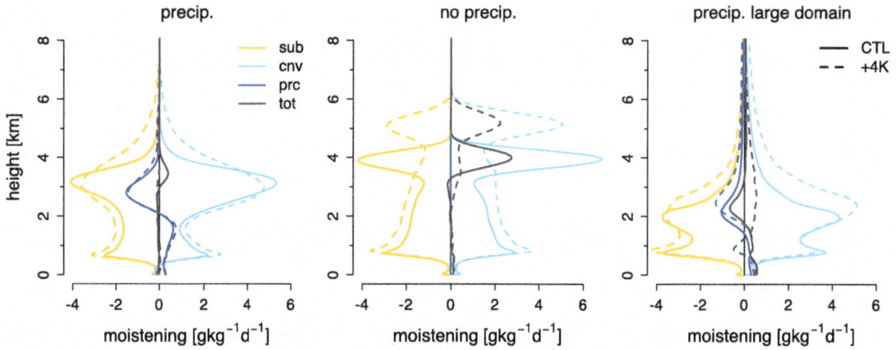

Fig. 5 Domain-averaged vertical profiles of moisture tendencies: *sub* for the large-scale subsidence (*orange*), *cnv* for convection (diffusive and advective processes—in *cyan*), *prc* for precipitation (*blue*) and *tot* for the total moisture tendency (*black*). From left to right is for precipitating and non-precipitating simulations on a small domain ($\sim$13 km), and the precipitating simulation over a larger domain ($\sim$50 km). The same experimental setup as for Fig. 4 is used. Results are shown for the present-day climate (*solid*) and as a response to a uniform warming of +4K at constant relative humidity (*dashed*). Figure adapted from Vogel et al. (2016)

is larger. Stationarity implies a drier boundary layer so as to induce a sufficiently large moisture flux to balance this increased rate of entrainment drying (Fig. 5). But a larger moisture flux also implies deeper mixing and more drying, enhancing these effects further so that the equilibrium in a warmer atmosphere evolves to a deeper and drier cloud layer, with a reduction in cloudiness above $\sim$1.5 km (Fig. 4).

This is a typical view of shallow cloud feedbacks, which is similar to the thermodynamic mechanism for stratocumulus cloud reduction reviewed in Bretherton (2015), and similar to the thermodynamic response in many climate models (Sect. 2). However, the above arguments neglect precipitation, which introduces a new process in the balance of the water budget. Precipitation also affects the assumed structure of the boundary layer and the spatial organization (e.g., Seifert and Heus 2013). Motivated by these findings, Vogel et al. (2016) performed LES experiments to study the response of trade-wind cumulus clouds to warming for non-precipitating and precipitating shallow cumulus clouds. They also performed simulations on a large domain of about $50 \times 50\,\mathrm{km}^2$ to better understand the role of organization. How these processes change our view of the balances determining cloudiness in the trade-wind layer are discussed below.

4.1.1 The Role of Precipitation

Studies of precipitating shallow cumulus (e.g., Blossey et al. 2013; Bretherton et al. 2013; Vogel et al. 2016) suggest that the main effect of precipitation is to restrain the deepening of the trade-wind layer, as explained in Stevens and Seifert (2008). Secondary effects arise from changes in the inversion and subcloud layer. With more precipitation, the cloud layer is more stable, but the inversion layer is less stable, so that clouds tend to detrain more continuously, leading to less stratiform cloudiness at the top of the cloud layer (compare left and middle panels in Figs. 4, 5). In addition, evaporation of precipitation in the lower part of the cloud layer induces a moistening and cooling, which yields an increase in cloud fraction near cloud-base compared to non-precipitating simulations. In a related study, Seifert and Heus (2013) explored the response of clouds to precipitation amount, rather than the differences between precipitating and non-precipitating simulations. They found that increasing precipitation leads to a reduction in cloud fraction over the whole trade-wind layer, including at cloud-base. Not withstanding that many of the responses to precipitation make physical sense, the magnitudes of the changes are not straightforward to assess. This is because, as a growing literature suggests, these are sensitive to the details of how the simulations are set up, ranging from the choice of microphysical schemes (Bretherton et al. 2013; Seifert and Heus 2013) to the effects of mean wind and resolution (Stevens and Seifert 2008; Matheou et al. 2011; Seifert and Heus 2013)—this point is further discussed in Sect. 4.2.

For precipitating layers, the response to warming is complicated by what are, at times, very strong changes in precipitation. For instance, in the warmer climate state of Vogel et al. (2016), increased surface fluxes with warming lead to congestus clouds developing with tops up 7 km. These dramatically change the structure of the boundary layer, weakening the inversion associated with more trade-wind-like clouds and limiting stratiform cloud formation. More compensating subsidence also leads to a shallowing and drying of the cloud layer, reducing cloud amount near cloud-base (Fig. 4).

4.1.2 *The Role of Organization*

Larger domain simulations ($\sim$50 kilometers as in Seifert and Heus 2013; Vogel et al. 2016) allow shallow convection to organize in clusters of variable depth (depending on the domain vertical extension). The reasons for this organization are still being debated, but phenomenologically it shares similarly with convective self-aggregation as seen in simulations of deep convection (Wing et al. 2017). Clouds organized in clusters tend to produce larger amounts of precipitation, which generates evaporative downdrafts and initiates cold pools that spread out and trigger new convective cells at the cold pool boundary, where subsequent shallow cumulus clouds form. Because most of the precipitation remains concentrated in the convective clusters that populate the moist regions of the domain (e.g., in Vogel et al. 2016), evaporation of precipitation is reduced although the cloud layer is overall drier. These processes can also influence the response of clouds to warming.

The greater precipitation efficiency that accompanies mesoscale organization leads to a more stable and drier trade-wind layer. In addition, and with the help of compensating subsidence in the drier area, this effect keeps the trade-wind clouds in the rest of the domain shallow. Therefore, in the presence of organized convection, the trade-wind boundary layer is drier and more stable, and trade-wind cumulus clouds are shallower, compared to when shallow cumulus clouds are more randomly distributed in space (in smaller domain simulations). In a warming climate, upward convective transport of moisture strengthens in the large domain simulations, comparable to the small domain simulations (Fig. 5). Because the amount of deep cloud clusters is enhanced with warming, precipitation, as mentioned above, increases much more strongly with warming than in the small domain simulations (Vogel et al. 2016). Thereby, the drying due to precipitation tends to replace the drying due to large-scale subsidence, which cannot efficiently balance the enhanced convective moistening in the presence of a few deep cloud clusters and an otherwise very shallow and dry trade-wind layer (Vogel et al. 2016). As a result, different changes in cloud fraction and vertical distribution occur in the larger domain: clouds become deeper with a reduced cloud fraction near cloud-base—a feature that is not captured in the smaller domain (Fig. 4), yet is reminiscent of the dynamics seen in parameterizations (Sect. 2).

Does the reduction in cloud-base cloud fraction with warming imply that the presence of organized clusters in larger domain simulations can trigger mechanisms that overcome the internal cumulus-valve mechanism? Recently, Neggers (2015b) has shown that a mass flux framework that takes into account the spatial distribution of cumulus horizontal sizes can introduce interactions between convective plumes of different sizes (see also Seifert et al. 2015). In particular, if large cumulus clouds are more abundant than small clouds, the vertical convective fluxes tend to dry at low levels and transport moisture to higher levels. This low-level drying is compensated by the smaller cumulus plumes that detrain at levels where larger plumes remove moisture (Neggers 2015b). More study on the role of spatial organization and the influence of the cumulus size distribution on trade-wind shallow cumulus cloud variability and feedback appears important to have a more complete understanding of shallow cumulus cloud mechanisms. Note that the effect of spatial organization on larger domains and cumulus size distribution might be related to each other, as larger domains lead to organized clusters and therefore a larger proportion of cumulus with larger cloud-base area. The main point being that the constancy of cumulus base cloud fraction is not necessarily something that can be taken for granted.

4.2 Robustness and Uncertainties of LES Studies

There is a tendency to view LES as surrogate of the truth, as able to fully represent the observed characteristics of the marine boundary layer. To some extent, this may be warranted by the robustness of simulated behavior across different LES. Simulated vertical distributions of cloud fraction and, to a slightly lesser extent, of projected cloud cover, tend to show relatively good agreement across different LES models in the Barbados Oceanographic and Meteorological EXperiment (BOMEX) and Rain In Cumulus over the Ocean (RICO) intercomparison cases of typical shallow trade-wind cumulus conditions (Siebesma et al. 2003; Van Zanten et al. 2011). An intercomparison case of the diurnal cycle of shallow cumulus over land also shows good model-to-model agreement (Brown et al. 2002). The cloud distributions of the above three intercomparison cases show a strong peak in cloud fraction at cloud-base, a rapid decrease in cloud fraction above cloud-base, and relatively small cloud fractions near the tops of cumulus clouds under the trade inversion. Total cloud cover ranges between about $13 \pm 6\%$ for BOMEX (Siebesma et al. 2003) and $19 \pm 9\%$ for RICO (Van Zanten et al. 2011), with the simulated cloud cover for RICO comparing favorably with corresponding lidar data. In simulations of an intermediate regime between stratocumulus and trade-wind cumulus, representative of the Atlantic Tradewind EXperiment (ATEX) field campaign and marked by a stronger inversion, the vertical distribution of cloud fraction has its maximum near the inversion instead of near cloud-base (Stevens et al. 2001). In this ATEX intercomparison case, there is more spread in simulated total cloud cover among the participating LES (total cloud cover ranges between 20 and 80% (mean $\pm 2\sigma$)), with the spread related to the representation of stratiform cloud amount under the inversion (Stevens et al. 2001). Also the CGILS intercomparison case of the response of shallow cumulus to climate change perturbations (location S6) shows the most apparent differences in the simulated cloud fraction profile near the top of the cloud layer under the trade inversion (Blossey et al. 2013). Whereas stratiform outflow layers are observed frequently at Barbados (Nuijens et al. 2014, 2015a), LES apparently have difficulties to properly simulate detrained layers of stratiform cloud. This difficulty is likely related to a poor representation of tight feedbacks between such outflow layers with radiation and subsidence, and to the fact that a very high vertical resolutions is necessary to resolve sharp inversions (Stevens et al. 2001). On the other hand, the range of cloud-base cloud fractions is quite consistent among the various intercomparison cases, with inter-model differences lying between 4.5 and 8% (Brown 1999; Stevens et al. 2001; Siebesma et al. 2003; Van Zanten et al. 2011; Blossey et al. 2013).

The comparison of the cumulative cloud fraction—the cumulative contributions to total cloud cover from the top down to the bottom of the cloud layer—estimated from LES and measured by a lidar indicates that the LES may not represent the full spectrum of cloud top height distributions present in nature (Figure 7 of Van Zanten et al. 2011). LES on large domains of $\sim 50 \times 50\,\mathrm{km}^2$ (about 16-times to 32-times larger than the domain sizes used for the intercomparison cases) can represent cloud populations with a wide range of cloud top heights, but cloud fractions in the upper cloud layer tend to be underestimated (Vogel et al. 2016). This underestimation is likely due to numerical diffusion, which is strongly related to the choice of advection scheme, the subgrid-scale model and the grid spacing. A thorough investigation of the impact of such model choices showed that cloud cover strongly decreased when a more dissipative monotone advection scheme was used instead of a centered differences scheme, or when a more dissipative subgrid formulation was used (relative decreases in cloud cover of up to 30%) (Matheou et al. 2011). Matheou et al. (2011) also find a relative decrease in cloud cover of up to 70% when the uniform horizontal and vertical grid spacing is increased from 20 m to 80 m. These results are in qualitative agreement with sensitivity studies presented in Stevens et al. (2001) and Siebesma et al. (2003) and show that one has to be careful when comparing absolute values of cloud cover between different LES studies, and between LES and observations. The strong decrease in cloud cover with larger grid spacing in Matheou et al. (2011) is partly due to reductions in cloudiness under the inversion, which cannot be resolved well at a vertical grid spacing of 80 m (see the liquid water specific humidities in their Figure 11). This again highlights that cloudiness near the top of shallow cumulus under the trade inversion is still poorly constrained by LES.

Studying how fields of shallow cumuli change in response to climate change perturbations and how they affect the planetary albedo and equilibrium climate sensitivity is also challenging using the current LES experimental setup. In this respect, LES yield an important limitation for climate studies, since they usually have to be run over small domains (10 to 50 kilometers) and therefore cannot realistically represent their variability under the wide range of conditions observed in nature, and especially their interactions with the large-scale circulation. For the same reason, LES have to be run over limited periods of time (a few days) and under simplified configurations (e.g., prescribed radiative cooling rate and sea surface temperature (SST)), and therefore all the process-scale interactions in the trade-wind layer and with the underlying ocean surface are not represented—for instance, the local interplay between cloud radiative forcing and turbulence as in Vial et al. (2016) and SST feedbacks on the trade-wind layer as in Tan et al. (2017).

Increasing computational resources now makes it possible to consider LES over larger domains, over larger timescales and under increasingly "realistic" configurations. For instance, simulations at 100 m resolution over the entire tropical Atlantic on timescales of months are now becoming possible with the ICOsahedral Non-hydrostatic (ICON) atmospheric model (Zängl et al. 2015; Heinze et al. 2016). Combined with observations of the planned field campaign EUREC^{4}A (Elucidating the Role of Cloud-Circulation Coupling in Climate) over this region (discussed next), these simulations will provide new opportunities to study in more details the key factors controlling the cloud responses to warming, including the interactions between the trade-wind boundary-layer processes and the large-scale mesoscale organization.

5 Connecting LES and GCM Interpretations of Shallow Cumulus Cloud Feedback Mechanisms

In order to better compare low-cloud changes and mechanisms between LES and GCM (or similarly SCM), a common interpretation framework is needed. Unlike in GCM and SCM, where turbulence and convection are usually represented by separate parameterization schemes (cf. Fig. 1), in LES these processes are part of a continuous spectrum of motions ranging from turbulent eddies (diffusive processes) to convective vertical drafts (advective processes). Convection is generally represented by the advective and diffusive flux divergence of the resolved and subgrid-scale flow and acts as a source of moisture over the whole trade-wind boundary layer (cf. Fig. 5). By adding the turbulent and convective tendencies in a GCM or SCM, tendency profiles comparable to LES can be generated (see Fig. 6 for a SCM example). In both LES and GCM, these turbulent and convective motions accomplish the vertical transports of heat and moisture that is supplied by surface sensible and latent heat fluxes, respectively.

Here we use the Vial et al. (2016) framework described in Sect. 2.2 to better interpret the contrasting model behaviors described in the preceding sections and provide a broader perspective:

- In a warming climate, all numerical models (LES, GCM, SCM) tend to simulate a more vigorous convective mixing due to increased latent heat flux (e.g., Rieck et al. 2012; Blossey et al. 2013; Vogel et al. 2016; Brient et al. 2015; Vial et al. 2016), yielding $d\mu > 0$ in Eq. 6. But this is not necessarily associated with changes in cloud fraction. For a given latent heat flux, the efficiency of the convective mixing at desiccating low-level clouds (i.e., $\mathcal{C}$) depends on where the convective drying maximizes with respect to the cloud layer (Vial et al. 2016). In most LES and according to the cumulus-valve mechanism, the convective mass flux originates in the subcloud layer where there is no cloud to desiccate. So $\mathcal{C}$ is likely to be very small. However, this appears to be different

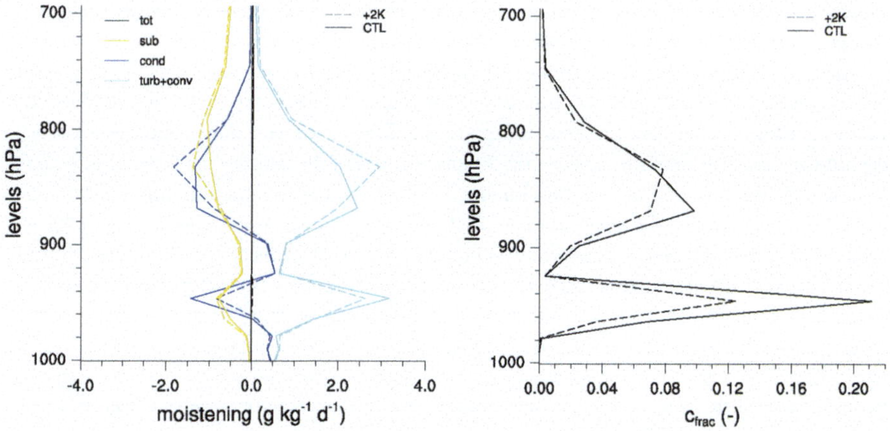

Fig. 6 Vertical distribution of moisture tendencies (*left*) and cloud fraction (*right*) for a SCM experiment under CGILS framework (shallow cumulus regime, s6) using the IPSL-CM5A-LR GCM. Moisture tendencies include: turbulence and convection (*cyan*), grid-average net condensation (*blue*, which includes cloud formation, precipitation and evaporation), large-scale subsidence (*orange*) and the sum of all moisture tendencies (*black*). Results are shown for the present-day climate (*solid*) and as a response to a surface warming of +2K (*dash*). Adapted from Vial et al. (2016)

in LES on larger domains, when convection organizes into more vigorous and deeper clusters. It has recently been shown that some larger convective plumes may originate within the cloud layer, and that these convective fluxes alone could dry the lower troposphere, just above cloud-base (Neggers 2015b; Seifert et al. 2015). Therefore, in larger domain simulations $\mathcal{C}$ might be stronger. On the other hand, in GCM and SCM, $\mathcal{C}$ is likely to depend on how the bulk convective mass flux is formulated, as shown in Vial et al. (2016).

- In all models, increased latent heat flux moistens the lower troposphere through turbulent diffusion and convective transport and favors cloud formation, yielding $\mathcal{T} > 0$.
- In all the models considered, λ (the change in surface evaporation per unit change in the convective mixing) is positive: more convective mixing dries the subcloud layer and increases surface evaporation. However, there might be some disagreement on how effective this is, as this depends on how efficiently increased convection brings dry air to the surface, and to what extent it is accompanied by increased warming. This effect would influence low-level stability and eventually the cloud response to warming (Brient et al. 2015).
- In both Rieck et al. (2012) and Vogel et al. (2016), the LES experimental setup prescribes uniform radiative cooling, and therefore the interaction between latent heat flux and cloud radiative cooling is neglected in these studies (i.e., $\lambda_r = 0$). The coupling between cloud radiative forcing and latent heat flux has been recently identified in Vial et al. (2016) in one model; its robustness across models and in observations remains to be shown.

6 Observational Support for Trade-Wind Shallow Cumulus Cloud Feedbacks

As discussed in the previous sections, the primary source of uncertainties in trade-wind cloud responses to warming in numerical models relies in how strong subgrid-scale vertical transports of heat and moisture affect cloudiness near cloud-base in response to changes in the large-scale environment. To this end, numerical experiments made it possible considerable progress on the understanding of the key processes that couple convection, turbulence and cloudiness in trade-wind boundary layers. In addition, several studies have used observations to test our physical understanding on a wide range of timescales and to constrain uncertainties of the simulated cloud changes in a warming climate (e.g., Clement et al. 2009; Qu et al. 2014; Brient and Schneider 2016). However, our confidence in low-level cloud feedbacks remains fairly low as the primary factors controlling low-level cloud variability in nature appear to be strongly dependent on the time and/or spatial scales at which the mechanisms are evaluated. Put differently, observational analyses of the factors controlling the trade-wind cloud responses to warming have not yet helped to clarify the inconsistency of the results that we have found between large-scale numerical models (e.g., GCM and SCM) and high-resolution simulations (e.g., LES).

On large domains and long timescales (interannual, decadal or climatological timescales), sea surface temperature explains a large part of the subtropical low-level cloud variability (e.g., Clement et al. 2009; Qu et al. 2014, 2015; Brient and Schneider 2016), with a reduced cloud cover when the sea surface is warmer. Note, however, that these observational studies consider both stratiform and cumuliform types of low-level clouds.

Nevertheless, this present-day relationship appears to be consistent with the lower-tropospheric mixing mechanism that controls the low-level cloud response to a warming climate (Sect. 2.2) in GCM and SCM simulations, as the cloud sensitivity to warming was found to be correlated with enhanced latent heat flux and vertical gradient of moisture between the boundary layer and the free troposphere (Qu et al. 2015), and both contribute to enhance the mixing of dry free tropospheric air into the boundary layer that leads to the reduction in low-level cloudiness (Rieck et al. 2012; Bretherton et al. 2013; Brient and Bony 2013; Vial et al. 2016). The observational studies cited above suggest that models that simulate a stronger cloud decrease in a warming climate (and thus a stronger cloud feedback) are more consistent with observations than models that simulate a weaker cloud feedback, and thus that high climate sensitivities are maybe more credible than low climate sensitivity estimates. This is in line with other studies that have related the low-level cloud feedbacks and/or climate sensitivity estimates with climatological indicators of the present-day lower-tropospheric mixing (such as vertical gradients in temperature and relative humidity, large-scale vertical velocity and shallowness of low clouds): models with a stronger lower-tropospheric mixing in the present-day climate are more efficient in depleting boundary-layer moisture as the climate warms, yielding a stronger low-cloud feedback and ECS ; these models tend to be more consistent with observations than models that simulate a weaker lower-tropospheric mixing in the present-day climate (Sherwood et al. 2014; Brient et al. 2015).

Measurements from the Barbados Cloud Observatory—a facility established on a windward promontory on Barbados to study factors controlling cloudiness in the trades (Stevens et al. 2016)—suggest that models can represent a fairly realistic climatology of the lower-tropospheric trade-wind layer on long-term means but through unrealistic variability on shorter timescales. Analysis of the Barbados data indicates that about 60% of observed cloud variance near cloud-base occurs on timescales smaller than a day (Nuijens et al. 2014, 2015a, b). These data suggest that cloudiness near cloud-base is more controlled by internal feedback processes on short timescales and is relatively independent of large-scale environmental factors (such as subsidence, surface temperature and vertical gradient of humidity) that vary on longer timescales (Bellon and Stevens 2013), consistent with the cumulus-valve mechanism (Sect. 3) and LES results on small domains (Sect. 4.1.1).

In their evaluation of climate model output as compared to the Barbados data, Nuijens et al. (2015b) argue that climate models (1) lack this cloud-base regulation mechanism associated with turbulence and convection that appears to be important in nature on sub-daily timescales and (2) are too sensitive to variations of the large-scale environment (lower-tropospheric relative humidity and thermal stratification) on timescales longer than a day (Nuijens et al. 2015a, b). Furthermore, observed cloud fraction at the inversion dominates the total variance in boundary-layer clouds and explains the seasonality of low-level cloudiness, with larger cloud cover (in winter) when surface winds and trade-wind inversions are stronger (Nuijens et al. 2014, 2015b). These relationships are not captured in climate models either. Because the Nuijens et al. studies point to relationships that are relevant for the interpretation of the simulated shallow cumulus cloud feedbacks in climate models, they raise the question of whether or not these models can simulate realistic changes in trade-wind cloudiness in a warming climate. However, one may also question whether these observed relationships capture all the interactions between the trade-wind boundary layer and the larger scale mesoscale organization that might be necessary to interpret the low-level cloud changes on a sufficiently large domain, which is also what matters for the climate sensitivity problem.

This question could be addressed with the planned field campaign EUREC^{4}A that will take place on January–February 2020 over a large oceanic area east of Barbados (Bony et al. in revision). Featured with its large experimental domain, and by linking profiles of cloudiness to large-scale fluxes of moisture and energy, including estimates of the cumulus mass flux, this campaign will make it possible to quantify macrophysical properties of shallow cumulus clouds as a function of the large-scale environment and thus to assess the existence of the model-based mechanisms that were discussed here under a wide range of large-scale conditions: (1) the vertical distribution of trade-wind cumulus clouds and its relation to convective mixing, latent heat flux and cloud radiative forcing (Vial et al. 2016) the non-uniformity in the spatial distribution of cloud-base area and its impact on the dynamics of trade-wind boundary layers and associated clouds (Neggers 2015b); and (3) the role of organized convection on trade-wind clouds (Seifert et al. 2015; Vogel et al. 2016). More information on the EUREC^{4}A field campaign and scientific goals are provided in Bony et al. in revision.

7 Synthesis

Fair-weather cumulus clouds, covering large areas of the tropical and subtropical oceans in the trade-wind regions, play a central role in the tropical cloud feedback uncertainties in climate models. Climate models predict different low-level cloud changes in response to warming, which results in a large dispersion in model-based estimates of cloud feedback and climate sensitivity. This large dispersion in model responses arises from differences in the balance of the key boundary-layer physical processes that are parameterized in climate models, especially convection and turbulence. Given the importance of low-level cloud feedbacks in climate change projections, understanding the factors controlling the low-level cloudiness across a wide range of temporal and spatial scales in a hierarchy of numerical models and in observations has emerged as an active research area.

Based on a review of past studies on this issue, we have identified three emergent topics for which further investigation would help understand and constrain shallow cumulus cloud feedbacks:

- the vertical distribution of shallow cumulus cloud layers and its relation to convective mixing, surface fluxes and cloud radiative forcing,
- the impact of the probability distribution of cumulus cloud-base areas on the dynamics of trade-wind boundary layers and associated clouds,
- the role of mesoscale organization, and accompanying episodes of deeper convection, in the trade cumulus variability and feedbacks.

These emergent topics would strongly benefit from combined analyses of high-resolution modeling and field experiments on large domains such as those discussed in this review.

Acknowledgements This paper arises from the International Space Science Institute (ISSI) workshop on Shallow clouds and water vapor, circulation and climate sensitivity. This work is supported by the European Research Council (ERC) Grant 694768 and the French National Research Agency (ANR) Project MORDICUS (ANR-13-SENV-0002-01). Thanks to the two reviewers for their constructive comments and helpful suggestions on an earlier version of the manuscript.

References

Albrecht BA, Betts AK, Schubert WH, Cox SK (1979) Model of the thermodynamic structure of the trade-wind boundary layer: Part I. Theoretical formulation and sensitivity tests. J Atmos Sci 36(1):73–89

Bellon G, Stevens B (2012) Using the sensitivity of large-eddy simulations to evaluate atmospheric boundary layer models. J Atmos Sci 69(5):1582–1601

Bellon G, Stevens B (2013) Time scales of the trade wind boundary layer adjustment. J Atmos Sci 70(4):1071–1083

Betts AK (1976) Modeling subcloud layer structure and interaction with a shallow cumulus layer. J Atmos Sci 33(12):2363–2382

Betts AK, Ridgway W (1989) Climatic equilibrium of the atmospheric convective boundary layer over a tropical ocean. J Atmos Sci 46(17):2621–2641

Blossey PN, Bretherton CS, Zhang M, Cheng A, Endo S, Heus T, Liu Y, Lock AP, Roode SR, Xu KM (2013) Marine low cloud sensitivity to an idealized climate change: the cgils les intercomparison. J Adv Model Earth Syst 5(2):234–258

Bony S, Dufresne JL (2005) Marine boundary layer clouds at the heart of tropical cloud feedback uncertainties in climate models. Geophys Res Lett 32(20):L20806

Bony S, Stevens B, Ament F, Bigorre S, Chazette P, Crewell S, Delanoë J, Emanuel K, Farrell D, Flamant C, Gross S, Hirsch L, Karstensen J, Mayer B, Nuijens L, Ruppert Jr JH, Sandu I, Siebesma P, Speich S, Szczap F, Totems J, Vogel R, Wendisch M, Wirth M (2017) EUREC4A: a field campaign to elucidate the couplings between clouds, convection and circulation. Surv Geophys (in revision)

Boucher O, Randall D, Artaxo P, Bretherton C, Feingold G, Forster P, Kerminen VM, Kondo Y, Liao H, Lohmann U, et al (2013) Clouds and aerosols. In: Climate change 2013: the physical science basis. Contribution of working group I to the fifth assessment report of the intergovernmental panel on climate change, vol 5, Cambridge University Press, Cambridge, pp 571–657

Bretherton CS (2015) Insights into low-latitude cloud feedbacks from high-resolution models. Philos Trans R Soc A 373(2054):20140415

Bretherton CS, Blossey PN, Jones CR (2013) Mechanisms of marine low cloud sensitivity to idealized climate perturbations: a single-les exploration extending the cgils cases. J Adv Model Earth Syst 5(2):316–337

Brient F, Bony S (2012) How may low-cloud radiative properties simulated in the current climate influence low-cloud feedbacks under global warming? Geophys Res Lett 39(20):L20807

Brient F, Bony S (2013) Interpretation of the positive low-cloud feedback predicted by a climate model under global warming. Clim Dyn 40(9–10):2415–2431

Brient F, Schneider T (2016) Constraints on climate sensitivity from space-based measurements of low-cloud reflection. J Clim 29:5821–5835

Brient F, Schneider T, Tan Z, Bony S, Qu X, Hall A (2015) Shallowness of tropical low clouds as a predictor of climate models response to warming. Clim Dyn 47(1):433–449

Brown A (1999) Large-eddy simulation and parametrization of the effects of shear on shallow cumulus convection. Bound Layer Meteorol 91(1):65–80

Brown A, Cederwall R, Chlond A, Duynkerke P, Golaz JC, Khairoutdinov M, Lewellen D, Lock A, MacVean M, Moeng CH et al (2002) Large-eddy simulation of the diurnal cycle of shallow cumulus convection over land. Q J R Meteorol Soc 128(582):1075–1093

Clement AC, Burgman R, Norris JR (2009) Observational and model evidence for positive low-level cloud feedback. Science 325(5939):460–464

Heinze R, Dipankar A, Carbajal Henken C, Moseley C, Sourdeval O, Trömel S, Xie X, Adamidis P, Ament F, Baars H, et al (2016) Large-eddy simulations over Germany using ICON: a comprehensive evaluation. Q J R Meteorol Soc 143(702):69–100

Kamae HY, Shiogama M, Watanabe T, Ogura T, Yokohata T, Kimoto M (2016) Lower tropospheric mixing as a constraint on cloud feedback in a multi-parameter multi-physics ensemble. J Clim. doi:10.1175/JCLI-D-16-0042.1

Klein SA, Hall A (2015) Emergent constraints for cloud feedbacks. Curr Clim Change Rep 1(4):276–287

Matheou G, Chung D, Nuijens L, Stevens B, Teixeira J (2011) On the fidelity of large-eddy simulation of shallow precipitating cumulus convection. Monthly Weather Rev 139(9):2918–2939

Medeiros B, Stevens B, Bony S (2015) Using aquaplanets to understand the robust responses of comprehensive climate models to forcing. Clim Dyn 44(7–8):1957–1977

Neggers R (2015a) Attributing the behavior of low-level clouds in large-scale models to subgrid-scale parameterizations. J Adv Model Earth Syst 7(4):2029–2043

Neggers R (2015b) Exploring bin-macrophysics models for moist convective transport and clouds. J Adv Model Earth Syst 7(4):2079–2104

Neggers R, Stevens B, Neelin JD (2006) A simple equilibrium model for shallow-cumulus-topped mixed layers. Theor Computat Fluid Dyn 20(5–6):305–322

Nuijens L, Serikov I, Hirsch L, Lonitz K, Stevens B (2014) The distribution and variability of low-level cloud in the North Atlantic trades. Q J R Meteorol Soc 140(684):2364–2374

Nuijens L, Medeiros B, Sandu I, Ahlgrimm M (2015a) The behavior of trade-wind cloudiness in observations and models: the major cloud components and their variability. J Adv Model Earth Syst 7(2):600–616

Nuijens L, Medeiros B, Sandu I, Ahlgrimm M (2015b) Observed and modeled patterns of covariability between low-level cloudiness and the structure of the trade-wind layer. J Adv Model Earth Syst 7(4):1741–1764

Qu X, Hall A, Klein SA, Caldwell PM (2014) On the spread of changes in marine low cloud cover in climate model simulations of the 21st century. Clim Dyn 42(9–10):2603–2626

Qu X, Hall A, Klein SA, DeAngelis AM (2015) Positive tropical marine low-cloud cover feedback inferred from cloud-controlling factors. Geophys Res Lett 42(18):7767–7775

Rieck M, Nuijens L, Stevens B (2012) Marine boundary layer cloud feedbacks in a constant relative humidity atmosphere. J Atmos Sci 69(8):2538–2550

Seifert A, Heus T (2013) Large-eddy simulation of organized precipitating trade wind cumulus clouds. Atmos Chem Phys 13(11):5631–5645

Seifert A, Heus T, Pincus R, Stevens B (2015) Large-eddy simulation of the transient and near-equilibrium behavior of precipitating shallow convection. J Adv Model Earth Syst 7(4):1918–1937

Sherwood SC, Bony S, Dufresne JL (2014) Spread in model climate sensitivity traced to atmospheric convective mixing. Nature 505(7481):37–42. doi:10.1038/nature12829

Siebesma AP, Bretherton CS, Brown A, Chlond A, Cuxart J, Duynkerke PG, Jiang H, Khairoutdinov M, Lewellen D, Moeng CH et al (2003) A large eddy simulation intercomparison study of shallow cumulus convection. J Atmos Sci 60(10):1201–1219

Stevens B (2006) Bulk boundary-layer concepts for simplified models of tropical dynamics. Theor Comput Fluid Dyn 20(5–6):279–304

Stevens B (2007) On the growth of layers of nonprecipitating cumulus convection. J Atmos Sci 64:2916–2931

Stevens B, Seifert R (2008) Understanding macrophysical outcomes of microphysical choices in simulations of shallow cumulus convection. J Meteorol Soc Japan 86:143–162

Stevens B, Ackerman AS, Albrecht BA, Brown AR, Chlond A, Cuxart J, Duynkerke PG, Lewellen DC, Macvean MK, Neggers RA et al (2001) Simulations of trade wind cumuli under a strong inversion. J Atmos Sci 58(14):1870–1891

Stevens B, Farrell D, Hirsch L, Jansen F, Nuijens L, Serikov I, Brügmann B, Forde M, Linne H, Lonitz K et al (2016) The barbados cloud observatory-anchoring investigations of clouds and circulation on the edge of the ITCZ. Bull Am Meteorol Soc 97:787–801

Tan Z, Schneider T, Teixeira J, Pressel KG (2017) Large-eddy simulation of subtropical cloud-topped boundary layers: 2. Cloud response to climate change. J Adv Model Earth Syst 9(1):19–38

Taylor KE, Stouffer RJ, Meehl GA (2012) An overview of cmip5 and the experiment design. Bull Am Meteorol Soc 93(4):485

Tiedtke M (1989) A comprehensive mass flux scheme for cumulus parameterization in large-scale models. Monthly Weather Rev 117(8):1779–1800

Tomassini L, Voigt A, Stevens B (2014) On the connection between tropical circulation, convective mixing, and climate sensitivity. Q J R Meteorol Soc 141(689):1404–1416

Van Zanten MC, Stevens B, Nuijens L, Siebesma AP, Ackerman A, Burnet F, Cheng A, Couvreux F, Jiang H, Khairoutdinov M et al (2011) Controls on precipitation and cloudiness in simulations of trade-wind cumulus as observed during RICO. J Adv Model Earth Syst 3(2):M06001

Vial J, Bony S, Dufresne JL, Roehrig R (2016) Coupling between lower-tropospheric convective mixing and low-level clouds: physical mechanisms and dependence on convection scheme. JAMES. doi:10.1002/2016MS000740

Vogel R, Nuijens L, Stevens B (2016) The role of precipitation and spatial organization in the response of trade-wind clouds to warming. J Adv Model Earth Syst 8:843–862

Watanabe M, Shiogama H, Yokohata T, Kamae Y, Yoshimori M, Ogura T, Annan JD, Hargreaves JC, Emori S, Kimoto M (2012) Using a multiphysics ensemble for exploring diversity in cloud-shortwave feedback in GCMs. J Clim 25(15):5416–5431

Webb M, Andrews T, Bodas-Salcedo A, Bony S, Bretherton C, Chadwick R, Chepfer H, Douville H, Good P, Kay J, et al (2016) The cloud feedback model intercomparison project (CFMIP) contribution to CMIP6 . Geoscientific Model Development Discussions 2016: in-open

Webb MJ, Lock AP (2013) Coupling between subtropical cloud feedback and the local hydrological cycle in a climate model. Clim Dyn 41(7–8):1923–1939

Webb MJ, Senior C, Sexton D, Ingram W, Williams K, Ringer M, McAvaney B, Colman R, Soden B, Gudgel R et al (2006) On the contribution of local feedback mechanisms to the range of climate sensitivity in two GCM ensembles. Clim Dyn 27(1):17–38

Webb MJ, Lock AP, Bretherton CS, Bony S, Cole JN, Idelkadi A, Kang SM, Koshiro T, Kawai H, Ogura T, Roehrig R, Shin Y, Mauritsen T, Sherwood SC, Vial J, Watanabe M, Woelfle MD, Zhao M (2015) The impact of parametrized convection on cloud feedback. Philos Trans R Soc A 373(2054):20140414

Wing AA, Emanuel KA, Holloway CE, Muller C (2017) Convective self-aggregation in numerical simulations: a review. Surv Geophys Rev. doi:10.1007/s10712-017-9408-4

Wyant MC, Bretherton CS, Blossey PN (2009) Subtropical low cloud response to a warmer climate in a superparameterized climate model. Part I: regime sorting and physical mechanisms. J Adv Model Earth Syst. doi:10.3894/JAMES.2009.1.7

Zängl G, Reinert D, Rípodas P, Baldauf M (2015) The ICON (ICOsahedral Non-hydrostatic) modelling framework of DWD and MPI-M: description of the non-hydrostatic dynamical core. Q J R Meteorol Soci 141(687):563–579

Zelinka MD, Zhou C, Klein SA (2016) Insights from a refined decomposition of cloud feedbacks. Geophys Res Lett 43(17):9259–9269

Zhang M, Bretherton CS, Blossey PN, Bony S, Brient F, Golaz JC (2012) The CGILS experimental design to investigate low cloud feedbacks in general circulation models by using single-column and large-eddy simulation models. J Adv Model Earth Syst. doi:10.1029/2012MS000182

Zhang M, Bretherton CS, Blossey PN, Austin PH, Bacmeister JT, Bony S, Brient F, Cheedela SK, Cheng A, Genio AD et al (2013) Cgils: Results from the first phase of an international project to understand the physical mechanisms of low cloud feedbacks in single column models. J Adv Model Earth Syst 5(4):826–842

Zhao M (2014) An investigation of the connections among convection, clouds, and climate sensitivity in a global climate model. J Clim 27(5):1845–1862

Surv Geophys (2017) 38:1355–1369
https://doi.org/10.1007/s10712-017-9427-1

Importance Profiles for Water Vapor

Brian Mapes[1] · Arunchandra S. Chandra[1] · Zhiming Kuang[2] · Paquita Zuidema[1]

Received: 20 March 2017 / Accepted: 13 September 2017 / Published online: 7 October 2017

Abstract Motivated by the scientific desire to align observations with quantities of physical interest, we survey how scalar importance functions depend on vertically resolved water vapor. Definitions of importance begin from familiar examples of water mass I^{m} and TOA clear-sky outgoing longwave flux I^{OLR}, in order to establish notation and illustrate graphically how the sensitivity profile or "kernel" depends on whether specific humidity *S*, relative humidity *R*, or ln(*R*) are used as measures of vapor. Then, new results on the sensitivity of convective activity I^{con} to vapor (with implied knock-on effects such as weather prediction skill) are presented. In radiative-convective equilibrium, organized (line-like) convection is much more sensitive to moisture than scattered isotropic convection, but it exists in a drier mean state. The lesson for natural convection may be that organized convection is less susceptible to dryness and can survive and propagate into regions unfavorable for disorganized convection. This counterintuitive interpretive conclusion, with respect to the narrow numerical result behind it, highlights the importance of clarity about *what is held constant at what values* in sensitivity or susceptibility kernels. Finally, the sensitivities of observable radiance signals I^{sig} for passive remote sensing are considered. While the accuracy of *R* in the lower free troposphere is crucial for the physical importance scalars, this layer is unfortunately the most difficult to isolate with passive remote sensing: In high emissivity channels, water vapor signals come from too high in the atmosphere (for satellites) or too low (for surface radiometers), while low emissivity channels have poor altitude discrimination and (in the case of satellites) are contaminated by surface emissions. For these reasons, active ranging (LiDAR) is the preferred observing strategy.

✉ Brian Mapes
bmapes@miami.edu

1 Rosenstiel School of Marine and Atmospheric Sciences, University of Miami, 4600, Rickenbacker Causeway, Miami, FL 33149, USA

2 Department of Earth and Planetary Sciences, School of Engineering and Applied Sciences, Harvard University, 20 Oxford St., Cambridge, MA 02138, USA

Keywords Water vapor · Humidity · OLR · Convection · Organized convection · Sensitivity · Susceptibility · Functional derivative

1 Introduction

Water vapor in the atmosphere is important for several reasons reviewed well in Sherwood et al. (2010). It carries latent heat that is released upon phase changes. Also, both vapor and condensed water interact strongly with radiation, with at least three major classes of consequences:

a. Direct contributions to global top-of-atmosphere (TOA) radiative energy budgets;
b. Dynamical effects, via both radiative and latent energy budgets in air layers;
c. Observational implications for remote sensing.

Interactions among these effects offer many motivations to deepen our understanding, and many possibilities for research. For instance, mid-level vapor affects low cloudiness, with its strong impacts on global shortwave TOA radiation budgets (Adebiyi et al. 2015; Stevens et al. 2017). Global atmospheric cooling importantly governs precipitation and the hydrologic cycle (e.g., Stephens and Ellis 2008; Previdi 2010; DeAngelis et al. 2015). Convective clouds depend on the vapor field, which in turn is shaped by convective fluxes as well as by latent heating-induced winds that shape transport and surface flux patterns. These couplings make water an intimate active tracer in the atmosphere.

One especially fruitful nexus of interests that can motivate remote sensing is the scientific desire to align observations with quantities of physical interest—or, to turn that around, to cast physical theory in terms of quantities that can be well estimated or strongly constrained from observables. For this reason, item (c) above is kept in context of water's other physical science importances, and is not viewed as a standalone motivation toward measurement for measurement's sake.

2 Simplest: The Profile of I^{m} = Mass Errors from Vapor Measurement Errors

To pursue these ideas quantitatively, it helps to define *importance* (something one cares about) in mathematical terms as a scalar–valued function I with complicated inputs. Such a dimension-reducing function is called a *functional*, and its sensitivities are called *functional derivatives* (see http://www.physicspages.com/2014/11/08/functionals-and-functional-derivatives/). Here we develop notation for how I is distributed over altitude in the atmosphere. From such a framework, progress toward the Grand Challenges motivating this volume can hopefully be steered and assessed.

This paper shows profiles (vertical distributions) of 4 different importances of water vapor at low latitudes. First, I^{m} = column-integrated vapor mass is used to clarify notation and fix ideas. As a second familiar example, I^{OLR} = Outgoing Longwave Radiation (OLR) illustrates the "kernel" approach of Soden et al. (2008). New results are then presented for I^{con} = rainfall (latent heating) by deep convective cloud systems. Finally, I^{sig} = radiative signal in low and high emissivity channels is discussed in the context of observing system design.

For each of these definitions of importance I, we want to know the impact of an increment (or a measurement error) of vapor, as a function of altitude. Such an increment might be expressed as mass (specific humidity S), or relative humidity R, or of some function thereof (for example, increments $\delta\ln(R)$ are relevant to some radiative quantities). In any partial derivative, it is necessary to specify what is held constant, so we will make that part of the mathematical notation where needed for clarity.

In this elementary case, we define an importance scalar or *functional* as I^{m} = column-integrated water vapor mass (CWV, units: kg m^{-2}, or mm of liquid equivalent), sometimes called precipitable water (PW):

$$I^m = \int_0^{p_s} \frac{\mathrm{d}p}{g} \tag{1}$$

where the profile of specific humidity S may be expanded as

$$S(p) = R(p) \times S_{\mathrm{sat}}(T(p), p) \tag{2}$$

Here we ignore any distinction between total pressure p versus the (p) notation for hydrostatic pressure, our monotonic mass coordinate of the vertical domain.

For tropical applications, consider a reference $T(p)$ that is a pseudo-adiabat characterized by its equivalent potential temperature $\theta_e = 350$ K. The corresponding S_{sat} profile is shown in Fig. 1a. The curve to a good approximation decreases linearly with p from 18 g/kg at 1000 hPa to near zero at about 300 hPa.

If a vapor concentration or specific humidity measurement had an error δS, which is constant in height, (1) shows that this error δS would contribute equally from all pressure levels to the error in I^{m}. We denote the accompanying kernel, the sensitivity of vapor mass m to an increment of S, as a functional derivative:

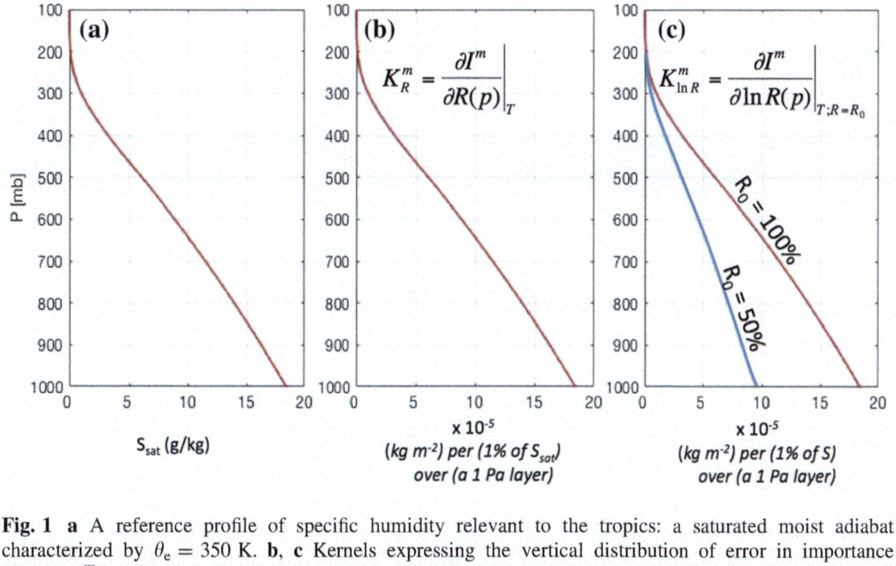

Fig. 1 **a** A reference profile of specific humidity relevant to the tropics: a saturated moist adiabat characterized by $\theta_e = 350$ K. **b**, **c** Kernels expressing the vertical distribution of error in importance function I^{m} = column vapor mass, incurred by measurements with a constant profile of measurement error δR (**b**) or $\delta\ln R$ (**c**)

$$K_S^m \equiv \frac{\delta I^m}{\delta S(p)} \tag{3}$$

Continuous function $S(p)$ notation is retained, even though all calculations below are on discrete levels, to spotlight the distinction between the layer depth appearing in the *units* of the estimated K, versus the vertical resolution of particular discrete calculations used in the estimate.

The functional derivative is defined as the sensitivity when $S(p)$ is held constant at all other altitudes. In our case, S also varies with time, latitude, and longitude, but these need not be specified as being held constant since they are orthogonal to the vertical domain (indexed by hydrostatic p in the usual way). The *value* of (3) is constant in pressure. Its units are (units of Importance) (units of denominator)$^{-1}$ (units of p)$^{-1}$ and the value is equal to 1.02 (kg m^{-2} *of vapor mass*) *per* (1 kg/kg *of S*) *over* (*a* 1 hPa *layer*), since an increment $\delta S = 1$ g/kg over a 10^5 Pa atmosphere equates to $10/9.8 = 1.02$ kg m^{-2} of column vapor mass (mm of liquid equivalent).

Now instead suppose we have a measurement with constant in height *relative* humidity error δR, like the relative humidity sensor in a radiosonde. What is the profile of its contribution to the importance error δI^{m}? That kernel can be notated as:

$$K_R^m \equiv \left.\frac{\delta I^m}{\delta R(p)}\right|_{T=\mathrm{pseudoadiabat@350K}} \tag{4}$$

Here the essential quantity held constant in the partial derivative (T) is specified overtly, as is necessary for clarity in any non-orthogonal "phase space" (Nolte 2010) of abstract variables such as thermodynamic quantities. The units are again (units of Importance) (units of denominator)$^{-1}$ (units of p)$^{-1}$, or (kg m^{-2}) *per* (1% *of* S_{sat}) *over* (*a* 1 Pa *layer*).

Since the contribution of an error δR to $S(p)$, given a fixed $T(p)$, is simply proportional to S_{sat} by Eq. (2), the shape of this kernel (Fig. 1b) is identical to the S_{sat} curve in Fig. 1a. To figure out its value, we can utilize the value of (3) quoted above, noting that where $S_{\mathrm{sat}} = 10$ g/kg (about 650 hPa), a 1% error will equal 10^{-4} kg/kg. Any real instrument's error profile $\delta R(p)$ can be multiplied by the kernel and integrated to give its total importance (in this case, mass error).

Finally, consider the importance of a measurement error in *fractional* R, $\delta\ln(R)$, which is constant with altitude. For saturated spectral bands, the radiative impacts of vapor are sometimes approximated as proportional to $\ln(R)$ (Spencer and Braswell 1997, Pierrehumbert et al. 2007). The desired kernel

$$K_{\ln R}^m \equiv \left.\frac{\delta I^m}{\delta \ln R(p)}\right|_{T=\mathrm{pseudoadiabat@350}K;R} \tag{5}$$

will have units of (kg m^{-2}) *per* (1% *of the existing vapor*) *over* (*a* 1 Pa *layer*). Its derivation can be performed using the chain rule as follows:

$$K_{\ln R}^m \equiv \left.\frac{\delta I^m}{\delta \ln R(p)}\right|_{T;R} = \left.\frac{\delta I^m}{\delta S(p)}\frac{\partial S}{\partial \ln R}\right|_{T;R=R_0} = \frac{\delta I^m}{\delta S(p)} R_0 \left.\frac{\partial S}{\partial R}\right|_T = R_0 K_R^m \tag{6}$$

A new point of notation has been introduced here: the semicolon in the subscript. Items to the right of the semicolon are not *held constant* in the differentiation, they are *evaluated at a given value of* $R(p)$, in this case $R_0(p)$. In this derivation, the core (definitional) dependence of I^{m} on S from (3) is converted into the desired dependence on $\ln(R)$ through a chain rule (first equality). Then, using a calculus fact (second equality), the R dependence is exposed as a multiplication by $R_0(p)$. At that point, the subscript ;R can be dropped from the last partial derivative.

The final equality simply notes that this result (Fig. 1c) is related to an earlier result (kernel (4), Fig. 1b). Again, we can leverage the values and units above to build this kernel for Fig. 1c.

With these bookkeeping notations and mathematical tools, we can express subtler sensitivities of physical scalars to vapor profiles as functional derivatives.

3 Radiative Kernels: Sensitivity of OLR to Humidity

One familiar climate importance for water vapor is its impact on I^{OLR} = outgoing longwave radiation (OLR) at top of atmosphere. The dependence of I^{OLR} on the water vapor profile is not a straightforward integral like I^{m}, but rather a result of sophisticated columnar radiative transfer computations. Following Soden et al. (2008), we may characterize this computation's dependencies as OLR($T_{\mathrm{em}}(p)$, $S(p)$, $C(p)$), where T_{em} is the emitting temperature of the air, and $C(p)$ is the radiation scheme's own optical measure of cloudiness.

The negative of the zonal and time mean of the TOA OLR kernel for water vapor (Fig. 2 of Soden et al. 2008) is reproduced here as Fig. 2. The negative sign was introduced by Soden et al. (2008) because their application was climate feedback, and *increased* OLR

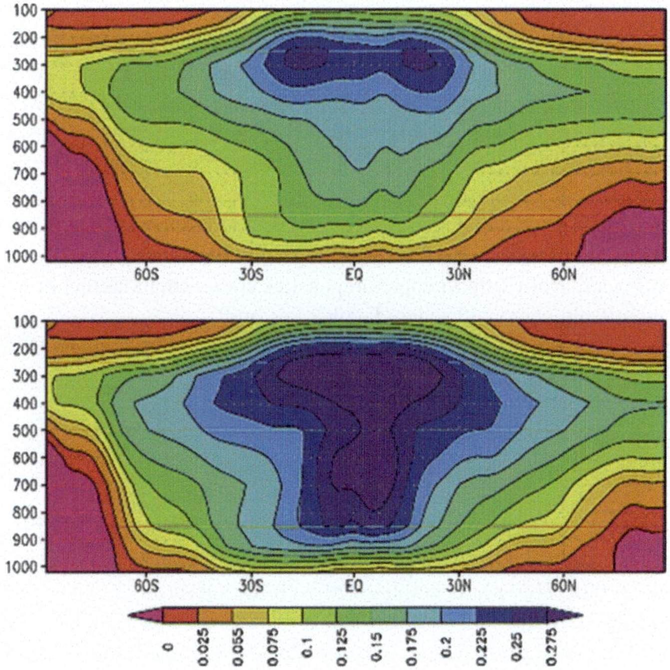

Fig. 2 Soden et al.'s (2008) "kernel" for OLR with respect to water vapor (their Fig. 2), averaged over time and longitude. The top plot is averaged over all-sky conditions (based on a climate model-predicted cloudiness field C) while the lower plot is with cloud radiative effects disabled. Units of the indicated numbers are (W m^{-2}) per (K in the $T_{\mathrm{sat}}(S, p)$ inversion of $S_{\mathrm{sat}}(T, p)$) over (a 100 hPa layer)

is a *negative* contribution to Earth's heat budget. Here we instead use $I^{\mathrm{OLR}} = +\,\mathrm{OLR}$, to express the impact of moisture on the *physical process* of upwelling longwave radiation.

In our notation, the two quantities plotted in Fig. 2 are:

$$K_{T_{\mathrm{sat}}}^{\mathrm{OLR}} = \left.\frac{\delta I^{\mathrm{OLR}}}{\delta T_{\mathrm{sat}}(\phi,p)}\right|_{\{T_{\mathrm{em}},S,C\}=\mathrm{sim}} \tag{7}$$

and

$$K_{T_{sat}}^{OLR_{clr}} = \left.\frac{\delta I^{OLR}}{\delta T_{sat}(\phi,p)}\right|_{\{T_{em},S\}=sim,C=0} \tag{8}$$

where T_{sat} indicates the inversion of the saturation specific humidity function $S(T, p)$ and ϕ is latitude. The set notation $\{\} = \mathrm{sim}$ is used to denote that multiple variables are utilized at climate model-simulated values. For brevity, we have ignored the fact that the kernel computations were done at different longitudes and times, then averaged. The actual computations behind Fig. 2 were performed by probing $\mathrm{OLR}(T_{\mathrm{em}}, S, C)$ with small increments δS at each altitude, with T_{em} and C held constant. Two separate calculations were performed, with C held constant at both realistic model-simulated values ("all-sky") and $C = 0$ ("clear sky" computations). Results of those computations were then recast by those authors, using the chain rule:

$$\left.\frac{\delta I^{\mathrm{OLR}}}{\delta T_{\mathrm{sat}}(\phi,p)}\right|_{T_{\mathrm{em}},S,C} = \left.\frac{\delta I^{\mathrm{OLR}}}{\delta S(\phi,p)}\right|_{T_{\mathrm{em}},C;S} \left.\frac{\partial S_{\mathrm{sat}}}{\partial T}\right|_{p} \tag{9}$$

It is important to notice that the *emission* temperature T_{em} was held constant, while T_{sat} was varied for Fig. 2, as the notation emphasizes. Allowing both to vary consistently leads to the well-known cancelation between the *lapse rate feedback* and *water vapor feedback* in climate sensitivity estimation, as discussed in Soden et al. (2008) and elsewhere. Here we wish to characterize I^{OLR}'s dependence on moisture profiles in more direct units of S, R, or $\ln(R)$, holding actual T constant. To do this, we must convert the kernels of Fig. 2 into other units as follows. For definiteness, line plots of the kernel profiles at specific longitudes and months are shown in Fig. 3a (using digital kernel data for two different climate models, downloaded from http://people.oregonstate.edu/~shellk/kernel.html and http://www.rsmas.miami.edu/personal/bsoden/data/kernels.html).

The easiest conversion is simply to reverse the post-processing done by Soden et al., by rearranging Eq. (9) above:

$$K_{S}^{\mathrm{OLR}} = \left.\frac{\delta I^{\mathrm{OLR}}}{\delta S(\phi,p)}\right|_{T_{\mathrm{em}},C=0;S} = \left.\frac{\delta I^{\mathrm{OLR}}}{\delta T_{\mathrm{sat}}(\phi,p)}\right|_{T_{\mathrm{em}},S,C=0} \left(\left.\frac{\partial S_{\mathrm{sat}}}{\partial T}\right|_{p;T}\right)^{-1} \tag{10}$$

The first factor on the right is kernel (8), the quantity in the bottom panel of Fig. 2 and the top row of Fig. 3. However, the conversion profile (second factor) must be evaluated at some reference profile $T(\phi, p)$, as emphasized with the subscript $;T$. For simplicity, since Fig. shows a tropical point, we use the 350 K pseudo-adiabat again as this reference T profile, a reasonable approximation. We can already anticipate that, since the slope of $S_{\mathrm{sat}}(T)$ increases with T, this factor (the inverse of that slope) will reduce the value of the kernel in the lower troposphere relative to the upper troposphere. In other words: molecule for molecule (or kg for kg), high-altitude vapor is much more important to OLR than low-

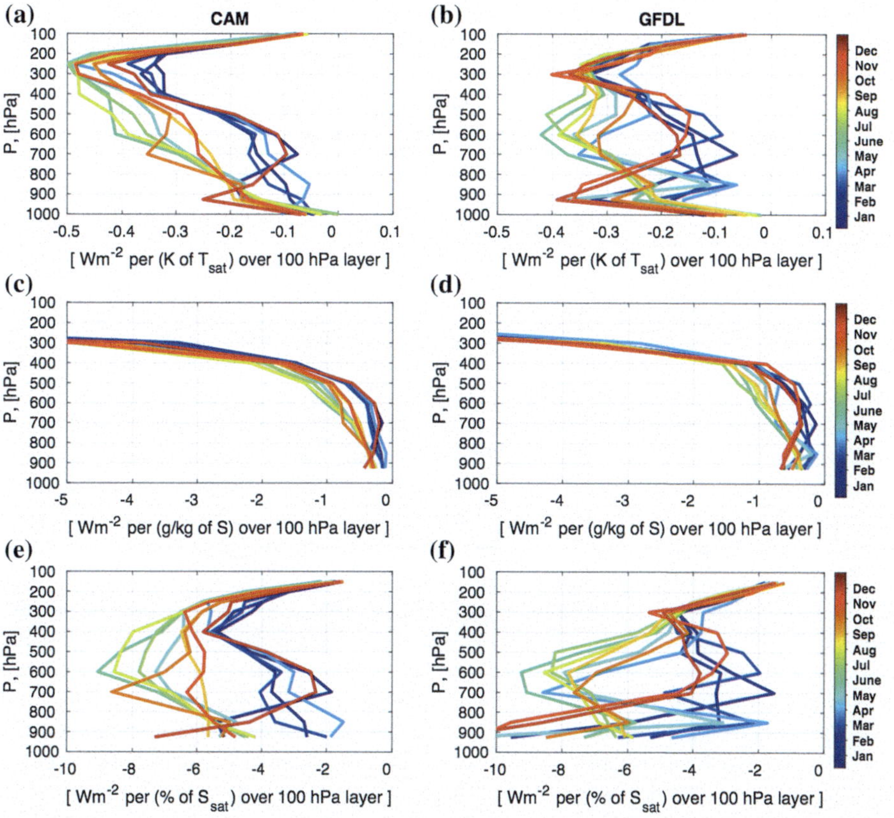

Fig. 3 Top row: The clear-sky kernel of Fig. 2b at a point in the Arabian Sea from **a** CAM and **b** GFDL climate models, using kernel datasets downloaded from the web sites of B. Soden and K. Shell. The twelve colored curves indicate 12 calendar month averages, with dry winter and moist summer monsoon conditions yielding distinctly different sensitivities to increments of moisture. Middle: kernel expressed in terms of increments δS instead using Eq. (10). Bottom: Kernel expressed in terms of increments δR using Eq. (11)

altitude vapor (Fig. 3, middle row), as emphasized in climate literature (e.g., Held and Soden 2000, Allan 2012).

Converting the denominator to reflect the sensitivities of OLR to *relative* humidity increments (δR) is then straightforward, following the steps in Sect. 2:

$$K_R^{\mathrm{OLR_{clr}}} = \left.\frac{\delta I^{\mathrm{OLR}}}{\delta R(p)}\right|_{\{T,S\},C=0} = \left.\frac{\delta I^{\mathrm{OLR}}}{\delta S(p)}\right|_{\{T,S\},C=0} \left(\left.\frac{\partial S}{\partial R}\right|_{T;p}\right) = K_S^{\mathrm{OLR_{clr}}} S_{\mathrm{sat}} \tag{11}$$

Further converting the sensitivity to increments $\delta \ln R$ involves multiplying by a R_0 profile as in (6), which would have to be obtained from these models' climatology of the Arabian Sea:

$$K_{\ln R}^{\mathrm{OLR_{clr}}} = R_0 K_R^{\mathrm{OLR_{clr}}} \tag{12}$$

This final conversion is not shown here, although the seasonality can be imagined in this region with its wet-summer monsoon climate. Even though (as noted above) *molecule for molecule, high-altitude vapor is especially important to OLR*, the bottom row of Fig. 3 indicates that *an increment of RH in the lower troposphere is approximately as consequential as an increment at other levels* (Held and Shell 2012; Vial et al. 2013). For some measurement systems, the *attribution of altitude to detected water molecules* might be another way to express the important uncertainties incurred in a retrieval process.

4 Importance of Vapor for Deep Convection

Deep convection's sensitivities to water vapor increments (or observation errors) are much less straightforward to estimate than the radiative sensitivities, because "convection" is not an instantaneous, local column process—or even a well-defined one (as elucidated in the Introduction of Sherwood et al. 2010). Simple parcel buoyancy arguments (as embodied in algorithms for convective available potential energy (CAPE) and other simple parcel indices) embody the assertion that only low-level (parcel source level) humidity matters. Indeed, if virtual temperature effects are considered, free-tropospheric moisture *reduces* naïvely computed CAPE. Meanwhile, in reality, mid-level moisture clearly impacts convection positively, as illustrated by the steep dependence of conditional rainfall on column water vapor (Bretherton et al. 2004; Neelin et al. 2009).

The next complexity of a theoretical account is to allow for mixing (entrainment) into updrafts. The buoyancy of a small-scale updraft then depends on humidity in the source layer, as well as in the layers that it has traversed and mixed with. For instance, the steep dependence on column vapor mentioned above may be usefully interpreted as a threshold-like dependence on parcel source-level humidity, plus a linear dependence on humidity in the free troposphere above that (Muller et al. 2009). Microphysical processes such as precipitation shedding during the finite time of parcel ascent are another ambiguity in determining bulk density and thus parcel buoyancy.

Even such a dressed-up (mixing-allowing, microphysics-respecting) parcel theory remains badly incomplete, however. The "importance" of convection as a process lies not in the fate of one hypothetical lone updraft, but rather in the time-integrated, net behavior of *ensembles* of convective *circulations* (perhaps in a scale-truncated sense) that must obey mass continuity in a particular geometrical configuration. Essentially, the kernel we seek is a subset of the cumulus parameterization problem (Arakawa 2004). For this reason, explicit convection-permitting models (CPMs) are needed to fill the gap between parcel theory and reality in estimating our desired sensitivity kernel, or even defining it with sufficient specificity.

In a convecting atmosphere, a humidity increment provokes a complex time-dependent and multi-scale response that includes thermally driven large-scale ascent as well as local cloud-scale overturnings. Observationally, all this is further superposed with noise. Simple regression of local convection measures on humidity observations thus gives us only a faint and distorted glimpse of moisture's true impact on convection (as discussed in Mapes et al. 2017). Active probing of models is required to even begin to interpret observations properly.

Early computation-limited studies used short integrations over small domains, in an *initial-value* approach with newly triggered convection, in 2D (Nicholls et al. 1988) or 3D (Takemi and Satomura 2000). Mapes (2017) discusses in more detail how domain constraints translate into function or importance measures. As computers have advanced, fully developed convection fields in larger and longer runs have been probed, for instance using humidity field relaxation experiments (Derbyshire et al. 2004 in that special issue) or transient impulses (Tulich and Mapes 2010; Kuang 2010). Long-time integrations make cyclic boundary conditions appealing for conservation reasons. The unreality of cyclicity is compensated by forcing applied through time, which subtly but profoundly reshapes the questions being asked and addressed. Taken together, all these approaches indicate that GCM convection parameterizations almost universally have inadequate sensitivity to mid-level moisture, rooted foreseeably in their use of simplistic updraft instability indices. However, offering a better quantitative estimate is more difficult than offering a critique.

Because deep convection is a "noisy" process, with lots of internal free variability, *ensemble* runs of cyclic CPMs (CCPMs), each with its internal ensemble of convective circulations, are required to find the systematic response. But although the response of convection to humidity is not very *deterministic*, its expectation value is very *linear* (Tulich and Mapes 2010). Unfortunately, probing such an ensemble produces large numbers of time-dependent responses that are difficult to summarize. However, the mathematical linearity of responses opens another avenue.

That linearity was exploited by Kuang (2010) to construct a time-invariant linear response matrix **M** through an elegant matrix inversion procedure. The time-dependent responses mapped by Tulich and Mapes (2010) were shown to be merely a facet of **M**, specifically $\exp(\mathbf{M}t)$, as illustrated in his Appendix. A finite-time propagator matrix $\mathbf{G} = (\exp(\Delta t \cdot \mathbf{M}) - \exp(0 \cdot \mathbf{M}))/\Delta t$ is another facet of **M**, and such matrices for $\Delta t = 4$ h are depicted in Fig. 8 of Kuang (2012).

Here let us define I^{con} = *rainfall anomalies over the subsequent* 3 h in a convecting patch of atmosphere of size $o(10^2–10^3)$ km. Our desired kernel or sensitivity profile can be obtained by integrating one quadrant of **G** over atmospheric mass. But what has been held constant in such a computation? An important part of the answer is *domain-averaged vertical motion* $[w] = 0$ *at all altitudes*, by the cyclic boundary conditions of a CCPM. All the other parameters specified in running the CCPM are also important, such as domain geometry and size. The kernel we seek can be expressed as:

$$K_S^{\text{con}} = \left.\frac{\delta I^{\text{con}}}{\delta S(p)}\right|_{RCE:\{[w]=0,\text{geometry},\text{SST},Q_{\text{rad}},\ldots\}} \tag{13}$$

where the set of conditions {…} for the radiative-convective equilibrium (RCE) reference state includes all of the parameters of the CCPM's forcing and configuration.

To better appreciate the role of domain geometry, consider two RCE configurations: 128 km × 128 km isotropic flow with no wind shear to break symmetry, and 2048 km × 64 km, also unsheared. Column water vapor maps for a 2048 km square sample of these two unbounded (but cyclic) domain symmetries are shown in Fig. 4. Domain-mean profiles are inset. Two differences are obvious: (1) convection is "organized" in the sense of *quasi-two-dimensional* in the long-domain run, and (2) the long-domain average is warmer and drier, because a large area with thermally capped (stable) dry air contributes a lot toward the domain average. One way to look at this is that the ensemble of convective circulations defining I^{con} has all the descending branches concentrated and reinforcing each other in the long domain, rather than distributed randomly and canceling each other as they do in the isotropic domain. This is a symmetry condition as in crystallography, not a domain "size" issue per se, since both atmospheres are really horizontally unbounded.

The kernel (13) differs substantially between these two cases, as seen in the right-hand panels of Fig. 5b, d. The left panels of Fig. 5a, c also show the sensitivity to temperature

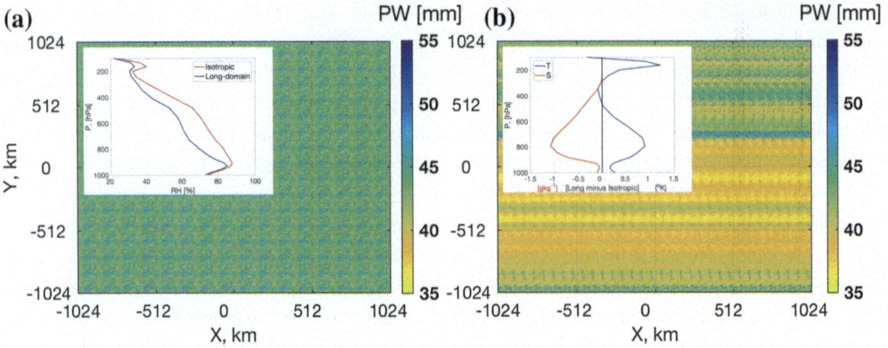

Fig. 4 Column water vapor maps for 2048 × 2048 km samples of the unbounded atmospheres embodied by cyclic cloud-permitting model (CCPM) simulations with **a** 128 × 128 km grid and **b** 2048 × 64 grid. Grid mesh size is 2 km in all cases. Inset line plots show that the domain average of the long domain with "organized" convection is warmer and drier in the lower free troposphere, as well as warmer near the tropopause

for completeness (discussed briefly below, as this is a novel result). The plot titles emphasize Unorganized (isotropic) versus Organized (long) geometries as the nature of the difference, but interpretation must be considered more carefully than that. The units in Fig. 5 should be self-explanatory after the discussions above.

Consider first the Unorganized sensitivity profiles in Fig. 5a, b, derived from the isotropic CCPM of modest size (128 km) with no wind shear. Convection consists of intermittent scattered cumulonimbus (Cb) clouds. As predicted by parcel notions of buoyant moist convection, an increment of water vapor δS in the lowest kilometer or two has the biggest effect, but the same δS of vapor added to other levels still has a positive effect. All levels above about 700 hPa are about equally important to rainfall production (panel 5b), in this geometry and RCE base state. Temperature sensitivity can be similarly understood in parcel buoyancy terms (panel 5a). There is positive sensitivity to temperature in the low-level parcel source layer, while ambient environment warmth from 900 to 500 hPa acts to reduce parcel buoyancy, an "inhibition" effect, but one that is much deeper than naïve undiluted parcel computations would suggest.

By contrast, Organized convection (long domain, panels 5c, d) is much more sensitive to water vapor increments at all levels, especially in the free troposphere (panel 5d). In this elongated domain (sometimes called "bowling-alley" geometry), convective circulations necessarily take the form of squall-like "layer overturning" (Kuang 2012) rather than sporadic buoyant parcel ascent. Interpretation therefore must recognize such layer overturning as the system whose response to a horizontally uniform increment δS is being measured. Such a horizontally uniform perturbation may seem like a fiction, but could perhaps be viewed as being generated by advection by a much larger-scale adiabatic vertical motion.

In the long geometry, a deep *layer* of lower-tropospheric air is rising, so enhanced humidity or temperature in that layer can enhance domain-mean precipitation for reasons that do not involve horizontal mixing (entrainment) into a convective-scale buoyant updraft. Apparently for this reason, humidity and temperature sensitivities are strong and positive up to 700 hPa. Upper-tropospheric δS increments also have a very strong impact on domain precipitation (panel 5d), presumably by enhancing stratiform precipitation from upper-level stratiform cloud (Houze 1997) and not from mixing effects on buoyancy in the upper levels of Cb updrafts. Upper-level temperature has a negative impact. Might that also be interpreted as an effect involving the precipitating upper-level stratiform cloud, or is it

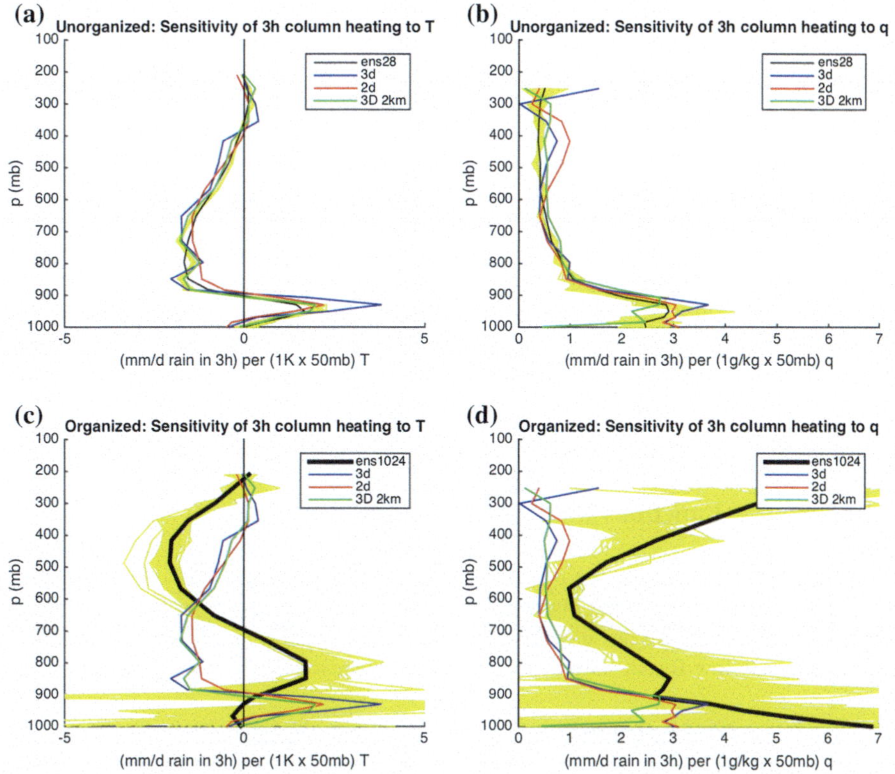

Fig. 5 Profiles of the importance of an increment of temperature (left) or water vapor (right) at any given altitude to rainfall rate (expressed in mm/day) averaged over the subsequent 3 h. Results are for a cyclic convection-permitting model at equilibrium with a forcing that produces a rainrate of about 4 mm/day. Top row: Unorganized refers to convection in small cyclic domains with no wind shear. Black curve and yellow band show estimates from ensembles of simulations. Early estimates made with smaller domains and a 2-dimensional computation (blue, green, red) are repeated in both panels for reference. Bottom row: Organized refers to convection in a 2048 × 64 km elongated cyclic domain. An ensemble approach has been used to estimate uncertainty (yellow) around a lightly vertically smoothed mean estimate (heavy black curve)

again an "inhibition" effect on convection—a reduction of the updraft buoyancy that is the ultimate energy source for the convective circulation (through the buoyancy flux b'w' source term in the kinetic energy equation)?

Our take-away lessons about moisture sensitivity must notice that the domain-mean condition is very much drier in the "Organized" (long domain) case (insets to Fig. 4), or to put it more sharply, *organized convective systems can exist in drier mean conditions than ordinary convection* (Takemi and Satomura 2000). In these CCPMs, background $S(p)$ is not imposed, and cannot easily be changed without breaking the equilibrium condition of RCE, so this is a result in itself. In nature, the asymptotically long-time equilibrium of the moisture field enforced in steady CCPM runs is rarely or never observed on the scales of Fig. 4b. In the burgeoning literature on radiative-convective equilibrium (RCE) (Wing et al. 2017; Mapes 2016), the >20 mm dynamic range of column water vapor (CWV) in Fig. 4 would take many days to develop, making it an artifact of cyclic boundaries that may not directly correspond to nature (Holloway et al. 2017).

If the most important result of these experiments is that organized convection can survive in drier environments, the functional lesson for nature might be that it is *less sensitive to dryness* than ordinary convection—opposite in sense from the face value of Fig. 5 (enhanced sensitivity). For instance, the ability of quasi-2D squalls to survive hostile environments can widen the time-mean tropical rainfall belt relative to its treatment only as vertical plume convection (Nolan et al. 2016). Sahelian Africa may similarly be the beneficiary of rain from organized storms, in environments too dry to support local vertical precipitating convection (e.g., Section 5 of Nicholson 2013).

To re-express the denominator of these kernels in terms of δR would involve multiplying them by $S_{sat}(T)$, making them much more bottom-heavy, as in the middle to bottom row differences of Fig. 3. This is a useful insight: it actually takes a very large or even unrealizable RH perturbation to account for 1 g/kg of δq in the middle or upper troposphere, as postulated in the units of Fig. 5, so the large kernel values at upper levels in Fig. 5b. may be somewhat irrelevant in practical terms.

5 Importance Functions for Passive Remote Sensing

To extend the radiative reasoning of Sect. 3, consider another Importance function: I^{sig} = signal detected by some instrument. Figure 6 sketches kernel profiles for I^{sig} = -brightness temperature for down-looking microwave instruments in spectral regions where water vapor has low (cyan, purple) and high (blue) emissivity. In this figure (following

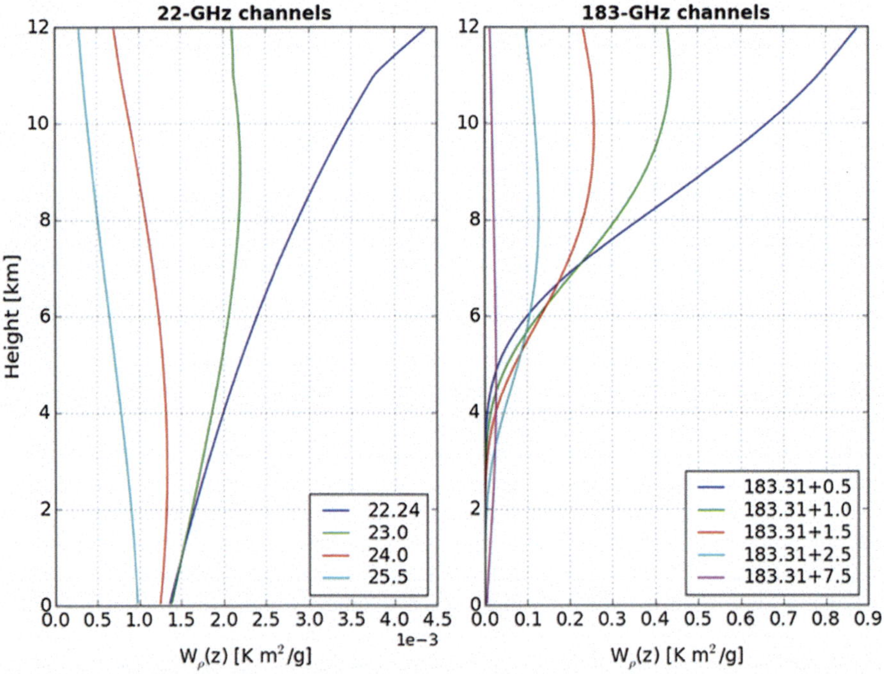

Fig. 6 Contributions of water vapor to brightness temperature seen from above, as a function of altitude, in **a** low and **b** high emissivity channels, with a radiatively cold background (water, with its high reflectivity and thus low emissivity). Conventions as in Mech et al. (2014)

Mech et al. 2014), our "kernel" K is depicted as a "weighting function" W, whose denominator is calibrated to increments not of S or R but of vapor density ρ.

$$K_{\rho}^{\mathrm{sig}} = \left.\frac{\delta I^{\mathrm{sig}}}{\delta \rho_{v}(z)}\right|_{\{T,\mathrm{backdrop}\};\rho} \equiv W_{\rho}(z) \tag{14}$$

The units can be unpacked as (K of signal) per (1 g m^{-3}) over a (1 m layer), while the small 1e−3 label clarifies that the axis numbers are really over a 1 km layer. Perhaps more clearly, it can be read as (K of signal) per (1 kg m^{-2} of a local *layer* of vapor mass, centered at each altitude).

In the limit of low emissivity in the measurement band, every molecule of vapor sends photons to the detector with equal efficiency, with no blockage by the intervening vapor. The cyan curve is near this limit and is greater at low levels because the emitting temperature is greater there, not because distant molecules are more efficacious than close ones at contributing to the detected signal. Meanwhile, the higher the emissivity in the band being observed, the more the photons come solely from the vapor nearest the detector, as indicated by blue curves for high emissivity bands. Using high emissivity bands, multichannel passive sensors can achieve fine vertical resolution, but only in a robust manner in areas close to the sensor.

Unfortunately, the only way to achieve fine resolution far from a passive sensor is through the error-prone differencing of highly overlapping kernels. In the downward-looking case, surface emission introduces further errors and ambiguities to such an exercise. As a result, no combination of kernels like Fig. 6 will yield good profile resolution and accuracy at low-middle levels, as desired based on the motivating physical process sensitivity kernels above. Upward-looking passive sensors could utilize high emissivity channels to achieve high resolution near the surface, but vertical discrimination in the lower free troposphere is harder, and made harder still by the decrease in both S and the emitting temperature with altitude. In practice, conventional microwave sensors achieve only about 2–3 distinct degrees of freedom in whole-troposphere moisture profiling (Brogniez et al. 2016; Zhang et al. 2017; see also Fig. 1 of Pincus et al. 2017, this volume).

While hyperspectral passive methods with very high sensitivity and precision can improve on existing technology, it seems clear that the best way to probe vertically resolved vapor in the lower free troposphere is with active sensors (Nehrir et al. 2017, see also Fig. 10 of Stevens et al. 2017; both in this volume). Light Detection And Ranging (LiDAR) and its radio cousin (RADAR) systems use time (range) gating to achieve the desired vertical resolution. Limb occultation uses angular geometry (see also Fig. 3 of Pincus et al. 2017, this volume). One challenge of LiDAR is to achieve sensitivity across a wide dynamic range: to best serve our needs, radiation must get through intervening layers, twice, and still interact to an accurately measurable degree with vapor in the desired mid-tropospheric layers. These may be quite dry, and yet the physical importance of accuracy may grow with background dryness, for instance in their radiative importance (Spencer and Braswell 1997; Pierrehumbert et al. 2007), raising the demands on laser power and/or detector sensitivity, especially when looking upward from the surface through optically thick intervening layers. Spaceborne active platforms have an advantage in this regard, and the use of multiple frequencies helps break up the extreme dynamic range challenge into more tractable chunks (Fig. 1 of Nehrir et al. 2017, this volume).

Radiative cooling of clouds at the top of the planetary boundary layer (PBL) is crucial to their dynamics, longevity, vigor, albedo, and thereby to the climate impact they exert in the visible spectrum (rejecting sunlight). Low-middle troposphere vapor is thus especially important to these clouds, by modulating cloud-top cooling, as discussed in Stevens et al. (2017) and elsewhere. Yet these are precisely the altitudes and dry conditions for which vapor measurement physics poses the greatest bind for both passive and active measurements, as sketched in the paragraphs above. Here lies an especially enticing opportunity for technological glory.

Designing such glorious systems will require a clear view of "importance" functions of vapor for both physical climate system impacts (radiation and convection) and detectability. It is hoped that this paper may help to clarify the challenge and the hopes for surmounting it.

6 Summary and Conclusions

We have explored the sensitivity of important processes (measured by scalars) to water vapor as a function of height. In every case, lower-middle troposphere is important. Unfortunately, passive measurements are poor for this region of the atmosphere. Active sensing (see Nehrir et al. 2017) is advocated as a crucial technology approach to improve measurements and, thereby, our understanding of important processes. Synthesis of all observations within atmospheric analysis systems will be a final challenge, with modeling as well as detection aspects (Pincus et al. 2017, this volume), but is a path to demonstrate such understanding as a powerful estimation and prediction capability in regions beyond the observed.

Acknowledgements This article is based on work supported by U.S. NOAA Grant NA13OAR4310156 and NASA Grant NNX15AD11G. Mario Mech kindly contributed the unpublished Fig. 6, related to a similar figure in Mech et al. (2014). We are grateful for comments by Bjorn Stevens, Robert Pincus, and anonymous reviewers whose efforts greatly improved the manuscript.

References

Adebiyi AA, Zuidema P, Abel SJ (2015) The convolution of dynamics and moisture with the presence of shortwave absorbing aerosols over the southeast Atlantic. J Clim 28:1997–2024. doi:10.1175/JCLI-D-14-00352.1

Allan RP (2012) The role of water vapour in Earth's energy flows. Surv Geophys 33(3–4):557–564

Arakawa A (2004) The cumulus parameterization problem: past, present, and future. J Clim 17:2493–2525

Bretherton CS, Peters ME, Back L (2004) Relationships between water vapor path and precipitation over the tropical oceans. J Clim 17:1517–1528

Brogniez H, Fallourd R, Mallet C, Sivira R, Dufour C (2016) Estimating confidence intervals around relative humidity profiles from satellite observations: application to the SAPHIR sounder. J Atmos Ocean Technol 33:1005–1022. doi:10.1175/JTECH-D-15-0237.1

DeAngelis AM et al (2015) An observational radiative constraint on hydrologic cycle intensification. Nature 528:249–253

Derbyshire S, Beau I, Bechtold P, Grandpeix J-Y, Piriou J-M, Redelsperger J-L, Soares P (2004) Sensitivity of moist convection to environmental humidity. Q J R Meteorol Soc 130:3055–3079. doi:10.1256/qj.03.130

Held IM, Shell KM (2012) Using relative humidity as a state variable in climate feedback analysis. J Clim 25(8):2578–2582

Held IM, Soden BJ (2000) Water vapor feedback and global warming. Annu Rev Energy Environ 25:441–475

Holloway CE, Wing AA, Bony S, Muller C, Masunaga H, L'Ecuyer TS, Turner DD, Zuidema P (2017) Observing convective aggregation. Surv Geophy. doi:10.1007/s10712-017-9419-1

Houze RAJ (1997) Stratiform precipitation in regions of convection: a meteorological paradox? Bull Am Meteorol Soc 78:2179–2196

Kuang Z (2010) Linear response functions of a cumulus ensemble to temperature and moisture perturbations and implications for the dynamics of convectively coupled waves. J Atmos Sci 67:941–962

Kuang Z (2012) Weakly forced mock-Walker cells. J Atmos Sci 69:2759–2786

Mapes BE (2016) Gregarious convection and radiative feedbacks in idealized worlds. J Adv Model Earth Syst. doi:10.1002/2016MS00065

Mapes BE (2017) Toward form-function relationships for mesoscale structure in convection: a review. J Meteorol Soc Jpn (**submitted**)

Mapes BE, Chandra A, Kuang Z, Song S, Zuidema P (2017) Estimating convection's moisture sensitivity: a model-observation synthesis using DYNAMO data. J Atmos Sci (**in preparation**)

Mech M, Orlandi E, Crewell S, Ament F, Hirsch L, Hagen M, Peters G, Stevens B (2014) HAMP the microwave package on the high altitude and long range research aircraft HALO. Atmos Meas Tech 7:4539–4553. doi:10.5194/amt-7-4539-2014

Muller CJ, Back LE, O'Gorman PA, Emanuel KA (2009) A model for the relationship between tropical precipitation and column water vapor. Geophys Res Lett 36:L16804. doi:10.1029/2009GL039667

Neelin JD, Peters O, Hales K (2009) The transition to strong convection. J Atmos Sci 66:2367–2384. doi:10.1175/2009JAS2962.1

Nehrir AR et al (2017) Emerging technologies and synergies for airborne and space-borne measurements of water vapor profiles. Surv Geophy. doi:10.1007/s10712-017-9436-0 **(this volume)**

Nicholls ME, Johnson RH, Cotton WR (1988) The sensitivity of two-dimensional simulations of tropical squall lines to environmental profiles. J Atmos Sci 45:3625–3649. doi:10.1175/1520-0469(1988)045<3625:TSOTDS>2.0.CO;2

Nicholson SE (2013) The West African Sahel: a review of recent studies on the rainfall regime and its interannual variability. ISRN Meteorol, vol. 2013, Article ID 453521. doi:10.1155/2013/453521

Nolan DS, Tulich SN, Blanco JE (2016) ITCZ structure as determined by parameterized versus explicit convection in aquachannel and aquapatch simulations. J Adv Model Earth Syst 8:425–452. doi:10.1002/2015MS000560

Nolte DD (2010) The tangled tale of phase space. Phys Today 63:33–38

Pierrehumbert RT, Brogniez H, Roca R (2007) On the relative humidity of the Earth's atmosphere. In: Schneider T, Sobel A (eds) The global circulation of the atmosphere: phenomena, theory, challenges. Princeton University Press, Princeton

Pincus R et al (2017) The representation of tropospheric water vapor over low-latitude oceans in (re-) analysis. Errors, impacts, and the ability to exploit current and prospective observations. Surv Geophys. doi:10.1007/s10712-017-9437-z **(this volume)**

Previdi M (2010) Radiative feedbacks on global precipitation. Environ Res Lett 5:025211. doi:10.1088/1748-9326/5/2/025211

Sherwood SC, Roca R, Weckwerth TM, Andronova NG (2010) Tropospheric water vapor, convection, and climate. Rev Geophys 48(2):RG2001. doi:10.1029/2009RG000301

Soden B, Shell Karen M, Kiehl Jeffrey T, Shields Christine A (2008) Quantifying climate feedbacks using radiative kernels. J Clim 21:3504–3520. doi:10.1175/2007JCLI2110.1

Spencer RW, Braswell William D (1997) How dry is the tropical free troposphere? Implications for global warming theory. Bull Am Meteorol Soc 78:1097–1106. doi:10.1175/1520-0477

Stephens GL, Ellis TD (2008) Controls of global-mean precipitation increases in global warming GCM experiments. J Clim 21:6141–6155

Stevens B, Brogniez H, Kiemle C, Lacour J-L, Crevoisier C, Kiliani J (2017) Structure and dynamical influence of water vapor in the lower tropical troposphere. Surv Geophys. doi:10.1007/s10712-017-9420-8

Takemi T, Satomura T (2000) Numerical experiments on the mechanisms for the development and maintenance of long-lived squall lines in dry environments. J Atmos Sci 57:1718–1740

Tulich S, Mapes BE (2010) Transient environmental sensitivities of explicitly simulated tropical convection. J Atmos Sci 67:923–940. doi:10.1175/2009JAS3277.1

Vial J, Dufresne J-L, Bony S (2013) On the interpretation of inter-model spread in CMIP5 climate sensitivity estimates. Clim Dyn 41:3339–3362

Wing AA, Emanuel K, Holloway CE, Muller C (2017) Convective self-aggregation in numerical simulations: a review. Surv Geophys. doi:10.1007/s10712-017-9408-4

Zhang J, Zuidema P, Turner D, Cadeddu M (2017) Tropical humidity vertical structure inferred from a microwave radiometer over the equatorial Indian Ocean. J Appl Meteorol

Surv Geophys (2017) 38:1371–1397
https://doi.org/10.1007/s10712-017-9420-8

Structure and Dynamical Influence of Water Vapor in the Lower Tropical Troposphere

Bjorn Stevens[1] · **Hélène Brogniez**[2] · **Christoph Kiemle**[3] · **Jean-Lionel Lacour**[4,5,6] · **Cyril Crevoisier**[7] · **Johannes Kiliani**[1]

Received: 12 February 2017 / Accepted: 4 July 2017 / Published online: 26 July 2017

Abstract In situ, airborne and satellite measurements are used to characterize the structure of water vapor in the lower tropical troposphere—below the height, z_*, of the triple-point isotherm, T_*. The measurements are evaluated in light of understanding of how lower-tropospheric water vapor influences clouds, convection and circulation, through both radiative and thermodynamic effects. Lower-tropospheric water vapor, which concentrates in the first few kilometers above the boundary layer, controls the radiative cooling profile of the boundary layer and lower troposphere. Elevated moist layers originating from a preferred level of convective detrainment induce a profile of radiative cooling that drives circulations which reinforce such features. A theory for this preferred level of cumulus termination is advanced, whereby the difference between T_* and the temperature at which primary ice forms gives a 'first-mover advantage' to glaciating cumulus convection, thereby concentrating the regions of the deepest convection and leading to more clouds and moisture near the triple point. A preferred level of convective detrainment near T_* implies relative humidity reversals below $z*$ which are difficult to identify using retrievals from satellite-borne microwave and infrared sounders. Isotopologues retrievals provide a hint of such features and their ability to constrain the structure of the vertical humidity profile merits further study. Nonetheless, it will likely remain challenging to resolve dynamically

✉ Bjorn Stevens
bjorn.stevens@mpimet.mpg.de

1 Max Planck Institute for Meteorology, 20146 Hamburg, Germany

2 Laboratoire Atmospheres, Milieux, Observations Spatiales, 78280 Guyancourt, France

3 Deutsches Zentrum für Luft- und Raumfahrt, Institute of Atmospheric Physics, 82234 Oberpfaffenhofen, Germany

4 UPMC Univ. Paris 06, Paris, France

5 LATMOS-IPSL, Universit Versailles St-Quentin, Paris, France

6 Institute of Earth Sciences, University of Iceland, Reykjavík, Iceland

7 Laboratoire de Météorologie Dynamique, CNRS, IPSL, Ecole Polytechnique, 91128 Palaiseau Cedex, France

important aspects of the vertical structure of water vapor from space using only passive sensors.

Keywords Water vapor · Convection · Atmospheric circulation · Ice initiation · Remote sensing · Atmospheric measurements · Clouds

1 Introduction

If it did not have to condense to become visible, water vapor would fuel the fascination of many more scientists. Imagine seeing with the naked eyes how elevated layers of water vapor, and its radiative effects, engender shallow circulations, or how pockets of humidity surround and socialize cumulus convection. Imagination is indeed necessary because water vapor's mysteries arise as much from its visible transparency as from the opulence of its infrared opacity (Stevens and Bony 2013).

Water vapor couples to atmospheric circulations directly, through water vapor's thermodynamic effect on the development of clouds and convection, and indirectly, through water vapor's non-local influence on infrared irradiances. The thermodynamic effect has been extensively studied over the past decades (Sherwood et al. 2010), the radiative effect—especially in the lower troposphere—has not. The purpose of this article is to review our understanding, and to evaluate our ability, to remotely sense important features of the lower-tropospheric water vapor distribution. We do so with the help of a series of flight campaigns in the tropical north Atlantic. In so doing we identify a new mechanism which, we hypothesize, contributes to the prevalence of mid-level clouds, near the triple-point temperature ($T_* = 0.01\,^\circ$C).

This paper is organized as follows. In Sect. 2 we fix terminology and provide an overview of the field measurements and satellite data-products used in this study. To provide context for the measurements, in Sect. 3 we review how water vapor's thermodynamic properties influence clouds and convection and how its radiative properties influence circulation. In Sect. 4 the ability of recent measurements, including state-of-the-art active and passive remote sensing, to delineate the humidity structure of the lower-troposphere, are evaluated. The measurements motivate a new hypothesis (Sect. 5) as to how water vapor radiative effects combine with barriers to ice initiation to help explain a preponderance of mid-level convection in the vicinity of deeper convection. The main points of the article are summarized in Sect. 6.

2 Data and Context

Terminology is fixed with the help of Fig. 1, which illustrates important features of the distribution of water vapor in the tropical troposphere. The *planetary boundary layer,* or PBL, is here defined as the layer below the lifting condensation level. It is assumed to be well mixed by turbulence, so that it is meaningful to talk about values of conserved quantities within the PBL in terms of single numbers, for instance the PBL specific humidity, denoted q_{PBL}. A *shallow cumulus layer,* whose top is usually between 1.5 and 3.0 km, can often be identified not only by the mean height of the clouds that form there, but also by a slight temperature inversion or, more commonly, a pronounced lapse in the humidity (hydrolapse) at its top. In Fig. 1 this height is denoted by z_{i} and placed at about 3 km. The layer between the average z_{i} and the height, z_*, of the triple-point isotherm T_* is

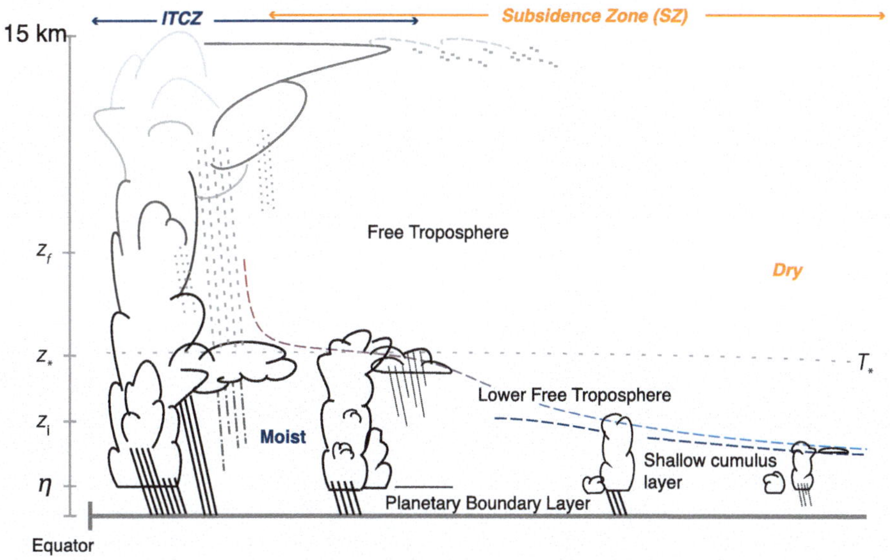

Fig. 1 Conceptual diagram illustrating layers of moisture, and some of the terminology, discussed in this article. The height of the sub-cloud layer, which delineates the planetary boundary layer or PBL, is denoted by η; the height of the trade-inversion or the hydrolapse delineating the shallow cloud layer is denoted by z_i; the height of the triple-point isotherm, T_*, is denoted by z_*, and the height of primary ice formation isotherm (T_f) is denoted by z_f. For the formation of ice by homogeneous nucleation $T_f = -38\,^\circ$C. The term lower free troposphere is used to denominate the region below the height of T_* and the average height of the fair-weather cumulus layer. In the present-day tropics η is at about 600 m, the z_i can be quite variable, but when evident is typically between 2 and 3 km. z_* locates at around 4.5 km

referred to as the *lower free troposphere*, LFT. We speak of the *triple-point level* rather than the 'melting level' or 'freezing level' as the latter are less well defined. In cases when the shallow cumulus layer is completely suppressed, the lower free troposphere will extend down to the top of the PBL. In disturbed conditions the shallow cumulus layer may extend to the triple-point level.

2.1 Airborne Measurements and the Barbados Cloud Observatory

The Next-Generation Airborne Remote Sensing for Validation Studies (NARVAL) field campaigns made use of the Germany Research Community's HALO (High Altitude Long Range) research aircraft (Stevens et al. 2016) to make measurements over the northern tropical Atlantic. NARVAL has had two phases: Phase I took place in the downstream winter trades in December of 2013; phase 2 took place in and around the vicinity of the Atlantic ITCZ in August 2016. NARVAL-1 consisted of eight flights, with about 40 h of flight time over the North Atlantic trades abutting the ITCZ. NARVAL-2 consisted of roughly twice as many flight hours distributed over ten flights in, around, and across the ITCZ. For NARVAL-1, the configuration of HALO is described by Stevens et al. (2016). For NARVAL-2, the configuration was similar. Most relevant for this study is the extensive deployment of dropsondes and continuous water vapor profiling using a downward staring differential absorption lidar as elaborated upon below.

Airborne measurements were taken east of the Barbados Cloud Observatory (BCO), over the ocean between 40–60°W and 5–20°N. The BCO comprises an extensive collection of advanced ground-based remote sensing instruments located on Deebles Point, a promontory on the windward side of Barbados at 59.43°W, 13.17°N. Measurements at the BCO were initiated by the Max Planck Institute for Meteorology and the Caribbean Institute for Meteorology and Hydrology, with support from the Museum of Barbados, to advance understanding of processes influencing the distribution of convective clouds in the lower troposphere. The BCO has been in continuous operation since 2010 (Stevens et al. 2016). During NARVAL-1, the western Atlantic ITCZ was located around 10°N; during NARVAL-2 it varied between 5–15°N. Following the terminology introduced in Fig. 1, NARVAL-2 sampled the broader ITCZ region, NARVAL-1 the subsidence zone.

2.1.1 Dropsonde Humidity Measurements During the NARVAL Campaigns

During NARVAL-1, roughly forty Vaisala RD94 Dropsondes (hereafter sondes) were launched, most from an altitude of about 14 km. During NARVAL-2, many more (roughly 200) were sondes, most from an altitude of 9 km. The Vaisala RD94 Dropsonde uses twin thin-film capacitors to measure humidity with a stated accuracy of 2% (absolute relative humidity). To reduce sampling biases and help remove outliers associated with sondes launched from the ferry flights, only sondes with a near-surface air-temperature greater than a threshold (25 °C for NARVAL-1 and 27 °C for NARVAL-2) are analyzed. On those flights where more than ten sondes satisfied this criterion, the list of sondes was sub-sampled by taking a uniform stride through the sonde-list, to ensure a relatively homogeneous coverage of eight to ten sondes per flight. In Fig. 2 the location of all sondes analyzed in this paper is presented and demarcates the area of flight operations as indicated earlier. All dropsondes were processed using the ASPEN V3.3-270 software, and the data were further visually inspected for possible biases or malfunction before being incorporated into the analysis.

Differences between the composite moisture sounding for NARVAL-1 versus NARVAL-2 measure differences between the winter trades (subsidence zone, Fig. 1) and the atmosphere near and around the ITCZ. In the NARVAL-1 soundings, the lower free troposphere is very dry, almost devoid of water vapor (Fig. 3). The shallow cumulus layer is well defined in terms of a moist layer extending from just above the top of a yet moister PBL to a height of 2 km, where it terminates in a strong hydrolapse. The principal mode of

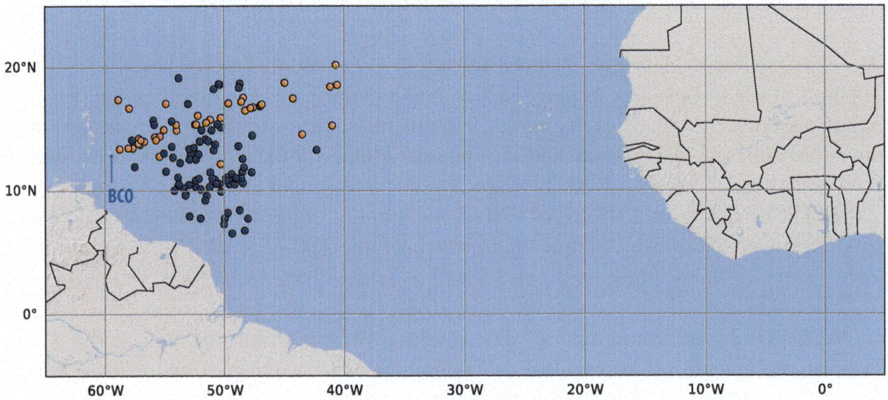

Fig. 2 Location of NARVAL-1 (*orange*) and NARVAL-2 (*blue*) sondes (42 and 81 sondes, respectively) used in the analysis and the location of the Barbados Cloud Observatory (BCO) is indicated

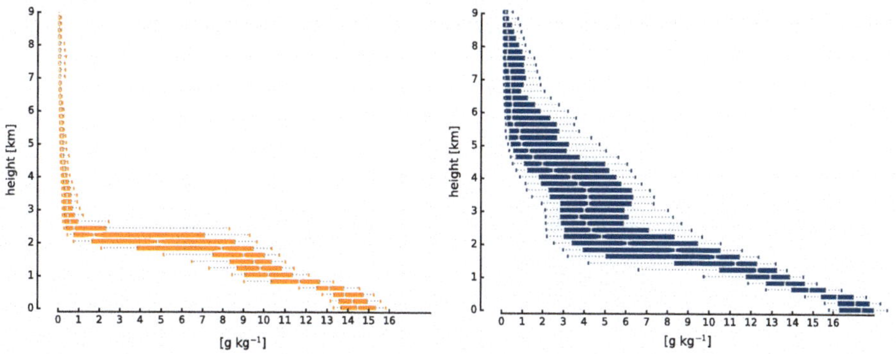

Fig. 3 NARVAL-1 (*left*) and NARVAL-2 (*right*) humidity profiles composited from dropsondes shown in Fig. 2. The box and whisker plot shows the distribution of humidity values measured within a vertical height interval in terms of the values at the 10th and 90th percentile (*whiskers*), the interquartile (*box*) and the median (*notch*) of the distribution

variability is thus expressed by variations of z_i, i.e., the height of the top of the shallow cumulus layer. This is evident in the large range of values of specific humidity, q, around 2 km. During NARVAL-2, where sondes were dropped in and across the ITCZ as well as in dry regions in the vicinity of deeper convection, a shallow cumulus layer is less distinct. This is because the height of the shallow cumulus layer is more variable as is the humidity in the lower free troposphere, with median humidity values near 5 g kg^{-1}. Surface temperatures are warmer in the air-masses sampled during NARVAL-2, as indicated by more elevated values of q_{PBL}. In both NARVAL-1 and NARVAL-2, the PBL is relatively well mixed, as shown by the relative constancy in the humidity within the lowest two (NARVAL-2) to three (NARVAL-1) levels of the humidity distribution.

2.1.2 WALES

Water vapor is continuously profiled by Water vapor Lidar Experiment in Space (WALES; Wirth et al. 2009), a differential absorption lidar designed to actively sense water vapor. WALES was operated on board HALO in a nadir viewing mode. Measurements at four wavelengths sample three water vapor lines of varying strength and allow WALES to measure water vapor over the enormous range of values found within the tropical troposphere. The water vapor molecule number density is proportional to the difference in absorption between the three "online" pulses and a reference "off-line" wavelength, as function of distance from the lidar. Weaker lines are better suited to measuring high-concentrations, where the signal from stronger lines would saturate, while stronger lines provide more signal to measure low concentrations. The differential absorption retrieval is applied with an effective vertical resolution of 200 m. To reduce instrument noise, the individual on- and off-line profiles are accumulated to 12 s averages, which results in a horizontal resolution of 2.5–2.9 km, depending on the aircraft speed. As a consequence, small clouds are detected only in the higher-resolved lidar backscatter (not shown here), and gaps $> \simeq 3$ km between clouds permit water vapor lidar profiles down to 200 m above sea level. Comparisons with the dropsondes yield agreement within a few percent, which is on the order of the water vapor spectroscopic and sonde measurement uncertainties. Details on WALES and its NARVAL-1 measurements are found in Kiemle et al.

(2017). More information on emerging technologies for better (tropical) water vapor measurements has been compiled by Nehrir et al. (2017).

2.2 SAPHIR and Megha-Tropiques

The Indo-French Megha-Tropiques satellite has been observing the tropical troposphere since October 2011. Megha-Tropiques carries three radiometers dedicated to the energy budget and the atmospheric water: Scanner for Radiation Budget (ScaRaB), Sounder for Atmospheric Profiling of Humidity in the Intertropics by Radiometry (SAPHIR) and Microwave Analysis and Detection of Rain and Atmospheric Structures (MADRAS). Because of mechanical issues, operation of MADRAS ceased after 15 months of observations (on January 26, 2013). A novelty of the Megha-Tropiques platform is its low inclination of 20 ° around the equator, which translates into a higher revisit frequency: The satellite orbits the tropics about fourteen times per day yielding between two to six observations per day at a given location between 25°N and 25°S. Further information about Megha-Tropiques is provided by Roca et al. (2015) and Capderou (2009).

We analyze humidity retrievals based on measurements by SAPHIR, which samples the 183.31 GHz water vapor absorption line with six channels. The use of microwave radiometry for water vapor profiling allows retrieval in cloud-covered scenes, within limits determined by the extent of scattering by large hydrometeors (Greenwald and Christopher 2002; Hong et al. 2005). The six channels all measure the same water vapor absorption line but with different radiometric sensitivity. Channel 1 measures the line within ±0.2 GHz, Channel 6 within ±11 GHz, (Eymard et al. 2002). Channel 1 is most sensitive to water vapor in the upper troposphere, near 300 hPa, Channel 6 is more sensitive to water vapor in the lower-troposphere, near 600 hPa (Roca et al. 2015). By virtue of their differences, signals from these channels can be used to estimate relative humidity (in % with respect to liquid water) in six unevenly spaced layers of the atmosphere, as indicated in Table 1. SAPHIR is a cross track scanning instrument with 130 non-overlapping (10 km diameter at nadir) footprints across a 1700-km scan.

We use water vapor retrievals from a multivariate regression scheme that provides the parameters of conditional distributions chosen a priori (Brogniez et al. 2016). Compared to the relative humidity estimation scheme described by Brogniez et al. (2016), which relied on a normal (Gaussian) distribution, here a Beta distribution is used as its compact support and allowance for asymmetries about the mean are more favorable for describing the relative humidity. The estimation of parameters also provides a characterization of the

Table 1 Layers in which relative humidity is retrieved based on six channels of the SAPHIR instrument aboard Megha-Tropiques

Level	Pressure range (hPa)	Description
1	100–200	Tropopause region
2	250–350	Upper troposphere
3	400–600	Congestus layer
4	650–700	Lower free troposphere
5	750–800	Shallow cumulus layer
6	850–950	Top of PBL and bottom of cumulus layer

retrieval uncertainty for every observing pixel of SAPHIR. Limits of such statistical schemes arise mainly from the training phase performed over a sample of tropical profiles, and from the signature of non-scattering clouds in the scene, as these cannot be accounted for in the radiative transfer computations that are a pre-requisite to the design of the retrieval technique. However, some studies have already shown that the three additional channels of SAPHIR compared to the configuration of the operational 183 GHz radiometers such as AMSU-B or MHS improve the estimation of the relative humidity of the upper part of the troposphere and in its lowest layers (Brogniez et al. 2013).

2.3 Infrared Atmospheric Sounding Interferometer (IASI)

The Infrared Atmospheric Sounding Interferometer (IASI) is a Fourier transform spectrometer based on a Michelson interferometer coupled to an integrated imaging system. IASI provides 8461 spectral samples, aligned in three bands between 645.00 and 2760.00 cm^{-1} (15.5 and 3.63 μm), with a spectral resolution of 0.50 cm^{-1} after apodization (filtering to remove diffraction rings) and a spectral sampling interval of 0.25 cm^{-1}. IASI scans across-track with a ground-swath width of approximately 2200 km and a nadir footprint of about 12 km diameter. One IASI was launched in October 2006 onboard the Metop-A satellite and another in September 2012 onboard Metop-B. This enables twice daily (09:30 and 21:30 Local time) global coverage.

Both the temperature and moisture of the troposphere and lower stratosphere are derived under cloud-free conditions, at the pixel level, with a vertical resolution of 1–2 km in the lower troposphere, and an accuracy of 1 K and 10%, respectively. Roughly ten degrees of freedom are estimated as determining the moisture profiles. The IASI weighting functions for water vapor are generally maximum above 700 hPa, so that larger errors can be expected below 3 km (Chazette et al. 2014).

IASI spectra also contain information on the isotopic composition of water vapor (ratio of HDO to H_2O) expressed as

$$\delta D = 1000\left(\frac{\mathrm{HDO}}{\mathrm{H_2O}|_{\mathrm{Vsmow}}} - 1\right), \tag{1}$$

where HDO denotes the relative amount of Hydrogen–Deuterium–Oxygen, and is normalized to the amount found in a standard, here taken to be Vienna Standard Mean Ocean Water. Because of the fractionation of the isotopologues during phase changes of water, δD observations are useful tracers of the hydrological cycle (Galewsky et al. 2016) and can provide additional constraints on the origin of water (e.g., González et al. 2016) and/or on the processes that give rise to its presence (e.g., Worden et al. 2007; Risi et al. 2008). δD can be retrieved after an inversion procedure specially adapted to reach the precision requirements of such observation (Schneider and Hase 2011; Lacour et al. 2012). In this study, we also present estimates of δD associated with the two NARVAL campaigns. We used the IASI δD retrievals developed at ULB/LATMOS (Lacour et al. 2012). The retrieved profiles have limited vertical information content (between 1 and 2 degrees of freedom) with a maximum of sensitivity in the free troposphere around 4.5 km, which is near z_*. These profiles have been cross validated with an error of about 38‰ in the free troposphere on a single measurement basis (Lacour et al. 2015).

3 How Lower-Tropospheric Humidity Influences Clouds, Convection and Circulation

Lower-tropospheric humidity is important because it determines the strength and depth of convection and influences circulations and cloudiness, especially in the tropics. Water vapor's influence can be described in terms of a few quantities. The specific humidity within the PBL, which we denote by q_{PBL}, the humidity within the lower free troposphere, q_{LFT}, as well as the height of the shallow (trade-wind, or fair-weather) cumulus layer, z_{i}. A fourth piece of information, describing the bottom or top-heaviness of the relative humidity profile in the lower free troposphere, may also be important.

3.1 Humidity in the Planetary Boundary Layer

PBL humidity, q_{PBL}, sets the potential for deep convection, as well as influencing cloud amount and surface fluxes. The role of humidity variations in determining cloud amount is discussed by Vial et al. (2017), and our focus is on how q_{PBL} influences convection. This can be understood with the help of the strict quasi-equilibrium ansatz for moist convection.

Strict quasi-equilibrium (Emanuel 1986) gives a name to the idea that convection instantaneously adjusts the temperature profile in the free troposphere to render convective clouds rising out of the PBL neutrally buoyant with respect to their environment. Roughly speaking, in the adiabatic limit the convective parcel rises isentropically from the surface, in which case neutral stability is associated with the saturated moist entropy (or the saturated moist static energy, h_{s}) in the free atmosphere equalling the moist entropy (or moist static energy) within the well mixed sub-cloud layer (Fig. 4). Hence if the boundary layer moist static energy, h_{PBL}, rises to a value that is larger than h_{s} in the free troposphere, then the atmospheric h_{s} is immediately adjusted to h_{PBL}—perhaps corrected for convective

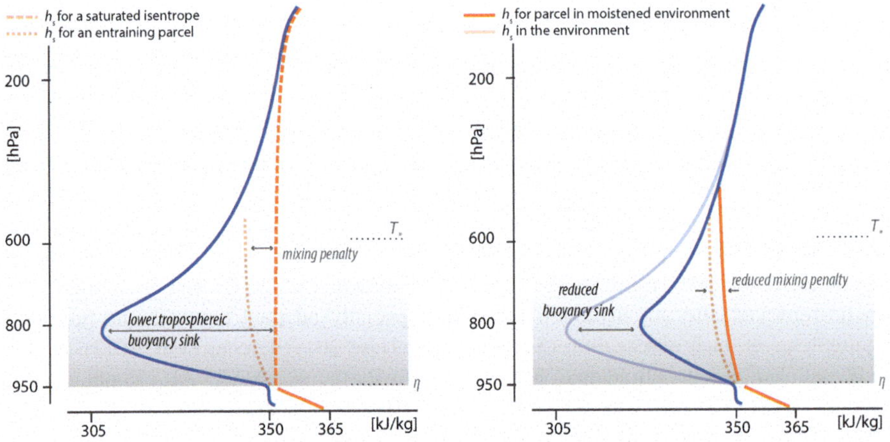

Fig. 4 Diagrammatic depiction of how lower-tropospheric humidity affects moist convective stability. On the *left*, the *red lines* show the saturated moist static energy for a parcel lifted isentropically from the PBL (*dashed*) or affected by cooling associated with lower-tropospheric mixing (*dotted*) versus the mean moist static energy (*blue*). The lower-tropospheric buoyancy sink is expected to be proportional to the environmental moisture deficit, which can be largest at the warmer temperatures of the lower troposphere. *Right panel* shows how reducing the moisture deficit in the lower troposphere reduces the mixing penalty and allows parcels rising from the PBL to more closely approach their saturated isentrope

downdrafts, which reduce h_{PBL}. In strict quasi-equilibrium, the thermal structure of the tropical atmosphere everywhere is set by the value of h_{PBL} wherever it maximizes.

The adjustment of the free-tropospheric h_s is made manifest by fast-moving gravity waves that rapidly act to align density surfaces with geopotential surfaces over areas much larger than that associated with the convection itself. Hence, if convection arises in one region, it will suppress convection in neighboring regions with smaller values of h_{PBL}. An unstable density stratification (convective instability) can thus be generated by reducing the free tropospheric value of h_s, for instance by radiative cooling, or by increasing the value of h_{PBL} where it maximizes, for instance through surface fluxes. During NARVAL-2, the variation of h_{PBL} within and around the ITCZ region varied by about 8 kJ kg^{-1}, variations in q_{PBL} alone (which amount to 3 g kg^{-1}, Fig. 3) contribute about 7.5 kJ kg^{-1} of variation. This type of thinking puts a premium on understanding processes that control, h_{PBL}, and by extension q_{PBL}. Despite its importance for deep convection, q_{PBL} is poorly quantified over the tropics.

In practice, air does not rise isentropically, as h_s is reduced by the mixing of convective parcels with environmental air as they ascend. Hence, the entire profile of h can influence convection, not just h_{PBL}. Because moisture anomalies are not communicated globally, values of h in the free troposphere vary more strongly on regional scales than does h_s, whereby the largest differences are concentrated in the lower free troposphere, where temperatures are warmer and differences between h_s and h_{PBL} can be larger. Mixing with low moist static energy air in the environment reduces h_s of the convecting air, acting as a buoyancy sink. Importantly, increasing the ambient moisture in the lower free troposphere reduces the lower-tropospheric buoyancy sink (Fig. 4), and increases the value of h_s for a test parcel. In such a situation, the temperature of the test parcel will more closely approximate the saturated isentrope, for which its temperature (and buoyancy) will be a maximum. This scenario (right panel of Fig. 4) emphasizes the importance of processes that control not just h_{PBL} but also the moisture deficit of the lower troposphere (strength of the lower-tropospheric buoyancy sink), and explains why convection starting with the same value of h_{PBL} is favored in a more humid lower free troposphere.

3.2 Column Water Vapor (Thermodynamic Effects)

There is a long literature discussing the very strong relationship between convective precipitation and column water vapor (see, for instance the discussion of the Kuo scheme in Anthes 1977). Bretherton et al. (2004) show that precipitation picks up exponentially with the increase in column relative humidity, defined as W/W_* where W is the water vapor path, and W_* is its saturated value given the temperature profile of the column. Initially found just over the tropical oceans this relationship has, with some caveats, been extended to other regions (Holloway and Neelin 2009; Schiro et al. 2016; Bergemann and Jakob 2016) and may also correlate with variations in shallow precipitation (Nuijens et al. 2009). The scale height of the saturation humidity following a moist adiabat is about 2 km. This means that column water vapor is largely determined by water vapor in the lower troposphere, between the top of the PBL, η, and z_* (Raymond et al. 2003). In other words, the observed relationship between column water vapor and precipitation is really a relationship between lower free troposphere humidity and precipitation.

The humidity of the lower free troposphere influences convection because it encodes the potential of mixing to evaporatively cool saturated air—the buoyancy sink (Fig. 4). Such sensitivities are readily evident in cloud-resolving model simulations, but are not always evident in convective parameterizations (Derbyshire et al. 2004; Kuang and Bretherton 2006; Wu et al. 2009). The failure of convective parameterizations to adequately capture this coupling may explain many of the difficulties in representing large-scale circulation features, such as the Madden-Julian Oscillation (MJO) (Bony and Emanuel 2005), processes of convective aggregation (Craig and Mack 2013; Becker et al. 2017), and the structure of the tropical convergence zones (Möbis and Stevens 2012; Oueslati and Bellon 2013).

3.3 Column Water Vapor (radiative effects)

Column water vapor, W, also strongly influences the distribution of radiative heating within the lower troposphere. In a nearly saturated atmosphere, the boundary layer cools at a rate of about 2 K d^{-1}, commensurate with the cooling rate of the free troposphere.[1] But as the humidity in the free troposphere is reduced, the cooling rate increases roughly sixfold, inversely with the free troposphere humidity, but increasingly nonlinearly as the free troposphere becomes very dry.

The effect of column water vapor on cooling rates within the PBL is illustrated by considering only the clear-sky cooling rates within the PBL, as a function of free-tropospheric moisture. Cooling rates for three different profiles are calculated and plotted in Fig. 5 (three left panels). In addition, pairs of cooling rates, for the PBL and the free-troposphere, are calculated and plotted versus W (right-most panel) for two further scenarios: one with a PBL depth of 600 m the other with an 1800 m deep PBL, in both cases W is varied by varying the free tropospheric relative humidity. The dry (orange) and moist (blue) profiles in the left panel correspond to special cases of the first scenario and were chosen in analogy to the mean soundings during NARVAL-1 and NARVAL-2 (Fig. 3). Clear-sky cooling rates are peaked near the top of the PBL in each case, but are much stronger for the drier free troposphere as the downwelling infrared radiance is roughly proportional to the number of water molecules above a particular layer (e.g., the cooling to space approximation, cf., Mapes and Zuidema 1996; Thomas and Stamnes 1999). These effects, though less pronounced, remain in force also for the scenario of a deep (1800 m) boundary layer (Fig. 5, right panel). When clouds top the moist layer, they can be expected to increase the outgoing irradiance from the top of the layer, thereby enhancing its cooling rate, but not its sensitivity to the overlying water.

Averaged over the PBL clear-sky cooling rates can be greater than 10 K d^{-1} or not differ substantially from the cooling rates within the free troposphere (Fig. 5). The degree of cooling can be expected to influence the state of the boundary layer, and hence surface fluxes, and may drive low-level circulations (Naumann et al. 2017). For instance, circulations associated with very strong radiative cooling rates are thought to be responsible for a net transport of moist entropy into convective regions and out of non-convective regions, thereby reinforcing convective regions at the expense of non-convective regions. Muller and Bony (2015) show that in cloud-resolving model simulations of radiative-convective equilibrium, cooling profiles similar to those shown for the case of the very dry free troposphere (blue line in Fig. 5), extract energy from the boundary layer, leading to the formation of radiatively driven cold pools. By neutralizing this cooling, their simulations no longer aggregate. The importance of these processes for the organization of convection is also emphasized in the review by Wing et al. (this issue). Naumann et al. (2017) recently

[1] We highlight the effect of infrared irradiances; absorption of visible radiation by water vapor mutes these effects, but scales with the effect of the infrared irradiance (Mapes and Zuidema 1996).

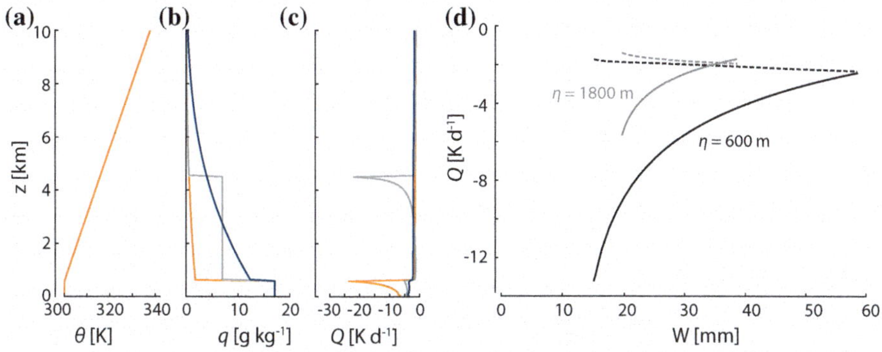

Fig. 5 Potential temperature (**a**), and humidity (**b**), used to calculate the infrared cooling rates in (**c**). The three profiles illustrate a very dry free troposphere (*orange*), with a nearly saturated free troposphere (*blue*) and with an elevated moisture layer (*gray*) between the top of the boundary layer and the 0 °C isotherm. **d** Cooling rates as a function of precipitable water, which is varied by varying the humidity above the boundary layer. *Solid lines* represent the radiative cooling rate (thermal-infrared only) averaged over the boundary layer and *dashed lines* the average cooling in the layer (with twice the depth of the boundary layer) above the boundary layer. Two pairs of lines are shown. The one with a larger dynamic range in precipitable water (and more cooling) corresponds to a boundary layer of 600 m in depth and the other to a layer of 1800 m in depth

developed a theoretical framework for studying how efficiently such cooling drives circulations in the lower troposphere. They show that infrared cooling driven by vertical gradients in lower-tropospheric moisture is as effective as differences in sea-surface temperatures, which following Lindzen and Nigam (1987) is the typical paradigm for understanding larger-scale circulations in the tropics.

3.4 Elevated Moist Layers

During NARVAL-2, there is the hint of a layer, or shelf, of constant absolute humidity within the lower free troposphere. Examples of an elevated moist layer are pronounced on individual NARVAL-2 research flights (RF), for instance, on RF03, which took place on August 12, 2016, and for which the moisture profile is plotted in Fig. 6. The sharp jump in q_{ft} around z_* (which locates at about 4.5 km) may be associated with moisture originating from convective outflow that has been advected away from a more remote source or simply mark the remains of now dissipated local convection, something that an analysis of isotopologues might be able to differentiate. Regardless of its origins it has a strong influence on the infrared irradiance.

Radiative heating responds not just to the total column water vapor, but also how it is distributed within the vertical. Unlike in the case of the dry-tongues analyzed by Mapes and Zuidema (1996), moisture shelves do not lead to a heating-cooling couplet across the top of the melting level, but rather the strong cooling at the top of the moist layer is offset by a reduction of the radiative cooling in the PBL (see Fig. 5). Whereas the cooling in the stably stratified free troposphere can be directly expected to induce subsidence, the reduction of cooling in the PBL will be partly compensated by a reduction in surface enthalpy fluxes (Naumann et al. 2017). However, the maximum of the cooling at the top of the moist layer is as strong as the PBL maximum in the very dry case (orange line) and may supplant very shallow circulations on the scale of the PBL by deeper circulation on the scale of z_* (Nishant et al. 2016). In contrast, the humid case (blue line) suggests that for a

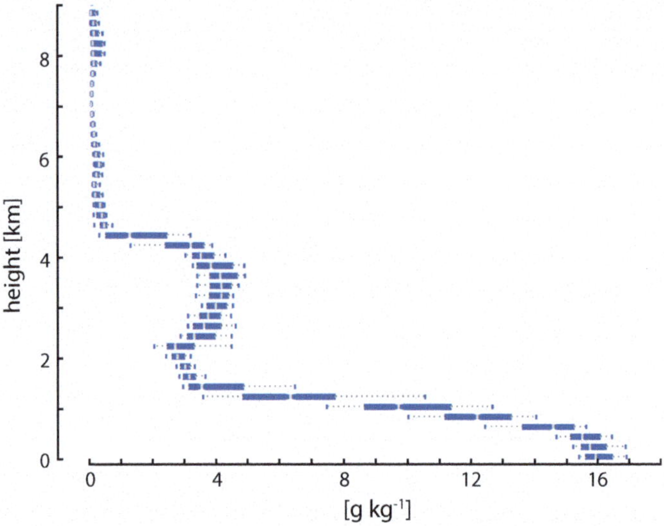

Fig. 6 Moisture profile from fifty dropsondes released in the *box*, with corners (52°W, 9°N) and (55°W, 14°N), during the third research flight of NARVAL-2 (August 12, 2016)

case with the same amount of precipitable water, but with q more bottom heavy so that it smoothly decreases with height, the structure of the resultant heating profile differs markedly. Thus, a fuller understanding of how water vapor in the lower troposphere affects clouds, convection and circulation also requires an understanding of its vertical distribution, as the gradient of absolute humidity is what determines the vertical profile of radiative heating.

4 Remotely Sensed Humidity Variations During NARVAL-1 and NARVAL-2

To better assess the structure of water vapor in the tropical atmosphere, we evaluate to what extent satellite retrievals capture important forms of water vapor variability as measured during the NARVAL series of campaigns. To do so, we compare water vapor retrievals from the SAPHIR and IASI instruments with dropsonde and water vapor lidar data collected during NARVAL-1 and NARVAL-2 research flights. Because of its better temporal sampling and its ability to extract water vapor information also in the presence of clouds, we focus on SAPHIR.

4.1 General Structure of Humidity Retrievals from SAPHIR

SAPHIR retrievals capture well the main, climatological, features of the humidity field as seen in the dropsonde data across the entire NARVAL-1 and NARVAL 2 periods. During NARVAL-1, an exceptionally dry free atmosphere (Fig. 7, left panel), with humidities well below 20%, is consistent with the dropsonde data presented in Fig. 3. A broader distribution of relative humidities between 750 and 800 hPa (SAPHIR layer 5) is consistent with the top of the shallow cumulus (trade-wind) layer being near this level—drier conditions being indicative of days with a shallower cloud layer. The August NARVAL-2 flights

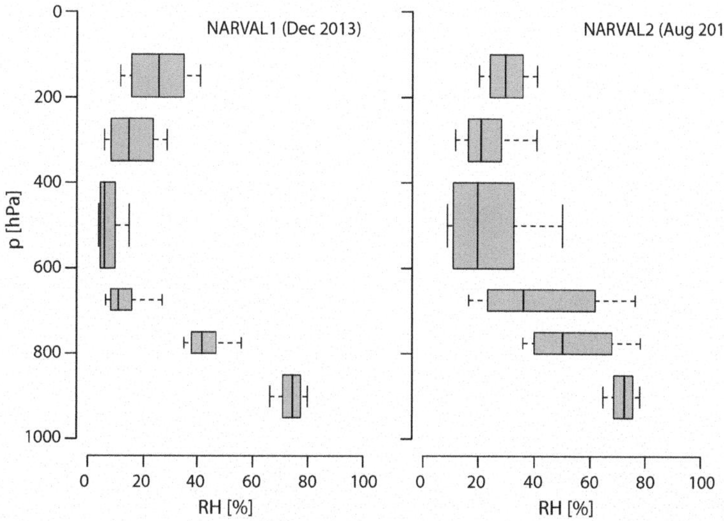

Fig. 7 Vertical profiles of relative humidities from SAPHIR for the flight area of NARVAL-1 and NARVAL-2

predominantly differ through a moister, and more variable, free troposphere as compared to the December NARVAL-1 flights. Figure 7 shows these differences to be evident at every level, but most pronounced between 400 hPa and 800 hPa, consistent with the dropsonde data in Fig. 3.

Our further evaluation of SAPHIR focuses on two flights: one (RF07 on December 19, 2013) from NARVAL-1 and the other (RF03 on August 12, 2016) from NARVAL-2. The humidity retrievals over the six SAPHIR layers are presented for these two flights in Fig. 8. The area of flight operations, for the case of NARVAl-2, and the flight track, for the case of NARVAL-1, are overlain. The main features of the humidity field are the sharp decline of humidity with height on both swaths, and clear meridional excursions of moist tropical air in the lower free troposphere—easterly waves. On 19 December (NARVAL-1), an easterly wave is evident in the eastern third of the swath, and the humidity in the lower troposphere is in general confined to a shallower layer. On 12 August, two easterly waves are evident, one just east of the area of flight operations at 50°W and 12°N, the other extending further to the north and locating east of 30°W. Neither of these meridional extended moist features is as deep, or as condensate laden, as the feature sampled on December 19, 2013 (NARVAL-1). To the extent convection can be associated with regions where retrievals (due to hydrometeor scattering) are not possible, for instance near (50°W, 10°N) it forms as expected in regions of exceptionally high near-surface humidities, not to mention a deeper layer of moisture, as anticipated by some of the arguments reviewed in Sect. 3.2 above. There is also some evidence of drier air near the surface in the vicinity of pixels where hydrometeor scattering resulted in failed retrievals. This would be consistent with downdrafts bringing entropy-poor (i.e., drier) air to the surface in the vicinity of deep convection. There also appears to be an imprint of the drier layers aloft also on the structure of the near-surface relative humidities; this is evident, for instance, in the large crescent-shaped features on December 19, 2013, and the dry feature aligned with the southern end of the track, features that are also evident in the form of the banded features in the cloud imagery from that day (Fig. 9).

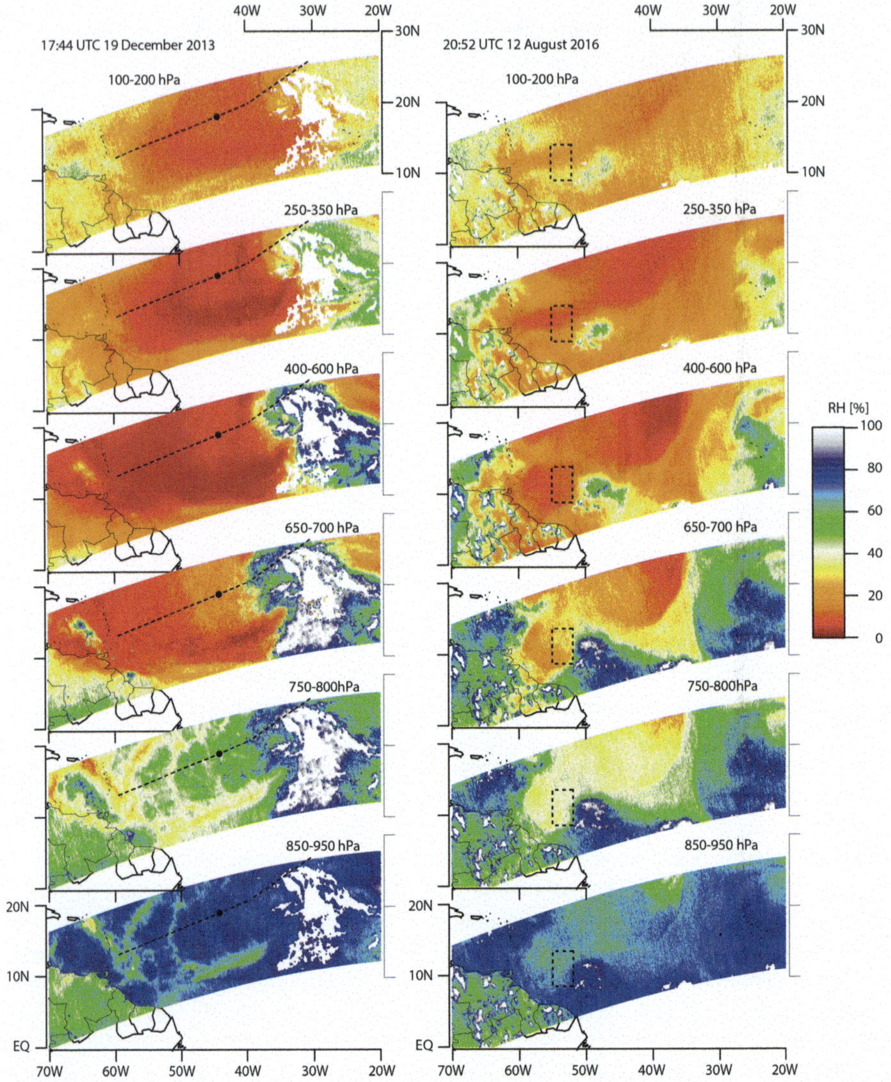

Fig. 8 Example of SAPHIR retrievals along orbit swaths coincident with NARVAL-1 RF07 (*left*) and NARVAL-2 RF03 (*right*). Shown are the relative humidities in the six layers over which humidity retrievals are performed. The HALO flight track and approximate time of the Megha-Tropiques overpass is indicated by the *dashed line* and *circle* in the *left panel* (NARVAL-1). In the case of NARVAL-2, the approximate HALO flight area is shown as a *rectangle* with a *dashed perimeter*

4.2 Evaluation of SAPHIR Retrievals by WALES

To get a better idea of what types of features can be reliably detected by the SAPHIR retrievals, and whether the apparent coincidence between cloudiness and features in the lower-tropospheric humidity field are real, the humidities in the five lower SAPHIR layers are compared to the WALES water vapor profiles. No comparison is made in the first (uppermost) layer because WALES at most sampled the lower 10 hPa of this level, and

Fig. 9 NASA Worldview MODIS (Aqua) visible imagery for NARVAL-1 RF07 (*upper*) and NARVAL-2 RF03 (*lower*). The swaths span the area from 20°W–70°W and from 0°N to 20°N identical to the bounding area for the SAPHIR swaths in Fig. 8

then only during NARVAL-1. Temperature fields from ECMWF analyses interpolated in space and time to the flight path are used to convert water vapor densities measured by the lidar into relative humidities. A comparison of the SAPHIR-retrieved relative humidities with those from WALES is presented along the cross section over the Atlantic for December 19, 2013 (NARVAL-1) in Fig. 10. Figure 11 shows the comparison of the SAPHIR humidity retrievals with WALES over the area of flight operations for August 12, 2016 (NARVAL-2).

For the 19 December comparison, SAPHIR retrievals of relative humidity agree well with WALES at most levels, especially in capturing the transition between the deep moist layer in the eastern part of the domain (Fig. 10) as compared to the much drier air to the west. Some differences are evident in fine-scale features, but these may arise due to a lack of coincidence. During NARVAL-1, SAPHIR also captures the vertical gradient in humidity, particularly when one compares layer 4 (relative humidities below 5%) with layer 6, where relative humidities are near 80%. Most remarkable, Fig. 10 suggests that the variations in the SAPHIR layer 6 retrievals, which correlate with patterns of cloudiness (Fig. 9), are also measured by WALES. Here our eye is drawn to what appears to be a long downward intrusion of dry air along a north–south-oriented line near 51°W (Fig. 8) in the SAPHIR data and sampled at around 1840 UTC by WALES. A second, somewhat broader, feature is also apparent at the very end of the record, near 51°W. These features are also evident in the layer 5 (2–2.6 km) retrievals, but much less pronounced than in WALES.

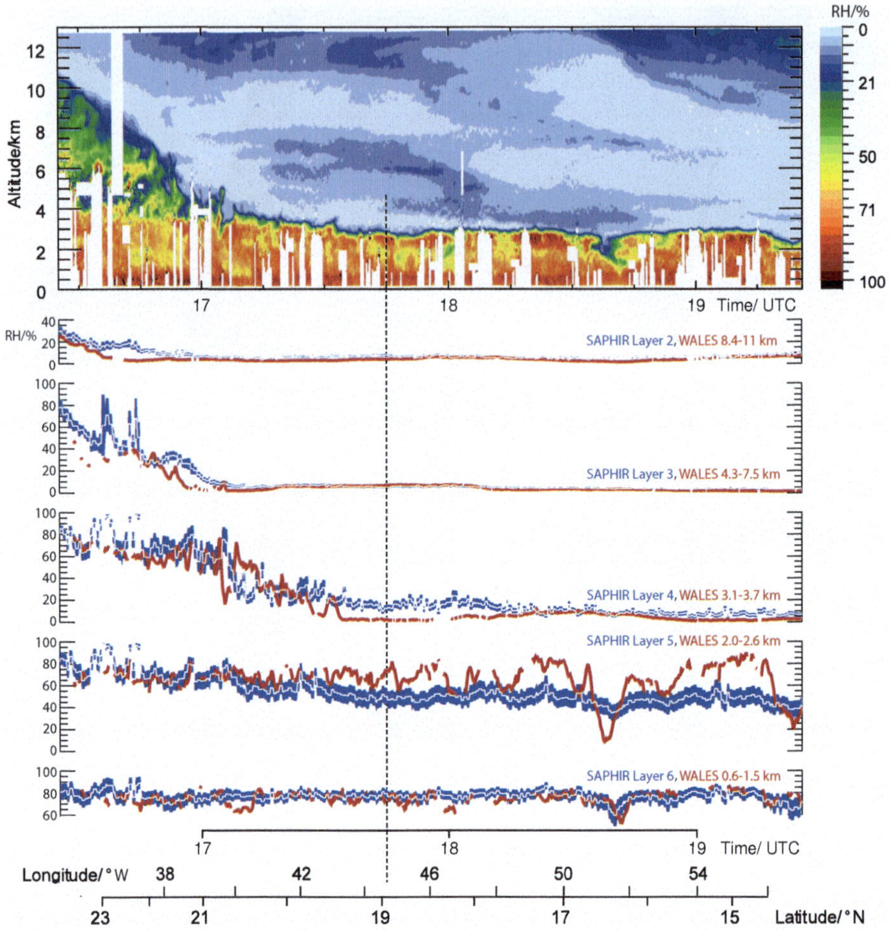

Fig. 10 Comparison of relative humidity profiles on December 12, 2013. Shown are measurements from the WALES instrument aboard HALO as part of NARVAL-1 and the Megha-Tropiques profiles along the same track. The time of the Megha-Tropiques overpass is indicated by the *vertical dashed line*. SAPHIR retrievals are shown (*saphire-blue*) bounded by the first and third interquartile uncertainty, WALES estimates show in *ruby-red*

These differences illustrate the challenges that retrievals from passive sensors face in the vicinity of a strong humidity gradient (80–10%), where the interpretation of the upwelling irradiances is not clear cut. In this case, the breadth of the sensitivity functions probably also leads to some of the moisture from the lower layers being interpreted as having arisen higher up. Earlier in the flight (before 1730 UTC), where the moisture is more homogeneous, the SAPHIR retrievals match the measurements by WALES quite well.

Considering their lack of coincidence, the SAPHIR humidity retrievals on 12 August also appear to capture the variability in relative humidity in the upper troposphere, and the gravest features of the vertical structure. In the lowest level (layer 6) near 1 km, across which the humidity profile varies as one expects climatologically, the humidities retrieved by SAPHIR are similar to that observed by WALES, with commensurate variability. On a layer-by-layer basis, particularly within the lower free troposphere, the retrievals have less

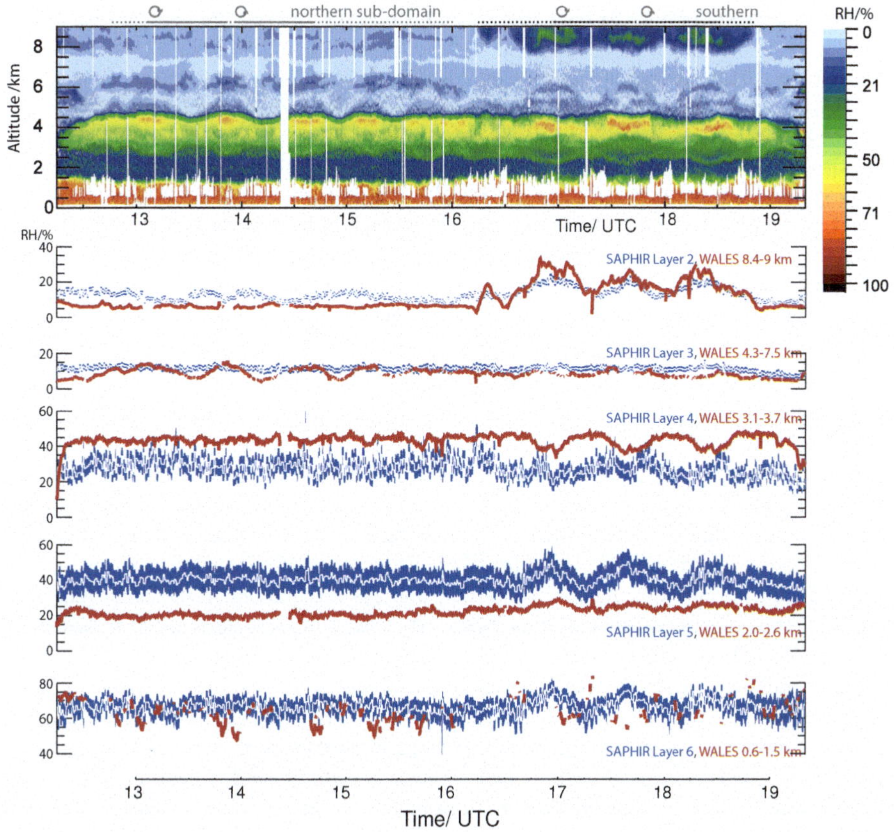

Fig. 11 Comparison of relative humidity profiles from the WALES instrument aboard HALO as part of NARVAL-2 RF03 (on August 12, 2016) and the Megha-Tropiques profiles along the same track. The overpass time of the Megha-Tropiques satellite was at 20:52 UTC. Color scale as in previous figure

skill. This appears to largely reflect the non-trivial vertical structure in the water vapor field, wherein a more humid layer, with relative humidities peaking near 4 km, overlays a relatively drier layer. The humidity structure measured by WALES is consistent with measurements from the dropsondes (Fig. 6), with absolute humidities constant or slightly increasing between about 1.5 and 4.5 km. These features are not well differentiated by the layer 4 and layer 5 SAPHIR retrievals (Fig. 11). SAPHIR under estimates the humidities (by about 15% absolute) around 3.4 km (its level 4) and compensates with commensurately higher humidities one kilometer lower (its level 5). Here again the broadness of the SAPHIR sensitivity functions makes it difficult to unravel the moist layer near 4 km from the absolutely drier layer aloft and a relatively drier layer below. When layers 4 and 5 are taken together, the SAPHIR retrievals capture the horizontal humidity variations seen in the WALES data below 3 km, but we interpret them in terms of a more climatological vertical structure, wherein relative humidity is strictly decreasing with height.

4.3 Comparison with IASI and the Added Value of Retrieving Isotopologues

Because the area of aircraft operations on August 12, 2016, was relatively cloud-free, we can explore whether IASI humidity retrievals are, by virtue of more independent information, better able to measure the vertical structure of water vapor in the lower free troposphere. Despite better vertical resolution, IASI retrievals are biased similarly to the retrievals using SAPHIR measurements. Neither instrument resolves the relative humidity inversion (relative humidity increasing with height between 2 and 4 km, Fig. 6) seen on August 12. In contrast to SAPHIR, IASI also provides retrievals within the PBL (below 1 km). But the IASI humidity retrievals are indicative of a substantially moister (10% absolutely) and shallower PBL as compared to what is measured in situ, by the sondes (Fig. 12).

One would expect different isotopic signatures for water subsiding downward, versus mixing upward. The former, for instance following from convective detrainment near z_*, would be associated with a subsiding moist layer and is consistent with the vertical humidity structure between 2 km and 4 km on 12 August. If this interpretation is correct, then the HDO/H_2O ratio (as measured by δD, Eq. 1) should follow the Rayleigh distillation value associated with water detrained at T_*. In the latter case, where water is mixed upward out of the boundary layer, δD would be relatively enriched. Differences in the pattern of retrieved specific humidity near z_* (660 hPa) and δD are evident in the IASI retrievals (Fig. 13), allowing us to test this hypothesis, or at least investigate to what extent measurements of δD might help constrain the vertical structure of water vapor.

We examine co-variability between retrieved humidity, q, and δD we focus again on the December 19, 2013, and August 12, 2016, cases, as shown in Fig. 14. In this example, different mixing processes are illustrated by lines in q–δD space. We expect water originating from condensate evaporation aloft to follow the Rayleigh distillation curve, where as mixing of moister air from the lower atmosphere with much drier air that has subsided

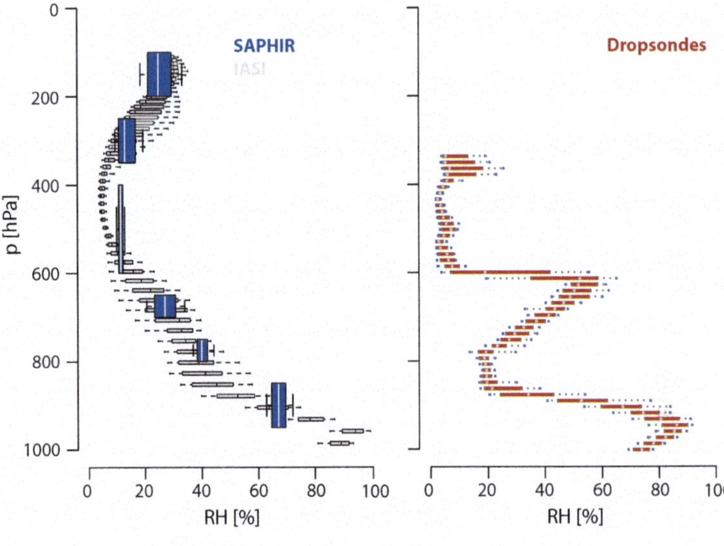

Fig. 12 Vertical profiles of relative humidities from SAPHIR and IASI (*left*) and from the dropsondes (*right*) for the flight area of RF03

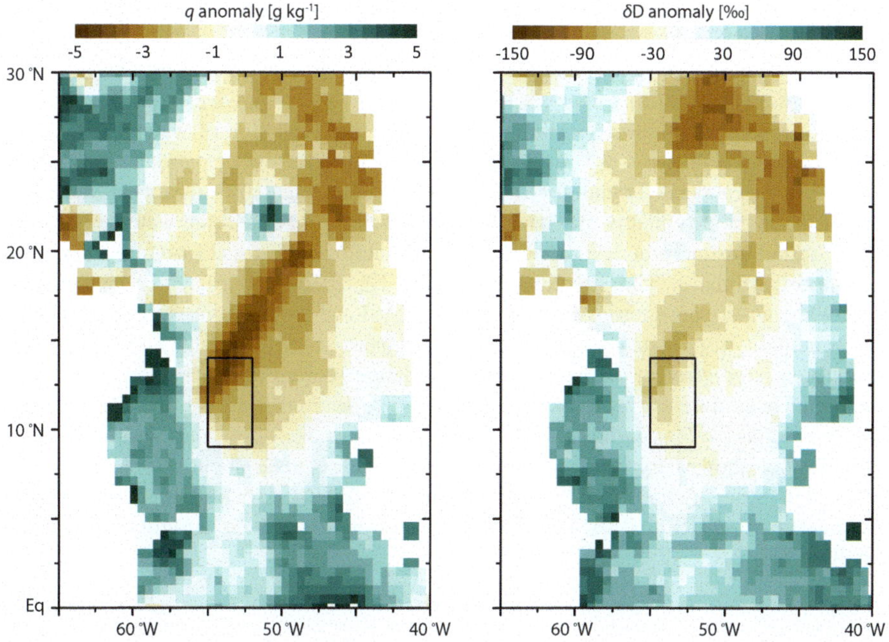

Fig. 13 Retrieved humidity (*left*) and δD anomaly (from domain mean) at 660 hPa for August 12, 2016, 21:30 Local Time IASI overpass

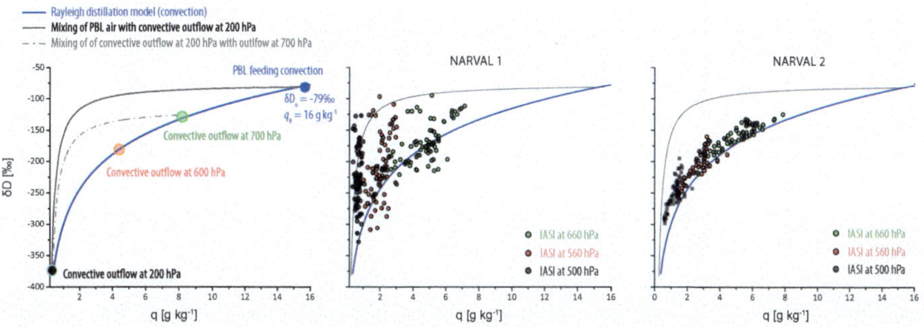

Fig. 14 q–δD diagram showing the variation of specific humidity, q and δD from retrievals at different heights. *Left panel* shows a schematic of how q and δD would vary following specific processes. Retrievals from NARVAL-1, for December 19, 2013, between 12°N–20°N and 40°W and 60°W, (*middle*) and NARVAL-2 for August 12, 2016, in the same area (*right*)

from a great altitude would be relatively enriched and follow the mixing curve (see also Risi et al. 2008; Galewsky et al. 2016). The August retrievals more closely follow the Rayleigh distillation curve than do the December profiles (Fig. 13), increasingly so as one approaches T_*. The diagrams are consistent with the idea that the humidity profiles originate more from preferred detrainment near T_* in the August period, whereas in the December period the humidity profiles are more characteristic of mixing up of moist PBL air into much drier free tropospheric air. In this latter (December) case, one would not expect relative humidity reversals. Whether this information can be more effectively used to robustly constrain the retrievals merits further study.

Despite a hint that retrieved isotopologues might help improve humidity retrievals, the overall impression from this analysis is that it is difficult for passive remote sensing to accurately capture non-trivial variability in the vertical distribution of water vapor in the lower troposphere. In the lower troposphere, above the PBL (where variations in water vapor are largest and also crucial for a variety of processes) the retrievals appear to capture some of the variability that is measured by both the dropsondes and the water vapor lidar (WALES), but misinterpret deviations from the climatological vertical structure encoded in the a priori estimates. The situation on August 12, 2016, is particularly challenging, whereby the water vapor retrievals from both the IASI and SAPHIR measurements yield a relative humidity profile that decreases, rather than increases, with height. This type of bias will cause a large error in the calculation of radiative heating rates (as shown in Fig. 5) which influence inferences as to the nature of circulations in the lower-troposphere. These biases are not as evident in the relative humidity profiles taken from the European Centre Analyses (not shown) for reasons that we can only speculate about. Nonetheless, the challenge in retrieving non-trivial profiles of humidity from passive measurements, such as those made by SAPHIR and IASI, could bias lower free troposphere humidity estimates to be bottom heavy.

5 A Hypothesis for the Preponderance of Melting Level Convection

There is a substantial literature documenting the prevalence of moisture and clouds in the middle troposphere, near z_*. Typically this is thought to be concentrated in the vicinity of deeper convection, similar to what was observed during NARVAL-2, where measurements took place in and around the ITCZ. In an analysis of data from the Tropical Ocean Global Atmosphere Coupled Ocean Atmosphere Response Experiment (TOGA-COARE), Johnson et al. (1999) accumulated a wide array of evidence from historical measurements dating back to the middle of the last century to make the case that there was a preferred mode of cloudiness near z_*. To the extent that there is preferential detrainment near z_* it is usually attributed to enhanced stability at this level associated with the melting of ice at temperatures warmer than T_*.

We hypothesize that the energy required for initial ice formation[2], which makes it difficult to form ice until ice-supersaturation is very high or temperatures are much colder than T_*, combined with the radiative response to sharp moisture gradients (cf. Sect. 3.4) as may arise from a preferred level of convective detrainment, further favor convective outflow at temperatures between T_* and T_{f}, that required for primary ice formation (cf, Fig. 1).

This hypothesis is illustrated with the help of Fig. 15. In the first panel, we envision a scenario in which, for some reason, moist convection reaches a height (or temperature) where ice is initiated. The freezing of condensate releases the enthalpy of fusion; this raises h_{s} by allowing parcels to rise isothermally until all the existing condensate freezes, thereby augmenting the buoyancy of the rising air. In a second step (shown in the second panel of Fig. 15), secondary ice-formation processes (for instance by contact of supercooled water with descending ice crystals) contribute to the glaciation of the entire convective system down to T_*. This further augments the convection, because the cloud may also be more

[2] Roughly, the Gibbs free energy cost of forming the nucleus, as arises in Classical Nucleation Theory.

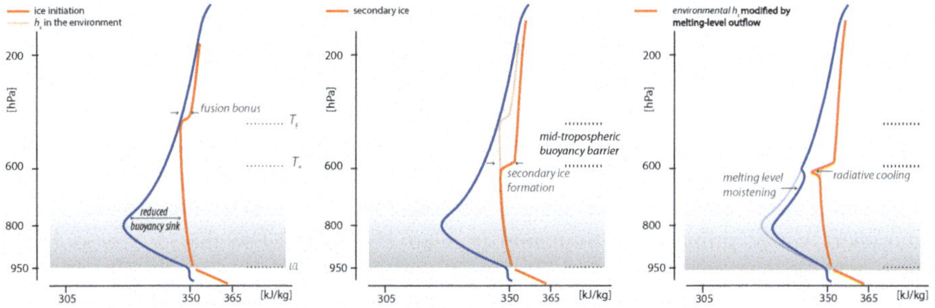

Fig. 15 Adaptation of Fig. 4 to illustrate how ice processes can act to stabilize the atmosphere near the melting level, so that the production of first ice inhibits remote convection at locations where ice must form by primary processes, from penetrating much above the melting level. Moisture gradients and radiative processes as the melting level reinforce this process

condensate laden at the warmer temperatures near z_*. Subsequent convection, in so far as it develops in a region where ice has not been initiated, will then encounter an environment more hostile to convective development for $T_* > T > T_\text{f}$. This enhances the effective stability felt by subsequent convection reaching heights above z_* but below z_f. Hence for convection to penetrate through this region, it will have to have started from substantially more favorable conditions (higher h_PBL or a moister lower troposphere) than the convection that first reached the ice-initiation level.

Enhanced detrainment from convection not strong enough to initiate ice will be associated with enhanced cloudiness and moisture gradients at z_*. This in turn initiates a radiative driven circulation feedback that further enhances the mid-level termination of convection. Clear-sky heating rates more negative than $-20\ \text{K d}^{-1}$ arise in response to strong moisture gradients near z_* (see Sect. 3.4). Even after averaging over a 1 km deep layer, the net cooling is many times larger than the background cooling that one expects in

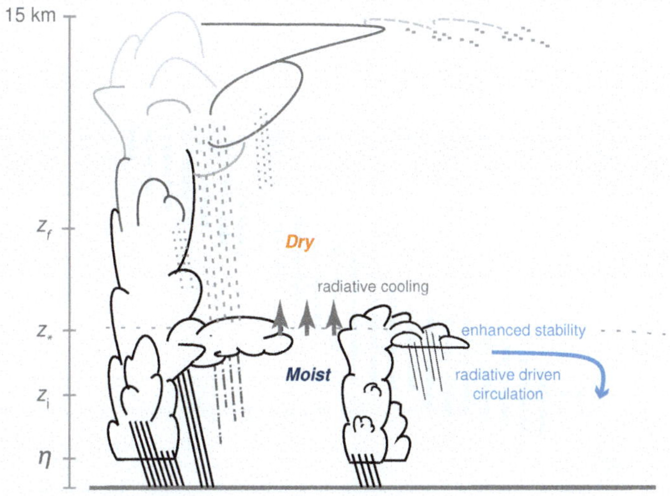

Fig. 16 Conceptual diagram illustrating how detrainment of moisture at the freezing level leads to moisture gradients, more radiative cooling from the top of this level, and a large-scale circulation that favors further detrainment

the absence of such a moisture gradient. As illustrated schematically in Fig. 16, this leads to greater detrainment at z_*: both directly, as a result of the enhanced mass divergence arising from this cooling (Bony et al. 2016); and indirectly, because of the stabilization with respect to the layer above z_*, which then makes it more difficult for convection to penetrate beyond this level. In analogy to the term used to explain the observed geographic preference for consumer brands that first occupy a market (Bronnenberg et al. 2009), we refer to this as the 'first-mover' advantage. We expect these ideas to be most pertinent in an atmosphere where primary ice formation is relatively difficult, so that $z_f - z_*$ is large, which could introduce a role for aerosol-cloud interactions.

The NARVAL-2 dropsondes support the idea that there is a preferred moisture source at T_*. This is evident in Fig. 3 where the median humidity is nearly constant within the layer we refer to as the lower free troposphere, i.e., between the top of the shallow cloud layer and z_*. Another way to see this is to plot the saturation temperature of a given sonde measurement versus the temperature at which it is measured. This is shown in Fig. 17. If the humidity at a given temperature is controlled by the humidity that the sampled air had at last saturation (Sherwood et al. 2010), and there was no preferred temperature of the mid-tropospheric moisture source, we would expect T_{sat} to be distributed between T and some minimum T characteristic of the lowest value of temperatures in the troposphere. Because tropospheric mixing is more efficient along isentropes (roughly isotherms), and air that was last saturated at much colder temperatures must have travelled a greater distance through the troposphere, T_{sat} should be less than, but roughly follow T, as for instance for $T < 0\,°C$ in Fig. 17. A surprising feature in Fig. 17 is the pronounced range of temperatures, between approximately 2 and 14 °C, over which T_{sat} is roughly constant, with a value slightly below T_*. This is consistent with air in the lower free troposphere ($z_i < z < z_*$) being associated with a preferred moisture source at temperatures at or slightly below T_*, as would occur for convective clouds preferentially terminating between z_* and z_f.

An analysis of data from the Barbados Cloud Observatory, where water vapor is profiled by a Raman lidar, supports the idea of a preferred moisture source at temperatures near or slightly below T_*. This, we claim, can best be seen by looking at something we call the

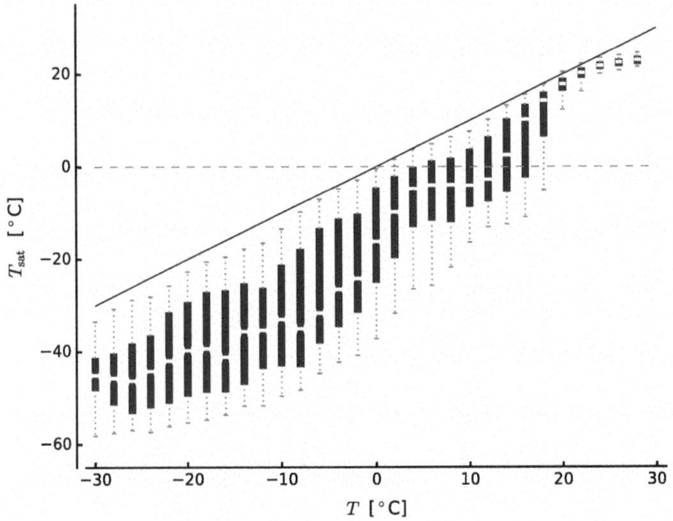

Fig. 17 NARVAL-2 dropsondes saturation temperature (dew-point) versus temperature

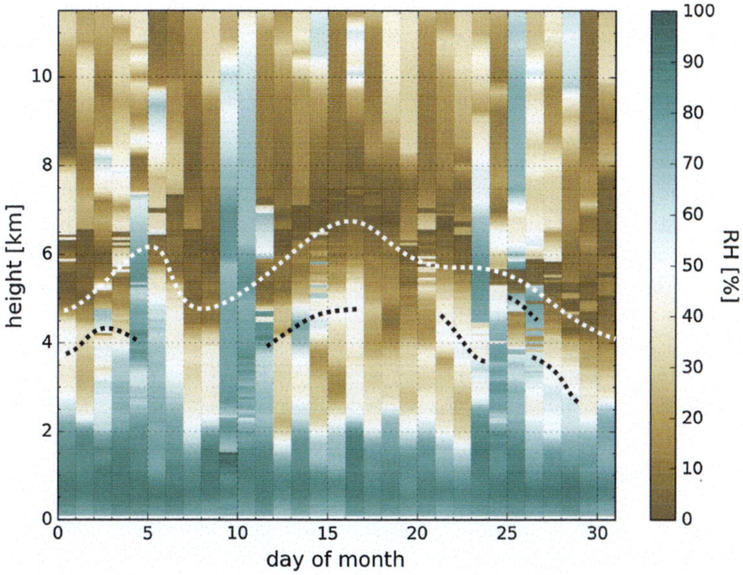

Fig. 18 Blended Raman and sounding night-time moisture (relative humidity) over the Barbados Cloud Observatory during August 2016, the period of NARVAL-2 operations. The *white* and *black dashed lines* are drawn free-hand to emphasize the base of the middle tropospheric dry layer (*white lines*) below which humidity increases to a secondary maximum (in places) *black lines*

tropospheric dry point, i.e., the levels within the troposphere where the humidity minimizes. On many days there is an additional relative humidity maximum below this dry point, but above the climatological cloud-base relative humidity maximum. In Fig. 18, this humidity minimum varies with its base between z_* (located near 4.5 km) and 6 km. Humidity profiles in other months, for which episodes of deep convection are present, also show similar features. In addition to a more quantitative evaluation of these data, these ideas could be tested by cloud-resolving models, for instance by varying T_f but allowing for secondary ice formation.

6 Conclusions

We reviewed mechanisms through which water vapor in the lower troposphere influences clouds, convection and circulation. For our purposes, the lower troposphere is defined as heights where the temperature is above the triple-point temperature, T_*. In the present-day tropics, this locates at a height $z_* \approx 4.5$ km. Three aspects of the water vapor are concentrated on: (1) the amount within the PBL, roughly the well-mixed layer extending from the surface to the cloud-base height, η at about $\eta \approx 0.6$ km; (2) the amount within the lower troposphere, between $\eta < z < z_*$; and (3) the way in which the water vapor is distributed over the lower-troposphere. The latter factor can be quantified by the depth of the shallow cumulus layer, and the amount of water vapor between the top of that layer (at about 2.5 km) and z_*. Boundary layer water vapor determines the convective potential, which can be thought of in terms of the potential energy that could be consumed by a convective overturning. The amount of lower-tropospheric water vapor determines the strength of the lower tropospheric buoyancy sink, which inhibits the vertical penetration of convection through the stabilizing effect of mixing with dry air. The amount and distribution of water vapor in the lower troposphere strongly influences the rate of radiative cooling of the PBL. Bulk cooling of the PBL can vary by a factor of three to five depending on the depth of the PBL and the amount of water vapor above. If the water vapor is distributed so that its relative humidity is relatively constant with height, the cooling within the lower free troposphere also varies little with height. Water vapor profiles for which the absolute humidity is approximately constant with height over a given layer are accompanied by large and sharp water vapor gradients at the top of this layer. These gradients concentrate the radiative cooling at the top of the moist layer, which may much more effectively drive circulations and help decouple the atmosphere in the vertical.

We analyzed in situ and remotely sensed measurements of lower-tropospheric water vapor measurements from two field campaigns (NARVAL-1 and NARVAL-2). The NARVAL missions were flown over the tropical Atlantic, between 45 and 60°W and 5 and 20°N. NARVAL-1 was flow in December 2013 and primarily sampled the winter trades north and adjacent to the ITCZ. NARVAL-2 flew in and around the ITCZ during August 2016. Water vapor was measured in situ by dropsondes, with more than 250 launched over the target region, and by a downward looking differential absorption lidar designed to measure water vapor throughout the troposphere. In terms of lower-tropospheric water vapor, the main difference between the campaigns is in the height of the shallow cumulus layer rising out of cloud base, and the amount of moisture above it, in what we call the lower free troposphere (Fig. 1). During the winter trades, the layer above the shallow cumulus layer is very dry and an extension of the free troposphere above. In air-masses within and near the ITCZ, the lower free troposphere is much moister and more continuous with the layers below. During NARVAL-2, most of the variation in boundary layer moist static energy can be explained by variations in boundary layer humidity, and elevated moisture layers, wherein absolute humidity was approximately constant.

We evaluated relative humidity retrievals from the SAPHIR instrument aboard the Megha-Tropiques satellite as well as from the IASI instrument aboard the Metop satellites. The SAPHIR retrievals are interesting because of their high temporal sampling of specific locations within the tropics, their ability to penetrate all but the thickest clouds, and their emphasis on the lower troposphere, with three of the levels retrieved corresponding to altitude ranges below z_*. In clear-sky conditions, the global coverage and richer spectral coverage of IASI better resolves the water vapor vertical structure. Despite an emphasis on the lower troposphere, the SAPHIR retrievals do not include a layer that overlaps with the PBL, leaving us relatively blind to variations of water vapor in this all important level. The SAPHIR retrievals capture the main differences between NARVAL-1 and NARVAL-2, and large-scale changes in the distribution of humidity associated with an Easterly wave. SAPHIR retrievals also show some skill in identifying large-scale dry filaments, interpretable as depressions in the height of the shallow cumulus layer (2–2.6 km) that are well correlated with meso-α scale patterns in cloudiness.

Non-trivial variations in the vertical structure of the humidity field in the lower troposphere (SAPHIR layers 4 and 5) are less reliably retrieved. SAPHIR-based retrievals of humidity fluctuations at the base of the shallow cumulus layer show relatively little correspondence to the lidar measurements. Relative humidity reversals, whereby in a layer of fixed absolute humidity the relative humidity increases rather than decreases with height, prove challenging to identify in the SAPHIR retrievals, which instead places the moisture in the lower levels, more consistent with relative humidity being constant through the layer. IASI retrievals are characterized by similar biases, although here the retrievals of isotopologues should be explored as a basis for differentiating between top and bottom heavy humidity profiles. The biases in the passive remote sensing that this study identifies could result in a substantial underestimation of the humidity within the lower free troposphere (between 3 and 5 km).

The frequency with which elevated layers of nearly constant humidity are found in the soundings of the NARVAL-2 flights is used to advance a hypothesis for a microphysical control on mid-level cloudiness. Whenever there is a large difference between the temperature, T_f, at which primary ice formation occurs, and the triple-point temperature, T_*, we hypothesize a 'first-mover' advantage for convection, whereby the first clouds sufficiently deep to form ice stabilize the atmosphere in a way that makes it more difficult for subsequent convection to reach the level at which primary ice forms. This leads to increased detrainment at levels above z_* but below the level of primary ice production. Detrainment or convective deposition of moisture at temperatures between T_* and T_f is self-reinforcing as strong radiative cooling associated with this moisture increases the atmospheric stability across the top of the developing moist tongues, and supports stronger subsiding circulations, which also contributes to the termination of convection at heights below z_f. The first-mover advantage for glaciating convection may help explain the preferred mode of cloudiness at temperatures near the triple point.

Acknowledgements Open access funding provided by Max Planck Society. This paper arises from the International Space Science Institute (ISSI) workshop on "Shallow clouds and water vapor, circulation and climate sensitivity". The NARVAL campaign was co-sponsored by the Max Planck Society, the Deutsche Forschungsgemeinschaft (German Science Foundation, project HALO-SPP 1294) and the DLR Institute of Atmospheric Physics. Jean-Lionel Lacour is grateful to the CNES for postdoctoral funding. P-E Kirstetter (NOAA NSSL) and C. Dufour (LATMOS) are acknowledged for their help on the SAPHIR data. The CNES and CNRS are gratefully acknowledged for the financial support to the scientific activity of the Megha-Tropiques mission. The ICARE group is also acknowledged for realizing the ground segment of the mission: The data are available at http://www.icare.univ-lille1.fr/mt. Brian E. Mapes and an anonymous reviewer are thanked for their constructive comments on an earlier version of this paper.

References

Anthes RA (1977) A cumulus parameterization scheme utilizing a one-dimensional cloud model. Mon Weather Rev 105(3):270–286

Becker T, Stevens B, Hohenegger C (2017) Imprint of the convective parameterization and sea-surface temperature on large-scale convective self-aggregation. J Adv Model Earth Syst. doi:10.1002/2016MS000865

Bergemann M, Jakob C (2016) How important is tropospheric humidity for coastal rainfall in the tropics? Geophys Res Lett 43(11):5860–5868

Bony S, Emanuel KA (2005) On the role of moist processes in tropical intraseasonal variability: cloud–radiation and moisture–convection feedbacks. J Atmos Sci 62(8):2770–2789

Bony S, Stevens B, Coppin D, Becker T, Reed KA, Voigt A, Medeiros B (2016) Thermodynamic control of anvil cloud amount. Proc Natl Acad Sci 113(32):8927–8932

Bretherton CS, Peters ME, Back LE (2004) Relationships between water vapor path and precipitation over the tropical oceans. J Clim 17(7):1517–1528

Brogniez H, Kirstetter PE, Eymard L (2013) Expected improvements in the atmospheric humidity profile retrieval using the Megha-Tropiques microwave payload. QJR Meteorol Soc 139:842–851. doi:10.1002/qj.1869

Brogniez H, Fallourd R, Mallet C, Sivira R, Dufour C (2016) Estimating confidence intervals around relative humidity profiles from satellite observations: application to the SAPHIR sounder. J Atmos Ocean Technol 33:1005–1022. doi:10.1175/JTECH-D-15-0237.1

Bronnenberg BJ, Dhar SK, Dubé J (2009) Brand history, geography, and the persistence of brand shares. J Polit Econ 117(1):87–115

Capderou M (2009) Sampling. Comparison with other Meteorological Satellites. Tech. Rep, Mega Tropiques Technical Report, IPSL

Chazette P, Marnas F, Totems J, Shang X (2014) Comparison of IASI water vapor retrieval with H_2O-Raman lidar in the framework of the Mediterranean HyMeX and ChArMEx programs. Atmos Chem Phys 14(18):9583–9596

Craig GC, Mack JM (2013) A coarsening model for self-organization of tropical convection. J Geophys Res Atmos 118(16):8761–8769

Derbyshire SH, Beau I, Bechtold P, Grandpeix JY, Piriou JM, Redelsperger JL, Soares PMM (2004) Sensitivity of moist convection to environmental humidity. QJR Meteorol Soc 130(604):3055–3079

Emanuel KA (1986) An air-sea interaction theory for tropical cyclones. Part I: steady-state maintenance. J Atmos Sci 43(6):585–605

Eymard L, Gheudin M, Laborie P, Sirou F, Gac CL, Vinson JP, Franquet S, Desbois M, Roca R, Scott N, Waldteufel P (2002) The SAPHIR humidity sounder. MEGHA-TROPIQUES 2nd Scientific Workshop, 2–6 July 2001, Paris, France

Galewsky J, Steen-Larsen HC, Field RD, Worden J, Risi C, Schneider M (2016) Stable isotopes in atmospheric water vapor and applications to the hydrologic cycle. Rev Geophys 54(4):809–865

González Y, Schneider M, Dyroff C, Rodríguez S, Christner E, García OE, Cuevas E, Bustos JJ, Ramos R, Guirado-Fuentes C, Barthlott S, Wiegele A, Sepúlveda E (2016) Detecting moisture transport pathways to the subtropical North Atlantic free troposphere using paired H_2O-δD in situ measurements. Atmos Chem Phys 16(7):4251–4269

Greenwald T, Christopher S (2002) Effect of cold clouds on satellite measurements near 183 GHz. J Geophys Res Atmos. doi:10.1029/2000JD000,258

Holloway CE, Neelin JD (2009) Moisture vertical structure, column water vapor, and tropical deep convection. J Atmos Sci 66(6):1665–1683

Hong G, Heygster G, Miao J, Kunzi K (2005) Detection of tropical deep convective clouds from AMSU-B water vapor channels measurements. J Geophys Res Atmos. doi:10.1029/2004JD004,949

Johnson RH, Rickenbach M, Rutledge SA, Ciesielski PE, Schubert WH (1999) Trimodal characteristics of tropical convection. J Clim 12:2397–2418

Kiemle C, Groß S, Wirth M, Bugliaro L (2017) Airborne lidar observations of water vapor variability in tropical shallow convective environment. Surv Geophys (in press)

Kuang Z, Bretherton CS (2006) A mass-flux scheme view of a high-resolution simulation of a transition from shallow to deep cumulus convection. J Atmos Sci 63(7):1895–1909

Lacour JL, Risi C, Clarisse L, Bony S, Hurtmans D, Clerbaux C, Coheur PF (2012) Mid-tropospheric δD observations from IASI/MetOp at high spatial and temporal resolution. Atmos Chem Phys 12(22):10,817–10,832

Lacour JL, Clarisse L, Worden J, Schneider M, Barthlott S, Hase F, Risi C, Clerbaux C, Hurtmans D, Coheur PF (2015) Cross-validation of IASI/MetOp derived tropospheric δD with TES and ground-based FTIR observations. Atmos Meas Tech 8(3):1447–1466

Lindzen RS, Nigam S (1987) On the role of sea surface temperature gradients in forcing low-level winds and convergence in the tropics. J Atmos Sci 44(17):2418–2436

Mapes BE, Zuidema P (1996) Radiative-dynamical consequences of dry tongues in the tropical troposphere. J Atmos Sci 53(4):620–638

Möbis B, Stevens B (2012) Factors controlling the position of the Intertropical Convergence Zone on an aquaplanet. J Adv Model Earth Syst 4(4):1–16

Muller CJ, Bony S (2015) What favors convective aggregation and why? Geophys Res Lett 42(13):5626–5634. doi:10.1002/2015GL064260

Naumann AK, Stevens B, Hohenegger C, Mellado JP (2017) A conceptual model of a shallow circulation induced by prescribed low-level radiative cooling. J Atmos Sci (in press)

Nehrir AR, Kiemle C, Lebsock M, Kirchengast G, Buehler SA, Löhnert U, Liu CL, Hargrave P, Barrera-Verdejo M, Winker D (2017) Emerging technologies and synergies for airborne and space-based measurements of water vapor profiles. Surv Geophys (in press)

Nishant N, Sherwood SC, Geoffroy O (2016) Radiative driving of shallow return flows from the ITCZ. J Adv Model Earth Syst 8(2):831–842

Nuijens L, Stevens B, Siebesma AP (2009) The environment of precipitating shallow cumulus convection. J Atmos Sci 66(7):1962–1979

Oueslati B, Bellon G (2013) Convective entrainment and large-scale organization of tropical precipitation: sensitivity of the CNRM-CM5 hierarchy of models. J Clim 26(9):2931–2946

Raymond DJ, Raga GB, Bretherton CS, Molinari J, López-Carrillo C, Fuchs Ž (2003) Convective forcing in the intertropical convergence zone of the eastern Pacific. J Atmos Sci 60(17):2064–2082

Risi C, Bony S, Fco Vimeux (2008) Influence of convective processes on the isotopic composition of precipitation and water vapor in the tropics: 2. Physical interpretation of the amount effect. J Geophys Res Atmos 113(D19):306

Roca R, Brogniez H, Chambon P, Chomette O, Cloché S, Gosset ME, Mahfouf JF, Raberanto P, Viltard N (2015) The Megha-Tropiques mission: a review after three years in orbit. Front Earth Sci 3:852

Schiro KA, Neelin JD, Adams DK, Lintner BR (2016) deep convection and column water vapor over tropical land versus tropical ocean: a comparison between the Amazon and the tropical western Pacific. J Atmos Sci 73(10):4043–4063

Schneider M, Hase F (2011) Optimal estimation of tropospheric H_2O and δD with IASI/METOP. Atmos Chem Phys 11(5):16,107–16,146

Sherwood SC, Roca R, Weckwerth TM, Andronova NG (2010) Tropospheric water vapor, convection, and climate. Rev Geophys 48(2):1481

Stevens B, Bony S (2013) Water in the atmosphere. Phys Today 66(6):29

Stevens B, Farrell D, Hirsch L, Jansen F, Nuijens L, Serikov I, Brügmann B, Forde M, Linné H, Lonitz K, Prospero JM (2016) The Barbados cloud observatory: anchoring investigations of clouds and circulation on the edge of the ITCZ. Bull Am Meteorol Soc 97(5):787–801

Thomas GE, Stamnes K (1999) Radiative transfer in the atmosphere and ocean. Cambridge atmospheric and space science series. Cambridge University Press, Cambridge

Vial J, Bony S, Stevens B, Vogel R (2017) Mechanisms and model diversity of trade-wind shallow cumulus cloud feedbacks: a review. Surv Geophys. doi:10.1007/s10712-017-9418-2

Wirth M, Fix A, Mahnke P, Schwarzer H, Schrandt F, Ehret G (2009) The airborne multi-wavelength water vapor differential absorption lidar WALES: system design and performance. Appl Phys B 96(1):201–213

Worden J, Noone D, Bowman K (2007) Importance of rain evaporation and continental convection in the tropical water cycle. Nature 445(7127):528–532

Wu CM, Stevens B, Arakawa A (2009) What controls the transition from shallow to deep convection? J Atmos Sci 66(6):1793–1806

Surv Geophys (2017) 38:1399–1423
https://doi.org/10.1007/s10712-017-9437-z

The Representation of Tropospheric Water Vapor Over Low-Latitude Oceans in (Re-)analysis: Errors, Impacts, and the Ability to Exploit Current and Prospective Observations

Robert Pincus[1,2] · Anton Beljaars[3] · Stefan A. Buehler[4] · Gottfried Kirchengast[5] · Florian Ladstaedter[5] · Jeffrey S. Whitaker[2]

Received: 20 March 2017 / Accepted: 6 October 2017 / Published online: 26 October 2017

Abstract This paper addresses the representation of lower tropospheric water vapor in the meteorological analyses—fully detailed estimates of atmospheric state—providing the wide temporal and spatial coverage used in many process studies. Analyses are produced in a cycle combining short forecasts from initial conditions with data assimilation that optimally estimates the state of the atmosphere from the previous forecasts and new observations, providing initial conditions for the next set of forecasts. Estimates of water vapor are among the less certain aspects of the state because the quantity poses special challenges for data assimilation while being particularly sensitive to the details of model parameterizations. Over remote tropical oceans observations of water vapor come from two sources: passive observations at microwave or infrared wavelengths that provide relatively strong constraints over large areas on column-integrated moisture but relatively coarse vertical resolution, and occultations of Global Positioning System provide much higher accuracy and vertical resolution but are relatively spatially coarse. Over low-latitude oceans, experiences with two systems suggest that current analyses reproduce much of the large-scale variability in integrated water vapor but have systematic errors in the representation of the boundary layer with compensating errors in the free troposphere; these errors introduce errors of order 10% in radiative heating rates through the free troposphere. New observations, such as might be obtained by future observing systems, improve the estimates of water vapor but this improvement is lost relatively quickly,

✉ Robert Pincus
Robert.Pincus@colorado.edu

[1] Cooperative Institute for Research in Environmental Sciences, University of Colorado, Boulder, CO 80309, USA

[2] Physical Sciences Division, NOAA Earth System Research Lab, Boulder, CO 80305, USA

[3] European Centre for Medium-Range Weather Forecasts, Shinfield Park, Reading RG2 9AX, UK

[4] Informatics and Natural Sciences Department of Earth Sciences, Meteorological Institute, Faculty of Mathematics, Universitt Hamburg, Bundesstrae 55, 20146 Hamburg, Germany

[5] Wegener Center for Climate and Global Change and Institute for Geophysics, Astrophysics, and Meteorology, Institute of Physics, University of Graz, 8010 Graz, Austria

suggesting that exploiting better observations will require targeted improvements to global forecast models.

Keywords Water vapor · Satellite · Microwave · Infrared · Radio occultation · Data assimilation · Tropospheric water vapor profiling

1 Tropospheric Water Vapor Over Low-Latitude Oceans

The vertical distribution of water vapor in the troposphere, through its influence on shallow and deep convection (see Mapes et al. 2017), controls many aspects of tropical climate. The distribution is tightly coupled to the processes underlying the self-aggregation of convection (Wing et al. 2017). Through its influence on radiative cooling rates the distribution of water vapor largely controls the amount of large-scale subsidence and hence the structure of thermal and moisture inversions to which the properties of low clouds are so sensitive, so the distribution plays an important role in determining low cloud feedbacks on climate change (Klein et al. 2017; Vial et al. 2017).

How can the distribution of water vapor be determined? Observations from field campaigns, including high-density dropsonde deployments (Stevens et al. 2017) or results from specialized remote sensing instruments on airborne platforms (Kiemle et al. 2017) provide the detail needed to spark and test hypothesis. Many studies, however, require observations with more uniform sampling across broader spatial and temporal scales (i.e., Lebsock et al. this issue and the studies reviewed by Klein et al. 2017). Retrievals based on measurements from microwave or infrared sounding instruments provide one possible source of observations for water vapor (see the use in Lebsock et al. this issue). An alternative is to include the same observations, or a subset thereof, in the construction of meteorological analyses. The two approaches are more similar than is immediately obvious (Parker 2016): both use radiative transfer models and prior knowledge of the likely state of the atmosphere to provide more refined estimates of water vapor concentrations. As we will show, analyses can be compromised by errors in the model used to produce short-term forecasts, but are enhanced by the use of a wide range of observations and by the influence of recent observations on the initial estimate of atmospheric state.

Here we assess the current state of routine knowledge of lower tropospheric water vapor over remote tropical oceans. We describe how observations are synthesized in time and space to produce meteorological analyses and discuss the two broad classes of observations—-multi- or hyperspectral sounding and radio occultation—that inform these estimates in current analysis systems. The representation of water vapor over remote low-latitude oceans in two analysis systems is compared to independent observations in regimes spanning a range of convective depth. In both systems the representation of boundary layer humidity is biased, resulting in compensating errors in the free troposphere; these errors lead to errors of order 10% in radiative heating rates in the free troposphere. Data assimilation experiments suggest that even substantially richer observational capabilities will need to be paired with systematic model improvements in order to realize more accurate estimates of boundary layer humidity.

2 Integrating Observations in Space and Time: Data Assimilation and (Re-)analysis

2.1 Producing Meteorological (Re-)analyses

In the context of this paper, an atmospheric analysis is an estimate of the instantaneous state of the atmosphere in which all state variables are defined at every location. The skill of current analyses is due to their use as initial conditions for numerical weather prediction (NWP): Accurate forecasts require accurate initial conditions, so NWP centers invest heavily in all aspect of analysis production including quality control, bias correction, and the use of as many observations and observation types as possible. Given our focus on lower tropospheric moisture analysis over tropical ocean our discussion focuses on global systems at operational centers. Examples in Sect. 4.1 are drawn from systems at the European Centre for Medium Range Weather Forecasts (ECMWF) and the US National Centers for Environmental Prediction (NCEP).

Meteorological analysis rely on data assimilation, a set of statistical techniques developed to make best use of a wide variety of observations that are irregular in time and space (see, e.g., Kalnay 2003). Data assimilation optimally combines short-range forecasts with new observations to produce an estimate of the state of the atmosphere at a given time (or the evolution of the atmosphere through the assimilation window). The skill of an analysis therefore depends on the forecast model, the assimilation system, and the available observations. The first two are discussed below; observations relevant to estimates of lower tropospheric water vapor are described in Sect. 3.

2.1.1 Global Forecast Models

Data assimilation uses a forecast model to propagate the atmospheric state in time and provide consistency between variables. Practical constraints, including a requirement to deliver timely forecasts, mean that model accuracy must be balanced against computational cost. This impacts model spatial resolution and so the maximum time step. At this writing operational models using equivalent grid spacing[1] of 10–20 km, time steps of around 10 min, and 50–150 levels in the vertical, with the highest between 0.2 and 0.01 hPa. The density of vertical levels is highest near the surface and decreases gradually with height. The vertical discretization especially is relevant to estimates of boundary layer humidity: At typical boundary layer inversion heights, the vertical level spacing varies from 200–400 m, so that sharp inversions can not be precisely located in the vertical.

Models must also include parameterizations for processes that are not explicitly resolved. In global NWP models the list of parameterized processes might include radiation, shallow and deep convection, cloud macro- and micro-physics, boundary layer turbulence, the impacts of subgrid orography and gravity waves, and the evolution of the land surface. Despite substantial progress, physical parameterizations remain a primary source of model error.

Model errors, whether arising from discretization, numerics, or parameterization deficiencies, impact the quality of the analysis both directly and indirectly, by reducing the

[1] Grid-point spacing is often equated with resolution but it is well known that, for numerical reasons, the true resolution is less. For instance, from a spectral analysis of ocean winds, Abdalla et al. (2013) conclude that, in the ECMWF model, 50% of the real variability is only achieved at scales of 3–5 times the grid point resolution.

effectiveness of observations. Model errors have a larger impact on fields that are only indirectly constrained by observations, e.g., moisture, cloud fields, precipitation and radiative/turbulent fluxes. Parameterization errors are of particular concern in the analysis of water vapor because the field is strongly impacted by heavily parameterized moist physics.

2.1.2 Data Assimilation Systems

Data assimilation uses observations to update the state of the atmosphere provided by a short-range forecast. Operational methods rely on the Kalman filter (Kalman 1960; Kalman and Bucy 1961) which minimizes the scaled root-mean-square difference between forecasts and all available observations while respecting the uncertainty in both. Covariances propagate information between variables and through space and time so that, for example, an observation of temperature at a given location can affect temperature, pressure, and winds in the surrounding region. The model-observation difference is computed in the space of the observations so that many observations types, especially the satellite observations described in the next section, rely on forward operators to map the model state to a synthetic (predicted) observation.

Analyses and forecasts are normally paired, with an analysis used to initialize a short (6–12 h) forecast that forms the background field or initial guess for the next analysis. This means that a given analysis is affected by not only the most recent observations but by all observations in the recent past. As a result, in most modern systems, the background forecast, which has the history of all previous observations, contains more information about the current state of the atmosphere than do the latest observations, and the difference between forecasts and the resulting analyses is subtle.

Because they rely on forecast models, analyses from continuous forecast/data assimilation cycling provide a self-consistent representation of the atmosphere in space and time. This is convenient; it is also normally more accurate than, for example, estimates for a single observing platform because the analysis exploits many different data sources (e.g., SYNOP stations, ships and buoys, radiosondes, aircraft, and satellite observations including the passive infrared and microwave and radio occultation described in Sect. 3). The impact of each observation on model state propagates in space and time via the forecasts, leading to better initial estimates of atmospheric state and an improved ability to use the observations (Eyre et al. 1993).

The Kalman filter on which analyses are based is optimal when errors are Gaussian, unbiased, and uncorrelated. All these conditions are violated to some extent in real systems with detrimental impacts on the analysis. Operational centers typically apply variational bias correction (Dee and Uppala 2009) to observations, especially those from satellite instruments, to ensure that the observations are unbiased with respect to the model, but this approach cannot distinguish between (possibly conditional) errors in the underlying observations, the forecast model, and the forward operator used to make model state to synthetic observations. Mischaracterization of observation and/or forecast errors can also degrade the analysis, as will be evident in Sect. 5.

2.1.3 Challenges in Assimilating Moisture

The assimilation of moisture poses special challenges because upper and lower bounds on absolute humidity mean that errors are often unlikely to be Gaussian. This is normally treated by representing humidity in the analysis generation with a specialized control variable that is more likely to be Gaussian: relative humidity, the logarithm of specific humidity, or variance scaled relative humidity (Hólm et al. 2002).

Systematic imbalances between observations and forecast models have historically been an issue for moisture. The ECMWF 40-year Re-analysis (ERA-40; Uppala et al. 2005) reanalysis, for example, contains small systematic positive increments of total column water vapor. This was due partly due to sampling bias in the observations, which were considered only in cloud-free areas, but also reflected errors in the forecast model and the forward operator as well as biases in the observations. Whatever the mix of causes, the model could not retain the excess moisture and precipitated heavily in the initial 12 h of each forecast, leading to an over-active hydrological cycle and a too-strong Hadley circulation. In more recent systems, which include variational bias correction and make greater use of satellite radiances in cloudy and precipitating areas (Bauer et al. 2010; Geer et al. 2010), the problem is less acute.

Humidity observations over remote tropical oceans consist almost entirely of satellite data (Andersson et al. 2005) from platforms such as those described in Sect. 3. The observational constraint on total column water vapor provided by passive microwave information is fairly accurate but the broad vertical weighting functions for passive sensors (see Fig. 1) mean that the vertical distribution of water vapor in the atmosphere is poorly constrained by observations. As a result the vertical distribution is very much controlled by the assimilating model. In the subtropics and stratocumulus areas with subsiding motion, an inversion is maintained as a balance between subsidence and boundary layer entrainment, both of which are controlled primarily by the model formulation (see Sect. 4.1.1).

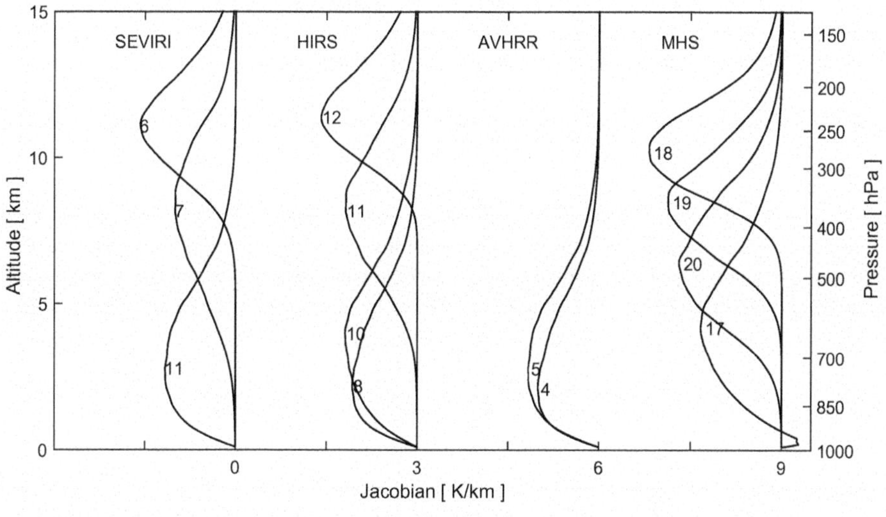

Fig. 1 Water vapor Jacobians for the SEVIRI (**a**), HIRS (**b**), and AVHRR (**c**) infrared and MHS (**d**) microwave instruments, calculated with the atmospheric radiative transfer simulator ARTS (Eriksson et al. 2011) for a tropical atmosphere (mean of radiosonde profiles for station Manus at 2.06°S, 147.42°E). Curves are labeled with the channel number for the respective instrument. The Jacobian is the derivative of the measurement with respect to changes in the atmospheric state; the figure shows brightness temperature change for nadir-viewing observations resulting from a fractional change in the water vapor concentration, normalized by the layer thickness. Jacobians identify the portion of the atmosphere to which the measurement is sensitive to changes in water vapor and are closely related to weighting functions

2.2 Analysis and Reanalysis

In operational NWP centers, all components of the analysis system including the forecast model, data assimilation system, and data handling (quality control and bias correction), are improved as opportunities present themselves. This sometimes results in systematic changes in analyzed fields. Reanalysis is the process of analyzing historic observations with a consistent system, normally one which has been improved since the initial analysis was produced. This removes the analysis system as a source of potential discontinuities in analyzed fields although changes in the kind, volume, accuracy, and spatial distribution of available observations can still introduce shifts.

Reanalysis inherits from operational analyses the optimal use of a wide range of observations and the convenience of a gridded, complete, self-consistent description of the atmosphere. There are several key limitations, however. Crucially for some applications is that there is no constraint that energy or mass in a reanalysis be conserved. More generally, estimates of quantities not directly constrained by observations, including the vertical distribution of water vapor over remote oceans, as well clouds, precipitation, and fluxes at the top and bottom of the atmosphere, are less certain and can only be assessed through careful evaluation (Trenberth et al. 2011). This issue is particularly relevant to estimates of water vapor in the oceanic boundary layer, as we show below.

An excellent resource for reanalysis is https://reanalyses.org. Some well-known recent global reanalysis projects are listed in Table 1. This list is likely to be out-of-date relatively quickly: ECMWF, for example, has already begun production of the fifth-generation reanalysis ERA5 (https://www.ecmwf.int/en/newsletter/147/news/era5-reanalysis-production).

3 What Measurements Inform Current Estimates?

Meteorological analyses, as described in Sect. 2, rely on observations to correct forecast errors and produce a more accurate estimate of the state of the atmosphere. In situ measurements such as radiosondes anchor the observational network but are few and far between over remote low-latitude oceans. Estimates of water vapor in this region rely primarily on three sources of information: sounding instruments that probe the vertical structure of water vapor using the spectral variation of emission by water vapor in the microwave and infrared regions, passive microwave measurements sensitive to column-integrated water vapor, and observations of radio occultation in limb geometry that resolve

Table 1 Global reanalyses covering the satellite epoch

Name	Epoch	References
ERA Interim	1979-present	Dee et al. (2011)
JRA-55 and variants	1958-present	Kobayashi et al. (2015)
MERRA-2	1980-present	http://gmao.gsfc.nasa.gov/pubs/tm/docs/Bosilovich803.pdf
NCEP/DOE Reanalysis II	1979-near present	Kanamitsu et al. (2002)

the density (temperature and humidity) structure with very high vertical resolution but coarse horizontal resolution.

3.1 Microwave and Infrared Sounding

3.1.1 Principles of Measurement

Measurements from microwave and infrared satellite sounders dominate the observations informing water vapor in meteorological analyses [see, e.g., Fig. 5 in Dee et al. (2011)], especially over remote oceans. These instruments observe radiation emitted by the atmosphere which originates near the altitude where optical depth, the integral of the absorption coefficient calculated from the satellite toward the atmosphere, reaches unity. Sounders exploit two facts. First, clear-sky atmospheric absorption in the troposphere is dominated by water vapor, so that additional moisture makes the atmosphere more opaque and shifts the emission level upward, to colder temperatures, reducing the measured radiance (see for example Buehler et al. 2004 Fig. 11) . Second, absorption and emission by water vapor depends strongly on frequency, so that measurements made at a range of frequencies with different amounts of absorption are sensitive to the moisture in different atmospheric layers and provide the ability to measure water vapor at different altitudes. This range of altitudes is illustrated by Fig. 1, which shows water vapor Jacobians for some widely used sensors (SEVIRI, HIRS, AVHRR, and MHS, summarized in Table 2). Each of these instruments has flown on multiple platforms; similar "one-off" instruments, such as the SAPHIR instrument aboard the Megha-Tropiques satellite (Brogniez et al. 2013), provide closely related information.

Passive satellite sounding with infrared imagers has a long history. Readily available infrared humidity data start with the HIRS instrument on the TIROS N satellite in 1978 (Shi and Bates 2011). Meteorological microwave satellite measurements also have a long history, but water vapor profile measurements (near 183 GHz) only became available with the SSM/T2 instrument on the DMSP F11 satellite in 1991, with data in the NOAA CLASS archive starting in 1994.

Legacy instruments have just a few (one to five) channels dedicated to water vapor; this provides a strict upper bound on the number of degrees of freedom in the vertical profile of water vapor that can be resolved. The strength of these sensors lies in their horizontal resolution (1–15 km , see Table 2) and the resulting ability to resolve horizontal structures.

Table 2 Four widely used infrared and microwave humidity sensors

Acronym	Spectral range	Orbit	Δt	Δx	Full name and references
SEVIRI	IR	GEO	15 min	3 km	Spinning enhanced visible and infrared imager (Schmetz et al. 2002)
HIRS	IR	LEO	12 h	10 km	High-resolution infrared radiation sounder (Klaes et al. 2007)
AVHRR	IR	LEO	12 h	1 km	Advanced very high-resolution radiometer (Klaes et al. 2007)
MHS	MW	LEO	12 h	16 km	Microwave humidity sounder (Klaes et al. 2007)

IR signifies infrared and MW microwave; GEO denotes geostationary and LEO for low Earth (polar) orbit; Δt and Δx approximately specify the temporal and spatial resolution, respectively

Figure 2 highlights the kind of information available from microwave and infrared sounding instruments. Here the image comes from channel 11 of the SEVERI instrument, which is sensitive to water vapor below about 6 km (Fig. 1). The measurement is expressed as a brightness temperature T_b i.e., the temperature at which a black body would produce the observed channel-integrated intensity. For SEVERI and similar instruments (downward-looking viewing geometry, in optically thick frequency regions dominated by water vapor absorption) T_b can be easily interpreted: Because the thermal source function for the radiative transfer (the Planck function) depends on temperature, while the atmospheric absorption depends on the water vapor content which is itself also strongly controlled by the temperature, the measured intensity is a good proxy for relative humidity averaged over the layer. High T_b indicates emission from lower in the layer, and hence low relative humidity.

These relationships may be used to develop scalings between observed T_b and layer-averaged relative humidity (e.g., Buehler and John 2005); in this image a 1 K change in T_b corresponds to approximately 9% relative difference in relative humidity. In practice, however, data assimilation systems use radiative transfer calculation using fast models (also called "forward operators") to predict the intensity that would be observed for a given distribution of temperature and humidity, and the data assimilation system works to minimize the difference between these predicted observations and the observations themselves (see Sect. 2.1.2).

3.1.2 Why Both Microwave and Infrared Observations are Useful

Figure 2 shows an image obtained from an infrared sounder on a geostationary satellite; instruments on such platforms can provide very high-time-resolution observations (15 m or less). It is not practical to deploy microwave sounders so far from the Earth because their spatial resolution would be greatly compromised. Spatial resolution is limited by diffraction, and thus infrared sensors inherently allow higher spatial resolutions than microwave sensors, because their apertures are much larger compared to the measured wavelength.

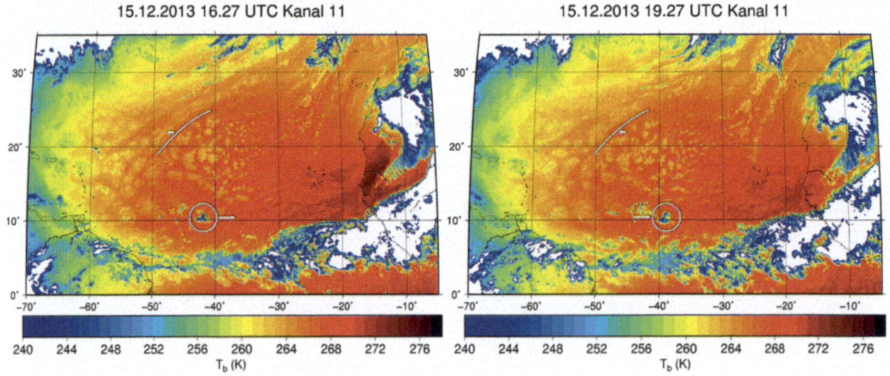

Fig. 2 Two different snapshots of water vapor brightness temperatures by SEVIRI Channel 11, separated by 3 h. The line indicates humidity and cloud structures below roughly 6 km, moving westward; the circle indicates an overlying high cold cloud, moving eastward. These images were taken in December 2013 over the tropical Atlantic. Brightness temperatures below the (somewhat arbitrary) threshold of 240 K are shown in white to indicate deep convection

But the figure also illustrates why microwave sounders provide such a useful complement to infrared observations. Brightness temperatures T_b below some threshold, as occur frequently in the image, correspond to relative humidity above saturation. This unphysical result indicates the presence of clouds. It is relatively easy to filter out the cloud contaminated data, but restricting humidity observations to clear skies along can lead to biases in average humidity and even its climatic trend (John et al. 2011). Microwave sensors are much less affected by clouds, and so continue to be central to meteorological analyses despite the relatively low horizontal resolution.

3.1.3 Prospects

The Jacobians of passive sensors are broad in the vertical (Fig. 1) because the radiation received at the sensor arises from a range of altitudes. Hyperspectral infrared instruments, including AIRS (Aumann et al. 2003) with 2378 channels and IASI (Clerbaux et al. 2009) with 8461 channels, offer the possibility of combining many high-accuracy channels to achieve higher vertical resolution, analogous to the sharpening of a blurred image in image processing (see for example Osher et al. 2005). (Hyperspectral microwave sensors do not yet exist but may be feasible in a few years, as discussed in the contribution by Nehrir et al. 2017).

The true information content from hyperspectral observations, however, is orders of magnitude lower than the number of channels because the measurements in individual channels are very highly correlated. Applying the Bayesian information content analysis described in more detail in Nehrir et al. (2017) to IASI, for example, shows that at most 16 independent pieces of information on the water vapor profile are theoretically possible, corresponding to a vertical resolution of approximately 1.5 km in the troposphere. In practice the information content of IASI is lower, with literature estimates ranging from 3 to 10 pieces of independent information (August et al. 2012; Lacour et al. 2012; Schneider and Hase 2011; Herbin et al. 2009). This may be firstly because often only a small subset of the channels is used, and secondly because in an NWP context better a priori information is available, which reduces the information content of the measurement relative to the a priori.

Ultimately, it is the physics of radiative transfer that limits the ability of passive sensors to probe the humidity structure of the lower troposphere.

3.2 Estimates of Precipitable Water from Microwave Observations

Constraints on total column water vapor come from observations in spectral regions where atmospheric absorption is weak enough that the surface is visible from space. This works particularly well in the microwave spectral region over ocean, because there the surface is radiatively cold and can be accurately modeled (the emissivity is around 0.6 and depends mostly on the wind speed). The atmosphere is then observed in emission in front of the cold background, and Jacobians are positive, instead of negative as in Fig. 1. The temperature dependence of the Planck function and the temperature control of absolute humidity then combine to make the measurement depend strongly on the total column water vapor.

Mears et al. (2015) validated such satellite-borne microwave total column water vapor measurements against ground-based GPS observations and found an overall bias of only approximately 1 kg/m^2 and standard deviations better than approximately 2 kg/m^2.

3.3 GNSS Radio Occultation: A Global Refractometer

An orthogonal source of water vapor information comes from observations of Global Navigation Satellite System (GNSS) radio occultation (RO) using the refraction of decimeter-wave GNSS signals near 1.2 and 1.6 GHz received at low Earth orbit (LEO) satellites.

3.3.1 Principles of Measurement

Figure 3a illustrates the GNSS RO observing geometry and summarizes the steps in estimating profiles of atmospheric state from phase measurements (see also Kursinski et al. 1997; Anthes 2011; Steiner et al. 2011). An occultation event occurs when a receiver, peering through the atmosphere toward a satellite with a GNSS transmitter in limb sounding geometry, observes the satellite setting behind or rising above the Earth's horizon. Profiles are assigned a horizontal position at point of closest approach to the Earth's surface ('tangent point') although the information derives from an along-path range of roughly 100–200 km around the tangent point in the troposphere with across-ray resolution of 1–2 km.

The fundamental measurement is of time delays of GNSS signals during occultation events, which can be directly related to excess phase paths. Because the measurement is based on time, which is uniquely well measured, RO measurements have unique long-term stability and narrow uncertainty. GNSS signals are not attenuated by clouds, so that sampling is unbiased. The vertical resolution of RO measurements, roughly 200–300 m (Gorbunov et al. 2004; Kursinski et al. 1997), is much higher than passive microwave and infrared sensors (Sect. 3.1). The primary disadvantage is that data are relatively sparse. The only system exploited to date for GNSS RO measurements is the U.S. Global Positioning System (GPS); with this constellation a single RO receiver in LEO acquires 300–600 RO events per day with global coverage.

Interpretation of RO refraction measurements is quite different than for brightness temperatures. Tracking and orbit data are used to compute excess phase profiles (Hajj et al. 2002; Schreiner et al. 2009; Anthes 2011) which are converted to Doppler shift profiles and then bending angle profiles (Ho et al. 2012; Steiner et al. 2013, and references therein). Bending angle profiles may be used in data assimilation or may form the basis of retrievals. The latter map bending angle to refractivity and dry-air profiles and finally to thermodynamic profiles of density, pressure, temperature, and tropospheric water vapor (Kursinski and Gebhardt 2014).

3.3.2 Vertical Resolution, Accuracy, and Limitations

Although meteorological analyses incorporate RO observations using their own forward operators (e.g., Healy and Thepaut 2006) the accuracy of the underlying observations is more clearly demonstrated by comparing retrievals of temperature and humidity to in situ observations, as in Fig. 4. Here the comparison is to Vaisala RS92 radiosonde profiles from the Global Reference Upper Air Network (GRUAN), which undertakes dedicated efforts to provide climate-quality processing of the data (Bodeker et al. 2016; Immler et al. 2010). Panels *a* and *b* show example profiles containing layered moisture structures (see also Stevens et al. 2017), illustrating the high vertical resolution. The reference radiosonde profiles (GRUAN RS92-GDP.2) are used at $\sim$ 300 m vertical resolution, while the RO

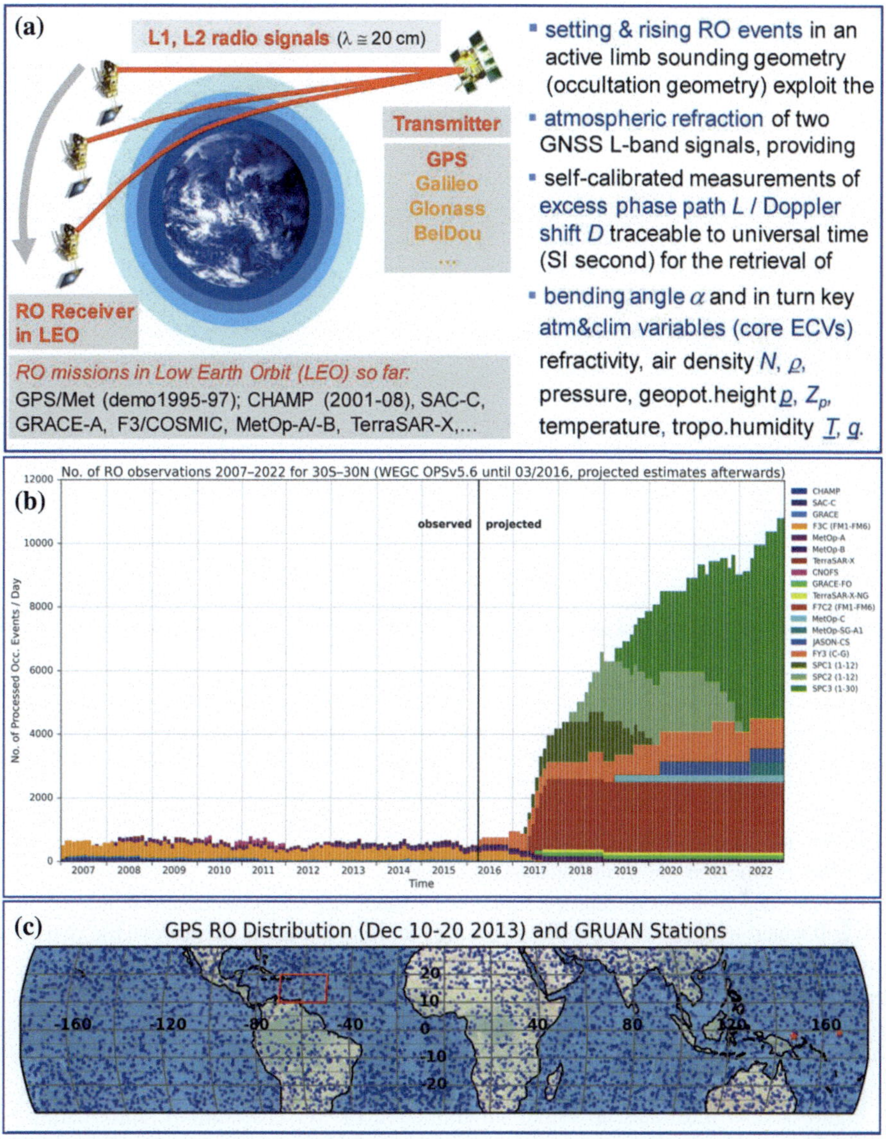

Fig. 3 **a** Schematic view of the GNSS RO active limb sounding observation geometry and explanation of measurement principles; the acronyms for RO missions are the common names in the RO literature. **b** Number of RO events per day over 2007–2022 at low latitudes (30S–30N), with actual numbers until 2016 and with projected numbers afterwards, where new RO missions currently prepared are successively deployed. **c** Coverage by RO events (blue dots) at low latitudes (30S–30N) during 10–20 December 2013 (NARVAL- South campaign), with showing a zoom also for a cell in the Barbados area (red box). Two available tropical GRUAN radiosonde stations are marked (red asterisk symbols)

profiles, taken from the recent OPSv5.6 reprocessing of all 2001–2016 RO data at the Wegener Center, were not yet optimized for water vapor profiling and so are only available at ~ 800 m vertical resolution. Profiles from the operational analysis produced at

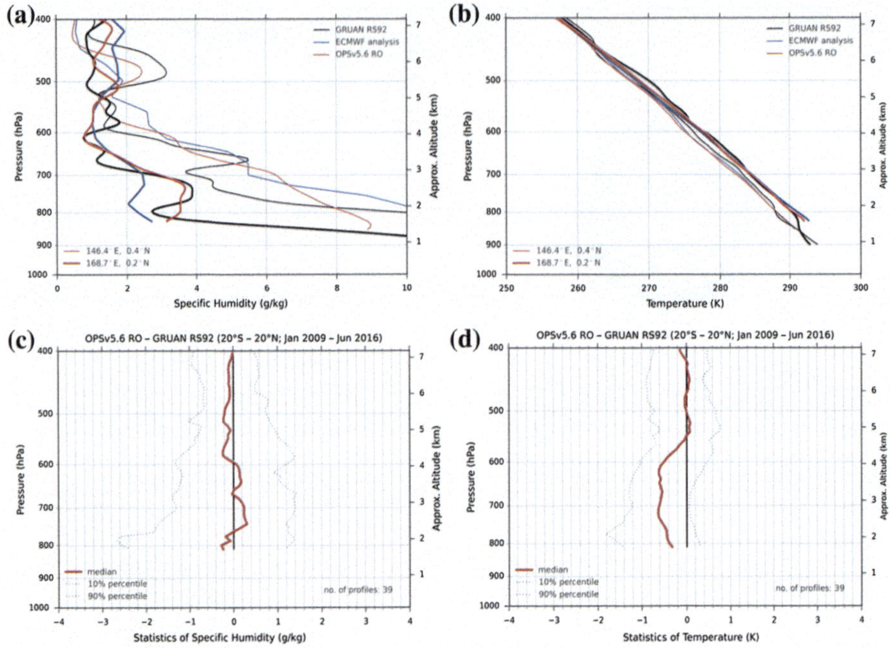

Fig. 4 Individual examples profiles of specific humidity (**a**) and temperature (**b**) from inter-comparing RO and ECMWF to collocated GRUAN data (about 200 km–2 h space–time difference in the two cases). Furthermore, statistics (median deviation and indicative percentiles) of RO specific humidity (**c**) and temperature (**d**) profiles are shown relative to GRUAN data, based on the available 300 km–3 h collocations over 2009–2016. The tropical GRUAN stations available are those marked in Fig. 4c

ECMWF are also shown. Even at reduced vertical resolution, RO captures moisture layering that is smeared out in the analysis profiles.

Panels *c* and *d* shows difference statistics between RO temperature and humidity profiles collocated with GRUAN profiles. Specific humidity is consistent between the two data sets to within 0.1–0.3 g/kg in the median, consistent with theoretical studies of accuracy (e.g., Kursinski and Gebhardt 2014; Ladstädter et al. 2015; Rieckh et al. 2016). Some amount of disagreement between radiosonde and RO profiles is expected due to imperfect collocation—sparse sampling by RO requires the use of relatively loose criteria for co-location ($\pm$ 300 km, $\pm$ 3 h) to obtain enough samples.

Figure 4 also illustrates an important limitation of RO observations: Information is available through the free troposphere, including the very valuable RO capacity of determining the boundary layer height (e.g., Anthes 2011; Ho et al. 2015), but is lacking within the atmospheric boundary layer. The retrieved RO profiles in Fig. 4a, b stop near the boundary layer top, which in one case is clearly indicated by the GRUAN profile. Though efforts are being made to exploit RO observations to determine boundary later water vapor (e.g., Sokolovskiy et al. 2014; Xie et al. 2012), the core strength of existing RO observations is in profiling of the free troposphere.

3.3.3 *Prospects*

Radio occultation measurements provide a valuable complement to observations by passive microwave and infrared sensors (Sect. 3.1) for characterizing water vapor. The observations play an important role in data assimilation systems (Sect. 2) because the high accuracy, itself a result of the ability to make the fundamental measurement of time with high accuracy, means that they can be incorporated as 'anchor observations' without bias correction, adding a globally distributed set of calibration points to the radiosonde network. When the measurements are used outside assimilation systems they are able to accurately determine the height of the boundary layer; this capability might be especially useful in process studies.

The record is short compared to passive imagers: Limited data are available from 2001 with the full existing system online only since 2007. Data are also relatively sparse horizontally. This sparsity means that RO observations are often more useful over the long term, say via their incorporation into (re-)analyses, then for limited-domain field campaigns. This is evident in Fig. 3c, which shows all low-latitude RO event locations during the NARVAL-South campaign between 10 and 20 December 2013 (Stevens et al. 2016). This field campaign focused on the region around Barbados (red 10° × 20° box). The entire low latitude band received 5759 RO events during this 11-day period; just 67 events occurred in the Barbados area.

The amount of GNSS data is posed to increase rapidly, however. Initial results have been obtained from the Chinese BeiDou system (Liao et al. 2016) and use of the European Galileo and Russian Glonass systems is expected in the next years. This will vastly expand the frequency of RO events. Figure 3b illustrates the total numbers available over the last decade at low latitudes within 30°S–30°N, around 600 events per day, and the projected strong increase by about an order of magnitude over the coming years. This leads to more than 8000 events per day as of 2020, which corresponds to more than 20,000 events per day globally.

Current observations rely on GNSS radio occultation. One promising technological advance is LEO–LEO occultation (Kirchengast and Schweitzer 2011; Kursinski et al. 2002; Liu et al. 2017), described as one of the emerging water vapor observation techniques by Nehrir et al. (2017).

4 Errors in Water Vapor Distributions and the Resulting Impacts

Satellite observations (Sect. 3) are used in analysis systems (Sect. 2) to produce estimates of the state of the atmosphere including the distribution of water vapor. The accuracy of this estimate over remote oceans, especially in the lower troposphere, has implications for the ability to use observations to test and generate hypotheses. In this section, we use observations to spot-check this distribution in two circumstances.

4.1 Assessing Errors in the Analyzed Distribution of Water Vapor

Because analyses are constructed by minimizing the difference between the state and all available observations, the accuracy of an analysis can be assessed only by comparison with independent observations. This is harder than it might seem—to produce the best possible analysis, forecasting centers go to great lengths to use all available high-quality observations, going so far as to facilitate the real-time availability of data from field campaigns. This means that opportunities to assess specific aspects of analysis, such as lower tropospheric water vapor over low-latitude oceans, are few and far between.

Below we exploit two sets of radiosonde measurements made during field campaigns to assess the accuracy of water vapor analysis. Sect. 4.1.1 compares radiosonde observations off the coast of Peru to fields in the ERA Interim reanalysis, which did not assimilate these observations. Section 4.1.2 exploits measurements made in the equatorial Pacific to assess a reduced-resolution version of the current NCEP data assimilation using data-denial experiments in which the observations were deliberately withheld. Both reanalyses make use of the full range of satellite observations as described in Sect. 3; the precise observations used will vary from day to day. Because the comparisons involve different analysis systems from different epochs, assessed in very different meteorological regimes, we do not expect the errors identified here to be consistent with each other, or representative in any broad sense. Our intent is rather to highlight similarities in the character and magnitude of errors and, in the next section, to ask to what extent these might be ameliorated with better observations of lower tropospheric water vapor.

4.1.1 Assessment in Subsiding Regions

The first example comes from radiosondes launched as part of the EPIC (Eastern Pacific Investigation of Climate Processes in the Coupled Ocean-Atmosphere System) campaign. This experiment took place from 16–21 October 2001 at 20°S/85°W in the East Pacific, a little over 800 km off the coast of Peru and Chile (Bretherton et al. 2004). We use 3-hourly radiosondes launched from the research vessel Ron Brown to assess the widely used ERA Interim reanalysis (ERA-I; Dee et al. 2011). The comparison exposes analysis error because the soundings were not used in the reanalysis.

In the EPIC domain, subsidence and relatively cold sea-surface temperatures frequently give rise to stratocumulus clouds under strong thermal and moisture inversions. These clouds are notoriously difficult to simulate in global models, partly because the inversions are very sharp and so poorly resolved on a relatively coarse numerical grid. As over much of the low-latitude oceans, few in situ observations were available for assimilation, so it is of interest to see how an analysis system copes.

Figure 5a shows the profile of specific humidity averaged over the entire period. A sharp jump in specific humidity is observed at the top of the boundary layer. ERA-I reproduces an inversion but it is too low (at 1147 and 918 m, by about one model level) and not sharp enough. The profile of potential temperature (not shown) indicates a consistent picture. There are no observations in ERA-I that impose sharp temperature and moisture gradients. Instead, the inversion is maintained by the model as the result of a subtle balance between weak subsidence and the entrainment process in the model physics.

The evolution of moisture in the observations and the reanalysis is illustrated in time-height cross section in Fig. 6. The mixed layer is more moist than observed. This is tied quite closely to the overly shallow boundary layer: since total column water vapor is well constrained by satellite microwave observations, the too-low inversion is compensated by excess boundary layer moisture. Temporal variability is also more muted in the analysis than in the observations, although the diurnal cycle is more regular than is observed. The observed variability includes a component of sampling or representativeness noise, but much of the observed variability is coherent over a period of 3 h, suggesting that the reanalysis does indeed miss some amount of variability. Finally, a moist layer descends from 4000 m on 16 October to about 2000 m on 20 Oct. Such a structure is related to the

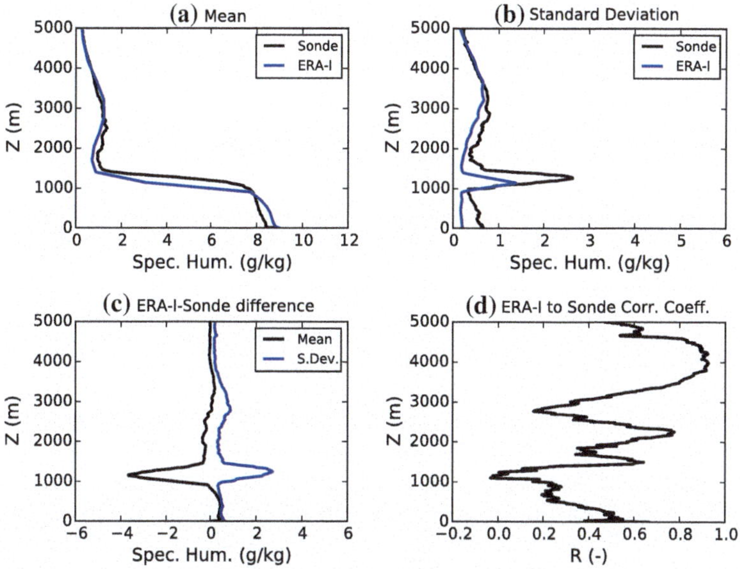

Fig. 5 **a** Mean radiosonde and ERA-I humidity profiles, **b** standard deviation of sonde and ERA-I humidity profiles, indicating temporal variability, **c** mean and standard deviation of difference in humidity between ERA-I and sonde, and **d** correlation coefficient of humidity between ERA-I and sonde

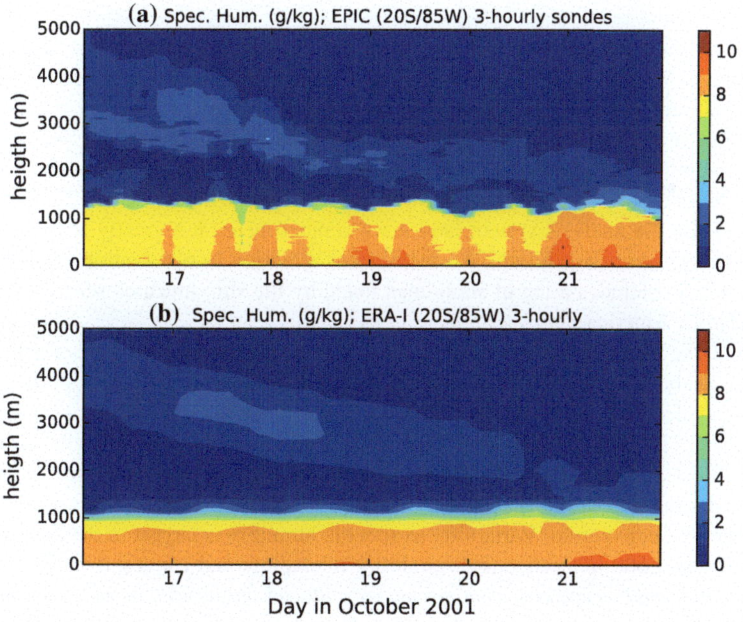

Fig. 6 Time-height cross section of 3-hourly radio sonde observations of specific humidity (top panel) and 3-hourly ERA-I reanalysis data. The time axis covers the EPIC campaign from 11 to 16 Oct 2001 at the "stratus buoy" location in the Eastern Pacific (20°S/85°W)

large-scale flow and ERA-I seems to capture the large scales of this feature rather well, but the smaller scale variability is smeared out.

The statistics of the variability are summarized in Fig. 5b–d. The correlation between ERA-I moisture and sonde observations varies strongly with height. The relatively low correlation between observations and analysis is due to the narrow range of conditions experienced at this single point, lack of variability in the forecast model (evident in panel b), and representativeness noise in the observations. The correlation drops to zero at a height of 1000 m, but this is due to the mismatch in inversion height: the 1000 m level is above the inversion in ERA-I and just below the inversion in the observations, so the time series are uncorrelated.

4.1.2 Assessment in Convecting Regions

A second opportunity for assessment, this one focused on regions of deeper convection, arises from a NOAA field campaign aimed at studying deep convection associated with sea-surface temperature anomalies. The El Niño Rapid Response field program (ENRR, see https://www.esrl.noaa.gov/psd/enso/rapid_response/) included dropsondes from the G4 aircraft and radiosonde launches from the island station at Kiribati and the research vessel Ron Brown, which was moving. Here we explore the impact of these sondes in a global assimilation with the currently operational NCEP hybrid ensemble-variational data assimilation system (Wang et al. 2013; Kleist and Ide 2015) using 80 members. The assimilation is similar to operational analyses in using all available observations including conventional observations, GPS radio occultation, satellite radiances, etc. but is run at reduced horizontal resolution (T254).

We assess the representation of lower tropospheric water vapor by using "data-denial" experiments in which the ENRR soundings were evaluated (i.e., the error between forecast and observations was computed) but did not affect the analyzed state. (A parallel set of assimilations in which the sondes did contribute to the analysis is used in Sect. 5.1). The forecast/assimilation system was cycled through all of February and March 2016. The original observations have quite high vertical resolution (c.f. Stevens et al. 2017) but were introduced into the analysis system, and are examined here, at a greatly reduced vertical resolution to prevent over-fitting. We restrict our attention to observations made west of 139° W, which excludes a leg of soundings made by the ship much closer to the western coast of North America.

Figure 7, the analog to Fig. 5, provides an overview of water vapor as observed (in blue) and in the analysis (red). Here all soundings are considered independently although many more dropsondes were launched from the G4 (529) than from Kiribati (54) or from the Ron Brown (69). The domain is deeper than in Fig. 5, extending to 250 hPa or roughly 10.5 km. No sharp inversion is evident in the mean sounding, partly as a result of sampling a much wider range conditions and partly because weaker subsidence in this region leads a wider variety of convection. Variability in specific humidity is largest in the layer between roughly 800 and 600 hPa reflecting variability in shallow and congestus convection; this variability is roughly captured, if at somewhat reduced amplitude, in the analysis (upper right). The analysis is too dry in the boundary layer and too moist in the free troposphere, with modest values in absolute terms (less than 0.5 g/kg) that become large when expressed as relative humidity at lower pressures. The relatively high correlation between analysis and observations (lower right) indicates that the analysis is able to reproduce regional and synoptic variations reasonable well.

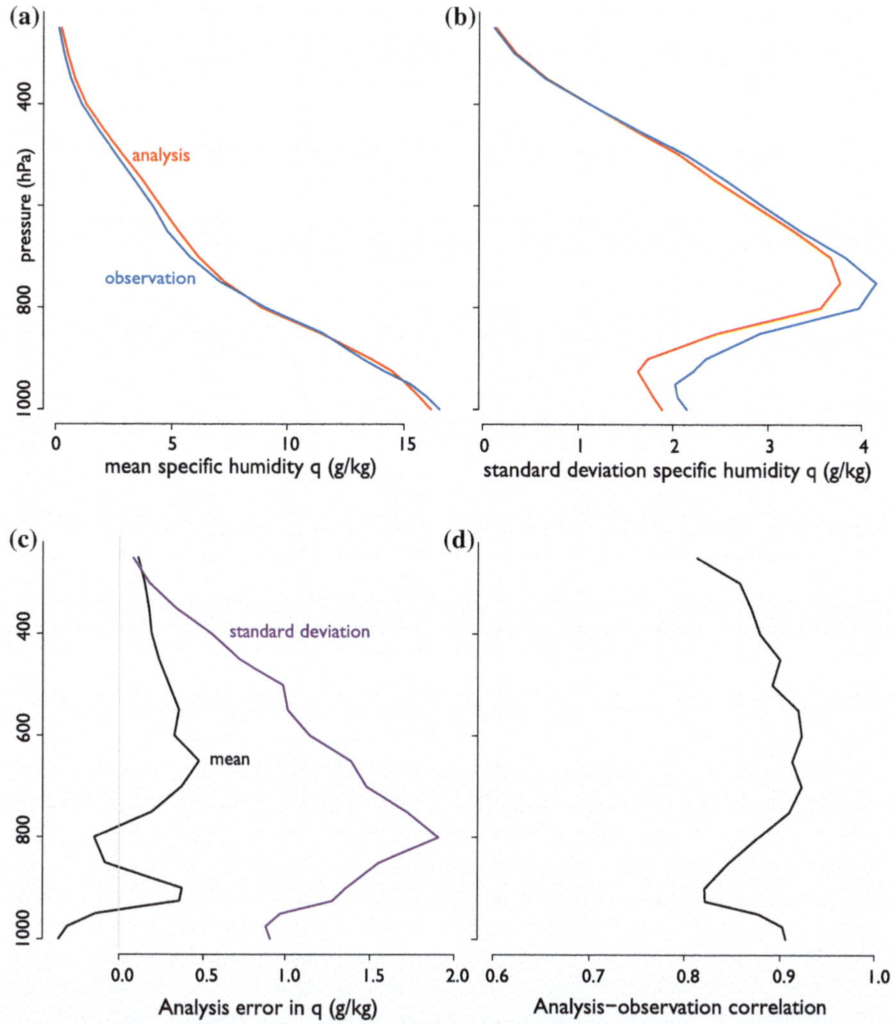

Fig. 7 As in Fig. 5 but for observations in the Tropical Pacific obtained during the El Niño Rapid Response Campaign. Observations include dropsondes from an aircraft and radiosondes launched at an island station and by a ship traversing the domain. Assimilation uses a reduced-resolution version of the operational system at the US National Centers for Environmental Prediction. **a** Mean sonde and analyzed specific humidity profiles, **b** standard deviation of sonde and analyzed specific humidity profiles, indicating temporal and spatial variability, **c** mean and standard deviation of difference in humidity between analysis and observations, and **d** correlation coefficient of humidity between analysis and observations

In practice the three platforms from which soundings are available sampled somewhat different environments (Fig. 8). Soundings from the fixed station at Kiritimati Island in the Republic of Kiribati, just off the Equator, are systematically wetter and less variable than soundings from the ship and aircraft, which sampled a much wider range of geographic locations as far as 20° from the Equator. Soundings from the Ron Brown are systematically moister than from the G4 with the region of maximum variability perhaps 50 hPa higher, suggesting greater sampling of congestus convection; this difference is at least partly the

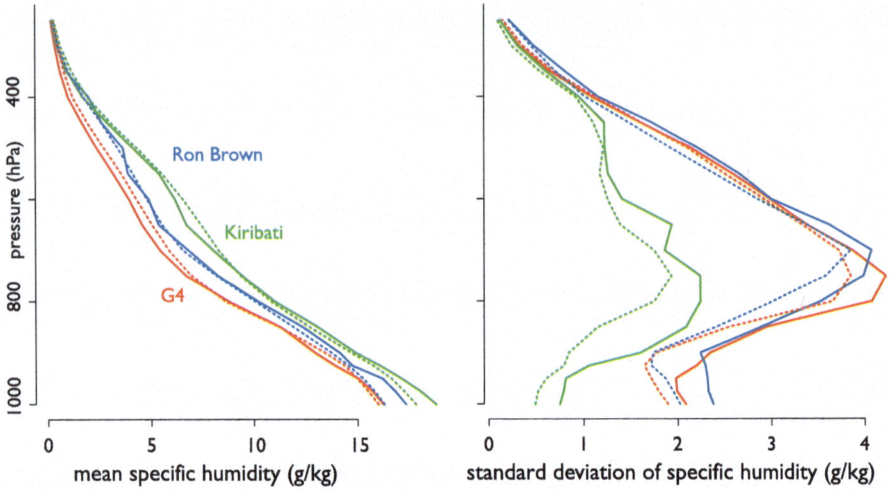

Fig. 8 Mean (left) and standard deviation of specific humidity in the El Niño Rapid Response soundings available from the G4 dropsondes (red) and from radiosondes launched from the Ron Brown (blue) and from Kiritimati Island in Kiribati (green). Solid lines show the observations and dashed lines the colocated analyses in experiments neglecting the sondes. Soundings from Kiribati, near the Equator, are systematically warmer and wetter than those obtained from the traveling platforms. The analysis captures the overall differences in the sampling of the large-scale environment but underrepresents the variability

result of the flight paths on the G4 being constructed to avoid hazardous large-scale regions of deep convection.

The analysis captures the broad-brush and synoptic differences among the sources of the soundings. Temporal variability is somewhat smaller in the analysis than in the soundings (compare the dashed and solid lines in Fig. 8, right panel) although some amount of the observed variability arises from sampling noise.

Because errors are relatively small, however, assessments of error can be sensitive to averaging assumptions. Figure 9 shows the analysis bias (left), standard deviation (right, dashed lines) , and root-mean-squared error (right, solid lines) in specific humidity. Estimates from all three platforms suggest that the analysis underestimates humidity in the boundary layer. The moist bias in the free troposphere (Fig. 7), however, is less robust. The bias is positive and largest in magnitude in soundings taken by the G4, which avoided regions of organized deep convection.

This suggests that the contrast in mid-tropospheric water vapor between regions of large-scale deep convection and the surrounding environment may be underestimated. To the extent that large-scale organization is influenced by convective self-aggregation (Wing et al. 2017) this implies that the observed (Tobin et al. 2012) and modeled (e.g., Bretherton et al. 2005) contrast in humidity associated with convective organization may not be fully captured in the analysis.

Soundings from the G4 are far more numerous than from the other sources and so dominate the estimates of analysis error in Fig. 7. Figure 10 repeats the lower left panel of this figure giving equal weight to the G4 and Ron Brown soundings. This more geographically representative view makes it clear that the analysis systematically underestimates boundary layer humidity and overestimates humidity in the free troposphere. The excursion toward more positive bias at 900 hPa is consistent with the boundary layer being deeper in the analysis than in the observations.

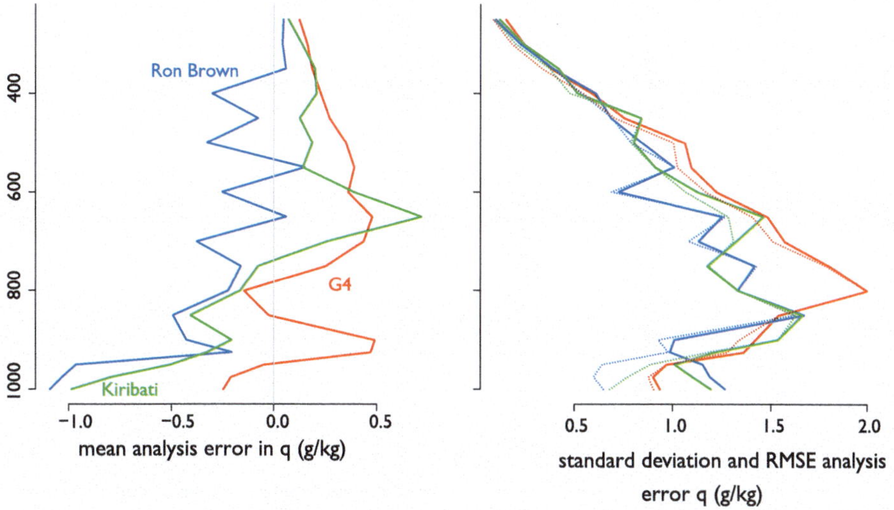

Fig. 9 Bias (analysis minus observations, left), standard deviation (right, dashed lines), root-mean-square error (solid lines) in specific humidity in the El Niño Rapid Response soundings. The G4 soundings, which preferentially sample outside extensive deep convection, also exhibit bias from roughly 700–500 hPa, suggesting that analysis underestimates the contrast between large-scale moist and dry regions

Fig. 10 Mean and standard deviation of difference in humidity between analysis and observations, computed using equal weights for soundings from the G4 and Ron Brown (and hence commensurate weights for all three sets of soundings). This reduces the impact of sampling biases in the G4 soundings on the overall estimate of error. The analysis systematically underestimates boundary layer humidity and overestimates humidity in the free troposphere and may underestimate the depth of the boundary layer

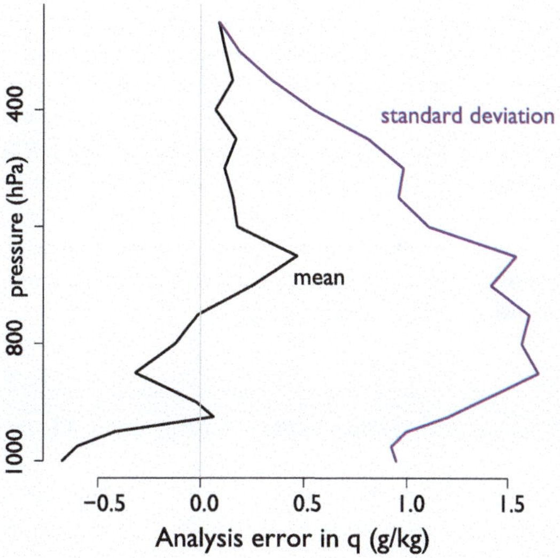

4.2 Assessing Impacts

As described in the introduction, the distribution of water vapor in the troposphere is intimately linked to convection and cloudiness because radiative cooling provides the destabilization needed to initiate convection and to determine large-scale vertical motion. Radiative cooling rates throughout the atmosphere are affected by the full vertical distribution of water vapor, so errors in water vapor abundance have non-local and non-obvious impacts on cooling rates. Figure 11 shows the impact of analysis errors in specific humidity

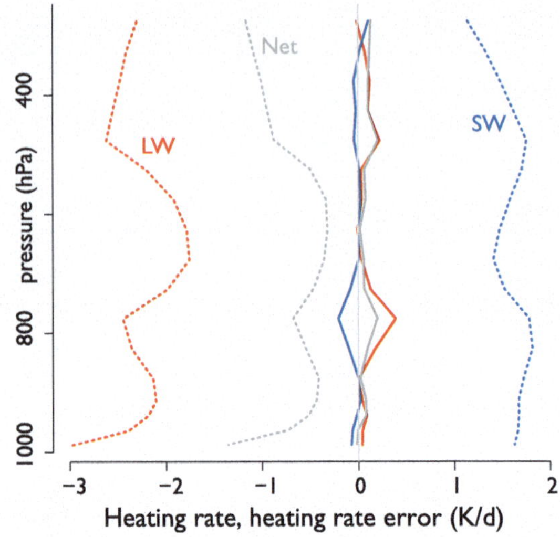

Fig. 11 Errors in longwave (red), shortwave (blue) and net (gray) heating rates arising from analysis errors in temperature and humidity. Dashed lines show mean heating rates

on clear-sky radiative cooling rates, computed by applying the SOCRATES radiative transfer model (a heavily revised version of the code described in Edwards and Slingo 1996) to the observed and analyzed temperature and humidity fields, then averaging across the heating rate differences. As in Fig. 10, each set of soundings is given roughly equal weight. Present-day concentrations of other well-mixed greenhouse gases are assumed. Water vapor abundance affects longwave (red) and shortwave (blue) fluxes in the opposite sense, i.e., increased humidity leads to more efficient longwave radiative cooling but also increased solar absorption. As a result net clear-sky cooling rates through the atmosphere are 0.5–1 K/d, with mean errors of 0.05–0.1 K/d, or roughly 10%, below 600 hPa, increasing to 0.15 K/d in the upper free troposphere where net cooling also increases.

Analysis errors in the radiative cooling (and resulting subsidence rates) to which low clouds are quite sensitive, especially errors that affect the ability of a reanalysis to identify regional and/or interannual contrast, can compromise observational studies of low cloud feedbacks (Klein et al. 2017).

5 Characterizing Water Vapor in a More Richly Observed World

As described in Sect. 3, present-day systems for routinely observing water vapor over remote oceans provide a relatively strong constraint on the vertical integral but a blurred view of the vertical distribution. Analyses of specific humidity that incorporate these loose constraints into imperfect models therefore exhibit errors in the vertical distribution (Figs. 5, 7). In particular, the lack of an observational constraint on boundary layer humidity allows analysis systems to make compensating errors that preserve column-integrated water vapor. In these very limited comparisons the ERA Interim system produces a boundary layer roughly 10% too shallow and 10% too moist, while errors in the current NCEP system (too dry in the boundary layer, too moist above) result in an artificially reduced moisture contrast between the boundary layer and the free troposphere. The analyses do capture the bulk of the large-scale variability, although the NCEP system may underestimate the contrast in humidity between convective and non-convective regions.

Far more sophisticated technologies for the remote sensing of water vapor are emerging, however, (see Nehrir et al. 2017). Might much better observations lead to greatly improved analyses of water vapor?

5.1 Limited Observations and Model Error

Figure 12 explores the impacts of the two most important sources of analysis error, namely model error and a lack of constraining observations, in the NCEP system. The left panel illustrates the impact of observations on the analysis by comparing the bias in specific humidity (solid lines), computed separately for each set of soundings, with the analysis increment in humidity averaged over the same locations in a parallel set of assimilations in which the observations *do* impact the analysis (dashed lines). Increments measure the degree to which all available observations change the short-term forecast; the degree to which increments averaged over time and space are nonzero reflects systematic model errors (Klinker and Sardeshmukh 1992; Rodwell and Palmer 2007).

At pressures below roughly 800 hPa the increments track the bias and its variation among sounding sources, indicating that the observations counteract systematic model error. The increments are not as large as the bias because the assimilation strikes a balance between the uncertainties in the observations and in the forecast (measured here by the ensemble spread).

Increments are small, however, at pressures above 800 hPa, where bias in humidity is largest. This partly reflects the fact that observations are thought to be less certain

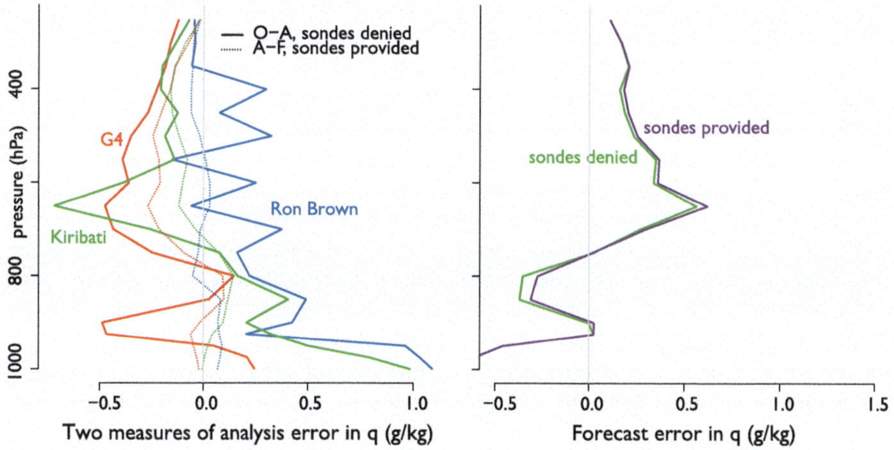

Fig. 12 Left: two views of systematic analysis error in specific humidity. Solid lines show the mean difference between withheld observations and analysis for each data source. (The same data are shown, with opposite sign, in the left panel of Fig. 9). Dashed lines show the mean analysis increment for specific humidity, i.e., the amount by which the forecast is changed by all observations, in data assimilation experiments in which the soundings are included. Increments are small below 850 hPa even though bias is large there, reflecting both larger observational uncertainty and more certain but erroneous forecasts. Right: mean forecast error in specific humidity, evaluated against evenly weighted sets of sondes, in assimilation experiments that include (purple) or neglect (green) the observations. Much of the improvement in the analysis of specific humidity is lost within the 6-hour forecast window, suggesting that analysis quality in humidity is strongly affected by model error

(specifically, less representative of large-scale conditions) closer to the atmosphere–surface interface. But it is also the result of limited variation in humidity across the ensemble in the lower troposphere, driven partly by the fact that all members of the ensemble see the same sea-surface temperature field. The result is that forecasts of lower tropospheric humidity are confident, so that low relative weight is assigned to the observations, but are nonetheless in error. The analysis is therefore unable to exploit the detailed information fully, and remains in error despite strong observational constraints from the soundings.

5.2 Exploiting Richer Observations

Observations that directly characterize boundary layer humidity, such as those obtained from field campaigns or from future observing systems, do improve the analysis of humidity throughout the atmosphere. The impact of the observations is limited, however, due to systematic model errors, lack of model variability, and errors in the background error estimates of global analysis systems. This applies especially to water vapor concentration, which is strongly affected by parametrization of processes with short time scales. Indeed, the observations collected during the El Niño Rapid Response campaign had little impact on even the 6-hour forecasts used as background states in the forecast/assimilation cycles. The right panel of Fig. 12 shows the mean difference between observed and forecast specific humidity in that neglect (green) or include (purple) the equally weighted sets of sondes. The forecast departures are quite similar in both experiments, indicating that improvements in the humidity analysis brought by the sondes don't last even as long as 6 h in the deep tropics.

Analysis systems provide the best currently available routine view of the distribution of water vapor over remote oceans, capturing much of the large-scale variability even as the vertical distribution is compromised by model error and observations that provide a relatively loose constraint on humidity near the surface. More detailed observations of the vertical structure of water vapor, especially in the boundary layer, would lead to an improved representation, especially if forecast model and assimilation systems are improved to better exploit the new observations. This will require better characterization of model error, as well as better characterization of error variances/covariances to allow analysis systems to represent vertical structures ranging from sharp inversions of the subtropics to more blended inversion structures of the deep tropics. High-resolution observations would, if made routinely, also provide the information needed to identify and correct systematic model errors.

In addition to the vertical structures caused by large-scale transport, detailed field observations also show fine-scale structures that have the signature of convection (for example as described by Stevens et al. (2017) and Kiemle et al. 2017). Such a mesoscale analysis will require both high-frequency and high-resolution observations as well as new assimilation techniques adapted to structures with a relatively short predictability horizon.

Acknowledgements This paper arises from the International Space Science Institute (ISSI) workshop on Shallow clouds and water vapor, circulation and climate sensitivity. We thank Oleksandr Bobryshev for preparing Fig. 1 and deriving scaling parameters and cloud thresholds in Sect. 3.1, Theresa Lang for preparing Fig. 2, and Lukas Kluft and Mareike Burba for IASI performance estimates. We thank the GRUAN project for tropical radiosonde data and M. Schwaerz and WGC's Occultation Processing System team for provision of OPSv5.6 RO data. K. Franklin Evans provided the SOCRATES radiative transfer calculations in Sect. 4. R.P. was supported by the U.S. National Science Foundation under grant ATM-1138394. Contributions by G. K. and F. L were funded by the Austrian Research Promotion Agency (FFG) under the projects OPSCLIMTRACE-OPSCLIMVALUE and by the Austrian Science Fund (FWF) under the project VERTCLIM (P27724-NBL).

References

Abdalla S, Isaksen L, Janssen P, Wedi N (2013) Effective spectral resolution of ECMWF atmospheric forecast models. ECMWF Newslett 137:19–22

Andersson E, Bauer P, Beljaars A, Chevallier F et al (2005) Assimilation and modeling of the atmospheric hydrological cycle in the ECMWF forecasting system. Bull Am Meteorol Soc 86(3):387

Anthes RA (2011) Exploring Earth's atmosphere with radio occultation: contributions to weather, climate, and space weather. Atmos Meas Tech 4(6):1077–1103

August T, Klaes D, Schlüssel P, Hultberg T, Crapeau M, Arriaga A, O'Carroll A, Coppens D, Munro R, Calbet X (2012) IASI on Metop-A: operational Level 2 retrievals after five years in orbit. J Quant Spectrosc Radiat Transf 113(11):1340–1371

Aumann HH, Chahine MT, Gautier C, Goldberg MD, Kalnay E, McMillin LM, Revercomb H, Rosenkranz PW, Smith WL, Staelin DH, Strow LL, Susskind J (2003) AIRS/AMSU/HSB on the aqua mission: design, science objectives, data products, and processing systems. IEEE Trans Geosci Remote Sens 41(2):253–264

Bauer P, Geer AJ, Lopez P, Salmond D (2010) Direct 4D-Var assimilation of all-sky radiances. Part I: implementation. Q J R Meteorol Soc 136(652):1868–1885

Bodeker GE, Bojinski S, Cimini D, Dirksen RJ, Haeffelin M, Hannigan JW, Hurst DF, Leblanc T, Madonna F, Maturilli M, Mikalsen AC, Philipona R, Reale T, Seidel DJ, Tan DGH, Thorne PW, Vömel H, Wang J (2016) Reference upper-air observations for climate: from concept to reality. Bull Am Meteorol Soc 97(1):123–135

Bretherton CS, Uttal T, Fairall CW, Yuter SE, Weller RA, Baumgardner D, Comstock K, Wood R, Raga GB (2004) The EPIC 2001 stratocumulus study. Bull Am Meteorol Soc 85(7):967–977

Bretherton CS, Blossey PN, Khairoutdinov M (2005) An energy-balance analysis of deep convective self-aggregation above uniform SST. J Atmos Sci 62(12):4273–4292

Brogniez H, Kirstetter PE, Eymard L (2013) Expected improvements in the atmospheric humidity profile retrieval using the Megha-Tropiques microwave payload. Q J R Meteorol Soc 139(673):842–851

Buehler SA, John VO (2005) A simple method to relate microwave radiances to upper tropospheric humidity. J Geophys Res 110(D2):D02,110

Buehler SA, Kuvatov M, John VO, Leiterer U, Dier H (2004) Comparison of microwave satellite humidity data and radiosonde profiles: a case study. J Geophys Res 109:D13103. doi:10.1029/2004JD004605

Clerbaux C, Boynard A, Clarisse L, George M, Hadji-Lazaro J, Herbin H, Hurtmans D, Pommier M, Razavi A, Turquety S, Wespes C, Coheur PF (2009) Monitoring of atmospheric composition using the thermal infrared IASI/MetOp sounder. Atmos Chem Phys 9(16):6041–6054

Dee DP, Uppala S (2009) Variational bias correction of satellite radiance data in the ERA-Interim reanalysis. Q J R Meteorol Soc 135(644):1830–1841

Dee DP, Uppala SM, Simmons AJ, Berrisford P, Poli P, Kobayashi S, Andrae U, Balmaseda MA, Balsamo G, Bauer P, Bechtold P, Beljaars ACM, van de Berg L, Bidlot J, Bormann N, Delsol C, Dragani R, Fuentes M, Geer AJ, Haimberger L, Healy SB, Hersbach H, Hólm EV, Isaksen L, Kållberg P, Köhler M, Matricardi M, McNally AP, Monge-Sanz BM, Morcrette JJ, Park BK, Peubey C, de Rosnay P, Tavolato C, Thepaut JN, Vitart F (2011) The ERA-Interim reanalysis: configuration and performance of the data assimilation system. Q J R Meteorol Soc 137(656):553–597

Edwards JM, Slingo A (1996) Studies with a flexible new radiation code. I: choosing a configuration for a large-scale model. Q J R Meteorol Soc 122(531):689–719

Eriksson P, Buehler SA, Davis CP, Emde C, Lemke O (2011) ARTS, the atmospheric radiative transfer simulator, version 2. J Quant Spectrosc Radiat Transf 112(10):1551–1558

Eyre JR, Kelly GA, McNally AP, Andersson E, Persson A (1993) Assimilation of TOVS radiance information through one-dimensional variational analysis. Q J R Meteorol Soc 119(514):1427–1463

Geer AJ, Bauer P, Lopez P (2010) Direct 4D-Var assimilation of all-sky radiances. Part II: assessment. Q J R Meteorol Soc 136(652):1886–1905

Gorbunov ME, Benzon H-H, Jensen AS, Lohmann MS, Nielsen AS (2004) Comparative analysis of radio occultation processing approaches based on Fourier integral operators. Radio Sci 39:RS6004. doi:10.1029/2003RS002916

Hajj GA, Kursinski ER, Romans LJ, Bertiger WI, Leroy SS (2002) A technical description of atmospheric sounding by GPS occultation. J Atmos Solar Terr Phys 64(4):451–469

Healy SB, Thepaut JN (2006) Assimilation experiments with CHAMP GPS radio occultation measurements. Q J R Meteorol Soc 132(615):605–623

Herbin H, Hurtmans D, Clerbaux C, Clarisse L, Coheur PF (2009) $H_2^{16}O$ and HDO measurements with IASI/MetOp. Atmos Chem Phys 9(24):9433–9447

Ho SP, Hunt D, Steiner AK, Mannucci AJ, Kirchengast G, Gleisner H, Heise S, von Engeln A, Marquardt C, Sokolovskiy S, Schreiner W, Scherllin-Pirscher B, Ao C, Wickert J, Syndergaard S, Lauritsen K, Leroy S, Kursinski ER, Kuo YH, Foelsche U, Schmidt T, Gorbunov M (2012) Reproducibility of GPS radio occultation data for climate monitoring: profile-to-profile inter-comparison of CHAMP climate records 2002 to 2008 from six data centers. J Geophys Res 117:D18111. doi:10.1029/2012JD017665

Ho SP, Peng L, Anthes RA, Kuo YH, Lin HC (2015) Marine boundary layer heights and their longitudinal, diurnal, and interseasonal variability in the southeastern pacific using COSMIC, CALIOP, and radiosonde data. J Climate 28(7):2856–2872

Hólm E, Andersson E, Beljaars A, Lopez P, Mahfouf JF, Simmons A, Thepaut JN (2002) Assimilation and modelling of the hydrological cycle: ECWMF's status and plans. Tech Memo 383:1–55

Immler FJ, Dykema J, Gardiner T, Whiteman DN, Thorne PW, Vömel H (2010) Reference quality upper-air measurements: guidance for developing GRUAN data products. Atmos Meas Tech 3(5):1217–1231

John VO, Holl G, Allan RP, Buehler SA, Parker DE, Soden BJ (2011) Clear-sky biases in satellite infrared estimates of upper tropospheric humidity and its trends. J Geophys Res 116(D14):1491

Kalman RE (1960) A new approach to linear filtering and prediction problems. Trans ASME J Basic Eng 82(Series D):35–45

Kalman RE, Bucy RS (1961) New results in linear filtering and prediction theory. Trans ASME J Basic Eng 83(Series D):95–108

Kalnay E (2003) Atmospheric modeling, data assimilation and predictability. Cambridge University Press, Cambridge

Kanamitsu M, Ebisuzaki W, Woollen J, Yang SK, Hnilo JJ, Fiorino M, Potter GL (2002) NCEP–DOE AMIP-II reanalysis (R-2). Bull Am Meteorol Soc 83(11):1631–1643

Kiemle C, Groß S, Wirth M, Bugliaro L (2017) Airborne lidar observations of water vapor variability in tropical shallow convective environment. Surv Geophys. doi:10.1007/s10712-017-9431-5

Kirchengast G, Schweitzer S (2011) Climate benchmark profiling of greenhouse gases and thermodynamic structure and wind from space. Geophys Res Lett 38(13):L13,701

Klaes KD, Cohen M, Buhler Y, Schlüssel P, Munro R, Engeln A, Clérigh E, Bonekamp H, Ackermann J, Schmetz J, Luntama JP (2007) An introduction to the EUMETSAT polar system. Bull Am Meteorol Soc 88(7):1085–1096

Klein SA, Hall A, Norris JR, Pincus R (2017) Low-cloud feedbacks from cloud-controlling factors: a review. Surv Geophys. doi:10.1007/s10712-017-9433-3

Kleist DT, Ide K (2015) An OSSE-based evaluation of hybrid variational-ensemble data assimilation for the NCEP GFS. Part I: system description and 3D-hybrid results. Mon Weather Rev 143(2):433–451

Klinker E, Sardeshmukh PD (1992) The diagnosis of mechanical dissipation in the atmosphere from large-scale balance requirements. J Atmos Sci 49(7):608–627

Kobayashi S, Ota Y, Harada Y, Ebita A, Moriya M, Onoda H, Onogi K, Kamahori H, Kobayashi C, Endo H, Miyaoka K, Takahashi K (2015) The JRA-55 reanalysis: general specifications and basic characteristics. J Meteorol Soc Japan 93(1):5–48

Kursinski ER, Gebhardt T (2014) A method to deconvolve errors in GPS RO-derived water vapor histograms. J Atmos Ocean Technol 31(12):2606–2628

Kursinski ER, Hajj GA, Schofield JT, Linfield RP, Hardy KR (1997) Observing earth's atmosphere with radio occultation measurements using the global positioning system. J Geophys Res 102(D19):23,429–23,465

Kursinski ER, Syndergaard S, Flittner D, Feng D, Hajj GA, Herman B, Ward D, Yunck TP (2002) A microwave occultation observing system optimized to characterize atmospheric water, temperature, and geopotential via absorption. J Atmos Ocean Technol 19(12):1897–1914

Lacour JL, Risi C, Clarisse L, Bony S, Hurtmans D, Clerbaux C, Coheur PF (2012) Mid-tropospheric δD observations from IASI/MetOp at high spatial and temporal resolution. Atmos Chem Phys 12(22):10,817–10,832

Ladstädter F, Steiner AK, Schwärz M, Kirchengast G (2015) Climate intercomparison of GPS radio occultation, RS90/92 radiosondes and GRUAN from 2002 to 2013. Atmos Meas Tech 8(4):1819–1834

Lebsock et al (this issue) An observational view of radiation-circulation feedbacks on convective aggregation. Surv Geophy (in press)

Liao M, Zhang P, Yang GL, Bi YM, Liu Y, Bai WH, Meng XG, Du QF, Sun YQ (2016) Preliminary validation of the refractivity from the new radio occultation sounder GNOS/FY-3C. Atmos Meas Tech 9(2):781–792

Liu CL, Kirchengast G, Syndergaard S, Kursinski ER, Sun YQ, Bai WH, Du QF (2017) A review of low Earth orbit occultation using microwave and infrared-laser signals for monitoring the atmosphere and climate. Adv Space Res. doi:10.1016/j.asr.2017.05.011

Mapes B, Chandra AS, Kuang Z, Zuidema P (2017) Importance profiles for water vapor. Surv Geophys. doi:10.1007/s10712-017-9427-1

Mears CA, Wang J, Smith D, Wentz FJ (2015) Intercomparison of total precipitable water measurements made by satellite-borne microwave radiometers and ground-based GPS instruments. J Geophys Res 120(6):2492–2504

Nehrir AR, Kiemle C, Lebsock MD, Kirchengast G, Buehler SA, Löhnert U, Liu C-L, Hargrave P, Barrera-Verdejo M, Winker D (2017) Emerging technologies and synergies for airborne and space-based measurements of water vapor profiles. Surv Geophys. doi:10.1007/s10712-017-9436-0

Osher S, Burger M, Goldfarb D, Xu J, Yin W (2005) An iterative regularization method for total variation-based image restoration. Multiscale Model Simul 4(2):460–489

Parker WS (2016) Reanalyses and observations: what's the difference? Bull Am Meteorol Soc 97(9):1565–1572

Rieckh T, Anthes R, Randel W, Ho S-P, Foelsche U (2016) Tropospheric dry layers in the Tropical Western Pacific: Comparisons of GPS radio occultation with multiple data sets. Atmos Meas Tech 10:1093–1110. doi:10.5194/amt-10-1093-2017

Rodwell MJ, Palmer TN (2007) Using numerical weather prediction to assess climate models. Quart J R Meterol Soc 133(622):129–146

Schmetz J, Pili P, Tjemkes S, Just D, Kerkmann J, Rota S, Ratier A (2002) An introduction to meteosat second generation (MSG). Bull Am Meteorol Soc 83(7):977–992

Schneider M, Hase F (2011) Optimal estimation of tropospheric H_2O and δD with IASI/METOP. Atmos Chem Phys 11(21):11,207–11,220

Schreiner WS, Rocken C, Sokolovskiy S, Hunt D (2009) Quality assessment of COSMIC/FORMOSAT-3 GPS radio occultation data derived from single- and double-difference atmospheric excess phase processing. GPS Solut 14(1):13–22

Shi L, Bates JJ (2011) Three decades of intersatellite-calibrated High-Resolution Infrared Radiation Sounder upper tropospheric water vapor. J Geophys Res 116(D4):D04,108

Sokolovskiy S, Schreiner W, Zeng Z, Hunt D, Lin YC, Kuo YH (2014) Observation, analysis, and modeling of deep radio occultation signals: effects of tropospheric ducts and interfering signals. Radio Sci 49(10):954–970

Steiner AK, Lackner BC, Ladstädter F, Scherllin-Pirscher B, Foelsche U, Kirchengast G (2011) GPS radio occultation for climate monitoring and change detection. Radio Sci 46(6):115

Steiner AK, Hunt D, Ho SP, Kirchengast G, Mannucci AJ, Scherllin-Pirscher B, Gleisner H, von Engeln A, Schmidt T, Ao C, Leroy SS, Kursinski ER, Foelsche U, Gorbunov M, Heise S, Kuo YH, Lauritsen KB, Marquardt C, Rocken C, Schreiner W, Sokolovskiy S, Syndergaard S, Wickert J (2013) Quantification of structural uncertainty in climate data records from GPS radio occultation. Atmos Chem Phys 13(3):1469–1484

Stevens B, Farrell D, Hirsch L, Jansen F, Nuijens L, Serikov I, Brügmann B, Forde M, Linne H, Lonitz K, Prospero JM (2016) The barbados cloud observatory: anchoring investigations of clouds and circulation on the edge of the ITCZ. Bull Am Meteorol Soc 97(5):787–801

Stevens B, Brogniez H, Kiemle C, Lacour JL, Crevoisier C, Kiliani J (2017) Structure and dynamical influence of water vapor in the lower tropical troposphere. Surv Geophys 105(3):270

Tobin I, Bony S, Roca R (2012) Observational evidence for relationships between the degree of aggregation of deep convection, water vapor, surface fluxes, and radiation. J Clim 25(20):6885–6904

Trenberth KE, Fasullo JT, Mackaro J (2011) Atmospheric moisture transports from ocean to land and global energy flows in reanalyses. J Clim 24(18):4907–4924

Uppala SM, Kållberg PW, Simmons AJ, Andrae U, Bechtold Vd, Fiorino M, Gibson JK, Haseler J, Hernandez A, Kelly GA et al (2005) The ERA-40 re-analysis. Q J R Meteorol Soc 131(612):2961–3012

Vial J, Bony S, Stevens B, Vogel R (2017) Mechanisms and model diversity of trade-wind shallow cumulus cloud feedbacks: a review. Surv Geophys 36(1):73

Wang X, Parrish D, Kleist D, Whitaker J (2013) GSI 3DVar-based ensemble-variational hybrid data assimilation for NCEP global forecast system: single-resolution experiments. Mon Weathr Rev 141(11):4098–4117

Wing AA, Emanuel K, Holloway CE, Muller C (2017) Convective self-aggregation in numerical simulations: a review. Surv Geophys 27(32):4391

Xie F, Wu DL, Ao CO, Mannucci AJ, Kursinski ER (2012) Advances and limitations of atmospheric boundary layer observations with GPS occultation over southeast Pacific Ocean. Atmos Chem Phys 12(2):903–918

Surv Geophys (2017) 38:1425–1443
https://doi.org/10.1007/s10712-017-9431-5

Airborne Lidar Observations of Water Vapor Variability in Tropical Shallow Convective Environment

Christoph Kiemle[1] · Silke Groß[1] · Martin Wirth[1] · Luca Bugliaro[1]

Received: 7 October 2016 / Accepted: 21 September 2017 / Published online: 4 October 2017

Abstract An airborne downward-pointing water vapor lidar provides two-dimensional, simultaneous curtains of atmospheric backscatter and humidity along the flight track with high accuracy and spatial resolution. In order to improve the knowledge on the coupling between clouds, circulation and climate in the trade wind region, the DLR (Deutsches Zentrum für Luft- und Raumfahrt) water vapor lidar was operated on board the German research aircraft HALO during the NARVAL (Next Generation Aircraft Remote Sensing for Validation Studies) field experiment in December 2013. Out of the wealth of about 30 flight hours or 25,000 km of data over the Tropical Atlantic Ocean east of Barbados, three $\sim$ 2-h-long, representative segments from different flights were selected. Analyses of Meteosat Second Generation images and dropsondes complement this case study. All observations indicate a high heterogeneity of the humidity in the lowest 4 km of the tropical troposphere, as well as of the depth of the cloud (1–2 km thick) and sub-cloud layer ($\sim$ 1 km thick). At the winter trade inversion with its strong humidity jump of up to 9 g/kg in water vapor mixing ratio, the mixing ratio variance can attain 9 $(g/kg)^2$, while below it typically ranges between 1 and 3 $(g/kg)^2$. Layer depths and partial water vapor columns within the layers vary by up to a factor of 2. This affects the total tropospheric water vapor column, amounting on average to 28 kg/m^2, by up to 10 kg/m^2 or 36%. The dominant scale of the variability is given by the extent of regions with higher-than-average humidity and lies between 300 and 600 km. The variability mainly stems from the alternation between dry regions and moisture lifted by convection. Occasionally, up to 100-km large dry regions are observed. In between, convection pushes the trade inversion upward, sharpening the vertical moisture gradient that is colocated with the trade inversion. In most of the water vapor profiles, this gradient is stronger than the one located at the top of the sub-cloud layer. Lidar observations in concert with models accurately reproducing the observed variability are expected to help evaluate the role these findings play for climate.

✉ Christoph Kiemle
christoph.kiemle@dlr.de

[1] DLR, Deutsches Zentrum für Luft- und Raumfahrt, Institut für Physik der Atmosphäre, 82234 Oberpfaffenhofen, Germany

Keywords Airborne lidar · Water vapor lidar · Shallow convection · Trade wind region · Cloud layer

1 Introduction

The WCRP (World Climate Research Programme) "Grand Challenge on Clouds, Circulation and Climate Sensitivity" endorses scientific progress in our understanding of the coupling between clouds, circulation and climate. In this context, the variability of water vapor in the trades, ubiquitous in our measurements, poses challenges to climate modeling because it modulates the low-cloud cover and eventually Earth's albedo (Bony et al. 2017). Besides being the precondition for cloud formation, the importance of water vapor in the trades is twofold. It lifts latent heat from the ocean surface via convection, and it is a key player in the radiation budget. Above the trade inversion, the almost absence of water vapor in the subsiding branch of the Hadley cell has the effect of an open window in a greenhouse, efficiently cooling the lower troposphere in the trades. Generally, the influence of water vapor on tropospheric heating rates depends on the vertical distribution of water vapor and on the strengths of its vertical gradients, both of which current water vapor observations, particularly from space, have difficulties in reproducing (Stevens et al. 2017). Secondary circulations between radiatively heated and cooled regions are supposed to occur (Zuidema and Torri 2017) which adds complexity to the situation. The interactions between shallow convection, circulation and radiation are at the heart of present scientific debate. They are held mainly responsible for the uncertainty in global climate sensitivity (see, e.g., Sherwood et al. 2014; Bony et al. 2015; WCRP 2016).

This case study focuses on water vapor lidar measurements obtained during the NARVAL (Next Generation Aircraft Remote Sensing for Validation Studies) field experiment around Barbados. From December 10, 2013, to December 20, 2013, the DLR airborne demonstrator WALES (Water vapor Lidar Experiment in Space; Wirth et al. 2009) was operated on board the German research aircraft HALO. A total of eight research flights were performed: four long-range transfer flights crossing the subtropical North Atlantic between Oberpfaffenhofen, HALO's home base in Southern Germany and Barbados, and four regional flights out of Barbados between 10–19°N and 39–59°W within the winter trades over the Tropical Atlantic Ocean. Dropsondes and a set of additional remote sensors were on board, including radar and microwave radiometers (Mech et al. 2014), as well as spectrometers measuring the up- and downwelling radiances. The ample and unique instrumentation evidently calls for a synergistic retrieval of the full suite of remote sensors that is pending. Meteosat Second Generation (MSG) satellite imagery and dropsonde profiles complement the lidar observations here. In August 2016, NARVAL was repeated with the same payload yet modified objectives and flight patterns (Stevens et al. 2017), generally located closer to the ITCZ than the measurements described here.

NARVAL lines up in a long list of scientific experiments in the trades. For example, water vapor and its variability were measured in situ onboard aircraft, e.g., LeMone and Pennell (1976), or ships, e.g., Zuidema et al. (2012). Nuijens et al. (2015) use ground-based lidar measurements of trade wind clouds and find that climate models have difficulties in reproducing the observed variability. To our knowledge, airborne water vapor lidar observations in tropical shallow convective environment have not been reported yet, despite the well-documented need for water vapor profiling with higher accuracy and higher vertical resolution than the current observations provide (Wulfmeyer et al. 2015; Pincus et al. 2017; Stevens et al. 2017). Earlier own work has demonstrated the value of airborne lidar observations to characterize the variability of humidity within the convective boundary layer over flat and mountainous terrain (Behrendt et al. 2007; Kiemle et al. 2011), as well as in the free troposphere (Flentje et al. 2005; Fischer et al. 2013), the tropical upper troposphere (Kiemle et al. 2008) and in cirrus clouds (Groß et al. 2014). In combination with a scanning Doppler wind lidar measuring horizontal winds, the advective tropospheric moisture transport along a curtain below the aircraft can be quantified (Schäfler et al. 2010). Eddy correlation of the Doppler lidar vertical velocity and of the water vapor lidar measurements provides profiles of the latent heat flux in a convective boundary layer (Kiemle et al. 2007, 2011). This dual-lidar combination did not exist for NARVAL because the mission goals focused on radar-lidar comparisons, unfortunately not leaving enough space for a wind lidar onboard HALO. The cross-Atlantic NARVAL transfer flights show dry layers and related synoptic-scale humidity variability in the mid- to upper troposphere, connected to tropopause folds that are dynamically linked to the subtropical jet, as has been observed, e.g., by Flentje et al. (2005) or Randel et al. (2016). This upper-level variability observed above the trade inversion extends equatorward yet is not topic of the paper. Cloud statistics as in Nuijens et al. (2015) and comparisons with satellite observations of clouds are addressed in a parallel study by Gutleben et al. (2017). This paper provides support to the preparation of future planned airborne (Bony et al. 2017) and spaceborne (Di Girolamo et al. 2008) water vapor lidar deployments. By showcasing selected measurement examples, it complements the more general overview on emerging technologies for measuring water vapor in the lower troposphere by Nehrir et al. (2017).

2 The DLR Airborne Water Vapor Lidar

A differential absorption lidar (DIAL) emits short and spectrally narrow laser pulses at a wavelength tuned to the center (in the case of WALES) or to a wing position of a molecular water vapor absorption line. The water vapor density can be derived from the difference in absorption between one or more “on-line” pulses and a reference “off-line” pulse as function of distance from the lidar. DLR’s airborne water vapor lidar WALES (Wirth et al. 2009) consists of two identical diode-pumped Nd:YAG lasers in a master oscillator power amplifier configuration with a pulse repetition rate of 100 Hz. Their frequency-converted 532-nm radiation pumps two optical parametric oscillators (OPOs). Each OPO generates two wavelengths in the 935-nm absorption band of water vapor. A water vapor absorption cell and a wave meter control and stabilize all four wavelengths, three on-line and one off-line. The pump light not converted to 935 nm is transmitted into the atmosphere for aerosol and cloud measurements, and is received with a high-spectral-resolution detector. The simultaneous use of three on-line wavelengths provides full vertical cover of tropospheric humidity with high accuracy also in the tropics. Due to its capabilities, the WALES airborne demonstrator has become a state-of-the-art instrument for climate and meteorological research. Over the past decade, it successfully completed several hundreds of flight hours in national and international field campaigns (Schäfler et al. 2010; Bhawar et al. 2011; Kiemle et al. 2011; Bielli et al. 2012; Fischer et al. 2013; Groß et al. 2014; Trickl et al. 2016).

During NARVAL, a typical HALO flight altitude of 13 km was selected, being the result of a compromise between radar operation constraints and maximum vertical coverage for lidar and dropsondes. The altitude is higher than during former experiments targeting lower tropospheric moisture which has implications for the lidar signal-to-noise ratio. In order to reduce instrumental noise to acceptable levels, the individual on- and off-line DIAL profiles are accumulated to 12-s averages, which results in a horizontal resolution of 2.8 km at a typical aircraft speed of 230 m/s. The vertical resolution is 290 m. The lidar backscatter signal is less affected by noise. Here the horizontal (vertical) resolution is 230 (15) m. Consequently, small clouds are detected only in the lidar backscatter, and gaps $< \sim 3$ km between clouds are too small to obtain water vapor lidar profiles down to the surface. Nevertheless, the mesoscale water vapor variability is sufficiently well resolved for the goals of this study. Differential absorption lidar measures molecule number density. Conversion into the mass mixing ratio, useful for thermodynamic relations, requires knowledge of dry air density, usually obtained from European Centre for Medium-range Weather Forecasts (ECMWF) analysis fields of atmospheric pressure and temperature that are interpolated in space and time to the lidar measurement positions. When available, a profile of air density is provided by the closest dropsonde.

On average over all available inter-comparisons, the typical deviations between lidar and dropsonde mixing ratio profiles amount to a few percent, which corroborates the results of lidar inter-comparisons by Bhawar et al. (2011) and Trickl et al. (2016). Total or partial columns of water vapor are obtained by vertical integration of absolute humidity, the product of mass mixing ratio and dry air density. Due to methodical constraints, water vapor lidar data below 250 m above sea level (asl) are not available. Given typical surface mass mixing ratios of around 13 g/kg (see Table 1), this amounts to about 4 kg/m^2 that are missing in the lidar-derived columns, which represents 14% of a typical total column of 28 kg/m^2. To solve this issue, we assume the unknown mixing ratios below 250 m to be identical to those measured at 250 m. We find agreement within a few percent between the resulting lidar-derived columns and the dropsonde results of Table 1, which is consistent with the above-mentioned lidar accuracy and consequently validates our approach.

3 The Meteosat Images

To obtain an overview of the general cloud situation at the time of the flight, false color composites have been produced using data from the imaging radiometer SEVIRI (Spinning Enhanced Visible and Infrared Imager; Schmetz et al. 2002) aboard the geostationary satellite Meteosat Second Generation (MSG) located at the equator above 0°E. SEVIRI combines a temporal resolution of 15 min with 12 channels in the solar and thermal spectral range. Its sampling distance amounts to 3 km in 11 channels at the subsatellite point, while the broadband high-resolution visible (HRV) channel has a sampling distance of 1 km. The colors red and green are produced by the superimposition of the low-resolution solar channels centered at 0.6 and 0.8 μm with the HRV channel to produce a high-resolution picture. The blue color stems from the low-resolution thermal channel centered at 10.8 μm such that the blue component is high for cold objects (such as ice clouds) and low for warm objects (such as low clouds). In more detail, the blue color ranges from 263 (maximum blue saturation) to 323 K (no blue component). In this color scheme, the sea surface appears deep blue/black since it shows very low solar reflection (apart from sun glint regions) and is warm. Very low clouds appear yellow because of their relatively high reflection (high R and G components) and comparatively high temperature (low B component). Thicker clouds appear white since they are both highly reflective and cold. Thin cirrus appears bluish/violet because it reflects little solar radiation and is very

Table 1 Main characteristics of the 3 selected lidar segments and of 7 related dropsondes (DS)

Subsection	4.2		4.3		4.4		
Topic/title	Strong heterogeneity in the cloud layer		Dry regions in the cloud and sub-cloud layers		Transport of moisture through the cloud layer		
Flight date	15 December		19 December		10 December		
UTC time	18:55–21:20		17:01–19:27		17:50–19:50		
Latitude	19–14°N		21–14°N		20–15°N		
Longitude	40–58°W		39–57°W		41–55°W		
Segment length	1982 km		2045 km		1571 km		
DS release times, UTC	19:07	19:52 Fig. 1a	17:01	18:25 Fig. 1b	18:20	18:50 Fig. 1c	19:22
DS latitude (°N)	18.5	17.0	20.9	17.1	18.8	17.6	16.1
Sfc. pressure (hPa)	1018.1	1014.6	1017.3	1010.0	1015.1	1013.2	1012.4
Sfc. temperature (°C)	25	26	23	25	25.2	25.7	26.3
Sfc. water vapor mixing ratio (g/kg)	12	13	12	14	13	15	14
Max. rel. humidity (%)	96	87	88	89	78	96	87
Altitude of max. rel. humidity (km)	2.0	0.8	1.0, 2.0	0.6, 2.6	0.6	0. 6	0.7
Wind speed (m/s)	14	16	11	11	11	12	13
Wind direction	70°	80°	70°	80°	60°	60°	70°
Total column water vapor, 0–13 km (kg/m^2)	26.5	23.6	30.5	32.5	23.5	32.1	23.8
Water vapor column below trade inversion (kg/m^2)	25	22	30	31	22	31	22
Altitude of trade inversion (km)	2.3	2.0	3.8	3.0	2.3	2.6	2.8
Humidity jump at trade inversion (g/kg)	8	7	5	8	4	9	2
Temperature jump at trade inversion (K)	6	4	0	2	2	4	1
Temperature lapse rate in cloud layer (K/km)	6.7	6.5	6.0	5.3	5.2	5.0	3.5

Wind speed and direction (0° is wind from north) are averaged over the dropsondes' lowest 3 km. All flights went from northeast to southwest in a straight line without turn, at ~ 13-km flight altitude. Local time is UTC—4 h. When the relative humidity exceeds 95%, the dropsonde likely passed through a cloud

cold. The different height of thick mid-level white clouds and high thin blueish cirrus clouds (sporadically present in the images presented in the following) can also be inferred from the distance between the clouds themselves and the shadow they cast onto the lower-lying objects (sea surface, lower clouds, aerosol layer) due to the low-lying sun: This exercise confirms that blueish objects are higher in the atmosphere than white objects. The clouds sensed by the airborne lidar are marked on these false color composites. A parallax correction has been applied under the assumption that the cloud top is at around 2 km height above the sea surface. The part of the flight path closest in time to the given SEVIRI slot, indicated by the start time of the scan in the top left corner of the figures, is plotted in cyan, while the flight route is red otherwise. The cyan cross denotes the point where the

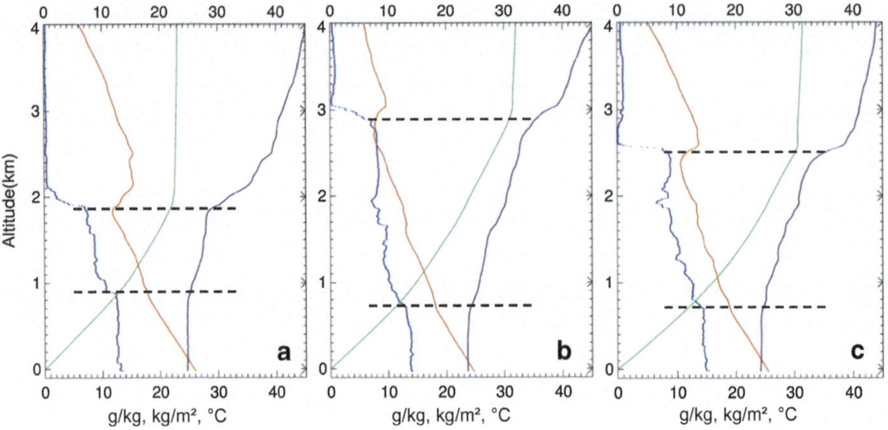

Fig. 1 Lower tropospheric profiles of 3 selected radiosondes dropped from the aircraft; **a** December 15, 2013, 19:52 UTC, **b** December 19, 2013, 18:25 UTC, **c** December 10, 2013, 18:50 UTC. Water vapor mixing ratio (blue), bottom-integrated water vapor column (green), temperature (orange) and potential temperature (violet). Dashed lines: heights of cloud layers from lidar observations. Table 1 lists additional data from these sondes

aircraft and SEVIRI are observing almost simultaneously. The corresponding UTC time is written in cyan next to it. All SEVIRI figures are in satellite projection.

4 Results

4.1 Overview

Out of the wealth of lidar data in the tropics, about 30 flight hours or 25,000 km, we select three 2.0- to 2.5-h-long, representative segments from different flights in the Western Atlantic Ocean, to the east of Barbados, between 21 and 14°N, north of the ITCZ. Our selection avoids aircraft turns, measurement interruptions, unavailability of dropsondes and fully overcast areas. Each segment showcases a topic of particular scientific interest. Table 1 gives an overview of the main characteristics of all segments, as well as of basic meteorological parameters obtained by the sondes dropped within the segments. We used the Vaisala RD94 GPS dropsondes that measure relative humidity with two independent sensors. After converting into mass mixing ratio, we find deviations between both sensors of < 0.5 g/kg, which roughly corresponds to the lidar accuracy. All sondes were processed using the ASPEN version 3.3–270 software, and the data were further visually inspected for possible biases or malfunction before being incorporated into the analysis. All sondes probed similar meteorological and thermodynamic conditions, owing to the trade wind regime and the ocean surface. Their lowest measurements in the atmosphere just above the ocean surface provide a good proxy of surface temperature and surface humidity. Both increase equatorward as expected, while surface pressure decreases. Total column water vapor, however, the vertically integrated absolute humidity, does not show a clear latitudinal dependency within our study area. It depends rather on the depth and humidity of an elevated moist layer with embedded clouds, termed cloud layer (Stevens 2005), which typically extends up to between 2 and 3 km asl in Fig. 1 and the lidar observations.

Figure 1, a sample of 3 sondes related to the lidar results, gives an overview of the thermodynamic situation in the lowest 4 km, mainly the trade wind layer. Since all segments lie within the downward subsiding branch of the tropical Hadley cell, we see a strong capping inversion at the cloud layer top, termed trade inversion (Stull 2015) accompanied by a strong gradient in water vapor mixing ratio. Below, the cloud layer extends down to about 0.8 km asl with conditionally unstable temperature lapse rates varying between 3.5 and 6.7 K/km in Table 1. Under such conditions, a saturated parcel of air is unstable to upward vertical displacements, and an unsaturated parcel is stable to small displacements. Slight variations in temperature and mixing ratio, as shown in Fig. 1, can suffice for saturation, or under-saturation, to occur. In the lowest level, the cloud-free, convective marine boundary layer is characterized by fairly zero vertical gradients in water vapor mixing ratio and potential temperature, indicating well-mixed conditions. Its temperature lapse rate in these 3 exemplary cases varies between 10.0 and 10.5 K/km, very close to dry adiabatic conditions. This sub-cloud layer is capped by a weaker inversion, accompanied by a smaller mixing ratio jump of roughly 4 g/kg. Table 1 shows that the relative humidity mostly peaks at the top of this sub-cloud layer. The average total column water vapor in Table 1 amounts to 28 kg/m^2. As typical for the winter trades, the humidity below the trade inversion accounts for $\sim$ 95% of that column, and the partial column above totals $\sim$ 1.4 kg/m^2. These dropsonde observations agree with the lidar results hereafter.

We dedicate each of the three following subsections to a different physical phenomenon or process, even if it is evident that the presented lidar examples show similarities, owing to the overall comparable meteorological conditions. Common features of the lidar observations include (a) the horizontal and vertical variability of humidity in the lowest 4 km, our key topic, (b) the omnipresence of an elevated moist layer with embedded clouds, the cloud layer, (c) the predominance of two altitudes for shallow clouds, namely the bottom and the top of the cloud layer, and (d) the absence of moist and aerosol layers in the free troposphere above the cloud layer. The latter is related to the winter trades with the lowest likelihood of cross-Atlantic Sahara dust transport (Prospero et al. 2014). The absence of aerosol from remote, continental origin eases the interpretation and the comparison between the lidar backscatter and water vapor measurements, as sea salt aerosol and evaporation, both originating at the sea surface, can be considered here the main sources for the observed backscatter and moisture, respectively.

4.2 Strong Heterogeneity in the Cloud Layer

On December 15, 2013, a local flight with start and landing in Barbados was performed in southwest–northeast–southwest sequence, interrupted by a north–south under-flight of the NASA A-Train satellites during the outbound flight leg, as displayed in Fig. 2. While the southwest part of the flight has few clouds, the northeast part shows large cloud clusters with smaller cloud-free zones in between. The overflown clouds can well be compared with the lidar backscatter in Fig. 3 which shows the way back to Barbados where the aircraft was heading straight southwest, without another A-Train under-flight, and nearly parallel with the trade wind blowing at $\sim$ 15 m/s (Table 1) yet 15 times faster. The resulting maximum deviations of the lidar measurements at 40 and 54°W, separated in time by $\sim$ 1 h from Fig. 2, are $\sim$ 54 km or $\sim$ 0.5° longitude. Not only do the large-scale patterns coincide, but also small clouds around 19:53 are visible in both satellite and lidar images at about 1 km asl. The smallest clouds though are difficult to distinguish in both figures. They appear as black dots in the lidar image and as an unresolved yellow–brown

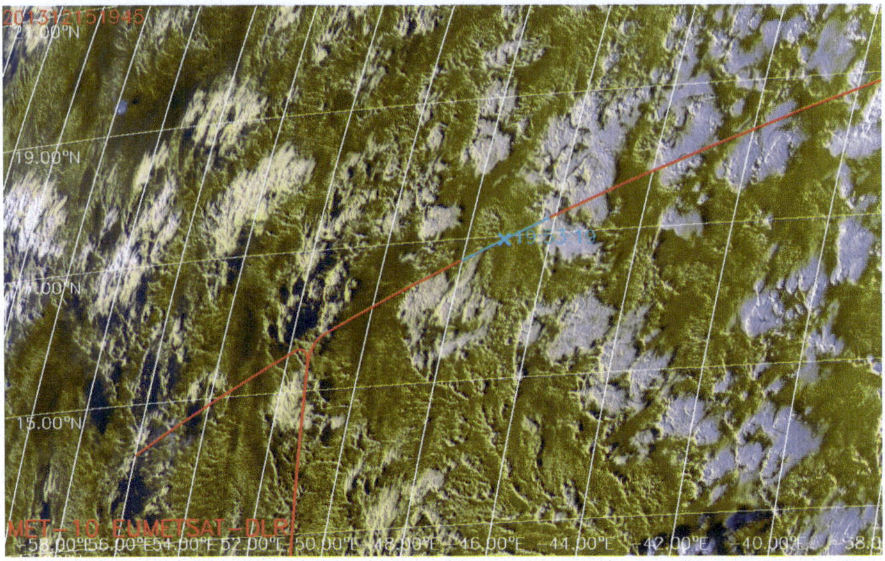

Fig. 2 MSG/SEVIRI false color composite for December 15, 2013, at 19:45 UTC using the 0.6-, 0.8- and 10.8-µm channels together with the high-resolution visible channel. The sea surface appears dark, low clouds yellow, thick clouds white. See Sect. 3 for details. Aircraft flight track superposed (red), between 18:55 (northeast) and 21:05 (southwest end of red line), with 15-min segment around 19:53 highlighted in blue, corresponding to the part of the flight path closest in time to this image. The north–south excursion at ~ 51 W was an under-flight of the NASA A-Train, yet during the earlier outbound flight part around 17:00 UTC. The distance between 2° in longitude is ~ 213 km, between 2° in latitude ~ 223 km. Barbados is situated approximately in the lower left corner, at 13°N, 59°W

background in the satellite image, where very low faint clouds in between the many little white (i.e., higher and thicker) cloud tops with horizontal extensions well below the satellite spatial resolution altogether result in a fairly uniform yellow–brown color. At 19:30, between two large cloud clusters, the top panel of Fig. 3 shows a cloud-free region ~ 70 km in width with maritime aerosol in the sub-cloud layer and less backscatter in the cloud layer. This corresponds in Fig. 2 to a uniform yellow–brown area at the intersection of the red flight path and 44.3°W. Comparing our two examples around 19:30 and 19:53, we conclude that it is nearly impossible to distinguish small, scattered low clouds from fully cloud-free regions in the MSG imagery because of the omnipresence of sea salt aerosol in the lowest layer which hides the expected darker ocean surface and because of the MSG/SEVIRI pixel size (large with respect to the lidar) that reduces the probability of observing a fully cloud-free area. An exceptionally dark area is shown in Fig. 2 at 54.2 W corresponding to Fig. 3 at 21:12. A region about 40 km wide, with backscatter coefficients two times lower than in the previous example in both the sub-cloud and the cloud layer, is observed by the lidar. Another dark region is seen in the next subsection, in Fig. 4 at 56.2 W and in Fig. 5 at 19:18 with additionally both layers thinner. These dark regions represent spots in the MSG images where reflectivity at 0.6 µm is as low as ~ 0.13 and reflectivity at 0.8 µm is below 0.08, indicating a cloud-free environment with low aerosol content. Aerosol variability is present and visible in detail in the lidar backscatter, yet difficult to quantify. Larger lidar backscatter can be due to larger aerosol sizes and concentrations. Hydrophilic sea salt particles swell with increasing relative humidity, observed

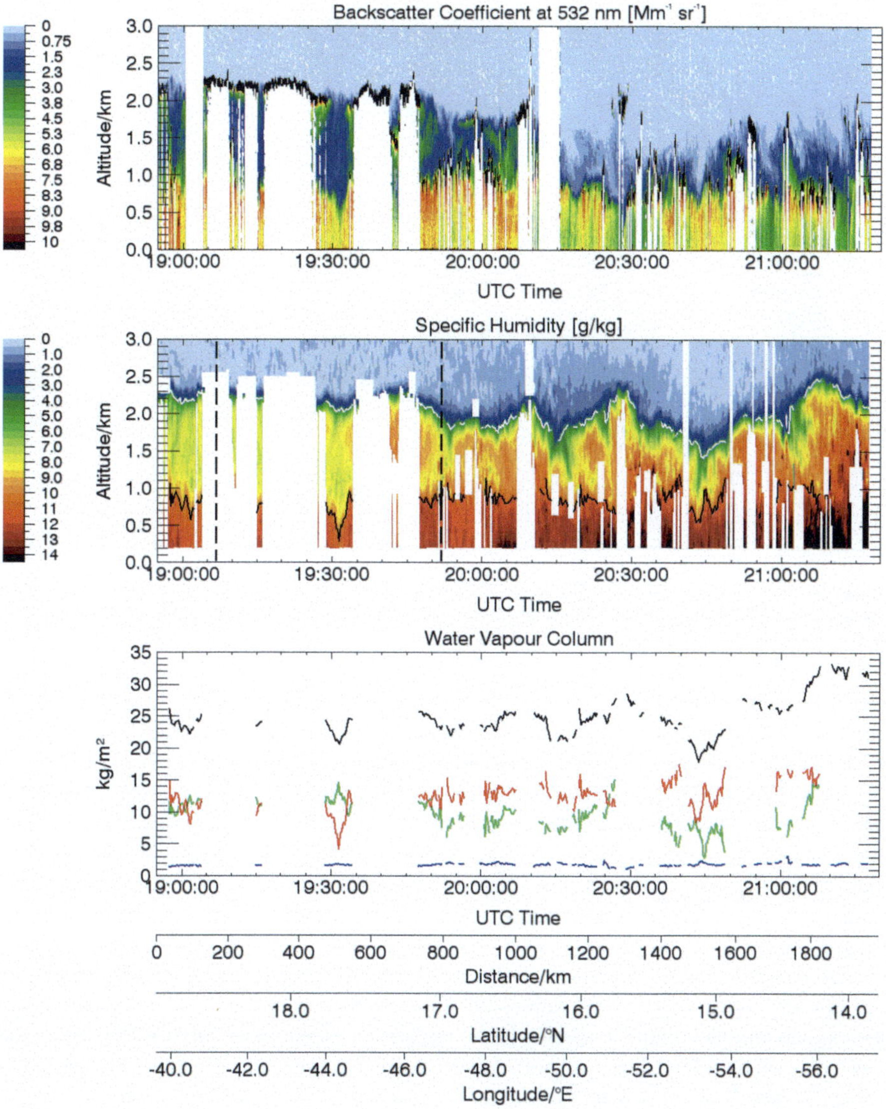

Fig. 3 Top: High-spectral-resolution lidar (HSRL) backscatter of the lower tropical troposphere on December 15, 2013; clouds black. Middle: Water vapor lidar cross section with dropsondes from Table 1 as dashed lines. The top of the cloud layer (white line) is defined by a threshold in mixing ratio of 4 g/kg, the top of the sub-cloud layer by one of 10 g/kg (black line; 11 g/kg after 19:45 UTC). Missing or too noisy data, mainly below clouds, are whitened. The lidar images are squeezed in the horizontal by a factor of $\sim$ 200 with respect to the vertical. Bottom: Total water vapor column between 0 and 13 km (black) and partial columns of the sub-cloud layer (red), the cloud layer (green) and the column aloft (blue). The columns are undefined when data are missing

at the top of the sub-cloud layer, while their concentrations increase with increasing surface winds, ocean roughness and intensity of turbulent mixing.

Figures 2 and 3 reveal a structural difference between the large upper-level cloud cover in the northeast and the more scattered, small and multi-level cloud occurrences in the

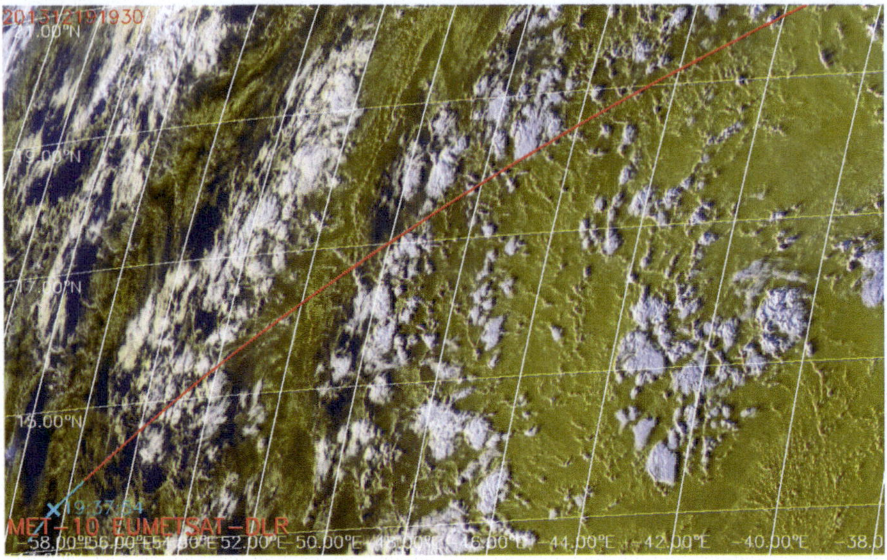

Fig. 4 As shown in Fig. 2 but for December 19, 2013, 19:30 UTC. Aircraft flight track starting 17:20 (northeast end of red line), with 15-min segment around 19:38 UTC highlighted in blue, corresponding to the part of the flight path closest in time to this image

southwest. The dropsonde of Fig. 1a is located at the transition between both cloud regimes. Moist convection is obviously responsible for lifting the cloud layer, e.g., at around 20:08, 20:28 and 20:54 UTC, and also the sub-cloud layer, e.g., at 20:00 and 20:32 UTC. The absence of high-resolution observations of vertical motion, however, hinders a more consolidated statement to this respect. Thresholds in the mixing ratio are more robust than maxima in vertical mixing ratio gradients to separate the layers. We tested both gradient and threshold methods. The gradient method works well for determining the upper layer limit, where it coincides within ~ 100 m with a threshold of 4 g/kg (the white line). The top of the sub-cloud layer, however, has a gradient so much weaker that the threshold gives more robust results, albeit not always satisfying. The black line in the middle panel of Fig. 3, outlining the top of the sub-cloud layer, mostly coincides well with the strong vertical aerosol backscatter gradients in the top panel, except after 20:50 UTC, where the lower part of the cloud layer is more humid. Here the threshold method fails, and data where the sub-cloud layer height would have exceeded 1.2 km have been removed. This is the maximum thickness of the sub-cloud layer in the lidar backscatter. The alternative to use the backscatter gradients fails due to disparities between the distributions of aerosol and humidity, and due to the frequent absence of aerosol at the cloud layer top.

Overall, the strong heterogeneity of layer depth, humidity, cloud and aerosol distribution of the cloud and sub-cloud layers marks the scenery. For example, at an altitude of 1.5 km, in the middle of the cloud layer, the lidar-observed mixing ratio varies by a factor of 3, between about 3 to 10 g/kg, which influences a lot the water vapor columns. The partial columns in both layers (red and green lines for the sub-cloud and cloud layers, respectively) vary by up to a factor of 2 and the total column (black) by up to 10 kg/m^2 or 30%. The wavy variations in the total column have a size of about 300 km, corresponding roughly to the extent of regions with higher-than-average humidity, interrupted by smaller dry regions.

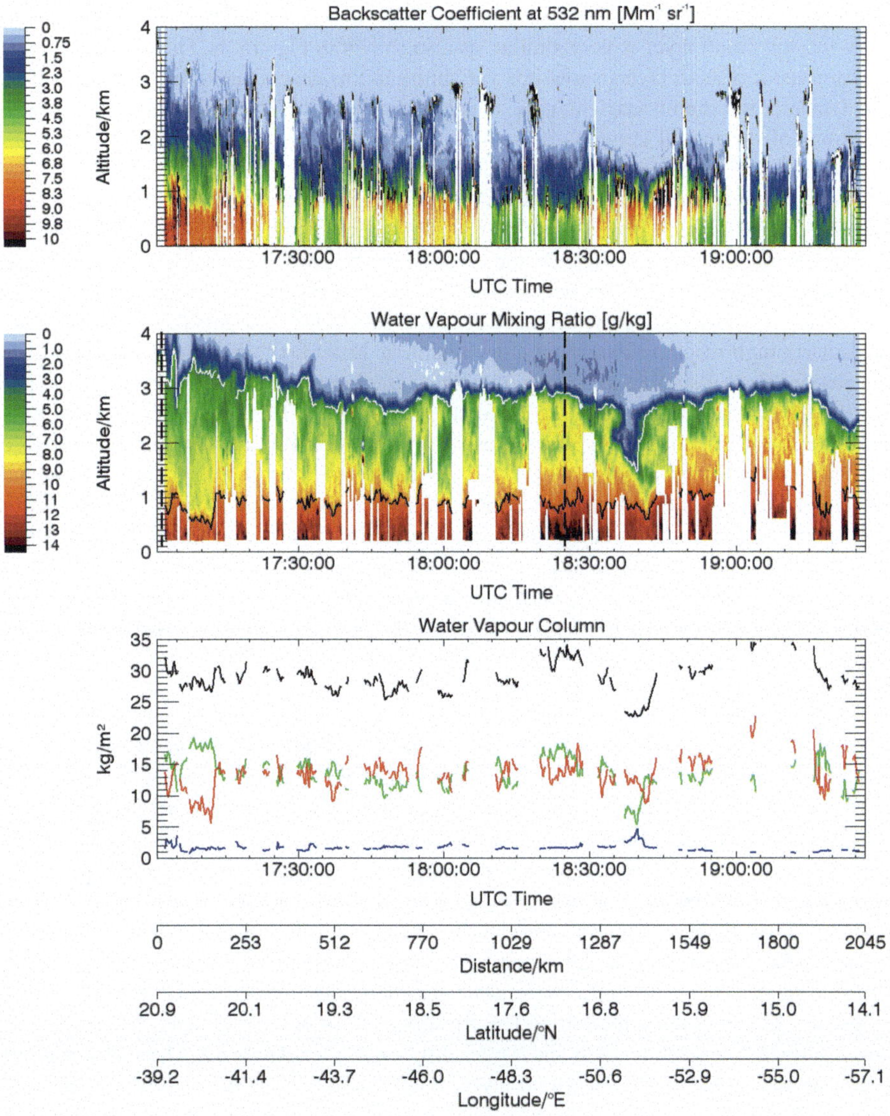

Fig. 5 As shown in Fig. 3 but for December 19, 2013. The top of the cloud layer is defined by a mixing ratio of 4 g/kg (white line), the top of the sub-cloud layer by 10 g/kg (black line; 11 g/kg after 18:20 UTC)

4.3 Dry Regions in Both Cloud and Sub-cloud Layers

This case has been observed 4 days later on December 19, 2013, along nearly the same flight track, with little more extent toward the northeast. In comparison, Figs. 2 and 4 show large cloud clusters aligned northeast–southwestward in the west and less cloudiness in the northeast. Indeed, in Table 1, the 17:01 UTC dropsonde has slightly lower relative humidity, compared to the 19:07 UTC dropsonde of the previous case. Again, in Fig. 5, we see humidity heterogeneity and the presence of two dominant cloud levels. However, the total water vapor is 20% higher, because the cloud layer is on average ~ 1 km thicker, while the sub-cloud layer is very similar, as also shown in Fig. 1a, b. The partial columns of water vapor of both layers show less variability in this case. Due to the thickness of the cloud layer, its column content now mostly equals that of the sub-cloud layer. Still, variations of the partial columns of both layers occasionally exceed a factor of 2. Conditional instability in the cloud layer, documented in Fig. 1 and Table 1, leads to the observed patchiness of convection. Uplifted by local convergence due to, e.g., shallow secondary circulations, a parcel of air may saturate at any height within the cloud layer, as observed in the lidar backscatter. On the other hand, the growth of clouds may be stopped by weak temperature inversions within the cloud layer, as shown, for example, in Fig. 1b at 1.7 km asl, corroborated by the presence of cloud tops in Fig. 5 at this height. Besides the particularly high reaching humidity, a dry region at 18:40 UTC is eye-catching in Fig. 5. Dry regions were also observed in the previous case, but this event is nearly as dry as the free troposphere in the middle of the cloud layer in 2 km asl. The total water vapor column decreases by 10 kg/m^2 or ~ 30% across a width of ~ 100 km and the near-surface humidity decreases by about 2 g/kg (15%). Because the water column aloft is dry, and because upper tropospheric clouds are absent, such regions significantly cool the lower troposphere (Stevens et al. 2017). It is this variability of water vapor and of the trade inversion height that modulates the amount of outgoing longwave radiation and the related cooling rates. The amount of cooling and the occurrence of possibly induced secondary circulations are a focus of current research (Zuidema et al. 2012; Zuidema and Torri 2017). In addition, radiative cooling may promote convection and clouds, and may be at the origin of convective aggregation, another focus of current research (Stevens et al. 2017).

The high-humidity regions in the cloud layer in Fig. 5 span across ~ 500 km, which is larger than in the previous case. This scale is again best deduced from the variations of the total column water vapor (black line), where we find higher levels between 18:00–18:40, and 18:40–19:20. The cloud layer extends nearly up to 4 km height in the northeast. Less subsidence in this northernmost part is a likely reason, since the 17:01 dropsonde profile (not shown here) has a much weaker inversion and moisture gradient than the 18:25 dropsonde of Fig. 1b. We also find dry regions in the sub-cloud layer, also ~ 100 km in size. At 17:10, for example, the sub-cloud water column decreases by nearly 10 kg/m^2. Unlike their counterparts in the cloud layer these dry regions probably do not affect radiation as much since they lie underneath the moist, insulating cloud layer. The 17:01 dropsonde, although not quite hitting the driest part of the sub-cloud layer, measured a 2° lower surface temperature than the 18:25 dropsonde (see Table 1). These dry regions obviously occur where the sub-cloud layer is thinner, as Fig. 5 shows. Overall, we find high variability in the depth of the sub-cloud layer. In this segment, for example, it varies between ~ 700 m (at 17:37 or 19:18) and ~ 1300 m, which is nearly a factor of 2.

4.4 Transport of Moisture Through the Cloud Layer by Shallow Convection

Figure 6 shows clouds again organized on large scales and a darker filament with few or no clouds between about 20°N/43°W and 15°N/58°W meandering about the flight trajectory. The MSG movie (not shown here) reveals that this filament persisted over the whole day, embedded in the trade wind flow moving westward. Similar filaments existed on other days of the campaign. Its darkest part was overflown between 19:00 and 19:25, and Fig. 7 confirms the absence of clouds here. Earlier, the slimmer northeastern part of the filament was crossed at around 18:08 and 18:25, where the lidar backscatter also indicates cloud-free conditions. The observations show that the width of the filament's cloud-free core

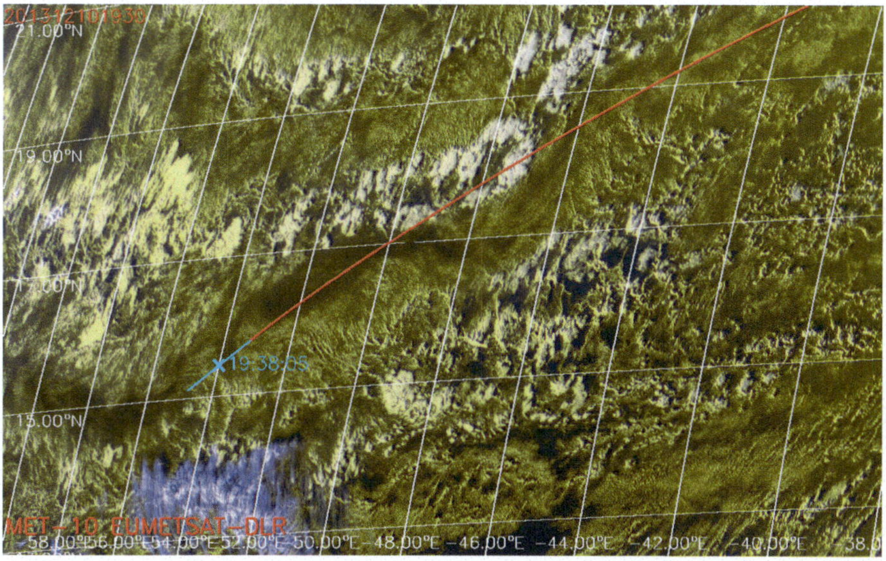

Fig. 6 As shown in Fig. 2 but for December 10, 2013, 19:30. In the lower left, cirrus emanating from deep convection at the ITCZ located more south is apparent in violet. Aircraft flight track starting 17:50 (northeast end of red line), with 15-min segment around 19:38 UTC highlighted in blue, corresponding to the part of the flight path closest in time to this image

varies between 30 and 100 km. All filament crossings show a significantly drier cloud layer in the water vapor cross section with around 7 instead of 9 g/kg at 1500 m asl.

This lidar segment displays an evident mechanism for upward transport, shallow convection. Figure 7 shows a heterogeneous cloud layer which the cloud convection has exceeded, giving a 200-km cloud topping between 2.5 and 2.8 km height and surrounded by a distinct elevated moist layer, vertically centered around 2.3 km asl. The higher humidity spans across 600 km, between 18:20 and 19:05. The availability of 3 dropsondes, the one displayed in Fig. 1c being just in the middle of the convective cluster, adds value to this case. They corroborate the lidar observation by indicating that convection has pushed the trade inversion upward by several 100 m and has significantly sharpened the moisture gradient at the height of the inversion. Table 1 also shows that the total column water vapor is augmented by 8.5 kg/m^2, or 30% in the 18:50 dropsonde, in agreement with the lidar water vapor columns (not shown here). Accordingly, the profiles of Fig. 1b, c are similar, except for the height of the inversion, whereas the 18:20 and 19:22 dropsondes (see Table 1) are more comparable to Fig. 1a. The observed elevated cloud cluster likely represents a moisture elevator, although evidence could, as mentioned before, only be provided by additional observations of vertical movements or of convergence. Again, the lidar backscatter reveals a predominance of two cloud top heights, with the smaller, scattered clouds topping at about 1 km asl. While the cirrus in the satellite image (Fig. 6) is well discernible, visual discrimination of the two different lower troposphere cloud top heights remains a challenge despite advanced imaging, mainly because the scattered clouds are too small and because Fig. 1c reveals that the temperature difference between the two heights is just a few degrees and consequently too low to generate a discernible color contrast in the satellite images.

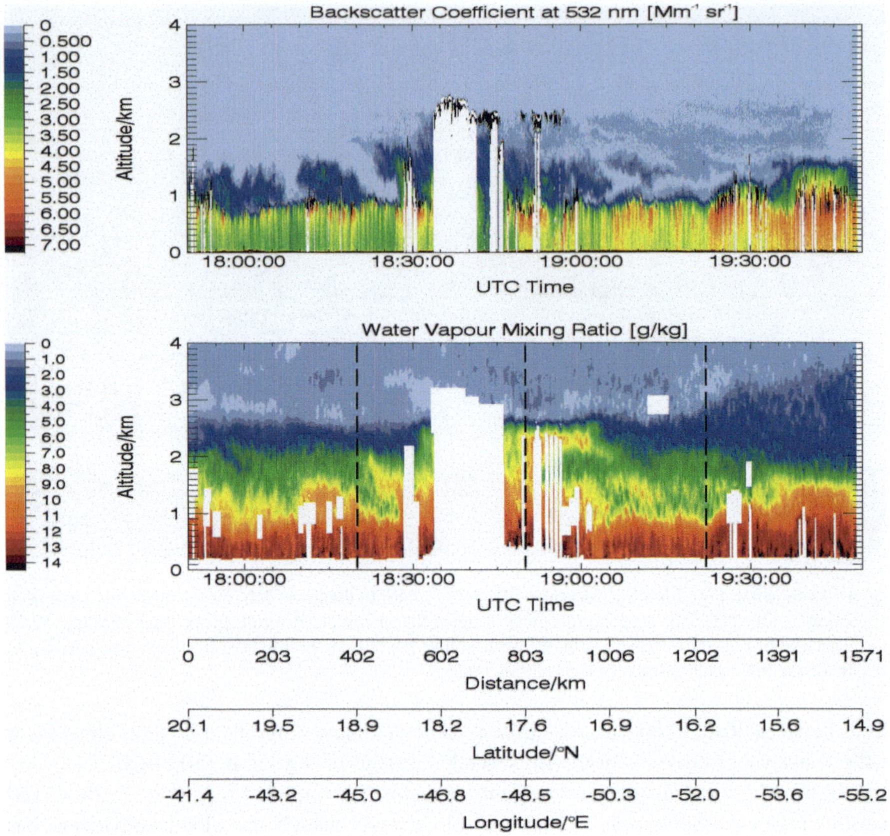

Fig. 7 As shown in Fig. 3 but for December 10, 2013

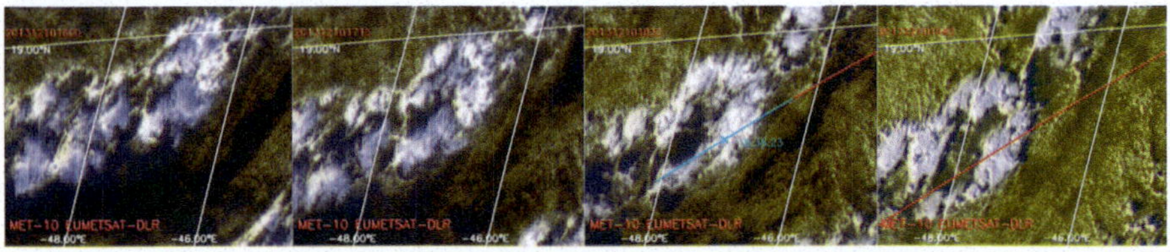

Fig. 8 Temporal evolution on December 10, 2013, from 16:00 to 19:45 UTC with 1:15 h increment of cloud cluster of Fig. 7 as observed by MSG. Third image (18:30) incudes HALO flight path with timestamp. Fourth image is 15 min later than Fig. 6. The background varies due to changing solar illumination and cloud temporal evolution

Figure 8 gives an impression how the convective cluster seen by the lidar in Fig. 7 evolves over time. It is part of a larger circle of clouds with size $\sim$ 200 km around an inner cloud-free region of $\sim$ 70 km diameter. The circle appears having organized itself from less-ordered cloud patches over the course of about 2.5 h. The whole cluster moves southwestward with the trade winds of $\sim$ 12 m/s, which results in a displacement of $\sim$ 160 km over the 3:45-h time span, in agreement with Fig. 8. Generally, lidar results from repetitive overflights of the same region (not shown here) provide evidence that meso-scale structures (variations) of the "background" humidity in the cloud layer remain discernible over at least some hours, similar to the cloud observations in Fig. 8.

4.5 Mean, Variance and Skewness Profiles of Water Vapor

Vertical profiles of statistical moments of horizontal water vapor lidar distributions between the clouds permit quantifying the variability for comparisons between the three presented cases and with former studies. Profiles for the horizontal domains of Figs. 3, 5 and 7 are shown in Fig. 9. In Fig. 9a, b that presents similarities, while the third case is different, the variance maxima coincide, as expected, with the heights of the strongest vertical moisture gradients located at the top of the cloud layer. The variance is proportional to the moisture jump across that boundary, with case a about twice as strong as case b. Below 1.5 km asl, the variances in a and b are similar. In contrast, the gradients at the top of the sub-cloud layer are so weak that the variances do not peak a second time there. Note that the mean profiles are smeared out by horizontal variability and that the vertical lidar resolution is 290 m, while the local gradients may be much steeper, as shown in Fig. 1. Case 3 is special. Due to the presence of two distinct convective regimes, one higher reaching in the middle of Fig. 7 and the other one surrounding it at lower levels, the mean profile in Fig. 9c shows three layers and the variance profile has two distinct peaks. In all three cases, the variance in the sub-cloud layer is lower than that in the cloud layer. A slight increase in the variance below 500 m asl may be indicative of instrumental noise that adds to the natural variance and increases with measurement range, i.e., augments the variance artificially toward the surface, as exemplified in Kiemle et al. (2007). Above its maximum at the trade inversion, the variance abruptly decreases to very low levels up to the flight altitude. In comparison with fluctuations of water vapor in a convective boundary layer (CBL) over land, with smaller spatial scales and different origins of variability, we find up to 10 times stronger moisture variances. Kiemle et al. (1997, 2007) derived mixing ratio variance profiles from CBL lidar curtains with maxima < 1 $(g/kg)^2$ at the top of the cloud-free CBL and mixing ratio jumps across the inversion comparable to those of the present case b. In the mid-CBL, their variances lie between 0.05 and 0.5 $(g/kg)^2$, after

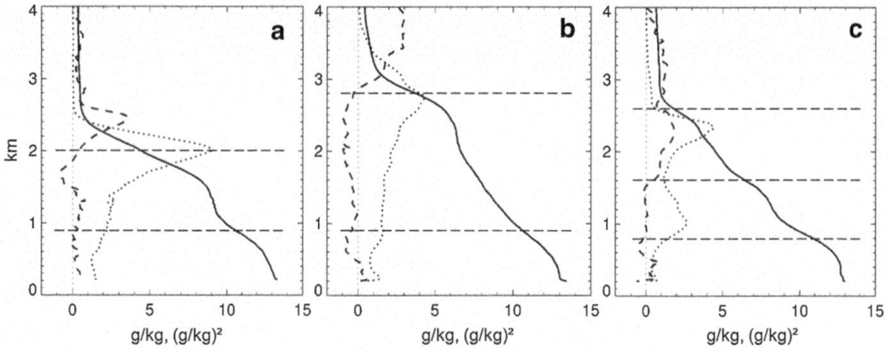

Fig. 9 Profiles of mean (solid), variance (dotted) and skewness (dashed) of horizontal water vapor lidar distributions between the clouds on **a** December 15, 2013, **b** December 19, 2013 and **c** December 10, 2013. Horizontal dashed lines: **a**, **b** average sub-cloud and cloud layer heights from black and white lines in Figs. 3 and 5; in **c** the cloud layer has two distinct top heights

filtering out variance on scales larger than ~ 10 km that were under-sampled. The comparison is consequently of limited value.

The skewness indicates the asymmetry of the water vapor distribution. A higher-order statistical moment, the skewness is more sensitive to outliers in the distribution, independent on whether they are real or measurement artifacts. Our interpretation consequently will be limited to the most prominent phenomena observed. In Fig. 9a, the skewness peaks at 2.5 km asl because of the presence of three upward protrusions of water vapor into the dry free atmosphere in Fig. 3 at 20:10, 20:30 and 21:10. Conversely, the minimum at 1.7 km is related to the presence of two downward protrusions of dry air into the cloud layer at 20:15 and 20:45. Above 2.5 and below 1.7 km, the skewness is insignificant, i.e., the distribution is symmetric about the mean. Maximum skewness in Fig. 9b is found above the trade inversion because of higher humidity in the very left of Fig. 5. More interesting is the region with negative skewness below, deeper than in case a, related to the prominent downward protrusion of dry air into the cloud layer at 18:40. The vertical extent of negative skewness corresponds to the depth of the protrusion. Skewness values between 0 and -1 correspond to values found below the top of a CBL by Kiemle et al. (1997) due to narrow entrainment protrusions of dry air, yet, as noted for the variance, just the morphology corresponds, but the spatial scales and the atmospheric processes responsible for the observed variability are very different. For example, the CBL protrusions are typically 1 km in width, while the one at 18:40 is 100 km wide. It is consequently appropriate to consider the present horizontal scales of variability to be about 100 times larger. The skewness is negative also between altitudes of 0.5 and 1.1 km, due to a suppressed sub-cloud layer around 17:10, again about 100 km large. Due to the two different regimes of convection and a consequently bimodal humidity distribution above 1.6 km, the skewness profile of Fig. 9c is only useful below this height, where we find relatively insignificant values again indicative of a balanced moisture distribution.

5 Conclusions and Outlook

Thanks to the capability of performing targeted, simultaneous measurements of aerosol or cloud backscatter and water vapor in regions of particular interest, an airborne water vapor lidar can help elucidate the complex interactions between water vapor, clouds and aerosol in the trades. During the HALO-NARVAL winter campaign in December 2013, high horizontal and vertical variability of humidity in the lower tropical troposphere, the omnipresence of an elevated moist layer with embedded clouds above the sub-cloud marine boundary layer and high variability in the depth of those two layers were observed. The mixing ratio variance within the layers is up to 10 times larger than within a cloud-free CBL over land, at horizontal scales of variance about 100 times larger. The variance mainly stems from the presence of dry regions in the cloud and sub-cloud layers, and the transport of moisture through the cloud layer by shallow convection. In the majority of lidar and dropsonde profiles, a strong temperature inversion and the strongest humidity gradient are located at the top of the cloud layer, while the sub-cloud layer top is characterized by weaker gradients of temperature and humidity. Conditional instability throughout the cloud layer creates an environment in which slight variations in temperature and humidity can have an impact on cloud depth and cover, thus creating the variability of cloud base and top heights observed within this layer.

Often, shallow convection is seen to precipitate, locally drying and cooling the atmosphere (Zuidema et al. 2012). We do not measure temperature by airborne lidar, but could, in the future, assess temperatures at the lifting condensation level when assuming that the lidar water vapor mixing ratio just at the top of the sub-cloud layer, below the base of optically thin clouds, corresponds to the saturation mixing ratio. Wherever the temperature lapse rate in the sub-cloud layer can be assumed dry adiabatic as shown in Fig. 1, we could extrapolate to the surface temperature and also obtain near-surface relative humidities using the near-surface lidar mixing ratios. Additionally, conditional sampling of the water vapor lidar profiles around precipitating versus non-precipitating clouds may shed more light on the implications of precipitating shallow convection. Furthermore, investigations of a possible correlation between cloud cover and ambient column water vapor may complement the analyses of water vapor distributions in the vicinity of clouds. In order to better quantify the observed heterogeneity, variance and spectral analyses of the broader humidity fields and of the depths of both cloud and sub-cloud layers are foreseen. Frequent gaps in the lidar data due to overlying clouds impede an easy approach, yet earlier work has opened ways to mitigate this issue (Kiemle et al. 2011). The observed intermittent character of convection in the cloud layer leads to water vapor distributions that may not be stationary enough for successfully applying Fourier analyses. For these cases, well-documented alternative approaches exist to characterize variability, as exemplified in Fischer et al. (2013). In contrast to the present case study, future work will use all available water vapor lidar measurements from both NARVAL winter and summer campaigns to give resulting probability distribution functions and related statistics a broader basis, similar to the comprehensive analyses performed by Randel et al. (2016).

With the absence of humidity and clouds above the trade inversion, the amount of outgoing longwave radiation and the related cooling rates are mainly governed by the trade inversion height and its variability. Lower heights increase the outgoing longwave radiation, yet the effects caused by such mesoscale variability on the trade wind layer are still fairly unknown. Future research could involve using the water vapor lidar profiles, together with auxiliary data as input for radiative transfer modeling to obtain estimates of the outgoing longwave radiation with high spatial resolution along the flight track. Recent model results suggest that low-level radiative cooling induces shallow circulations (Naumann et al. 2017). The presented lidar observations as well as models accurately reproducing the observed variability will help evaluate the role that trade wind water vapor and its variability play for climate.

Acknowledgements This paper arises from the International Space Science Institute (ISSI) workshop on "Shallow clouds and water vapor, circulation and climate sensitivity." Valuable support during the flight campaign was provided by Andreas Fix, Christian Büdenbender and Axel Amediek, all DLR. The NARVAL campaign was co-sponsored by the Max Planck Society, the Deutsche Forschungsgemeinschaft (German Science Foundation, project HALO-SPP 1294) and the DLR Institute of Atmospheric Physics. The dropsonde data were processed by Yanfei Gong, DLR, who also tested layer separation methods. We are grateful to Klaus Gierens, DLR, who provided an internal review, to Andreas Schäfler, DLR, for helpful discussions, and to an anonymous reviewer for many valuable comments.

References

Behrendt A, Wulfmeyer V, Kiemle C, Ehret G, Flamant C, Schaberl T, Bauer H-S, Kooi S, Ismail S, Ferrare R, Browell EV, Whiteman DN (2007) Intercomparison of water vapor data measured with lidar during IHOP_2002 Part II airborne-to-airborne systems. J Atmos Ocean Technol 24:22–39

Bhawar R et al (2011) The water vapour intercomparison effort in the framework of the convective and orographically-induced precipitation study: airborne-to-ground-based and airborne-to-airborne lidar systems. Q J R Meteorol Soc 137:325–348. doi:10.1002/qj.697

Bielli S, Grzeschik M, Richard E, Flamant C, Champollion C, Kiemle C, Dorninger M, Brousseau P (2012) Assimilation of water-vapour airborne lidar observations: impact study on the COPS precipitation forecasts. Q J R Meteorol Soc. doi:10.1002/qj.1864:10.1002/qj.1864

Bony S, Stevens B, Frierson D et al (2015) Clouds, circulation, and climate sensitivity. Nat Geosci 8:261–268. doi:10.1038/ngeo2398

Bony S, Stevens B, Ament F et al (2017) EUREC^{4}A: a field campaign to elucidate the couplings between clouds, convection and circulation. Surv Geophys. doi:10.1007/s10712-017-9428-0

Di Girolamo P, Behrendt A, Kiemle C, Wulfmeyer V, Bauer H, Summa D, Dörnbrack A, Ehret G (2008) Simulation of satellite water vapour lidar measurements: performance assessment under real atmospheric conditions. Remote Sens Environ 112(4):1552–1568

Fischer L, Craig GC, Kiemle C (2013) Horizontal structure function and vertical correlation analysis of mesoscale water vapor variability observed by airborne lidar. J Geophys Res 118:1–12. doi:10.1002/jgrd.50588

Flentje H, Dörnbrack A, Ehret G, Fix A, Kiemle C, Poberaj G, Wirth M (2005) Water vapour heterogeneity related to tropopause folds over the North Atlantic revealed by airborne water vapour differential absorption lidar. J Geophys Res 110:D03115. doi:10.1029/2004JD004957

Groß S, Wirth M, Schäfler A, Fix A, Kaufmann S, Voigt C (2014) Potential of airborne lidar measurements for cirrus cloud studies. Atmos Meas Technol 7:2745–2755

Gutleben M, Groß S, Wirth M, Ewald F, Schäfler A (2017) Applicability of spaceborne lidar measurements to study shallow marine convection over the subtropical North Atlantic Ocean. To be submitted to AMT

Kiemle C, Ehret G, Giez A, Davis KJ, Lenschow DH, Oncley SP (1997) Estimation of boundary-layer humidity fluxes and statistics from airborne differential absorption lidar (DIAL). J Geophys Res 102(D24):29189–29203

Kiemle C, Brewer WA, Ehret G, Hardesty RM, Fix A, Senff C, Wirth M, Poberaj G, LeMone MA (2007) Latent heat flux profiles from collocated airborne water vapor and wind lidars during IHOP_2002. J Atmos Ocean Technol 24:627–639. doi:10.1175/JTECH1997.1

Kiemle C, Wirth M, Fix A, Ehret G, Schumann U, Gardiner T, Schiller C, Sitnikov N, Stiller G (2008) First airborne water vapor lidar measurements in the tropical upper troposphere and mid-latitudes lower stratosphere: accuracy evaluation and intercomparisons with other instruments. Atmos Chem Phys 8:5245–5261

Kiemle C, Wirth M, Fix A, Rahm S, Corsmeier U, Di Girolamo P (2011) Latent heat flux measurements over complex terrain by airborne water vapour and wind lidars. Q J R Meteorol Soc 137:190–203. doi:10.1002/qj.757

LeMone MA, Pennell WT (1976) The relationship of trade wind cumulus distribution to subcloud layer fluxes and structure. Mon Weather Rev 104:524–539

Mech M, Orlandi E, Crewell S, Ament F, Hirsch L, Hagen M, Peters G, Stevens B (2014) HAMP—the microwave package on the high altitude and long range research aircraft (HALO). Atmos Meas Tech 7:4539–4553. doi:10.5194/amt-7-4539-2014

Naumann A-K, Stevens B, Hohenegger C, Mellado JP (2017) A conceptual model of a shallow circulation induced by prescribed low-level radiative cooling. J Atmos Sci. doi:10.1175/JAS-D-17-0030.1

Nehrir AR, Kiemle C, Lebsock M, Kirchengast G, Buehler SA, Löhnert U, Liu CL, Hargrave P, BarreraVerdejo M, Winker D (2017) Emerging technologies and synergies for airborne and space-based measurements of water vapor profiles. Surv Geophys (in press)

Nuijens L, Medeiros B, Sandu I, Ahlgrimm M (2015) Observed and modeled patterns of covariability between low-level cloudiness and the structure of the trade-wind layer. J Adv Model Earth Syst. doi:10.1002/2015MS000483

Pincus RA, Beljaars A, Kirchengast G, Buehler BS, Landstaedter F (2017) The distribution of water vapor over low-latitude oceans: current best estimates, errors, and impacts. Surv Geophys (in press)

Prospero JM, Collard F-X, Molinié J, Jeannot A (2014) Characterizing the annual cycle of African dust transport to the Caribbean Basin and South America and its impact on the environment and air quality. Glob Biogeochem Cycles 29:757–773. doi:10.1002/2013GB004802

Randel WJ, Rivoire L, Pan LL, Honomichl SB (2016) Dry layers in the tropical troposphere observed during CONTRAST and global behavior from GFS analyses. J Geophys Res Atmos 121:14142–14158. doi:10.1002/2016JD025841

Schäfler A, Dörnbrack A, Kiemle C, Rahm S, Wirth M (2010) Tropospheric water vapor transport as determined from airborne lidar measurements. J Atmos Ocean Technol 27:2017–2030. doi:10.1175/2010JTECHA1418.1

Schmetz J, Pili P, Tjemkes S, Just D, Kerkmann J, Rota S, Ratier A (2002) An introduction to meteosat second generation (MSG). Bull Am Meteorol Soc 83:977–992. doi:10.1175/1520-0477(2002)083<0977:AITMSG>2.3.CO;2

Sherwood SC, Bony S, Dufresne J-L (2014) Spread in model climate sensitivity traced to atmospheric convective mixing. Nature 505:37–42

Stevens B (2005) Atmospheric moist convection. Annu Rev Earth Planet Sci 33:605–643. doi:10.1146/annurev.earth.33.092203.122658

Stevens B, Brogniez H, Kiemle C, Lacour J-L, Crevoisier C, Killiani J (2017) Structure and dynamical influence of water vapor in the lower tropical troposphere. Surv Geophys. doi:10.1007/s10712-017-9420-8

Stull R (2015) Practical meteorology: an algebra-based survey of atmospheric science. University of British Columbia, p 938. https://www.eoas.ubc.ca/books/Practical_Meteorology/

Trickl T et al (2016) How stratospheric are deep stratospheric intrusions? LUAMI 2008. Atmos Chem Phys 16:8791–8815. www.atmos-chem-phys.net/16/8791/2016/. doi:10.5194/acp-16-8791-2016

WCRP (2016) http://wcrp-climate.org/index.php/gc-clouds

Wirth M, Fix A, Mahnke P, Schwarzer H, Schrandt F, Ehret G (2009) The airborne multi-wavelength water vapor differential absorption lidar WALES: system design and performance. Appl Phys B 96:201–213

Wulfmeyer V, Hardesty RM, Turner DD, Behrendt A, Cadeddu MP, Di Girolamo P, Schlüssel P, Van Baelen J, Zus F (2015) A review of the remote sensing of lower tropospheric thermodynamic profiles and its indispensable role for the understanding and the simulation of water and energy cycles. Rev Geophys 53:819–895. doi:10.1002/2014RG000476

Zuidema P, Torri G (2017) Precipitation-induced oceanic cold pools and their interactions with the larger-scale environment. Surv Geophys (in press)

Zuidema P et al (2012) On trade wind cumulus cold pools. J Atmos Sci 69:258–280. doi:10.1175/JAS-D-11-0143.1

Surv Geophys (2017) 38:1445–1482
https://doi.org/10.1007/s10712-017-9448-9

CrossMark

Emerging Technologies and Synergies for Airborne and Space-Based Measurements of Water Vapor Profiles

Amin R. Nehrir[1] · Christoph Kiemle[2] · Mathew D. Lebsock[3] · Gottfried Kirchengast[4] · Stefan A. Buehler[5] · Ulrich Löhnert[6] · Cong-Liang Liu[7] · Peter C. Hargrave[8] · Maria Barrera-Verdejo[9] · David M. Winker[1]

Received: 1 February 2017 / Accepted: 6 October 2017 / Published online: 21 November 2017

Abstract A deeper understanding of how clouds will respond to a warming climate is one of the outstanding challenges in climate science. Uncertainties in the response of clouds, and particularly shallow clouds, have been identified as the dominant source of the discrepancy in model estimates of equilibrium climate sensitivity. As the community gains a deeper understanding of the many processes involved, there is a growing appreciation of the critical role played by fluctuations in water vapor and the coupling of water vapor and atmospheric circulations. Reduction of uncertainties in cloud-climate feedbacks and convection initiation as well as improved understanding of processes governing these effects will result from profiling of water vapor in the lower troposphere with improved accuracy and vertical resolution compared to existing airborne and space-based measurements. This paper highlights new technologies and improved measurement approaches for measuring lower tropospheric water vapor and their expected added value to current observations. Those include differential absorption lidar and radar, microwave occultation between low-Earth orbiters, and hyperspectral microwave remote sensing. Each methodology is briefly

✉ Amin R. Nehrir
amin.r.nehrir@nasa.gov

1 NASA Langley Research Center, Hampton, VA 23681, USA

2 DLR, Institut für Physik der Atmosphäre, 82234 Oberpfaffenhofen, Germany

3 Jet Propulsion Laboratory, California Institute of Technology, Pasadena, CA 91109, USA

4 Wegener Center for Climate and Global Change (WEGC) and Institute for Geophysics, Astrophysics, and Meteorology/Inst. of Physics, University of Graz, Graz 8010, Austria

5 Center for Earth System Research and Sustainability (CEN), Meteorological Institute, Universität Hamburg, 20146 Hamburg, Germany

6 Institute for Geophysics and Meteorology, University of Cologne, Pohligstr. 3, 50969 Cologne, Germany

7 National Space Science Center (NSSC), Chinese Academy of Sciences, 100190 Beijing, China

8 School of Physics & Astronomy, Cardiff University, 5 The Parade, Cardiff CF24 3AA, UK

9 Forschungszentrum Jülich, Wilhelm-Johnen-Straße, 52428 Jülich, Germany

explained, and measurement capabilities as well as the current technological readiness for aircraft and satellite implementation are specified. Potential synergies between the technologies are discussed, actual examples hereof are given, and future perspectives are explored. Based on technical maturity and the foreseen near-mid-term development path of the various discussed measurement approaches, we find that improved measurements of water vapor throughout the troposphere would greatly benefit from the combination of differential absorption lidar focusing on the lower troposphere with passive remote sensors constraining the upper-tropospheric humidity.

Keywords Remote sensing · Water vapor profiles · Atmospheric science · Lidar · Differential absorption lidar · Radar · Differential absorption radar · Microwave occultation · Hyperspectral microwave · Emerging technology

1 Introduction

The dominant source of uncertainty in climate sensitivity estimates has been traced to model diversity in the response of clouds to climate change (Bony and Dufresne 2005; Vial et al. 2013). Narrowing this uncertainty and improving the confidence of climate projections represents one of the greatest challenges faced by the Earth science community, suggesting better constraints on the magnitude of cloud feedbacks are key. While we have had little success in reducing the range of estimated climate sensitivity, we have greatly improved our knowledge of the processes involved. Fundamental mechanisms behind longwave feedbacks due to deep tropical clouds are better understood (Hartmann and Larson 2002), although questions remain about the importance of mesoscale organizing processes (Wing et al. 2017; Wing and Emanuel 2014; Mauritsen and Stevens 2015; Mapes et al. 2017). Shallow clouds are mediated more by atmospheric circulation than by microphysical processes. Petters et al. (2013) performed large-eddy simulation (LES) studies on the sensitivity of overcast stratocumulus liquid water path (LWP) to the humidity of the air in the free troposphere just above cloud top, finding that radiative and evaporative cooling at the cloud top are highly sensitive to the moistness of the free troposphere above cloud top. Realistic variations in moisture were found to have effects on LWP similar to or larger than those expected from perturbations in cloud condensation nuclei. This parallels the known importance of humidity for the development of deeper convection (James and Markowski 2010).

Modeling across scales suggests that many aspects of circulation—the depth of the marine boundary layer (BL), the moisture gradient between the sea surface and the free troposphere, the rate of subsidence, etc.—will change in response to a warming climate. As the community gains a deeper understanding of the processes involved in regulating shallow clouds and their radiative effects, there is a growing appreciation of the critical role played by small fluctuations in water vapor in and above the marine boundary layer and the coupling of water vapor and atmospheric circulations.

In the tropics, water vapor has a maximum variability in the lower troposphere near 800 hPa (Holloway and Neelin 2009). Recent observations (Nuijens et al. 2009, 2017) and LES studies (Seifert et al. 2015) of trade cumulus regimes suggest that subtle variations in humidity in the lower troposphere play a dominant role in convective development and in buffering aerosol-cloud radiation interactions. Mixing processes acting to redistribute water vapor between the free troposphere and the marine boundary layer can impact the albedo and frequency of shallow marine clouds. Lower tropospheric mixing has been seen to scale strongly with feedbacks in models (Sherwood et al. 2014) but is poorly constrained by observations. Proxies for small- and large-scale mixing are given by the structure and depth of the moist layer over the tropical oceans. Better constraints on water vapor anomalies in the lower troposphere-requiring higher vertical ($\sim \leq 0.3$ km) and horizontal ($\sim \leq 10$ km) resolution and accuracy ($\leq 2\%$)—would improve understanding in all of these areas and constitute a tremendous advance (ISSI Workshop on Shallow clouds and water vapor, circulation and climate sensitivity).

In summary, progress on the role of atmospheric water and circulation requires improved understanding of underlying processes by better observations of water vapor, clouds, and winds, especially in the lower troposphere. A high vertical measurement resolution is needed because strong vertical moisture gradients at the top of the mixed layer and in the free troposphere aloft can strongly influence radiation and the development of both shallow and deep convection (Stevens et al. 2017). Generally, water vapor measurements are challenging, both because water vapor is highly variable in space and time on a large range of scales, and because the physics of the various measurement approaches used to date fundamentally limits each instrument's performance in different ways. The ideal hygrometer still does not exist, but interesting new technologies are emerging that show promise in overcoming limitations of current measurement approaches. It is therefore timely to explore their capabilities and synergies, and to discuss their advances with respect to the conventional measurement techniques.

In situ hygrometers, mainly deployed on land stations, buoys, ships, balloons or aircraft can be highly accurate but suffer from poor representativeness as they offer point or line measurements in space or time. They are very irregularly distributed around the globe: Their density is poor over remote regions and oceans, and vertical profiles are rare. Spectrometers on satellites provide global coverage yet are impeded by low measurement sensitivity in the lower troposphere due to blurring by the water vapor column aloft. Their complex retrieval is sensitive toward thin clouds or aerosol layers such that measurement biases may arise if clouds and aerosol remain undetected, or if they are accounted for at incorrect altitudes. Their data have coarse spatial resolution, particularly in the vertical, limb sounding being no option in the lower troposphere. Better spatial resolution, combined with the interesting possibility to trade-off vertical versus horizontal resolution versus measurement precision, in order to optimize the measurements to the needs of their users/communities, can be provided by active remote sensing such as differential absorption lidar (DIAL) and (in the future) radar (DAR). DIAL and DAR come at the cost of higher instrument complexity, and a certain loss of coverage since most active instruments have no scanning capability in order to keep complexity within reasonable limits. Clouds are obstacles for DIAL, yet the lidar sensors' field of view is sufficiently small that every cloud gap provides an opportunity for vertical profiling. An upcoming cost-efficient technique is radio/microwave (signal) occultation. Since the measurement geometry is similar to limb sounding, however, the benefit for profiling humidity in the planetary boundary layer is limited due to the inherent broad horizontal averaging kernels. The value of this new technique will grow with the number of exploitable occultation events. Lastly, new detector technologies open the possibility for hyperspectral microwave sensors that can support thousands of spectral channels with little to no sensitivity to cloud contamination. This paper highlights these new or improved technologies for measuring lower tropospheric water vapor presented at the ISSI Workshop on Shallow clouds and water vapor, circulation and climate sensitivity, and their expected added value to current observations. The potential synergies and added benefits of each measurement technique compared to existing observational methods are also discussed.

2 Differential Absorption Lidar

The differential absorption lidar (DIAL) technique was first used to measure lower tropospheric water vapor by Schotland (1966), just a few years after the discovery of the laser, and it has since been used to measure a number of trace gases in the atmosphere. So the technique is not new, yet currently benefits from significant technological advances. It utilizes atmospheric molecular (Rayleigh) and aerosol (Mie) backscattered return signals from a tunable single-frequency, pulsed laser to directly measure range resolved profiles of water vapor molecular number density N(R). Two spectrally close laser pulses (typical 10–100 ns pulse width) are transmitted to the atmosphere near simultaneously (0.2–1 ms separation) with one pulse tuned to the center or wing of a gas absorption line, called the online (λ_{on}), and the second pulse, called the off-line (λ_{off}), tuned to a less absorbing spectral location. The separation between the online and off-line wavelengths is typically < 1 nm to minimize biases resulting from differences in atmospheric backscatter and extinction coefficients. The retrieval of concentration profiles is achieved using the DIAL equation,

$$N(R) = \frac{1}{2\Delta R(\sigma_{\mathrm{on}} - \sigma_{\mathrm{off}})} \ln\left[\frac{P(\lambda_{\mathrm{on}}, R)P(\lambda_{\mathrm{off}}, R + \Delta R)}{P(\lambda_{\mathrm{off}}, R)P(\lambda_{\mathrm{on}}, R + \Delta R)}\right],$$

which exploits the differential attenuation of the backscattered return signals ($P_{\lambda\mathrm{on}}$, $P_{\lambda\mathrm{off}}$ (W)) between the online and off-line wavelengths along with a priori knowledge of the differential absorption cross section $\Delta\sigma$ (cm^2) over a range bin ΔR (m) (Schotland 1974).

Given the laser on–off pulse pair temporal and spectral separation are sufficiently small (≤ 1 ms and < 1 nm, respectively), it can be assumed that the atmospheric volume scattering and extinction coefficients are constant between the online and off-line atmospheric return signals. As a result, DIAL from aircraft and space have the potential for high accuracy measurements of trace gases, in particular, water vapor throughout the troposphere, as they are self-calibrating and not prone to bias resulting from aerosol and cloud contamination.

2.1 Measurement Capabilities

Water vapor DIAL has similar performance over land and ocean as it is insensitive to surface emissivity. Further, water vapor profiles can be retrieved during day and night, at all latitudes, and during all seasons. Multiple online wavelengths can be implemented to increase the dynamic range required for measurements in different climates as well as for full tropospheric profile measurements from the upper troposphere and lower stratosphere (UTLS) down to the tropical marine layer (Fig. 1). Generally, Lidar on aircraft can efficiently probe target regions of interest for the study of particular atmospheric processes in two or three (when using a scanner) spatial dimensions. When installed nadir-viewing, another advantage arises from the fact that air density, and partly also humidity and aerosol concentration, increases exponentially with range below the aircraft, thus partly compensating the decrease in backscatter signal intensity that is proportional to the range-squared. Airborne water vapor DIAL systems have been developed over the past 20 years with focus on tropospheric and UTLS measurements from local to synoptic scales for weather and climate studies (Higdon et al. 1994; Bruneau et al. 2001; Browell et al. 1997; Ehret et al. 1998, 1999; Poberaj et al. 2002; Wirth et al. 2009; Ferrare et al. 2004). Lidar technology developments have evolved at DLR, Germany, to enable advanced DIAL systems which operate in the strongly absorbing 935-nm water vapor band. These systems have been developed as airborne demonstrators for future satellite missions and have been deployed within many large-scale field experiments. They have demonstrated measurement capability in the UTLS region, as well as in cirrus clouds (Kiemle et al. 2008; Groß et al. 2014). Recently, airborne water vapor lidar observations focused on tropical shallow convective environments have been reported (Kiemle et al. 2017), while earlier work has demonstrated the value of airborne lidar to characterize the variability of humidity and of latent heat fluxes within the convective boundary layer over land (Kiemle et al. 2011).

Developed at NASA as an initial step toward a space-based water vapor DIAL, the airborne Lidar Atmospheric Sensing Experiment (LASE) has also demonstrated that profile measurements of water vapor can be made to an accuracy of better than 6% or 0.01 g/kg (whichever is larger) throughout the troposphere with 330 m and 14 km vertical and horizontal resolution, respectively (Browell et al. 1997; Ferrare et al. 2004). Figure 2 shows recent measurements from LASE, which has now been in operation for nearly two decades. These results are over the central Great Plains during the Plains Elevated Convection At Night (PECAN) mission and illustrate the high vertical resolution achievable and high accuracy relative to radiosonde profiles. Use of online/off-line surface reflection signals provides column water vapor and also for extending the water vapor profiles to within 100 meters above the surface, although not implemented in the figures shown here. Water vapor DIAL has the potential to provide high accuracy and high vertical resolution profiles from space, particularly near the surface where retrievals from passive and active microwave sensors struggle the most. Future prospects for Earth observing systems focused on atmospheric dynamics will greatly benefit from complementary measurements between active DIAL sensors which have greatest sensitivity to the near the surface

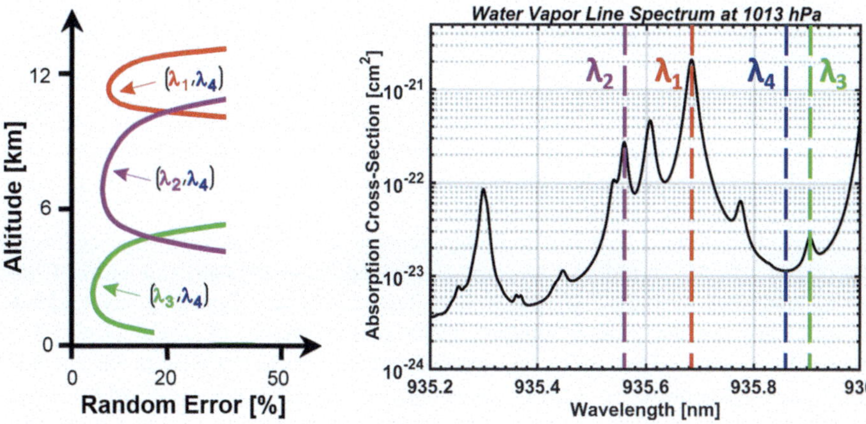

Fig. 1 A space-based DIAL transmitter will require transmission of two or more online wavelengths to adequately measure the large water vapor dynamic range from the upper troposphere down to the near-surface atmosphere. The different online wavelengths experience different water vapor optical thicknesses, resulting in different measurement sensitivities. For WALES and HALO (see Sect. 2.2), three combinations of signals using three wavelength pairs result in a composite water vapor profile throughout the troposphere, with expected relative random error profiles depicted in the left panel (ESA 2001). Hard target returns from the surface can be used to measure the full column at each wavelength using the integrated path DIAL approach (IPDA). Water vapor measurements from the lowest retrievable range bin can be extended down to the surface using IPDA. Measurements/missions focused on the mid-lower troposphere would benefit from the reduced number of required online wavelengths as well as from the added flexibility in the transmitter design by having additional access to weaker absorption lines in the 820 and 720 nm spectral bands

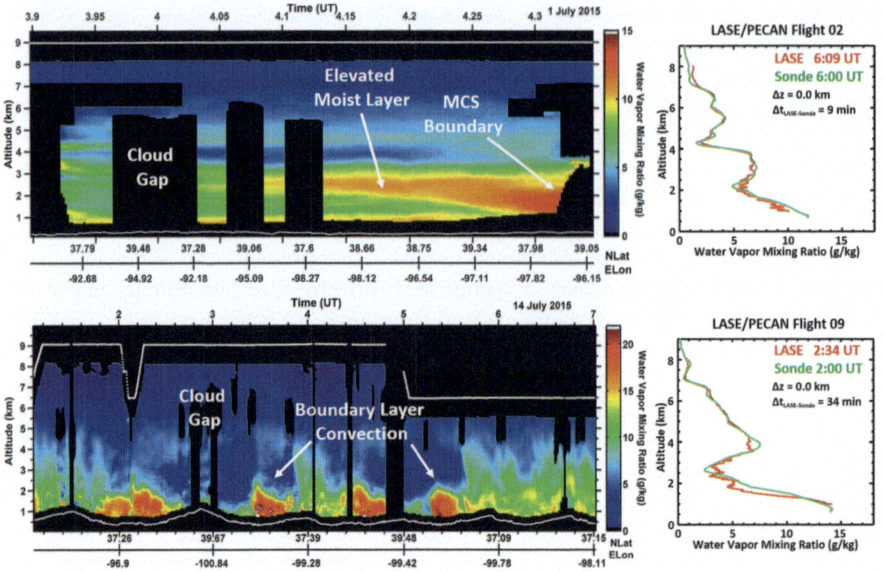

Fig. 2 Examples of water vapor profiles measured by LASE in advance of two separate mesoscale convective systems (MCS) during the 2015 PECAN mission over the US central Great Plains. The LASE time series data (left panels) reveal strong spatial variability ahead of high impact weather systems that are not resolved by passive systems. Overpasses of PECAN ground sites by the NASA DC-8 aircraft show good agreement between LASE and coincident radiosonde measurements (right panels). The sonde comparison shown for PECAN flight 02 occured outside of the corresponding time series chart to the left. The white lines at the top and bottom of the time series charts represent the aircraft flight path and surface height, respectively. The blacked out sections in the time series charts lack data due to lidar beam extinction by cloud. Δz and Δt represent the spatial and temporal separation between the sond launch and aircraft overpass time and location, respectively

atmosphere, and passive/active microwave and occultation approaches (discussed below) whose averaging kernels are weighted to the mid-upper troposphere. The combination of active and passive retrievals has the potential to result in high-resolution water vapor profiles from the surface to the stratosphere.

Lidar is characterized by small footprints relative to radars and radiometers (on the order of 100 m) and can profile above low clouds and below optically thin cirrus. Cloud masking and shot averaging allows probing between broken boundary layer clouds. Opaque clouds are the primary limiting factor to the number of useful observations retrieved from space-based active and passive remote sensors. Similar to the methods carried out by Kiemle et al. (2015), the number of useful water vapor DIAL profile measurements that extend down to the surface can be determined using the cloud fraction measurements from the Cloud-Aerosol Lidar and Infrared Pathfinder Satellite Observations (CALIPSO) mission. CALIPSO level-2 version 3-01 cloud layer products are aggregated into a 278×278 km^2 ($2.5° \times 2.5°$) grid where a cloud optical depth (OD) threshold of 1 (86.5% two-way attenuation) is applied to the grid as a means to differentiate between clear and cloudy (cloud OD > 1) conditions. Figure 3 shows the global map of CALIPSO-derived cloud-free fraction averaged over 12 consecutive months in 2007. The resulting cloud-free fraction map shows most favorable conditions along the northern and southern subtropics, a geographic area where water vapor has a maximum variability in the lower troposphere (Holloway and Neelin 2009) and subtle variations in lower tropospheric

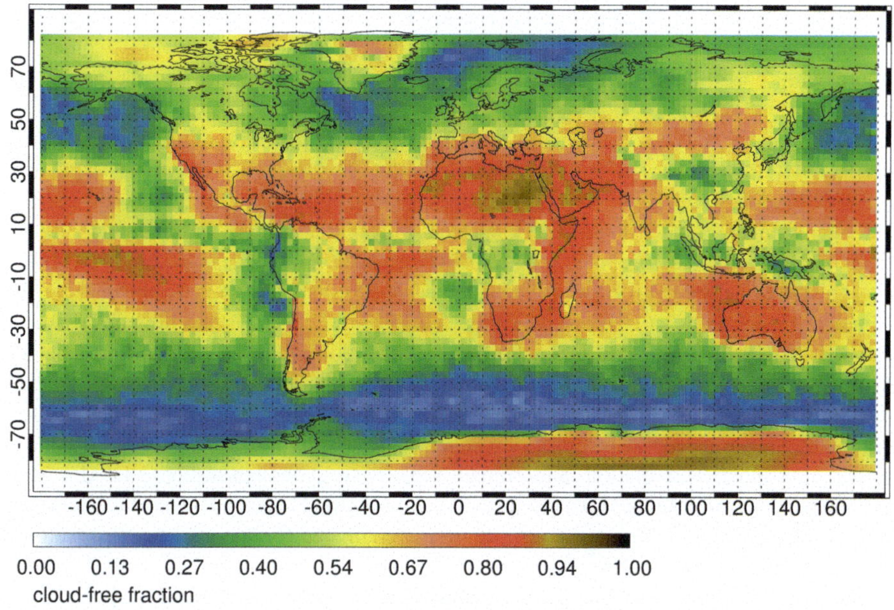

Fig. 3 Global map of annual mean CALIPSO-derived cloud-free fraction. A CALIPSO column is defined to be cloud-free whenever its total measured cloud optical depth is less than 1. The cloud statistics were compiled from data collected in 2007. In 2007, the annual global cloud-free fraction was 53.5%, suggesting that the retrieved water vapor profiles of a space-based DIAL could extend down to the surface for over half of the total retrieved profiles and thereby substantially add value to the next generation of atmospheric remote sensors targeting tropospheric dynamics

humidity play a dominant role in convective development and in buffering aerosol–cloud–radiation interactions. DIAL measurements are impacted by optically thin clouds with OD < 1; however, the signals are further attenuated, not fully obstructed. In consequence, more horizontal and/or vertical averaging can provide profiles with accuracy and precision similar to cloud-free profiles. The coarse cloud-mask blurs many of the small cloud-free regions that appear in higher-resolution space lidar analyses (see, e.g., Kiemle et al. 2015, Fig. 3 for CALIPSO cloud stats). The likely reduction in average OD using this blurred cloud mask is not quantified here, however, the effect is less detrimental to instruments with small field of view such as lidar by enabling profiling within small cloud gaps, and is expected to be most pronounced in regions with (deep) convective clouds and fair weather cumulus that have sharp edges, particularly in the tropics.

Simulations based on the experience gained with airborne demonstrators at NASA and DLR have demonstrated the potential for high-resolution water vapor DIAL profiling from a LEO satellite throughout the troposphere (Ismail and Browell 1989; Di Girolamo et al. 2008). The Water Vapour Lidar Experiment in Space (WALES) is a European Space Agency Earth Explorer mission proposed in the early 2000s targeting profile measurements of tropospheric water vapor using the DIAL technique (ESA 2001). Unfortunately, WALES was not selected for further development due to cost and technical risk associated with the laser under consideration; however, several feasibility studies were carried out and demonstrated the utility of such measurements for weather and climate applications. A concept study by Gerard et al. (2004) assessed the performance of a space-based water vapor DIAL relative to passive infrared measurements such as from IASI (Infrared

Atmospheric Sounding Interferometer). The results of the study indicated that for a fixed error threshold, the WALES water vapor DIAL concept under study would extend water vapor measurements to higher altitudes than for IASI, with improved vertical resolution near the surface (Gerard et al. 2004). Di Girolamo et al. (2008) also carried out an end-to-end simulation of a WALES like lidar to assess the systematic and random error performance under a wide range of atmospheric conditions. The results confirmed the capability of a space-based water vapor DIAL of accurately measuring the water vapor and aerosol structure from the UTLS down to the surface with better than ± 5% bias and ± 20% random error. The measurement requirements resulting from the space-borne WALES mission study (Gerard et al. 2004; DiGirolamo et al. 2008) are shown in Table 1. Although the concept study was carried out over a decade ago, the general sampling and accuracy requirements for a space-based DIAL for the application of improving Numerical Weather Prediction is still valid. The third column represents target measurement requirements (resulting from the 2016 Shallow clouds and water vapor, circulation and climate sensitivity Workshop) needed to advance our understanding of the interplay between shallow clouds, atmospheric circulation, and climate sensitivity. Although airborne DIALs have demonstrated the high spatial resolutions called out for cloud/climate studies, the improved vertical and spatial sampling requirements pose a challenge for space-based observations with current technologies.

The high accuracy and precision capability of a space-based water vapor DIAL, WALES, is further demonstrated in a model-simulation comparison shown in Fig. 4. A single global latitude slice from a Numerical Weather Prediction (NWP) model run at 122 W longitude on May 24, 2001, is used to define the water vapor field that is ingested into the WALES lidar simulator. Thin and thick clouds in the NWP model are depicted by the brown and black feature masks, respectively. These feasibility studies, including the model-simulation comparison in Fig. 4, demonstrate the ability of the DIAL technique to measure fine-scale water vapor features from the upper troposphere down to the near-

Table 1 Space-borne water vapor DIAL observational requirements adapted from the WALES lidar mission concept (Gerard et al. 2004)

Requirement		Threshold			Baseline			Cloud Climate Sensitivity Studies (ISSI Workshop)
		LT	MT	UTLS	LT	MT	UTLS	LT
Vertical resolution	(km)	1.0	1.0	2.0	0.5	1.0	2.0	0.3
Horizontal resolution	(km)	100	150	200	10	50	100	≤ 10
Spatial coverage		Global						Tropics
Timeliness/Revisit time		<3 hours (operational meteorology), < Monthly revisit (e.g. CALIPSO/ CloudSat nadir sampling)						
Dynamic range	(g/kg)	0.01-15			0.001-25			1-25
Random error (1σ)	(%)	20			5			3
Systematic error	(%)	< 5			< 2			< 2

Measurement requirements for cloud and climate sensitivity studies result from the 2016 Shallow clouds and water vapor, circulation and climate sensitivity workshop (Stevens et al. 2017; Bony et al. this issue). Dynamic range requirements for DIAL missions targeting lower tropospheric measurements can be relaxed to cover those ranges typically observed from the mid-troposphere (< 10 km) down to the surface. Narrowing the focus of a mission to measurements in the mid-lower troposphere opens the possibilities of new laser technologies with reduced complexity and hence risk. Lower troposphere (LT) ranges from ~ 0 to 5 km, mid-troposphere (MT) ranges from ~ 5 to 10 km, and upper troposphere/lower stratosphere (UTLS) ranges from ~ 10 to 17 km

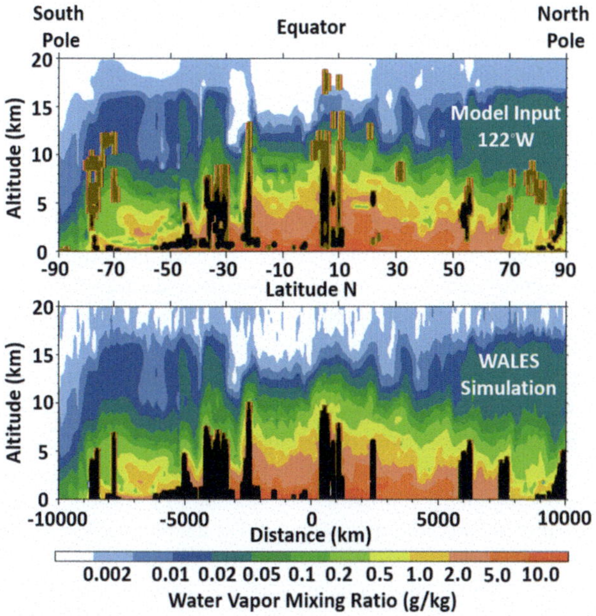

Fig. 4 Pole-to-pole water vapor cross section along 122°W (mainly over the Pacific Ocean and Canada) representing the "true" atmospheric state for input into the WALES simulation model (top), and output of this model, i.e., what this space lidar would be able to observe (bottom). Information below thick clouds is lost (black), yet the tropospheric water vapor distribution and variability is very well reproduced. In total, 83% of the simulated cloud-free observations below 15 km agree to within 20% with the model reference

surface atmosphere with high precision and low bias in clear and cloudy skies during both day and night conditions. These capabilities have been further realized via airborne demonstrators (airborne WALES and LASE) that match and often exceed the WALES space mission requirements (as well as those required for shallow cloud studies) on precision and accuracy, although over much shorter horizontal averaging scales. The requirements for space-based observations of the tropical marine boundary layer (with the resolution required to improve our understanding of cloud-climate feedbacks) are challenging and would require substantial technological investments in the near term. However, airborne DIAL measurements with capabilities that surpass the aforementioned measurement requirements can be used in the interim to conduct process oriented field studies that supplement space-borne observations that lack the required spatial resolution/sampling, but offer global coverage over long time series.

2.2 Technology Readiness

NASA and DLR have fielded airborne water vapor DIAL systems for targeted process studies over the last two decades and have successfully demonstrated key elements required for accurately measuring water vapor profiles from airborne platforms. While laser energy and spectral purity requirements for airborne and space-based DIAL measurements are fulfilled by LASE and WALES, the electrical efficiencies for both systems are unsuitable for scaling to space. Advances in laser and lidar technologies leveraged from previous successful missions such as CALIPSO as well as ongoing developments such as

EarthCare, ADM-Aeoulus, ICESAT-II, MERLIN, and CALIPSO follow-on mission concepts (Albert et al. 2015) have enabled a new class of airborne water vapor DIALs that serve as prototypes for future satellite missions. NASA is currently developing a new airborne DIAL to fly on a variety of mid-to-high altitude platforms for profile measurements of water vapor and aerosols and clouds from the surface to the lower stratosphere. This development will serve as a prototype for future space-based water vapor DIAL missions. Recent investments in airborne DIAL have led to reduction in size, weight, and power budgets commensurate with those required for future space-based water vapor DIAL missions. Further technology investments are required, however, in the areas of high-power lasers, optical receivers, and detectors to scale technology used in current airborne systems to be suitable for satellite DIAL missions with reasonable cost and reduced risks of failure.

The LASE and WALES DIAL systems operate at 820 and 935 nm, respectively. The 935 nm spectral band was chosen for WALES to allow for full tropospheric profile measurements from the UTLS down to the surface (Wirth et al. 2009). The 820 nm spectral band was chosen for the LASE system to increase sensitivity to the near-surface atmosphere and utilize mature Ti/Sapphire laser technology (Moore et al. 1997; Browell et al. 1997, 1998). Furthermore, this spectral band was the only region accessible by means of direct laser generation at the time, opposed to WALES which was developed in the mid-2000s and utilizes parametric conversion to generate coherent laser light in the 935 nm spectral band. The newest DIAL system under development at NASA is the High Altitude Lidar Observatory (HALO), planned for aircraft deployment on the NASA King Air platform in 2018. HALO is configured to fly on the NASA ER-2 aircraft, which will provide a viewing geometry able to simulate the performance of an eventual satellite instrument, but will also be capable of flying on smaller aircraft for targeted process studies. HALO is a multi-function lidar and is designed as a technology test bed for future space-based trace gas DIAL missions. To that end, HALO will have the capability to operate at the 935, 820, and 720 nm water vapor bands to allow for risk reduction of key water vapor DIAL technologies (namely the optical receiver filters, detectors, and transmitter technologies) required for future satellite missions. HALO is much more compact and electrically efficient compared to its predecessor LASE system and employs a more suitable set of technologies for scaling to space.

WALES operates at 935 nm, providing coverage throughout the troposphere. For a mission focused on the lower troposphere (below 8–10 km), the 820 or 720 nm spectral bands have advantages. Increased Rayleigh scattering at these shorter wavelengths increases the lidar signal and allows for use of more sensitive detectors with lower noise characteristics. Furthermore, direct generation of laser radiation at these wavelengths based on new laser materials is emerging and would significantly reduce the complexity and improve the efficiency of a laser transmitter required for space-based water vapor DIAL. A mission focused on lower tropospheric measurements would also reduce the number of required online wavelengths, significantly reducing the complexity and cost of a space-based DIAL transmitter, while at the same time increasing the signal to noise ratio due to an effective increase in DIAL wavelength-pair pulse repetition rate.

Targeted investments are needed in the area of single-frequency lasers to increase the technology readiness level (TRL) for space. The WALES airborne demonstrator has achieved the laser pulse energy and spectral purity required for space-based water vapor DIAL measurements; however, the size, weight, and power (SWAP) figures are not scalable to a space-based implementation. HALO laser pulse energies, while sufficient for aircraft applications need to be increased by a factor of 10 to be suitable for space, while maintaining reasonably high electrical wall plug efficiency (> 5–8%). Leveraging emerging technologies in single photon counting detectors could, however, reduce the requirement on laser pulse energy to those currently achieved with the HALO 935 nm laser, which exhibits the SWAP required for a space-based implementation.

Transitioning to a low energy and high pulse repetition rate laser architecture, however, requires the use of narrow field of view receivers and extremely narrow-band and frequency agile optical filters. Technology advancements in narrow-band (1–20 pm full width at half maximum) rapidly tunable (frequency agile) etalons are also required. The solar background noise in DIAL systems is proportional to the spectral width of the band-pass filter in the receiver. Online/off-line wavelength pairs used to sample the targeted water vapor absorption line(s) are typically separated by a few 100 pm (Fig. 1). Rather than accommodating both wavelengths within the receiver band-pass filter, narrow-band frequency agile etalons are needed which can be tuned between the online and off-line wavelengths on a shot-by-shot basis. The spectral width of the etalons can then be as low as 1–20 pm, and if successfully implemented has the potential to significantly increase the daytime measurement signal to noise ratio over high albedo scenes, allowing for boundary layer water vapor measurements with better than 5–10% precision using existing and higher TRL laser architectures. The transmitter and receiver subsystem TRLs using technologies developed for WALES, LASE, and HALO (as well as applicable technologies developed under existing space-programs) are at 4–5 and 5–6, respectively. TRL 4 is component/subsystem validation in a laboratory environment, TRL 6 is system/subsystem model or prototyping demonstration in a relevant end-to-end environment. The aforementioned systems and receiver and transmitter technologies could be matured to be space-flight ready within the next 5–7 years by targeted investments.

3 Differential Absorption Radar

Differential Absorption Radar (DAR) is the radar analogue of DIAL. The method exploits the difference in gaseous attenuation between two or more channels closely spaced in frequency. This differential attenuation can be related to the gas content of the atmosphere. A key difference between the DAR and DIAL is in the transmitted frequency and related atmospheric scattering properties. DIAL uses clear-sky molecular and aerosol scattering, whereas DAR uses cloud and precipitation scattering targets, thereby enabling sounding capabilities within clouds, highly complementary to DIAL.

In a general sense the DAR technique can be used to sense any species with an absorption feature in the microwave spectrum. For example, prior studies have proposed to observe surface pressure using channels in the 60 GHz oxygen absorption complex scattered off of the Earth surface (Lin and Hu 2005; Millán et al. 2014) and some initial observations have been made with an airborne prototype (Lawrence et al. 2011). Use of the 22 GHz absorption line has been proposed to observe water vapor within precipitation echoes (Meneghini et al. 2005). Additionally, exploiting the differences in absorption by water vapor in the continuum has been proposed to retrieve water vapor (Ellis and Vivekanandan 2010; Tian et al. 2007).

The application of DAR proposed here utilizes two radar channels in the G-Band adjacent to the 183 GHz water vapor absorption line (Millan et al. 2016; Lebsock et al. 2015). At range r, the ratio of the radar reflectivity (Z) at two close frequencies (ν) near the

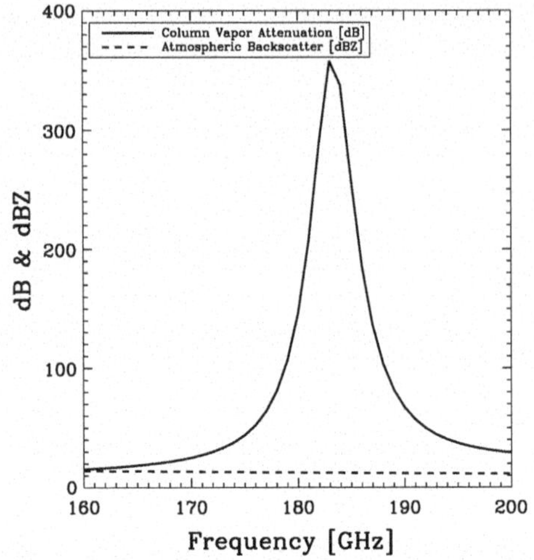

Fig. 5 Example of the spectral variation of the column-integrated attenuation due to water vapor and the backscatter due to clouds and precipitation for a trade wind boundary layer. Droplet backscatter variations are essentially negligible relative to water attenuation across the absorption line. Data taken from the simulations of Lebsock et al. (2015)

absorption line is approximately proportional to the water vapor path (u_{h_2o}) between the transceiver and the scattering volume,

$$\frac{Z(\upsilon_1, r)}{Z(\upsilon_2, r)} \propto u_{\mathrm{h_2o}}.$$

Here, it has been assumed that variation of extinction properties of the cloud with frequency is much smaller than the variation of the attenuation due to water vapor. In practice, there are small variations in the cloud extinction properties, which can be accounted for. To demonstrate the underlying physics, Fig. 5 shows the spectral variation of the atmospheric attenuation and particle backscatter across the 183 GHz line for a case from the cloud simulations in Lebsock et al. (2015). The column-integrated water vapor attenuation varies by several hundred dB while the column-integrated backscatter varies in a quasi-predictable manner by only 2.5 dBZ.

To provide profiling capabilities a DAR instrument might be implemented as either a range-gated or Frequency Modulated Continuous Wave (FMCW) radar. Water vapor profiles can be obtained by differencing the retrieved partial water vapor path between adjacent range bins and dividing by the range resolution, similar to DIAL retrieval of water vapor concentrations. As a result, the uncertainty in the derived water vapor profiles is an inverse function of the radar range resolution.

In addition to in-cloud profiling capabilities, a downward-looking DAR could provide spatially continuous observations of the column-integrated water vapor (CWV) from the surface reflection, which should be available over all surface types and nearly all atmospheric conditions, with the notable exception of heavy precipitation, which attenuates the signal too strongly.

3.1 Measurement Capabilities

The primary advantage of a water vapor DAR is that it specifically profiles water vapor within clouds. This is a unique measurement capability that cannot be obtained by existing technologies. For example, microwave sounders are sensitive to water vapor in cirrus clouds; however, they have broad weighting functions that encompass both cloud and surrounding clear atmosphere. The observations can further be contaminated by the presence of other clouds. Global Positioning System Radio Occultation (GPS-RO, here below) is also sensitive to water vapor within clouds; however, RO has extremely broad horizontal weighting functions, suffers non-uniqueness near the Earth surface, and has infrequent sampling. Methods that rely on infrared, near-infrared, and visible wavelengths do not penetrate water clouds at all.

Depending on channel selection, a downward-looking water vapor DAR can feasibly sound either high altitude cirrus clouds or low altitude boundary layer clouds. Channels near the absorption line center are required to sound high altitude clouds, whereas channels that attenuate less strongly in the wings of the absorption line are required to sound low altitude clouds. Sampling by a two-channel DAR will be limited by channel selection. The vertical sampling of the CloudSat Cloud Profiling Radar (CPR) may be considered as a best-case sampling capability provided by a DAR within the near future, including the capability to sample Earth's ubiquitous multi-layered cloud systems. It must be noted that a DAR matching the vertical sampling capability of CloudSat would require a high-power multi-frequency system. More likely, the first DAR will be targeted to a particular altitude range through channel selection.

Initial studies have documented expected uncertainty of a space-borne DAR using instrument simulators. Lebsock et al. (2015) specifically examine the potential for sounding water vapor within boundary layer clouds. They report an expected precision in CWV of 0.5–2 kgm^{-2} ($\sim$ 2–8% in the subtropics) with biases not exceeding 0.25 kgm^{-2}. They find that profiling uncertainties are scene-dependent, ranging between 1 and 3 gm^{-3} at a range resolution of 500 m. Millan et al. (2016) apply an instrument simulator to global observations from CloudSat. They report precision in CWV of 0.3 kgm^{-2} with biases rarely exceeding 2.6 kgm^{-2} (10% of the global mean CWV). These larger biases are only found in heavily precipitating scenarios. They report an average single-bin precision of 89% with biases generally lower than 38% above 3 km and 77% near the surface.

Table 2 provides proposed observational requirements for a space-borne DAR. The DAR provides both profiling and CWV capabilities. The premise for the profiling requirements is that a DAR provides in-cloud profiling with the precision and resolution of a state-of-the-art infrared sounder, thereby extending water vapor profiling to all cloud scenarios. The requirements for the CWV observation are based on the current precision of passive microwave observations at very low spatial resolution over water surfaces only. DAR could advance the science possible with CWV observations by substantially increasing the spatial resolution (due to the high frequency employed) and extend sampling capability to all surface types thereby enabling quantification of the spatial variance of water vapor on small scales.

3.2 Technology Readiness

DAR is a newly emerging technology. There have only been initial observations of surface pressure using a 60 GHz DAR (Lawrence et al. 2011). A prototype water vapor DAR

Table 2 Proposed space-borne water vapor DAR observational requirements for profiling and column-integrated water vapor (CWV) measurements

Requirement		Profile Below 100 hPa level	Column Integrated
Vertical resolution	(km)	2.0	N/A
Horizontal resolution	(km)	40	2
Spatial coverage		Global	
Temporal coverage		< Monthly revisit (e.g. CloudSat/CALIPSO nadir sampling)	
Dynamic range	(g/kg or kg/m^2)	0.1-20	1-80
Random error (1σ)	(% or kg/m^2)	20 %	0.5 kg/m^2
Systematic error	(% or kg/m^2)	25 %	1.0 kg/m^2

Requirements for profiling mirror those of the Atmospheric InfraRed Sounder (AIRS; Auman et al. 2003), which samples the clear-sky water vapor, while those for precision of the CWV follow from the advanced microwave scanning radiometer (AMSR) for Earth observing systems. Expected precision of a DAR should exceed this required precision, possibly achieving 0.1 kg/m^2 (i.e., Millan et al. 2016). The profiling capability would thus extend the clear-sky capability of state of the art infrared sounding into cloudy atmospheres. The column-integrated observations would provide an order of magnitude increase in spatial resolution over passive microwave and add capability over all surface types. Systematic error estimates are derived from the study of Millan et al. (2016)

transceiver is currently being built at the Jet Propulsion Laboratory under the NASA Earth Science Technology Office's Advanced Component Technology (ACT) and Instrument Incubator Program (IIP). Due to restrictions on radio-frequency transmission surrounding the 183 GHz line, the prototype instrument is being developed for the 167–174.8 GHz band, which is currently reserved for but unused for communications systems. This tunable, all-solid-state G-Band transceiver leverages recent innovations in extremely high frequency radar system architecture, Schottky diode frequency-multiplier power handling, and III-V semiconductor amplifiers. The transceiver's TRL began at 2 (technology concept) and is currently at 4 (component/subsystem validation in laboratory environment) as of 2017. For space-flight, TRL-6 (system/subsystem model or prototyping demonstration in a relevant end-to-end environment) must be achieved. Additional investments in system design and integration will be required to increase the TRL of the G-Band DAR to 6 for both airborne and space-borne applications. In particular, maximum transmit power is currently in the 0.5 W range which is suitable for close range aircraft observations but must rise by 2 orders of magnitude to enable a satellite DAR. Such increases in transmit power are not manifestly infeasible in the near future using vacuum electronic power sources like those employed by current space-borne radars (e.g., CloudSat).

4 Microwave Occultation

Based on the established and meanwhile widely successful technique of Global Navigation Satellite System (GNSS) radio occultation (RO), using the refraction of decimeter wave L-band signals near 1.2 and 1.6 GHz received at low-Earth orbit (LEO) satellites, temperature and humidity profiles in the troposphere can be retrieved only by cousing a priori temperature and/or humidity profile information (see Pincus et al. 2017, for a brief RO introduction in this special issue; for detailed review see, e.g., Kursinki et al. 1997; Anthes 2011; Steiner et al. 2011).

As an emerging technique advanced from RO, LEO–LEO microwave occultation LMO), using centimeter and millimeter wave signals between LEO transmitter and

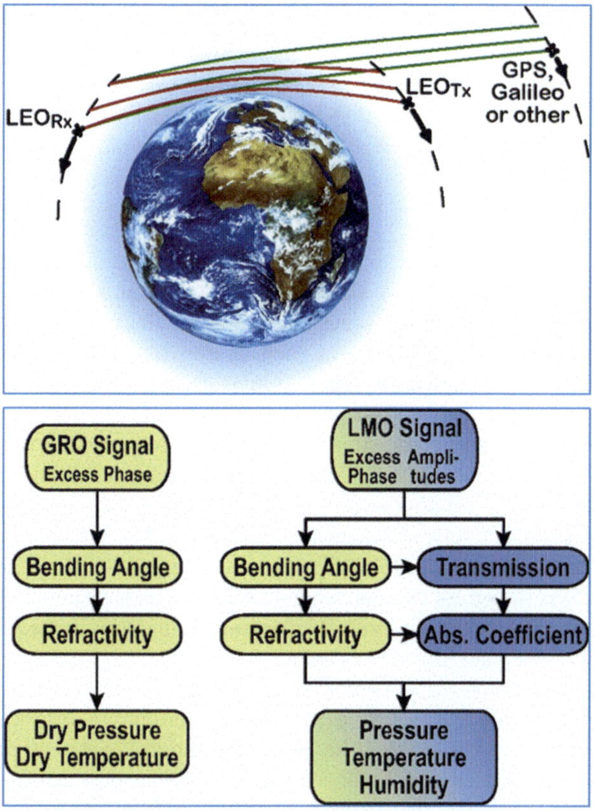

Fig. 6 Schematic view of the LMO (red signal paths) and GRO (GNSS RO, with GPS and Galileo systems named as example; green signal paths) occultation event geometry (top panel), and schematic view of the retrieval processing chain for LMO (right) and GRO (left) data (bottom panel). *Source* Schweitzer et al. (2011)

receiver satellites, exploits both refraction and absorption (water vapor) of signals in the X/K band within 8–23 GHz to overcome RO's temperature-humidity ambiguity in the troposphere. Figure 6 illustrates the LMO measurement technique compared to RO (for detailed LMO introduction see Kursinski et al. 2002, 2009; Kirchengast and Hoeg 2004; Schweitzer et al. 2011; for a recent review see Liu et al. 2017).

The LMO technique thus enables to retrieve pressure, temperature, and humidity profiles without a priori background information. It can also provide accurate profiling of these key thermodynamic state variables higher up over the full stratosphere, including water vapor if using also absorption signals near the 183 GHz water vapor absorption line, due to the insensitivity of the higher signal frequencies of LMO to ionospheric influences. Furthermore, stratospheric ozone profiles can be retrieved if additional absorption signals near the 195 GHz ozone absorption line are used. Liquid water and ice cloud properties as well as turbulence strength can be retrieved as by-products.

Since the late 1990s a series of LMO-related proposals and associated development work have been pursued, mainly in USA and Europe, which is reviewed by Liu et al. (2017). The current USA-led and European-led LMO mission concepts pursued toward implementation are ATOMMS—Active Temperature, Ozone, and Moisture Microwave

Spectrometer (Kursinski et al. 2009, 2012) and the LMO component of ACCURATE—Atmospheric Climate and Chemistry in the UTLS Region And climate Trends Explorer (Kirchengast and Schweitzer 2011; Schweitzer et al. 2011). The unique potential of LMO strongly suggests that this ongoing work toward demonstration missions should lead to launches of such first missions in the 2020–2022 timeframe.

4.1 Measurement Approach

The LMO measurement approach exploits the refraction and absorption of centimeter wave and optionally millimeter wave signals (within X/K band 8–23 GHz and optionally within 178–196 GHz) by performing accurate geodetic tracking measurements of the complex signal (phase and amplitude) at each chosen frequency during LEO-to-LEO occultation events, as illustrated in Fig. 6 (top panel) compared to GNSS RO. The atmospheric excess phase path due to refraction, relative to straight-line phase path in vacuum, and the amplitude attenuation due to water vapor (and optionally ozone) absorption, are derived from the raw tracking data of phase and amplitude and are the basis for subsequent retrieval of bending angle/refractivity profiles and transmission/absorption coefficient profiles, respectively, as schematically shown in Fig. 6 (lower panel) compared to GNSS RO. The well proven GNSS RO approach (Kursinski et al. 1997; Anthes 2011; Pincus et al. 2017) is shown for comparison with facilitate understanding that LMO works very similar in terms of geometry (with the transmitters in LEO instead of GNSS orbits near 20,000 km altitude) and fully shares the refractometric part of measurement principle.

The core algorithm steps for RO are: (1) complementing the "refractive" Abel transform for refractivity profile retrieval from bending angle profiles (Kursinski et al. 1997), an "absorptive" Abel transform yields differential absorption coefficient profiles retrieval from differential transmission (amplitude attenuation) profiles obtained from transmission differences between adjacent signal frequencies to eliminate broadband effects (Kursinski et al. 2002); (2) The joint availability of refractivity and differential absorption coefficient profiles then allows to unambiguously retrieve pressure, temperature, and humidity profiles by an optimal estimation scheme (Schweitzer et al. 2011). A detailed description of the LMO measurement approach and detailed algorithm descriptions of the steps in the retrieval processing chain are provided by Kursinsiki et al. (2002), Gorbunov and Kirchengast (2007), and Schweitzer et al. (2011).

4.2 Measurement Capabilities

In order to summarize the mission performance and information provided by LMO, and in this way concisely indicating LMO's capabilities and limitations, Table 3 summarizes observational requirements of a typical LMO mission concept (ATOMMS) based on the review by Liu et al. (2017). The measurement requirements are derived from a combination of flow down requirements needed to advance global climate monitoring and numerical weather prediction (Liu et al. 2017) as well from physical measurement limitations. Achieving this required performance for temperature and humidity profiling by suitable LMO instrumentation will at the same time ensure adequate performance also for the other variables such as pressure and other by-products. Four to five signals at X/K band frequencies, and optionally two to five signals within 178–196 GHz (max. three if without ozone capability), ensure the needed information content is provided for the retrieval.

Regarding the density of coverage in space and time with occultation events, the number of transmitter (Tx) and receiver (Rx) satellites is key. Generally designing Tx and

Table 3 LMO observational requirements (based on ATOMMS requirements table of Liu et al. 2017)

Requirement			Temperature		Humidity	
			Target	Threshold	Target	Threshold
Horizontal domain			Global			
Horizontal resolution		(km)	500	1000	500	1000
Time sampling		(hrs)	12	24	12	24
No. of profiles per grid box per month[a]			40	30	40	30
Vertical domain		(km)	1-80	3-50	1-50	3-18
Vertical sampling	LT	(km)	0.25	0.5	0.25	0.5
	UT	(km)	0.5	1	0.5	1
	LS	(km)	0.5	1	0.5	1
	US	(km)	1	2	1	2
RMS accuracy[b]	LT	(K or %RH)	1	2	5	10
	UT	(K or %RH)	0.5	1	10	20
	LS	(K or %RH)	0.5	1	10	20
	US	(K or %RH)	1	2	20	-
Long term stability			0.1	0.15	2	3
(per decade)			(K/decade)		(%RH/decade)	
Timeliness	Climate	(days)	7	14	7	14
	NWP	(hrs)	1.5	3	1.5	3
Mission lifetime		(years)	> 3			

[a]No. of profiles to be fulfilled in global average by all grid boxes but also any individual grid box shall receive at least 80% of this number. Grid box is here defined as square of the horizontal sampling requirement [box of size Horiz. sampling (km) × Horiz. sampling (km)] or any box of equivalent size with at least 500 km length of its smaller dimension

[b]Understood to be the accuracy for an individual occultation event over the required vertical domain (in the LT from Top of BL upward) at a vertical resolution consistent with the required sampling [i.e., a resolution of 2 × vertical sampling (km)]

Rx satellites into counter-rotating orbits for optimal event geometry (Fig. 6), initial demonstration missions typically aim at small 2 Tx and 2 Rx constellations (Kirchengast and Hoeg 2004; Kirchengast and Schweitzer 2011), leading to about 240 events per day, comparable to the initial GNSS RO demonstration mission CHAMP, Challenging Mini-Satellite Payload (Wickert et al. 2001). Subsequent operational LMO missions will intend to provide a much higher number, well meeting the sampling requirements. For example, a 6 Tx and 6 Rx constellation can provide about 3800 events per day (Kirchengast and Schweitzer 2011), more than the successful COSMIC six-satellite RO constellation (Anthes et al. 2008).

A range of feasibility studies and technology projects for LMO instruments, and on-ground demonstration experiments, have evidenced that also the required accuracy and stability is achievable (ESA 2004; Kursinski et al. 2012; Liu et al. 2017, and references therein). Figure 7 illustrates example performance results for pressure, temperature, and humidity profiling from quasi-realistic end-to-end simulations, accounting for the effects of instrumental errors, clouds, and scintillation noise from turbulence (Schweitzer et al. 2011; Kursinski et al. 2009). The profiles are generally found unbiased and within target accuracy requirements. The main limitation regarding lower troposphere water vapor profiling is degraded accuracy and poor spatial resolution in the atmospheric boundary layer similar to RO. The strength of the technique is in the free troposphere down to the top of the boundary layer, which itself can be detected with high accuracy (Anthes 2011; Ho et al. 2015) and allows other active and passive remote measurements to optimize sensitivity for the lower troposphere.

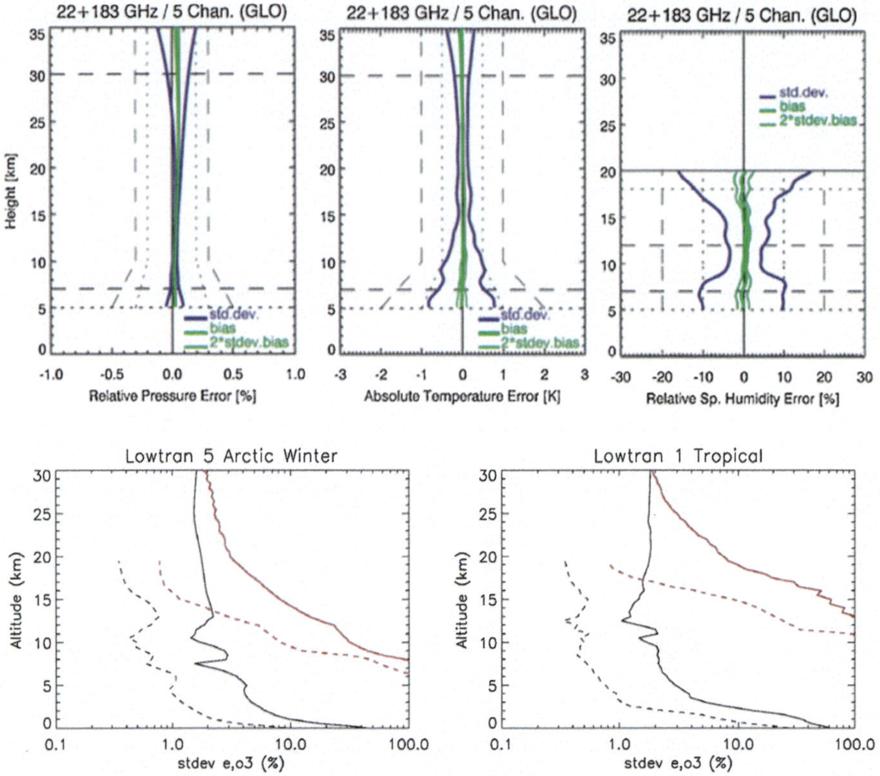

Fig. 7 Retrieval performance estimated for pressure, temperature and specific humidity profiles for LMO, using three K band channels and a 179/182-GHz channel pair [top panels; *Source* Schweitzer et al. (2011)], and retrieval performance estimated for water vapor (black) and ozone (red) profiles for LMO (solid) and aircraft occultations (dashed) in Arctic winter (left) and tropical (right) conditions [bottom panels; *Source* Kursinski et al. (2016); E.R. Kursinski, pers. communications (2016)]. Schweitzer et al. (2011) assessed the performance by employing end-to-end simulations of a full-day ensemble of a few hundred LMO events, using realistic orbits, ray tracing through an atmospheric NWP analysis field, and observational error models for obtaining simulated LMO observations. An atmospheric profiles retrieval processing chain was then run and its results compared to the "true" NWP analysis profiles at the LMO event locations, for estimating performance statistics such as those shown in the top panels. Kursinski et al. (2016) used a somewhat simpler but similar approach, with forward modeling toward reasonably realistic simulated observations, separately for LEO-to-LEO and aircraft-to-aircraft geometries, and then applying atmospheric retrieval error estimation to obtain results such as shown in the bottom panels

4.3 Technology Readiness

A series of technological feasibility studies and technology development projects for LMO instrumentation and related ground-based demonstration campaigns (ESA 2004; Kursinski et al. 2009, 2012; Kirchengast and Schweitzer 2011; Liu et al. 2017; and references therein) have prepared the basic technological readiness for now working toward LMO demonstration missions targeted for launch in the 2020–2022 timeframe. This applies in particular to the USA-led ATOMMS (Kursinski et al. 2009, 2012) and the Europe-led ACCURATE/LMO mission concepts (Kirchengast and Schweitzer 2011; Liu et al. 2017), where in various relevant studies also associated data processing chains have been developed to good maturity.

In China, a LMO pre-study has been carried out at the National Space Science Center (NSSC Beijing), which included design and performance analysis studies and prototype transmitter and receiver development (Liu et al. 2017). A pair of LMO transmitter and receiver has been manufactured and a related ground-based experiment is scheduled for later in 2017. Recent advances in the development of small satellites in the Microsat/Nanosat class now increasingly provide an economical way to realize LMO missions. This avenue is therefore pursued next as summarized by Liu et al. (2017).

In view of the promise of highly cost-effective potential LMO missions in the Microsat/Nanosat class, the most relevant next technology advancements required are those toward LMO instruments miniaturization, in order to become compliant with a Microsat/Nanosat-type LMO demonstration mission approach. Working in this direction, a recently commenced European–Chinese cooperation initiative toward an ACCURATE-1 LMO mission focuses mainly on initial X/K band technology and science demonstration, while a USA initiative toward an ATOMMS-1 mission intends both X/K band and 178–196 GHz band demonstrations.

Complementary to the activities related to LMO hardware, also further advancements of LMO data processing are required, as well as further demonstration campaigns. Hopefully, in the future, the LMO technique will combine with the growing observation density of GNSS RO to jointly profile the atmospheric thermodynamic state, and in particular water vapor in the (free) troposphere, with steadily increasing accuracy and resolution.

5 Hyperspectral Microwave

Current down-looking passive microwave humidity sensors are limited to only a few spectral channels, typically less than ten. This is comparable to the situation of infrared sensors before the arrival of hyperspectral sensors such as the Atmospheric Infrared Sounder (AIRS) and IASI, both summarized in a recent intercalibration article by Wang et al. (2015). Despite their spectral limitations, microwave sensors play a big role in operational meteorology, because of their key advantage of being only weakly affect by clouds.

Recently, a new detector technology has emerged, Kinetic Inductance Detectors (KIDs) (Day et al. 2003; Doyle et al. 2008; Mauskopf et al. 2014; Baselmans et al. 2008; Bueno et al. 2014; Griffin et al. 2015). Each detector is a superconducting resonator circuit element, built by lithography, and cooled to a sub-Kelvin temperature, i.e. at a fraction of 1° K. These detector circuit elements present an inductive load to a superconducting microwave stripline, with each detector element tuned to resonate at a slightly different frequency. Incident photons on a detector break superconducting Cooper pairs, which changes the population of quasi-particles in the superconducting film. This alters the kinetic inductance of the film, with the effect of modifying the resonant frequency of the circuit. This results in a shift of detector frequency, amplitude and phase, proportional to the incident photon flux, and this is the detection signal.

The key advantages of this new technology are that detectors are relatively straightforward to produce in large arrays, and they are very sensitive, with photon-noise limited performance demonstrated for a variety of applications (Griffin et al. 2015). KIDs also allow a high multiplexing ratio, such that up to 1000 pixels can be read out on one readout line, which greatly simplifies instrument design. Recent developments have demonstrated the ability to implement KIDs as on-chip spectrometers (Shirokoff et al. 2014; Endo et al. 2012). In this implementation, a broadband antenna feeds a filterbank, where each narrowband filter feeds and individual KID. We propose to optimize such a configuration (channel optimization) for applications to meteorology and climatology. Such a configuration allows photon-noise-limited sensitivity in the frequency region of approximately 100–1000 GHz, with individual spectral channel resolution (ν/Δν) of approximately 800. Studies are currently in progress to investigate alternate materials to demonstrate response in the region 50–60 GHz.

Combined, the large array capability and high sensitivity can be exploited to build a hyperspectral microwave detector. The high sensitivity requirement here is perhaps less obvious than the high pixel number one. But according to the radiometer formula

$$\Delta T = \frac{T_s}{\sqrt{\Delta\nu\Delta t}},$$

the noise of a spectral channel (ΔT) is proportional to the spectral bandwidth ($\Delta\nu$) to the power of minus one half, with the consequence that hyperspectral instruments with conventional detectors would be quite noisy. (The other two parameters in the above equation are a system-specific constant (T_s) and the integration time (Δt).) The measurement approach of the proposed instrument is to passively measure a large part of the millimeter/submillimeter spectral range with a moderately high spectral resolution. The initial instrument concept studied below by retrieval simulations had 2303 channels between 100 and 1000 GHz. The simulated brightness temperature spectrum for a tropical atmosphere is shown in the top plot of Fig. 8. The bottom plot of the same figure shows the humidity Jacobian, i.e., the derivative of the spectrum with respect to fractional changes in the water

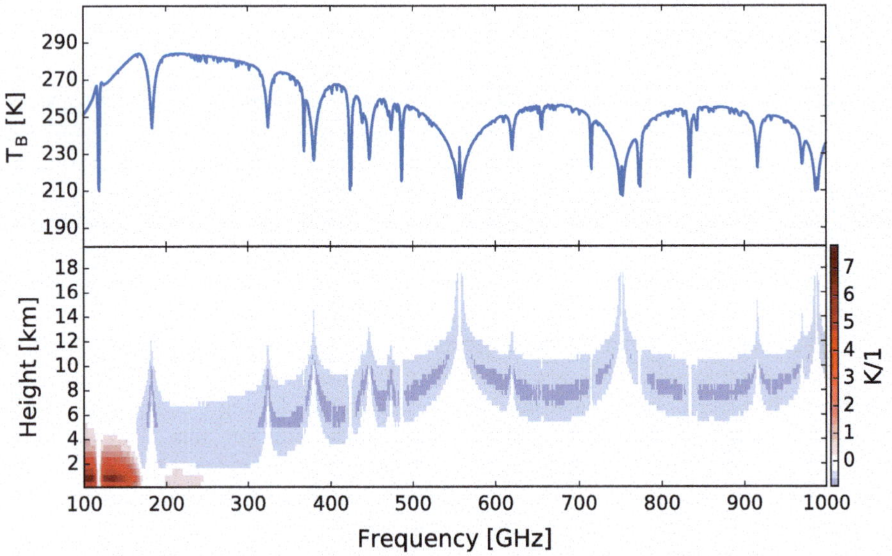

Fig. 8 Simulated brightness temperature spectrum for a tropical atmosphere (top) and associated humidity Jacobian (bottom). The Jacobian roughly shows the brightness temperature change in Kelvin for a local doubling of the water vapor volume mixing ratio on a 0.5 km vertical grid. Blue colors indicate that the Jacobian is negative, which is typically the case for frequencies above approximately 170 GHz

vapor volume mixing ratio (VMR) at different altitudes. The simulation was performed with the Atmospheric Radiative Transfer Simulator (ARTS) (Eriksson et al. 2011). The data analysis for such an instrument would benefit from various laboratory studies for spectroscopic parameters in the submillimeter spectral range that were carried out in the last years, for example in the context of planned limb sounding missions (Perrin et al. 2005; Verdes et al. 2005).

5.1 Measurement Capabilities

We have studied the capabilities of a hypothetical KID-based hyperspectral microwave sensor by optimal estimation (Rodgers 1990) retrieval simulations. As a reference, we also used the same retrieval setup on an idealized hyperspectral infrared sensor, emulating a future generation of IASI (the spectral resolution is taken from the current IASI instrument, but the noise is lower to allow for technical development and give a fairer comparison). The retrieval setup is summarized in Table 4. It closely follows the study by Schneider and Hase (2011). Besides temperature and water vapor, other relevant trace gases that significantly affect the spectrum are also included but currently not retrieved. For the KID microwave sensor this is mainly oxygen, but also ozone is quite important, because its lines are so ubiquitous (e.g., John and Buehler 2004). For simplicity, the KID microwave instrument was assumed to have regularly spaced channels with $\nu/\Delta\nu = 1000$, resulting in more than 2000 channels. But the number of channels could be reduced significantly without performance degradation if channels are chosen intelligently, since information content is not distributed equally. Initial tests show that approximately 100 channels would be enough in practice.

The left plot in Fig. 9 shows the retrieval error for water vapor on a 0.5 km retrieval grid for both sensors. One caveat is that surface emissivity here was assumed to be known perfectly, whereas in reality there is some uncertainty on this knowledge, especially over desert and arid surfaces. The very small retrieval errors close to the surface for both instruments therefore are overoptimistic. Overall, according to this simulation the KID

Table 4 Summary of retrieval simulation setup

Infrared	8534 Channels, 19–83 THz
Sensor	NeΔT 0.1–0.2 K
Microwave	2303 channels, 100–1000 GHz
Sensor	NeΔT 0.007–0.024 K
Retrieved	Temperature, retrieved in Kelvin
Species	Water vapor, retrieved in ln(VMR) coordinates
Water vapor a priori covariance matrix	Depending on altitude, mimicking Schneider and Hase (2011)
	Standard deviation 1–0.25 in ln(VMR) coordinates
	Correlation length 2.5–10 km
Atmosphere	FASCOD tropical atmosphere (Anderson et al. 1986)
Surface emissivity	Microwave: 0.6; infrared: 1.0

NeΔT indicates the instrument noise, which was assumed to vary across the band, based on typical instrument behavior, with the range indicated

VMR volume mixing ratio

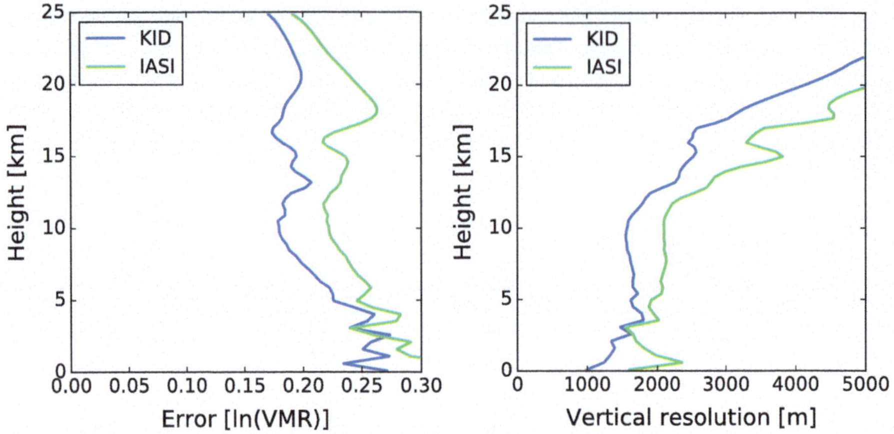

Fig. 9 Simulated water vapor retrieval error profiles for microwave and infrared instrument (left) and vertical resolution of the measurements (right). An ln(VMR) error of 0.1 roughly corresponds to a relative error of 10% on the VMR

microwave sensor would perform significantly better than the infrared sensor in the troposphere. The KID microwave sensor noise presented here is based on the assumption of a pushbroom geometry. An alternative (preferred) KID-based solution is a scanning geometry that uses the array only for the detection of different frequencies, not for different spatial pixels. For a sensor scanning over 100 spatial pixels, instead of measuring them simultaneously, the noise would increase by a factor of 10. Simulations were also carried out with the noise increased by a factor of 10, and as expected they show higher retrieval errors and a poorer vertical resolution, but still competitive with the IASI simulation.

Of course, a crucial question is also the vertical resolution of the measurement. The right plot in Fig. 9 shows the vertical resolution of the retrieved water vapor profiles for both sensors, computed as the full width at half maximum of the averaging kernel functions for retrieval on a 0.5 km vertical grid. Even here the KID sensor performs better than the infrared sensor according to our preliminary simulations. Since the inherent vertical resolution of the measurement is poorer than the chosen 0.5 km retrieval grid, part of the total error in the left plot is due to smoothing error. When retrieval grid resolution is reduced to 1 km in an alternative retrieval setup, retrieval error is reduced by a factor of 2. Comparing these preliminary KID sensor simulations to the expected enhanced performance of IASI next generation (IASI-NG) still shows promise. IASI-NG will have 16,923 spectral channels covering roughly the same bandwidth as IASI, providing twice the spectral resolution (0.25 cm^{-1}) and hence higher vertical resolution resulting from narrower weighting functions (Crevoisier et al. 2014). A major advancement of IASI-NG over IASI is the 2× improvement in radiometric noise, leading to reduction in retrieved water vapor uncertainty by $\sim$ 3–5%, although these improvements are realized between 800 and 200 hPa (Crevoisier et al. 2014). The vertical resolution and accuracy of IASI-NG water vapor retrievals below 800 hPa will remain approximately the same as IASI, thereby maintaining the relevance of preliminary instrument performance simulations for KID sensors presented here and detailed space-based water vapor DIAL simulations presented in Gerard et al. (2004).

These performance figures depend strongly on the spectral resolution as well as the assumed instrument noise (NeΔT in Table 4). It is therefore quite possible that a future cooled infrared instrument would achieve even better performance. However, a key advantage of the KID microwave sensor is that it would be able to see through most of the clouds that are opaque to the infrared sensor. Additionally, it would provide information on the cloud hydrometeors. The scientific background for that measurement is discussed in Buehler et al. (2012) and Jiménez et al. (2007). Retrieval simulations for cloud parameters in the submillimeter spectral range so far have been done only for sensors with a small number of channels. Exploiting a hyperspectral sensor for this purpose needs faster radiative transfer algorithms, a development that is already ongoing. It is likely that there will be a large benefit, mostly because measuring the full spectrum should allow for a much better separation of the cloud signal from the water vapor signal, so that humidity and clouds really could be retrieved independently, with only little error correlation. However, this is yet to be demonstrated with explicit retrieval simulations.

In summary, we have demonstrated that a KID sensor would deliver high resolution and precision water vapor profiles, comparable or better to a hyperspectral infrared sensor, and with much better global coverage, since it would work in most of the infrared-cloudy areas, i.e., through thin clouds with water paths below a few grams per square meter. This is a significant advantage for operational meteorology, since cloudy areas likely have a strong impact on forecast skill. It also is a significant advantage for climatology, because the clear-sky sampling of infrared sensors introduces a humidity dry bias (John et al. 2011). The capabilities of such a sensor to deliver hydrometeor profiles and water vapor profiles in areas with thick clouds still have to be demonstrated, but are likely to constitute a significant advancement over currently planned submillimeter sensors that are based on conventional technology. Threshold requirements for profiling water vapor in clear air for operational meteorology and climate studies would be similar to those of infrared sounders such as AIRS and IASI.

5.2 Technology Readiness

Kinetic inductance detectors are a new technology, first proposed by Day et al. (2003). But since then they have progressed rapidly. A recently completed European Union-funded project, SPACEKIDS (Griffin et al. 2015), delivered two laboratory prototype instruments, one of which was optimized for observing the Earth's atmosphere, a high radiative background case compared to typical applications in astronomy. Currently, array sizes of up to 1000 pixels can be realized in the laboratory with a single readout line and electronics set. In addition, full filterbank implementations have been demonstrated in a laboratory environment (Shirokoff et al. 2014; Endo et al. 2012).

The focal plane would need to be cooled to $\sim$ 300 mK, with an operational design lifetime goal of around 7 years. This requires high reliability mechanical cryocoolers, attached to a ^{3}He sorption cooling system. The ^{3}He system would be based upon proven heritage from Herschel PACS and SPIRE (Griffin et al. 2008) instruments, with the possibility of continuous 300 mK cooling, recently developed by Cardiff University, in collaboration with Chase Research.

The Planck JT compressors were supplied by EADS Astrium, based upon a licensed design from RAL, the compressors are first generation linear motor reciprocating mechanisms originally developed for Stirling cycle coolers which have now amassed around 15 years of in-orbit operational life between 24 mechanisms. The 4 K cooler requires pre-cooling to temperatures below the inversion temperature for helium (about 27 K). In practice, the lower the pre-cooling temperature (within limits), the better the cooling power. The RAL two-stage 15 K cooler has been in development for many years and is now able to deliver around 250 mW at 15.4 K and a no load base temperature of 9.8 K. The compressor input power under these conditions would be around 135 W. This could provide the basis of an excellent pre-cooler for our instrument. In summary, overall the technology is at a readiness level where the logical next step would be to build an airborne prototype instrument, or a small demonstration satellite mission. The remaining challenge is foreseen in the design of radiators with sufficient cooling power to suit the higher infrared radiation load from Earth (compared to applications for astronomy) that a meteorological mission in low-Earth orbit would be exposed to.

6 Synergies of Observing Systems

As the previous subsections have shown, promising technologies for improved water vapor profiling are emerging and may be operationally applicable within the time frame of a decade. However, no single remote sensing technology will ever deliver simultaneously full height coverage of water vapor in a highly, temporally and spatially resolving manner in all atmospheric conditions, which would be highly relevant for understanding and modeling atmospheric circulation patterns on a wide variety of scales. Profiling lidar systems are severely hampered by thick clouds, whereas future DAR systems will rely on the presence of backscattering particles within clouds. Passive remote sensing systems in the microwave region can retrieve water vapor in the presence of clouds, are, however, hampered by surface emission and rather low vertical resolution. Active limb sounding by LMO will feature high vertical resolution, no surface interference, and high accuracy and long-term stability but deliver comparatively sparse horizontal coverage and resolution unless large microsatellite constellations are deployed. Passive infrared spectrometers can accomplish a higher horizontal resolution, but are again disturbed by the presence of clouds.

Clearly, this calls for an optimized synergy of the different observation systems, complementing their respective strengths in different situations. Such methods have been developed successfully for classifying cloud type as well as for retrieving microphysical quantities from Cloudsat/CALIPSO cloud radar and lidar observations (Ceccaldi et al. 2013; Stein et al. 2011) but also using the Cloudnet instrument suite (Illingworth et al. 2007) from surface-based remote sensing observations.

6.1 Value of Ground-Based Profiling

The bulk of ground-based atmospheric profiling has historically been undertaken by routine upper-air-soundings, but worldwide the density of this network is under pressure. Clearly, upper-air-soundings state benchmark measurements when it comes to profiling the atmosphere in terms of humidity, temperature and winds. However, carrying out upper-air-soundings on a routine basis presents enormous cost and labor factors. Also, launches are typically on a 12-h basis, so that the diurnal cycle of the boundary layer development and other short-lived weather events will not be captured. Certainly, such temporal developments are crucial for the initialization and evaluation of atmospheric models. Additionally, single ascents need approx. 1 h to pass through the troposphere and also underlie wind drifts effects. Another drawback concerning upper-air-soundings is that no methods have been established to measure cloud macro- and microphysical properties on an operational basis.

Ground-based remote sensing has an enormous potential for accurately profiling the lower troposphere with high vertical and temporal resolution as shown by Wulfmeyer et al. (2015). It has been known for some time that passive microwave radiometers (MWR) offer the potential of filling the gap in retrieving the atmospheric thermodynamic state (Westwater 1997), i.e., the distribution of temperature, humidity and liquid water, in a quasi-continuous and instantaneous manner. Additionally ground-based infrared spectrometers, such as the commercially available atmospheric emitted radiance interferometer (AERI) instrument (Knuteson et al. 2004) also show high potential for retrieving temperature and humidity profiles from the surface (Turner and Löhnert 2014)—in addition to cloud and aerosol parameters. While technical developments of the last decades have led to passive systems, which can operationally profile the thermodynamic structure of the atmosphere, their vertical resolution is rather low showing the best retrieval performance in the lower part of the boundary layer. Typical vertical resolutions are on the order of 50 to 500 m in the lower troposphere. Active lidar systems (employing either DIAL or Raman technology) are able to detect sharp vertical gradients in humidity and/or temperature with much smaller uncertainties in general. Many ground-based lidar systems, however, do not provide full coverage of the troposphere, can show daytime-dependent noise characteristics and are not yet at a technical development stage to be called fully mature for 24/7 unattended operation. Exceptions to the aforementioned generalization, among others not cited here, are the fully operational DOE ARM Raman lidar (Turner et al. 2016), the NCAR field deployable micropulse DIAL (Spuler et al. 2015; Weckwerth et al. 2016; Nehrir et al. 2011a, b, 2012), the Cloud Observing Radar and Lidar at the Barbados cloud observatory (Stevens et al. 2016), and the Raman Lidar for Meteorological Observations at the MeteoSwiss Regional Center of Payerne (Brocard et al. 2013), all of which exhibit varying levels (spatial and temporal) of daytime dependent noise characteristics. Still, promising developments will enable many more operational ground-based lidars to come online within the next decade with improved capabilities for measurements in the mid-lower troposphere.

Ground-based remote sensing profiling can be a benchmark for accurately and continuously profiling the lower troposphere if synergetic methods are applied to exploit the complementary potential of the different measurement systems. Developments in this direction have just begun and require scientific expertise in different observation systems simultaneously as well as the employment of advanced retrieval methods solving the classical inverse problem of atmospheric remote sensing in multiple ways. Examples for such developments are given in the next section. Ebell et al. (2017) retrieve cloud, water vapor and temperature profiles simultaneously through synergy of ground-based active and passive remote sensing. Such observations are highly important for model evaluation, e.g., for the small-scale, cloud-resolving model ICON-LEM (Heinze et al. 2017) developed with the $HD(CP)^2$ (High Definition Cloud and Precipitation for Climate Prediction) research initiative as well as for process studies on clouds and precipitation.

One demanding question to be resolved is to what extent local ground-based remote sensing observations could complement existing measurements within the global observation system. Here, the combination of existing satellite and ground-based profiling methods can lead to an increase in local information content (see next section) for the simultaneously observed atmospheric column. However, research is also required to what extent a local observation with continuous and temporally highly resolved profile information can improve satellite observations within a certain radius of influence.

6.2 Examples of Synergetic Applications

In the following two examples of synergetic retrieval applications for atmospheric thermodynamic profiling are given. These are intended to show the potential of variational methods for optimally exploiting the information content contained in the respective observations resulting in a retrieval with an improved performance with respect to overall uncertainty and information content. Note, an objective way to analyze the information content of variational retrievals is to evaluate the number of degrees of freedom (DOF) for signal for the retrieved profile; i.e., the number of independent levels of humidity that can be determined.

6.2.1 Combining Lidar and Microwave Radiometer

During HOPE (HD(CP)^2 Observational Prototype Experiment, Macke et al. 2017), water vapor Raman lidar and microwave radiometer observations were carried out simultaneously at JOYCE (Jülich Observatory Cloud Evoultion, Löhnert et al. 2015) for profiling water vapor profiles. A variational retrieval was applied (Barrera-Verdejo et al. 2016) to optimally exploit both observation systems leading to the uncertainty estimates of the profile shown in Fig. 10. Lidar-only retrievals that are carried out in cloud-free conditions are, as expected from the physical basis of the problem, significantly more accurate than the MWR-only retrieval in terms of standard uncertainty (compare Fig. 10 left and center panels, red lines). The lidar signal is assumed to be attenuated at varying heights, i.e., due to clouds, the lidar-only retrieval uncertainty rapidly increases with height toward the a priori profile. However, the benefit of the synergy of both systems becomes clear in exactly these situations (Fig. 10 right panel). The retrieval uncertainty stays well below the lidar-

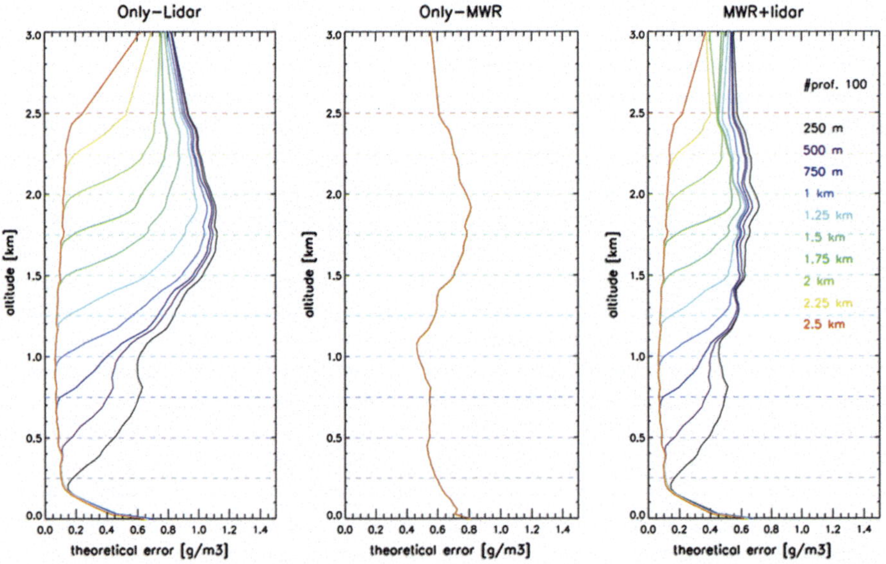

Fig. 10 Mean uncertainty of 100 clear-sky cases of absolute humidity retrieval during HOPE. Left: Water vapor lidar only, Center: MWR only, Right: Combination of both lidar and MWR. The dashed horizontal lines indicate levels of full hypothetical lidar attenuation, i.e., due to clouds. The corresponding colored lines (bold) show the uncertainty in case the lidar provides no signal above the dashed line

only and MWR-only values throughout all heights, with the highest impact visible in the heights above 1 km (up to 0.3 gm^{-3}). In these cases, the degrees of freedom for signal of the MWR ($\sim$ 2) are used to improve the profile above the height of full lidar extinction, where in the MWR-only case these DOF are used to retrieve the full profile leading to the observed large uncertainties on the order of 0.6 gm^{-3}.

Ground-based MWR observations (HATPRO) can also be combined with highly spectrally resolved infrared observations (AERI) to improve the DOF of the humidity profile to be $\sim$ 4 (Ebell et al. 2013). Here, the additional information is distributed throughout the whole atmospheric profile.

6.2.2 Combining Satellite and Ground-Based Observations

Satellite sounders typically only provide little information in the boundary layer wherefore the combination with ground-based sensors proves beneficial (Fig. 11). Combining AMSU-A/MHS microwave satellite sensors with ground-based MWR is less beneficial

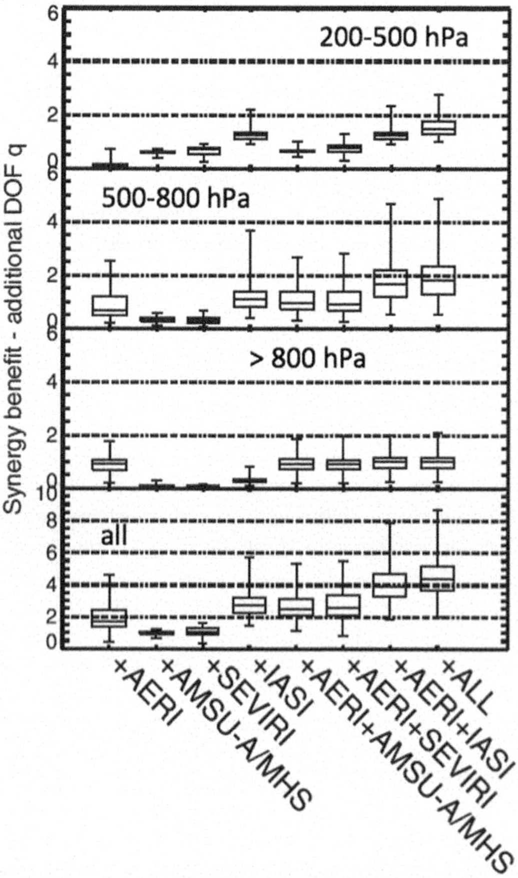

Fig. 11 Synergy benefit for absolute humidity profile in terms of additional DOF compared to HATPRO-only retrieval. Median (line in box), 0.25 and 0.75 quantiles (box boundaries), and minimum and maximum values (whiskers) of the profile sample are shown. From Ebell et al. (2013)

than the combination with IASI, the latter being a satellite-based spectrally highly resolving infrared spectrometer. Compared to the HATPRO-only retrieval, the DOF increases by 2.5. The main advantage of adding SEVIRI (geostationary satellite data) to ground-based MWR observations is obtained between 200 and 500 hPa. In this study, the additional DOF is roughly equal to the sum of the DOF of a single instrument so one can conclude that the information content of the sensors is optimally combined. Combining all sensors used in this study results in the largest gain of information content.

6.3 Future Perspectives

There is strong need for combining satellite and ground-based observations of wind, water vapor and clouds. The examples motivate that improvements in uncertainty reduction and information content gain are possible; however, the synergy even of existing observation systems is today far from optimally exploited. Physically consistent combinations of observations will become more important with the advent of advanced profiling systems, which will work under specific atmospheric conditions.

Passive instruments on geostationary satellites, but especially on current polar-orbiters are basically the "workhorses" for capturing the mid- and upper-tropospheric dynamic and thermodynamic environment (including clouds). It is imperative that the international atmospheric science community pursues the sustainment of such measurement systems in space (preferably with increased spectral resolution) in order to guarantee these vitally important data streams for global assimilation systems in the decades to come. In spite of their importance for large-scale flows, however, these sensors reveal limited retrieval performance in the boundary layer. Future satellite concepts could thus be comprised of a passive cross-track, high spectral resolution scanner in addition to a nadir pointing active system for vertically highly resolved profiles of water vapor in the column above the satellite track.

Feasibility studies of such a system could be carried out at ground-based remote sensing sites using manifold observations of the boundary layer (i.e., in critical regions of convective aggregation) through a campaign-based deployment of mobile and/or fixed ground-based profiling stations including Doppler lidar (BL winds), cloud radar (BL clouds), microwave radiometer (BL temperature and humidity) and DIAL and/or Raman (humidity profiles below clouds). A much more complete picture of the column could then be obtained by directly combining satellite and ground-based measurements within variational approaches. The representativeness for going from a point measurement to an area is an additional challenge, but could be resolved by overpasses (curtains) either from Cloudsat/CALIPSO or from dedicated aircraft (HALO, HIAPER, NASA DC-8 aircraft) observations. A complete picture of the region of interest would probably need to involve an advanced variational data assimilation scheme with these multiple types of sensors.

7 Summary and Conclusions

Water vapor is the most dominant greenhouse gas and plays a vital role in both the climate and weather. Water vapor profile measurements with improved accuracy and vertical resolution would benefit many major research programs including the World Climate Research Programs (WCRP), the World Weather Research Program (WWRP), the Stratosphere-troposphere Processes and their Role in Climate (SPARC) program, the

Network for the Detection of Atmospheric Composition Change (NDACC), and the Global Energy and Water Cycle Experiment (GEWEX) program. In addition to providing lower tropospheric profiles, the discussed measurement approaches can provide simultaneous high accuracy and high vertical resolution water vapor profiles in the UTLS that are needed to address both the chemical and radiative role of water vapor to changes in climate. Accurate, high vertical resolution water vapor profiles would also improve estimates of surface evaporation and surface fluxes of sensible and latent heat which are critical to numerous boundary layer processes important to both weather and climate, but also for constraining atmospheric heat budgets across scales. The recent NASA Weather Focus Area community workshop report (Zeng et al. 2015) calls for a continuing investment to improve satellite measurements of water vapor in the atmospheric boundary layer.

Current satellite water vapor products rely mainly on passive measurements of spectrally resolved microwave and infrared radiances at the top of the atmosphere, but such radiances are weighted toward the mid-troposphere and are not very sensitive to water vapor in the lower troposphere where it varies the most. The radiative kernel functions are broad and are unable to resolve either the marine boundary layer or the gradients in water vapor near the trade inversion, both of which are critical to processes that govern shallow cloud cover and convection initiation. Poor vertical resolution also mutes circulation patterns associated with the radiative response to large water vapor gradients. Passive infrared measurements have a vertical resolution of at best 1–2 km in the lower troposphere with relative errors of 10–20% (August et al. 2012). Microwave profilers can provide humidity measurements in regions of non-precipitating clouds; however, they can suffer from underlying biases resulting from uncertainties in the microwave surface emissivity. Tropospheric humidity retrievals from GNSS radio occultation show promise due to their favorable combination of high vertical resolution, accuracy, long-term stability, all-weather capability and global as well as local time coverage, but are limited in their horizontal resolution (especially near the surface) and by the need for a priori background temperature profiles required to retrieve humidity profiles.

Space-based water vapor lidars are ideal for high accuracy measurements of water vapor throughout the troposphere and lower stratosphere as they are self-calibrating and not prone to bias resulting from aerosol and cloud contamination. Furthermore, they can retrieve water vapor profiles during day and night, at all latitudes, through broken cloud fields and tenuous high altitude cirrus (with reduced spatial resolution), and during all seasons. In addition to profiles of water vapor, DIAL can provide high accuracy measurements of total precipitable water, cloud occurrence, cloud-top-height, and cloud water phase. Further, gradients in both water vapor and aerosol can be used to characterize the height of the (marine) boundary layer inversion. This capability is useful for studying vapor-cloud interactions, although to comprehensively address questions related to shallow cloud processes, DIAL observations should be augmented with observations of cloud optical depth, liquid water path, and droplet size from passive sensors, in addition to detection of drizzle and light precipitation from radar.

Airborne DIAL prototypes have been fielded over the past 20 years demonstrating key technologies and resulting science capabilities from active sensing platforms. A concept study was carried out by Gerard et al. (2004) to assess the performance of a space-based water vapor DIAL in relation to passive measurements such as IASI. The results of the study indicated that for a fixed error threshold, a water vapor DIAL would extend the water vapor measurement to a greater altitude as well as further down into the lower troposphere than IASI, would provide higher precision and accuracy in the lower troposphere and would have approximately twice the vertical resolution. The study also indicated that the water vapor measurements would extend from the lower stratosphere down to the surface which would have significant positive impacts in determining the background error covariance used in Numerical Weather Prediction (NWP) analysis, as well as for providing a stable absolute reference for calibrating passive sensors. Although the DIAL technique has been extensively demonstrated via airborne simulators and is also the most mature of the technologies described here, a comprehensive funding profile is still required to increase the technology readiness level of critical subsystems such as the laser, detectors, and filters, required for a space-based implementation.

Complementary to DIAL, the DAR technique provides profiles of water vapor within clouds. The DAR technique presented here samples the 183 GHz water vapor absorption line with two discrete RF channels, one that sits on the wing of the absorption line and one that is located away from the absorption line. Similar to the DIAL technique, the ratio of the radar reflectivity's at the online and off-line frequencies is approximately proportional to the water vapor path. The DAR technique is a powerful new emerging technology in that it will provide profile measurements of water vapor within clouds at different altitudes, an area which no satellite or airborne sensor currently has the capability to retrieve high accuracy measurements. A combined DIAL and DAR payload will serve as a powerful measurement suite providing both clear air and in-cloud profiles of water vapor, in addition to column measurements from both systems.

An extension of GNSS RO, LMO utilizes centimeter and millimeter wave signals between LEO satellites in the X and K bands to retrieve pressure, temperature, and humidity profiles using an optimal estimation framework. Together with pressure and temperature retrievals, the LMO water vapor retrievals show promise in providing high accuracy global profiles throughout the troposphere and stratosphere in clear and cloudy conditions. A primary limitation of LMO is an inherently limited horizontal resolution of approximately 100–200 km along the sounding paths, which is thus not able to resolve fine-scale features in water vapor fields in the near-surface atmosphere. An LMO observing system coupled with nadir pointing active and passive systems, which provide high spatial resolution measurements in the lower troposphere, would extend the vertical coverage of water vapor profiles from the stratosphere down to the surface. The first LMO demonstration missions, ACCURATE-1 and ATOMMS-1, are currently under development and expected to launch in the 2020–2022 timeframe.

Hyperspectral passive microwave KID sensors, as discussed above, would extend the capabilities of current hyperspectral infrared sensors to cloudy areas. This all-sky capability and the wide measurement swath would be advantages even compared to active DIAL measurements, although accuracy and horizontal and vertical resolution would be inferior to DIAL. In terms of technological readiness, hyperspectral KID sensors are the least mature of the techniques discussed here, but currently developing rapidly. Laboratory systems have already been realized, and the next step will be airborne prototypes, which could be ready within the next 5 years. A top level system/subsystem TRL breakdown based on the NASA Earth Science Technology Office definitions is presented for each measurement approach in Table 5. The TRL breakdown is intended to summarize the relative matureness of each measurement approach in order to estimate the potential readiness date for a future space-borne mission, rather than to elucidate the exact level of readiness of each instrument(s) subsystem(s), which is required when transitioning from ground/airborne demonstrators to a space-borne implementation.

In addition to profiles of water vapor, information on the origin of water vapor is also important. Stable naturally occurring isotopes of hydrogen (Deuterium–Oxygen, HDO) have been used in hydrology for decades to provide constraints on the origins of water. The

Table 5 System/subsystem technology readiness levels for the different airborne and space-based technologies presented here. For orientation, TRL 2 is a technology concept, TRL 4 is component/subsystem validation in a laboratory environment, TRL 6 is system/subsystem model or prototyping demonstration in a relevant end-to-end environment

Measurement approach	Aircraft TRL	Space-borne TRL	Comment
Differential absorption lidar			
Transmitter	9	4–5	Nd/YAG pump lasers developed for current space-based lidar missions (e.g., ADM Aeolus, EarthCARE, MERLIN, CALIPSO follow-on mission concept) can be adapted for use as a pump laser for a water vapor DIAL transmitter. TRL for DIAL single-frequency pulsed and continuous wave seed lasers reflect technologies developed under the LASE, WALES, HALO, and MERLIN programs
Receiver	9	5–6	TRL reflects use of highly sensitive single photon sensitive silicon and mercury cadmium telluride (HgCdTe) detectors. Telescope adapted from ADM Aeolus. Frequency agile filter not included in TRL assessment
Differential absorption radar			
Transmitter	4	2	TRL based on existing GaN W-band power amplifier and GaAs Schottky diode frequency multiplication to achieve > 0.5 W continuous transmit power at G-band
Receiver	4	4	TRL based on proven quasi-optical duplexer and antenna and mature back-end digital electronics
LEO–LEO microwave occultation			
ATOMMS airborne demonstrator	5	NA	ATOMMS hardware developed for demonstration on NASA WB-57 platform. TRL for ground-based (mountain top) operation is at 8
ACCURATE microsat transmitter and receiver	NA	5	TRL reflects that successful phase A studies have been conducted and that related space instrument transmitter and receiver breadboard simulators have been developed and validated via ground-based (lower troposphere) long-path experiments
Hyperspectral microwave			
KID sensor	3	3	The KID spectrometer system is operated under vacuum in a cryogenic environment. The only difference with a satellite-based implementation is the enhanced cosmic ray flux. TRL is at 3 for space-based implementation, as the full cooling chain (radiator design) for LEO needs to be demonstrated

ratio of HDO to H_2O (expressed as δD) is a tracer for water origin and lifetime and consequently sensitive to many key processes of the hydrological cycle such as mixing of air masses, convection, transport and cloud detrainment (Galewsky et al. 2016; Worden et al. 2007). Measurements of this water isotopologue therefore add value to measurements of the main isotope alone. Global HDO datasets have become available from satellites such as IASI, Tropospheric Emissions spectrometer (TES) for sensitivity in the upper troposphere or SCanning Imaging Absorption SpectroMeter for Atmospheric CHartographY (SCIAMACHY) and Greenhouse Gas Observing Satellite (GOSAT) for the full tropospheric column measurements (Lacour et al. 2012; Schneider and Hase 2011; Frankenberg et al. 2013). These existing satellites provide global observations of the distribution of water isotopes, providing a unique view on the processes that control the global water cycle. Despite recent advancements, passive remote sensing of HDO in the troposphere using reflected or emitted radiances is typically an ill-posed problem (Galewsky et al. 2016) that limits the accuracy in which vertical distributions can be retrieved in non-trivial heterogeneous scenes (Stevens et al. 2017). Measurements of water isotopes using the measurement approaches presented here are mostly out of the scope of this paper, however, future prospects exist for microwave and LMO.

Hyperspectral passive KID sensors presented here would also extend the capabilities of current infrared sensors for column HDO measurements to cloudy areas with potentially higher spectral resolution, thereby enabling sharper weighting functions in the mid-upper troposphere. LMO also has the potential for retrieving HDO profiles, but similar to the other retrieved products, suffers from poor spatial resolution near the surface. Although column HDO measurements using integrated path DIAL have been demonstrated to date, HDO profile measurements are largely out of scope for active remote sensing such as DIAL and DAR due to the very low atmospheric concentrations. For a HDO DIAL or DAR with similar precision as for H_2O, for example, one would require a HDO line with an absorption cross section larger by a factor n_H2O/n_HDO than the usual H_2O DIAL/DAR cross section, where n_x is the molecule number density of isotope x. Due to longer integration paths across the atmosphere, passive remote sensing is in a better position, albeit at the cost of coarse spatial resolution, which may jeopardize the overall benefit as most of the abovementioned hydrological cycle processes are small scale and reside in the mid-to-lower troposphere.

Despite the crucial role of water vapor, and particularly in the lower troposphere, in virtually every climate process (Stevens and Bony 2013), it is important to note that no satellite missions are currently planned which could overcome the limitations of the current deployed space-based sensor suite. The new and emerging technologies and analysis tools presented here have the potential to bridge this observational gap and provide high resolution and accurate water vapor profiles from the surface through the stratosphere. Future climate and dynamics focused Earth observing systems requiring high resolution water vapor profiles will require complementary measurements from DIAL sensors optimized for lower tropospheric measurements as well as from passive sensors whose weighting functions are optimized from the free troposphere up through the stratosphere. The addition of active DAR measurements for in cloud water vapor profiles would complete the portfolio, providing retrievals of water vapor in both clear and cloudy conditions.

In addition to satellite sensors, there is also a need for highly accurate ground-based and airborne remote and in situ measurements of tropospheric water vapor profiles for satellite validation, detailed process studies not achievable using space-borne sensors, and for use as transfer standard to combine satellite datasets. In parallel, the ground-based and airborne lidars and aircraft in situ measurements are vital for looking at detailed micro-/macrophysical processes associated with water vapor and its interaction with clouds. In the near term, airborne process studies such as the European NARVAL and EUREC^{4}A campaigns (Kiemle et al. 2017; Bony et al. 2017), will pave the way for next-generation space-based water vapor measurements required to overcome existing measurement deficiencies.

Acknowledgements This paper arises from the International Space Science Institute (ISSI) Workshop on "Shallow clouds and water vapor, circulation and climate sensitivity." The coauthors wish to acknowledge Lukas Kluft for his performance simulations of the hyperspectral microwave measurement approach, as well as Richard Ferrare and Susan Kooi for their contributions to the LASE data processing. Stefan Buehler was supported in part by the Cluster of Excellence "CliSAP" (EXC177), University of Hamburg, funded through the German Science Foundation (DFG). He also received support from the HD(CP)^2 project (Fkz. 01LK1505D; Fkz. 01LK1502B), funded by the German Federal Ministry of Education and Research (BMBF) as "Forschung für Nachhaltige Entwicklung" (FONA). The lidar synergies work presented by Maria Barrera Verdejo and Ulrich Löhnert has been supported by High Definition Clouds and Precipitation for advancing Climate Prediction HD(CP)^2 (FKZ 01LK1209A, 01LK1502E) funded by the German Ministry for Education and Research. The DAR related activities were carried out at the Jet Propulsion Laboratory, California Institute of Technology, under a contract with the National Aeronautics and Space Administration. The HALO program is supported by the National Aeronautics and Space Administration Earth Science Technology Office and Earth Science Division. The work at the Wegener Center has been supported by the European Space Agency (ESA) and the Aeronautics and Space Agency of the Austrian Research Promotion Agency (FFG-ALR). Cong-Liang Liu's work was supported by the Chinese Academy of Sciences (CAS) and the National Natural Science Foundation of China (NSFC). We thank the two anonymous reviewers for helpful comments and suggestions and also the UK Science and Technologies Facilities Council, under grant reference number ST/N000706/1, for providing support to enable open access to this article.

References

Albert M, Puffenburger K, Schum T, Fitzpatrick F, Litvinovitch S, Jones D, Rudd J, Hovis F (2015) UV lifetime laser demonstrator for space-based applications. Proc SPIE 9612, Lidar remote sensing for environmental monitoring XV. doi: 10.1117/12.2191188

Anderson GP, Clough SA, Kneizys FX, Chetwynd JH, Shettle EP (1986) AFGL atmospheric constituent profiles (0–120 km), AFGL, TR-86-0110

Anthes RA (2011) Exploring Earth's atmosphere with radio occultation: contributions to weather, climate and space weather. Atmos Meas Tech 4:1077–1103. doi:10.5194/amt-4-1077-2011

Anthes RA, Bernhardt PA, Chen Y, Cucurull L, Dymond KF et al (2008) The COSMIC/FORMOSAT-3 mission: early results. Bull Atmos Metorol Soc 89:313–333. doi:10.1175/BAMS-89-3-313

August T, Klaes D, Schlussel P, Hultberg T, Crapeau M, Arriaga A, O'Carroll A, Coppens D, Munro R, Calbet X (2012) IASI on Metop-A: operational level 2 retrievals after five years in orbit. J Quant Spectroscopy Rad Trans Iss 113:1340–1371

Barrera-Verdejo M, Crewell S, Löhnert U, Orlandi E, Di Girolamo P (2016) Ground-based lidar and microwave radiometry synergy for high vertical resolution absolute humidity profiling. Atmos Meas Tech 9:4013–4028. doi:10.5194/amt-9-4013-2016

Baselmans J et al (2008) Noise and sensitivity of aluminum kinetic inductance detectors for sub-mm astronomy. J Low Temp Phys 151(1–2):524–529

Bony S, Dufresne J-L (2005) Marine boundary layer clouds at the heart of tropical cloud feed-back uncertainties in climate models. Res Lett, Geophys. doi:10.1029/2005GL023851

Bony S, Stevens B, Ament F, Crewell S, Delanoe J, Farrell D et al (2017) EUREC4A: a field campaign to elucidate the couplings between clouds, convection and circulation. Surv Geophys. doi:10.1007/s10712-017-9428-0

Brocard B, Philipona R, Haefele A, Romanens G, Mueller A, Ruffieux D, Simeonov V, Calpini B (2013) Raman Lidar for meteorological observations, RALMO—Part 2: validation of water vapor measurements. Atm Meas Tech. doi:10.1175/BAMS-D-14-00247.1

Browell EV, Ismail S, Hall W, Moore A, Kooi S, Brackett V, Clayton M, Barrick J, Schmidlin F, Higdon N, Melfi S, Whiteman D (1997) LASE Validation Experiment. In: Ansmann A, Neuber R, Rairoux P, Wandinger U (eds) Advances in atmospheric remote sensing with lidar. Springer, Berlin, pp 289–295

Browell EV, Ismail S, Grant WB (1998) Differential absorption lidar (DIAL) measurements from air and space. Appl Phys B 67:399–410

Bruneau D, Quaglia P, Flamant C, Meissonnier M, Pelon J (2001) Airborne lidar LEANDRE II for water-vapor profiling in the troposphere. I. System description. Appl Opt 40:3450–3461

Buehler SA, Defer E, Evans F, Eliasson S, Mendrok J, Eriksson P, Lee C, Jimenéz C, Prigent C, Crewell S, Kasai Y, Bennartz R, Gasiewski AJ (2012) Observing ice clouds in the submillimeter spectral range: the cloudice mission proposal for ESA's earth explorer 8. Atmos Meas Tech 5:1529–1549. doi:10.5194/amt-5-1529-2012

Bueno J et al (2014) Anomalous response of superconducting titanium nitride resonators to terahertz radiation. Appl Phys Lett 105(19):192601

Ceccaldi M, Delanoë J, Hogan RJ, Pounder NL, Protat A, Pelon J (2013) From CloudSat-CALIPSO to EarthCare: evolution of the DARDAR cloud classification and its comparison to airborne radar-lidar observations. J Geophys Res Atmos 118:7962–7981. doi:10.1002/jgrd.50579

Crevoisier C, Clerbaux C, Guidard V, Phulpin T, Armante R, Barret B, Camy-Peyret C, Chaboureau, J-P, Coheur P-F, Crépeau L, Dufour G, Labonnote L, Lavanant L, Hadji-Lazaro J, Herbin H, Jacquinet-Husson N, Payan S, Péquignot E, Pierangelo C, Sellitto P, Stubenrauch, C (2014) Towards IASI-new generation (IASI-NG): impact of improved spectral resolution and radiometric noise on the retrieval of thermodynamic, chemistry and climate variables. Atmos Meas Tech 7:4367–4385. https://doi.org/10.5194/amt-7-4367-2014

Day PK et al (2003) A broadband superconducting detector suitable for use in large arrays. Nature 425(6960):817–821

Di Girolamo P, Behrendt Andreas, Kiemle Christoph, Wulfmeyer Volker, Bauer Heinz, Summa Donato, Dörnbrack Andreas, Ehret Gerhard (2008) Simulation of satellite water vapour lidar measurements: performance assessment under real atmospheric conditions. Remote Sens Environ 112(4):1552–1568. doi:10.1016/j.rse.2007.08.008

Doyle S et al (2008) Lumped element kinetic inductance detectors. J Low Temp Phys 151(1):530–536

Ebell K, Orlandi E, Hünerbein A, Löhnert U, Crewell S (2013) Combining ground and satellite based measurements in the atmospheric state retrieval: assessment of the information content. J Geophys Res 18:6940–6956. doi:10.1002/jgrd.50548

Ebell K, Löhnert U, Päschke E, Orlandi E, Schween JH, Crewell S (2017) A 1-D variational retrieval of temperature, humidity, and liquid cloud properties: performance under idealized and real conditions. J Geophys Res Atmos 122:1746–1766. doi:10.1002/2016JD025945

Ehret G, Fix A, Weiss V, Poberaj G, Baumert T (1998) Diode-laser-seeded optical parametric oscillator for airborne water vapor DIAL application in the upper troposphere and lower stratosphere. Appl Phys B 67:427–431

Ehret G, Hoinka K, Stein J, Fix A, Kiemle C, Poberaj G (1999) Low stratospheric water vapor measured by an airborne DIAL. J Geophys Res 104:31351–31359

Endo A et al (2012) Development of DESHIMA: a redshift machine based on a superconducting on-chip filterbank. In: SPIE astronomical telescopes + instrumentation. international society for optics and photonics

Ellis SM, Vivekanandan J (2010) Water vapor estimates using simultaneous dual-wavelength radar observations. Radio Sci. doi:10.1029/2009RS004280

Eriksson P, Buehler SA, Davis CP, Emde C, Lemke O (2011) ARTS, the atmospheric radiative transfer simulator, version 2. J Quant Spectrosc Radiat Transfer 112(10):1551–1558. doi:10.1016/j.jqsrt.2011.03.001

ESA (2001) The five candidate earth explorer core missions—WALES—WAter Vapour Lidar Experiment in Space. ESA SP-1257(2). ESA Publication Division, Nooredwijk

ESA (2004) ACE + : atmosphere and climate explorer, 4th report of reports for mission selection–the six candidate earth explorer missions. ESA SP-1279(4) 60 pp and Technical & Programmatic Annex 39 pp, ESA Publ Div, Noordwijk

Ferrare R, Browell E, Ismail S, Kooi S, Brasseur L, Brackett V, Clayton M, Barrick J, Linné H, Lammert A, Diskin G, Goldsmith J, Lesht B, Podolske J, Sachse G, Schmidlin FJ, Turner D, Whiteman D, Tobin D, Revercomb H, Demoz B, Girolamo P (2004) Characterization of upper troposphere water vapor measurements during AFWEX using LASE. J Atmos Oceanic Tech 21:1790–1808

Frankenberg C, Wunch D, Toon G, Risi C, Scheepmaker R, Lee J-E, Wennberg P, Worden J (2013) 2013: water vapor isotopologue retrievals from high-resolution GOSAT shortwave infrared spectra. Atmos Meas Tech 6:263–274

Galewsky J, Steen-Larsen HC, Field RD, Worden J, Risi C, Schneider M (2016) Stable isotopes in atmospheric water vapor and applications to the hydrologic cycle. Rev Geophys 54:809–865

Gerard E, Tan DGH, Garand L, Wulfmeyer V, Ehret G, Di Girolamo P (2004) Major advances foreseen in humidity profiling from WALES. Bull Am Meteor Soc 85:237–251

Gorbunov ME, Kirchengast G (2007) Fluctuations of radio occultation signals in X/K band in the presence of anisotropic turbulence and differential transmission retrieval performance. Radio Sci. doi:10.1029/2006rs003544

Griffin M et al (2008) Herschel-SPIRE: design, ground test results, and predicted performance. In: Society of photo-optical instrumentation engineers (SPIE) conference Series

Griffin M et al (2015) SPACEKIDS: kinetic inductance detector arrays for space applications. In: 40th international conference on Infrared, Millimeter, and Terahertz waves (IRMMW-THz). 2015. IEEE

Groß S, Wirth M, Schäfler A, Fix A, Kaufmann S, Voigt C (2014) Potential of airborne lidar measurements for cirrus cloud studies. Atmos Meas Tech 7:2745–2755. doi:10.5194/amt-7-2745

Hartmann DL, Larson K (2002) An important constraint on cloud-climate feedback. Res Lett, Geophys. doi:10.1029/2002GL015835

Heinze R, Dipankar A, Henken CC, Moseley C, Sourdeval O, Trömel S, Xie X, Adamidis P, Ament F, Baars H, Barthlott C, Behrendt A, Blahak U, Bley S, Brdar S, Brueck M, Crewell S, Deneke H, Di Girolamo P, Evaristo R, Fischer J, Frank C, Friederichs P, Göcke T, Gorges K, Hande L, Hanke M, Hansen A, Hege H-C, Hoose C, Jahns T, Kalthoff N, Klocke D, Kneifel S, Knippertz P, Kuhn A, van Laar T, Macke A, Maurer V, Mayer B, Meyer CI, Muppa SK, Neggers RAJ, Orlandi E, Pantillon F, Pospichal B, Röber N, Scheck L, Seifert A, Seifert P, Senf F, Siligam P, Simmer C, Steinke S, Stevens B, Wapler K, Weniger M, Wulfmeyer V, Zängl G, Zhang D, Quaas J (2017) Large-eddy simulations over Germany using ICON: a comprehensive evaluation. QJR Meteorol Soc 143:69–100. doi:10.1002/qj.2947

Higdon NS, Browell E, Ponsardin P, Grossmann B, Butler C, Chyba T, Mayo M, Allen R, Heuser A, Grant W, Ismail S, Mayor S, Carter A (1994) Airborne differential absorption lidar system for measurements of atmospheric water vapor and aerosols. Appl Opt 33:6422–6438

Ho S-P, Peng L, Anthes RA, Kuo Y-H, Lin H-C (2015) Marine boundary layer heights and their longitudinal, diurnal, and interseasonal variability in the southeastern Pacific using COSMIC, CALIOP, and radiosonde data. J Climate 28:2856–2872. doi:10.1175/JCLI-D-14-00238.1

Holloway C, Neelin J (2009) Moisture vertical structure, column water vapor, and tropical deep convection. J Atmos Sci 66:1665–1683. doi:10.1175/2008JAS2806.1

Illingworth AJ, Hogan RJ, O'Connor EJ, Bouniol D, Brooks ME, Delanoe J, Donovan DP, Eastment JD, Gaussiat N, Goddard JWF, Haeffelin M, Klein Baltink H, Krasnov OA, Pelon J, Piriou J-M, Protat A, Russchenberg HWJ, Seifert A, Tompkins AM, van Zadelhoff G-J, Vinit F, Willen U, Wilson DR, Wrench CL (2007) Cloudnet–continuous evaluation of cloud profiles in seven operational models using ground-based observations. Bull Am Meteorol Soc 88:883–898. doi:10.1175/BAMS-88-6-883

Ismail S, Browell E (1989) Airborne and spaceborne lidar measurements of water vapor profiles: a sensitivity analysis. Appl Opt 28:3603–3614

James RP, Markowski PM (2010) A numerical investigation of the effects of dry air aloft on deep convection. Mon Wea Rev 138:140–161. doi:10.1175/2009MWR3018.1

Jiménez C, Buehler SA, Rydberg B, Eriksson P, Evans KF (2007) Performance simulations for a submillimetre wave cloud ice satellite instrument. Q J R Meteorol Soc 133(S2):129–149. doi:10.1002/qj.134

John VO, Buehler SA (2004) The impact of ozone lines on AMSU-B radiances. Geophys Res Lett 31:L21108. doi:10.1029/2004GL021214

John VO, Holl G, Allan RP, Buehler SA, Parker DE, Soden BJ (2011) Clear-sky biases in satellite infrared estimates of upper tropospheric humidity and its trends. J Geophys Res 116:D14108. doi:10.1029/2010JD015355

Kiemle C, Wirth M, Fix A, Ehret G, Schumann U, Gardiner T, Schiller C, Sitnikov N, Stiller G (2008) First airborne water vapor lidar measurements in the tropical upper troposphere and mid-latitudes lower stratosphere: accuracy evaluation and inter comparisons with other instruments. Atmos Chem Phys 8:5245–5261

Kiemle C, Wirth M, Fix A, Rahm S, Corsmeier U, Di Girolamo P (2011) Latent heat flux measurements over complex terrain by airborne water vapour and wind lidars. Q J R Meteorol Soc 137:190–203. doi:10.1002/qj.757

Kiemle C, Ehret G, Kawa SR, Browell EV (2015) The global distribution of cloud gaps in CALIPSO data. J Quant Spectrosc Radiat Transf 153–101:95–101. doi:10.1016/j.jqsrt.2014.12.001

Kiemle C, Groß S, Wirth M, Bugliaro L (2017) Airborne lidar observations of water vapor variability in tropical shallow convective environment. Surv Geophys. doi:10.1007/s10712-017-9431-5

Kirchengast G, Hoeg P (2004) The ACE + mission: an atmosphere and climate explorer based on GPS, GALILEO, and LEO-LEO radio occultation. In: Kirchengast G, Foelsche U, Steiner AK (eds) Occultations for probing atmosphere and climate. Springer, Berlin, pp 201–220. doi:10.1007/978-3-662-09041-1_19

Kirchengast G, Schweitzer S (2011) Climate benchmark profiling of greenhouse gases and thermodynamic structure and wind from space. Geophys Res Lett 38:L13701. doi:10.1029/2011gl047617

Knuteson, R.O., et al., 2004: Atmospheric Emitted Radiance Interferometer. Part I: Instrument design. J. Atmos. Oceanic Technol., 21, 1763-1776/

Kursinski ER, Hajj GA, Schofield JT, Linfield RP, Hardy KR (1997) Observing Earth's atmosphere with radio occultation measurements using the Global Positioning System. J Geophys Res 102:23429–23465
Kursinski ER, Syndergaard S, Flittner D, Feng D, Hajj G, Herman B, Ward D, Yunck T (2002) A microwave occultation observing system optimized to characterize atmospheric water, temperature, and geopotential via absorption. J Atmos Oceanic Tech 19:1897–1914. doi:10.1175/1520-0426(2002)019<1897:Amooso>2.0.Co;2
Kursinski ER, Ward D, Otarola A, Frehlich R, Groppi C, Albanna S, Shein M, Bertiger W, Pickett H, Ross M (2009) The active temperature, ozone and moisture microwave spectrometer (ATOMMS). In: Steiner AK, Pirscher B, Foelsche U, Kirchengast G (eds) New Horizons in Occultation Research: Studies in Atmosphere and Climate. Springer, Berlin, pp 295–313. doi:10.1007/978-3-642-00321-9_24
Kursinski ER, Ward D, Stovern M, Otarola AC, Young A, Wheelwright B, Stickney R, Albanna S, Duffy B, Groppi C, Hainsworth J (2012) Development and testing of the active temperature, ozone and moisture microwave spectrometer (ATOMMS) cm and mm wavelength occultation instrument. Atmos Meas Tech 5:439–456. doi:10.5194/amt-5-439-2012
Kursinski ER, Ward D, Otarola AC, McGhee J, Stovern M, Sammler K, Reed H, Erickson D, McCormick C, Griggs E (2016) Atmospheric profiling via satellite to satellite occultations near water and ozone absorption lines for weather and climate. In: Proceeding of the SPIE 9881 earth observing missions and sensors: development, implementation, and characterization IV, 98810Z. doi:10.1117/12.2224038
Lacour J-L, Risi C, Clarisse L, Bony S, Hurtmans D, Clerbaux C, Coheur P-F (2012) Mid-tropospheric deltaD observations from IASI/MetOp at high spatial and temporal resolution. Atmos Chem Phys 12:10817–10832
Lawrence R, Lin B, Harrah S, Hu Y, Hunt P, Lipp C (2011) Initial flight test results of differential absorption barometric radar for remote sensing of sea surface air pressure. J Quant Spectrosc Radiat Transf 112:247–253. doi:10.1016/j.jqsrt.2010.06.001
Lebsock MD, Suzuki K, Millán LF, Kalmus PM (2015) The feasibility of water vapor sounding of the cloudy boundary layer using a differential absorption radar technique. Atmos Meas Tech 8:3631–3645. doi:10.5194/amt-8-3631-2015
Lin B, Hu Y (2005) Numerical simulations of radar surface air pressure measurements at O2 bands. IEEE Geosci Remote Sens Lett 2:324–328. doi:10.1109/LGRS.2005.848515
Liu C-L, Kirchengast G, Syndergaard GS, Kursinski ER, Sun YQ, Bai W-H, Du Q-F (2017) A review of low Earth orbit occultation using microwave and infrared-laser signals for monitoring the atmosphere and climate. Adv Space Res. doi:10.1016/j.asr.2017.05.011
Löhnert U, Schween JH, Acquistapace C, Ebell K, Maahn M, Barrera-Verdejo M, Hirsikko A, Bohn B, Knaps A, O'Connor E, Simmer C, Wahner A, Crewell S (2015) JOYCE: jülich observatory for cloud evolution. Bull Am Meteor Soc 96:1157–1174. doi:10.1175/BAMS-D-14-00105.1
Macke A et al (2017) The HD(CP)^2 Observational prototype experiment hope—an overview. Atmos Chem Phys 17:4887–4914. doi:10.5194/acp-17-4887-2017
Mapes B, Chandra AS, Kuang Z, Zuidema P (2017) Importance profiles for water vapor. Submitted to Surveys in Geophysics
Mauritsen T, Stevens B (2015) Missing iris effect as a possible cause of muted hydrogical change and high climate sensitivity in models. Nat Geosci. doi:10.1038/NGEO2414
Mauskopf P et al (2014) Photon-noise limited performance in aluminum LEKIDs. J Low Temp Phys 176(3–4):545–552
Meneghini R, Liao L, Tian L (2005) A feasibility study for simultaneous estimates of water vapor and precipitation parameters using a three-frequency radar. J Appl Meteorol 44:1511–1525. doi:10.1175/JAM2302.1
Millan L, Lebsock M, Livesey N, Tanelli S (2016) Differential absorption radar techniques: water vapor retrievals. Atmos MeasTech Discuss. doi:10.5194/amt-2016-72
Millán L, Lebsock M, Livesey N, Tanelli S, Stephens G (2014) Differential absorption radar techniques – Part 1: surface pressure. Atmos Meas Tech Discuss 7:5795–5827. doi:10.5194/amtd-7-5795-2014
Moore AS et al (1997) Development of the lidar atmospheric sensing experiment (LASE)–an advanced airborne DIAL instrument, advances in atmospheric remote sensing with lidar. Springer, Berlin, pp 281–288
Nehrir AR (2011a) Development of an eye-safe diode-laser-based micro-pulse differential absorption lidar (MP-DIAL) for atmospheric water-vapor and aerosol studies. Ph.D. thesis, Montana State University
Nehrir AR, Repasky KS, Carlsten JL (2011) Eye-safe diode-laser-based micropulse differential absorption lidar (DIAL) for water vapor profiling in the lower troposphere. J Atmos Oceanic Technol 28:131–147. doi:10.1175/2010JTECHA1452.1

Nehrir AR, Repasky KS, Carlsten JL (2012) Micropulse water vapor differential absorption lidar: transmitter design and performance. Opt. Express 20:25137–25151. doi:10.1364/OE.20.025137
Nuijens L, Stevens B, Siebesma AP (2009) The environment of precipitating shallow cumulus convection. J Atmos Sci 66:1962–1979
Nuijens L, Emanuel K, Masunaga H, L'Ecuyer T (2017) Implications of warm rain in shallow cumulus and congestus clouds for large-scale circulations. Submitted to Surveys in Geophysics
Perrin A, Puzzarini C, Colmont J-M, Verdes C, Wlodarczak G, Cazzoli G, Buehler S, Flaud J-M, Demaison J (2005) Molecular line parameters for the MASTER (millimeter wave acquisitions for stratosphere/troposphere exchange research) database. J Atmos Chem 50(2):161–205. doi:10.1007/s10874-005-7185-9
Petters JL, Jiang H, Feingold G et al (2013) A comparative study of the response of modeled nondrizzling stratocumulus to meteorological and aerosol perturbations. Atmos Chem Phys 13:2507–2529. doi:10.5194/acp-13-2507-2013
Pincus R, Beljaars A, Kirchengast G, Buehler S, Ladstaedter F (2017) The distribution of water vapor over low-latitude oceans: Current best estimates, errors, and impacts. Surv Geophys, in revision (status Aug 2017, to appear also in the ISSI-Shallow Clouds special issue)
Poberaj G, Fix A, Assion A, Wirth M, Kiemle C, Ehret G (2002) Airborne all-solid-state DIAL for water vapour measurements in the tropopause region: system description and assessment of accuracy. Appl Phys B 75:165–172. doi:10.1007/s00340-002-0965-x
Rodgers CD (1990) Characterization and error analysis of profiles retrieved from remote sounding measurements. J Geophys Res 95(D5):5587–5595
Schneider M, Hase F (2011) Optimal estimation of tropospheric H_2O and δD with IASI/METOP. Atmos Chem Phys 11(21):11207–11220. doi:10.5194/acp-11-11207-2011
Schotland RM (1966) Proc fourth symp remote sens environ, 273 p. ERIM, Ann Arbor
Schotland RM (1974) Errors in the lidar measurement of atmospheric gases by differential absorption. J Appl Meteor 13:71–77. doi:10.1175/1520-0450
Schweitzer S, Kirchengast G, Schwaerz M, Fritzer J, Gorbunov M (2011) Thermodynamic state retrieval from microwave occultation data and performance analysis based on end-to-end simulations. J Geophys Res 116:D10301. doi:10.1029/2010jd014850
Seifert A, Heus T, Pincus R, Stevens B (2015) Large-eddy simulation of the transient and near equilibrium behavior of precipitating shallow convection. J Adv Mod Earth Sys 7:1918. doi:10.1002/2015MS000489
Sherwood SC, Bony S, Dufresne J-L (2014) Spread in model climate sensitivity traced to atmospheric convective mixing. Nature 505:37–42
Shirokoff E et al (2014) Design and performance of superspec: an on-chip, kid-based, mm-wavelength spectrometer. J Low Temp Phys 176(5–6):657–662
Spuler SM, Repasky KS, Morley B, Moen D, Hayman M, Nehrir AR (2015) Field-deployable diode-laser-based differential absorption lidar (DIAL) for profiling water vapor. Atmos Meas Tech 8:1073–1087. https://doi.org/10.5194/amt-8-1073-2015
Stein T, Delanoë J, Hogan R (2011) A comparison among four different retrieval methods for ice-cloud properties using data from CloudSat, CALIPSO, and MODIS. J Appl Meteor Climatol 50:1952–1969. doi:10.1175/2011JAMC2646.1
Steiner AK, Lackner BC, Ladstädter F, Scherllin-Pirscher B, Foelsche U, Kirchengast G (2011) GPS radio occultation for climate monitoring and change detection. Radio Sci. doi:10.1029/2010RS004614
Stevens B, Bony S (2013) Water in the atmosphere. Phys Today 66(6):29. doi:10.1063/PT.3.2009
Stevens B et al (2016) The barbados cloud observatory-anchoring investigations of clouds and circulation on the edge of the ITCZ. Atmos Met Soc, Bull. doi:10.1175/BAMS-D-14-00247.1
Stevens B, Brogniez H, Kiemle C, Lacour J-L, Crevoiser C, Killiani J (2017) Structure and dynamical influence of water vapor in the lower troposphere. Surv Geophys. doi:10.1007/s10712-017-9420-8
Tian L, Heymsfield GM, Li L, Srivastava RC (2007) Properties of light stratiform rain derived from 10- and 94-GHz airborne Doppler radars measurements. J Geophys Res 112. doi:10.1029/2006JD008144. http://doi.wiley.com/10.1029/2006JD008144
Turner DD, Löhnert U (2014) Information content and uncertainties in thermodynamic profiles and liquid cloud properties retrieved from the ground-based atmospheric emitted radiance interferometer (AERI). J Appl Meteor Climatol 53:752–771. doi:10.1175/JAMC-D-13-0126.1
Turner DD, Goldmsith J, Ferrare RA (2016) Development and applications of the ARM Raman Lidar. Meteorol Monogr 57:18.1–18.5. doi:10.1175/AMSMONOGRAPHS-D-15-0026.1
Verdes CL, Buehler SA, Perrin A, Flaud J-M, Demaison J, Wlodarczak G, Colmont J-M, Cazzoli G, Puzzarini C (2005) A sensitivity study on spectroscopic parameter accuracies for a mm/sub-mm limb sounder instrument. J Mol Spectro 229(2):266–275. doi:10.1016/j.jms.2004.09.014

Vial J, Dufresne J-L, Bony S (2013) On the interpretation of inter-model spread in CMIP5 climate sensitivity estimates. Clim Dyn. doi:10.1007/s00382-013-1725-9
Wang L, Han Y, Jin X, Chen Y, Tremblay DA (2015) Radiometric consistency assessment of hyperspectral infrared sounders. Atmos Meas Tech 8:4831–4844. doi:10.5194/amt-8-4831-2015
Weckwerth T, Weber K, Turner D, Spuler S (2016) Validation of a water vapor micropulse differential absorption lidar (DIAL). J Atmos Oceanic Technol 33:2353–2372. doi:10.1175/JTECH-D-16-0119.1
Westwater ER (1997) Remote sensing of tropospheric temperature and water vapor by integrated observing systems. Bull Am Meteor Soc 78:1991–2006
Wickert J, Reigber C, Beyerle G, König R, Marquardt C, Schmidt T, Grunwaldt L, Galas R, Meehan T, Melbourne W, Hocke K (2001) Atmosphere sounding by GPS radio occultation: first results from CHAMP. Geophys Res Lett 28:3263–3266. doi:10.1029/2001GL013117
Wing AA, Emanuel KA (2014) Physical mechanisms controlling self-aggregation of convection in idealized numerical modeling simulations. JAMES 6:47–59. doi:10.1002/2013MS000269
Wing AA, Emanuel K, Holloway CE, Muller C (2017) Convective self-aggregation in numerical simulations: a review. Surv Geophys. doi:10.1007/s10712-017-9408-4
Wirth M, Fix A, Mahnke P, Schwarzer H, Schrandt F, Ehret G (2009) The airborne multi-wavelength H_2O-dial wales: system design and performance. Appl Phys B 96:201–213
Worden J, Noone D, Bowman K (2007) Importance of rain evaporation and continental convection in the tropical water cycle. Nature 445:528–532
Wulfmeyer V, Hardesty RM, Turner DD, Behrendt A, Cadeddu MP, Di Girolamo P, Schlüssel P, Van Baelen J, Zus F (2015) A review of the remote sensing of lower tropospheric thermodynamic profiles and its indispensable role for the understanding and the simulation of water and energy cycles. Rev Geophys 53:819–895. doi:10.1002/2014RG000476
Zeng et al (2015) Workshop report on scientific challenges and opportunities in the NASA weather focus area

Surv Geophys (2017) 38:1483–1508
https://doi.org/10.1007/s10712-017-9452-0

Observational Constraints on Cloud Feedbacks: The Role of Active Satellite Sensors

David Winker[1] · Helene Chepfer[2] · Vincent Noel[3] · Xia Cai[4]

Received: 7 August 2017 / Accepted: 17 November 2017 / Published online: 30 November 2017

Abstract Cloud profiling from active lidar and radar in the A-train satellite constellation has significantly advanced our understanding of clouds and their role in the climate system. Nevertheless, the response of clouds to a warming climate remains one of the largest uncertainties in predicting climate change and for the development of adaptions to change. Both observation of long-term changes and observational constraints on the processes responsible for those changes are necessary. We review recent progress in our understanding of the cloud feedback problem. Capabilities and advantages of active sensors for observing clouds are discussed, along with the importance of active sensors for deriving constraints on cloud feedbacks as an essential component of a global climate observing system.

Keywords Cloud feedback · Satellite lidar · Radar · Deep convection · Shallow clouds

1 Introduction

Equilibrium climate sensitivity (ECS)—the sensitivity of global-mean surface temperature to a doubling of CO_2—has become an organizing concept of climate science which reflects many aspects of climate change including changes of the hydrologic cycle and regional climate. ECS was first estimated in the late 1970s, based on clever inferences from two early global circulation models, with a range of 1.5–4.5 °C (Charney et al. 1979). Since

✉ David Winker
david.m.winker@nasa.gov

1 MS/475, NASA Langley Research Center, Hampton, VA 23681, USA

2 LMD/IPSL, CNRS, UPMC, University of Paris 06, 75252 Paris, France

3 Laboratoire d'Aérologie, CNRS, 31400 Toulouse, France

4 Science Systems and Applications, Inc (SSAI), Hampton, VA 23666, USA

then, our understanding of the climate system has improved markedly and much effort has been applied to reducing this range of uncertainty. However, in the most recent report of the IPCC, the ECS is still judged to be "likely in the range 1.5–4.5 °C" (IPCC 2013), illustrating that the uncertainties in ECS have proven to be remarkably resilient. Much progress has been made, however, in the identification and understanding of the processes that control climate sensitivity (Stevens et al. 2016a). A reduction in the range of ECS—particularly at the high end where the risks are greatest and the economic benefits of better information are largest—would be tremendously valuable in improving our abilities to better plan for climate change (Cooke et al. 2013; Neubersch et al. 2014) and would help determine the levels of greenhouse gas emissions compatible with a global warming target of 1.5 or 2 °C.

Estimates of ECS based on the historical surface temperature record tend to fall at the low end of the IPCC range (Otto et al. 2013; Lewis and Curry 2015; Forster 2016), while observational tests of processes relevant to cloud feedbacks tend to point toward the high end of the range (e.g., Sherwood et al. 2014; Su et al. 2014). This inconsistency has been a focus of research efforts in recent years, and questions have been raised as to whether the historical record is fundamentally unsuitable for constraining ECS (Knutti and Hegerl 2008). In spite of these difficulties, techniques of analyzing model output have become increasingly sophisticated over the last decade and it is now well established that the largest source of model diversity in ECS is due to cloud feedbacks (Bony et al. 2006; Webb et al. 2013). The large spread in model ECS can be traced to the complexity of the processes involving clouds and the difficulty of realistically representing these processes in global models. The occurrence and properties of clouds are the result of multiple competing processes taking place over a wide range of space and time scales. Many processes must be highly parameterized in the global models used for climate studies, and there are questions as to whether the key processes are captured sufficiently well. Much has been learned from high-resolution eddy-scale models, but interactions across spatial and temporal scales are also important and models are only beginning to reach the point where these scale interactions can be directly simulated.

Providing observational constraints on cloud feedbacks poses a significant challenge. Clouds have a large impact on Earth's radiation budget with global-mean shortwave (SW) and longwave (LW) cloud radiative effects estimated at -47.1 and $+26.5$ W/m^2, respectively, resulting in a net cooling of about 20 ± 4 W/m^2 (Loeb et al. 2009). Because the radiative effects of clouds are large relative to the radiative forcing of 3.7 W/m^2 expected from a doubling of CO_2 (Myhre et al. 2001), even small forced changes in clouds can act to significantly enhance or mitigate greenhouse gas warming. Observing these small changes requires an observing system which is highly accurate and stable over the decades required for trends to emerge from the noise of climate variability.

The international "A-train" constellation, built around a core consisting of the Aqua, CALIPSO, and CloudSat satellites, represents a major advance underway since the late 1990s in our ability to observe Earth's climate system. The A-train has provided an unprecedented, comprehensive set of observations of clouds, aerosols, and atmospheric state (L'Ecuyer and Jiang 2010) which have significantly advanced our understanding of climate. But more progress is needed. In this paper, we address the question of how observations can be used to reduce current uncertainties in cloud feedbacks. In the next section, we briefly summarize the current understanding of cloud feedbacks. Section 3 reviews some of the strengths and weaknesses of lidar for monitoring cloud changes. In Sect. 4, we review some of the approaches used to determine cloud feedbacks from satellite sensors, difficulties posed by non-ideal instruments, and discuss new opportunities presented by the emerging record from active satellite sensors. In particular, what could be learned from extending the current record obtained from the active sensors in the A-train constellation to multidecade time scales. Finally, Sect. 5 presents an outline of the satellite observations beyond the A-train necessary to continue advancing our understanding of cloud processes and feedbacks.

2 What Do We Know? What Do We Still Need to Know?

Cloud feedbacks, and tropical low-cloud feedbacks in particular, have been identified as the dominant source of uncertainty in model estimates of climate sensitivity (Bony and Dufresne 2005; Vial et al. 2013). Reducing these uncertainties has proven difficult, in part due to the difficulty of quantifying the factors controlling clouds and their response to a warming climate. But rapid progress has been made in developing techniques to interrogate model simulations and in delineating the cloud types and factors responsible for both feedbacks and their uncertainties. Plausible physical mechanisms have been proposed. Ultimately, climate models must be tested against observations and model parameterizations must capture the key physical mechanisms operating in the real world.

The effect of clouds on outgoing radiative fluxes is referred to as the cloud radiative effect (CRE). This can be computed as the difference between all-sky fluxes and fluxes from cloud-free scenes, or between cloudy and cloud-free scenes:

$$\mathrm{CRE} = F_{\mathrm{clear}} - F_{\mathrm{all-sky}} = \mathrm{CF}(F_{\mathrm{clear}} - F_{\mathrm{cloud}}) \quad (1)$$

where CF is the total cloud fraction and F refers to the net upwelling minus downwelling radiative fluxes at the top of the atmosphere (TOA). A "cloud radiative feedback" is the perturbation of outgoing radiative fluxes due to the changes of clouds in response to climate warming. While the change in CRE over time is equal to the change in clear-sky minus all-sky radiative fluxes, the cloud radiative feedback is the change in outgoing radiation due to clouds alone. Clouds have a masking effect, in that they reduce the contribution of changes in lower-level temperatures, surface albedo, and moistening to the outgoing radiative fluxes from those which would occur if the clouds were not present. Therefore, cloud feedbacks are estimated by adjusting the change in CRE for cloud masking of the temperature and water vapor feedbacks. This is commonly done using the radiative kernel approach pioneered by Soden et al. (2008). In climate models, longwave (LW) cloud feedbacks are driven by changes in mid- and high clouds (Zelinka et al. 2016). The largest feedback uncertainties, however, are associated with the shortwave (SW) feedbacks of shallow marine clouds (Bony et al. 2006; Soden and Held 2006; Vial et al. 2013).

2.1 What Observable Changes Do We Expect Under Climate Change?

Cloud radiative feedbacks result from changes in cloud fraction, height, or optical depth in response to rising greenhouse gas concentrations. Changes in optical depth can arise from changes in cloud liquid water path or by changes in thermodynamic phase. Zelinka et al. (2016) developed a decomposition of cloud feedback in terms of changes in macroscopic cloud properties. Their analysis of the feedbacks predicted by an ensemble of climate models shows low-cloud feedbacks (> 680 hPa) are dominated by a robust positive net cloud amount feedback arising from SW effects, with essentially zero altitude feedback from low clouds. A negative low-cloud optical depth feedback is found, primarily at high latitudes, which may involve changes of cloud phase from ice to water as the atmosphere warms. In contrast, non-low clouds (mid-level and high clouds above 680 hPa) show a robust and positive altitude feedback due to LW CRE. The LW and SW components of the cloud amount and cloud optical depth feedbacks of non-low clouds exhibit relatively large model diversity but tend to compensate each other. The cloud altitude, amount, and optical depth feedbacks vary geographically in complex patterns, compensating each other some places and reinforcing in others.

Determining the sensitivities of these cloud responses to warming surface temperatures, in the presence of large natural variability, represents one of the main challenges in characterizing and constraining cloud feedbacks. The many uncertainties associated with model estimates of cloud feedbacks motivate approaches based on observations. These can be roughly divided into studies analyzing observed trends over the satellite era and those which try to derive constraints on long-term cloud feedbacks from observations of inter-annual variability.

2.2 How Might We Diagnose These Changes?

Models robustly predict that high clouds will rise in a warming climate such that cloud top temperatures remain nearly constant, producing a positive LW feedback (Hartmann and Larson 2002; Zelinka and Hartmann 2010). The rise in cloud height occurs at all latitudes and in the extratropics is associated with increasing height of the tropopause. This appears to be a consequence of the fundamental mechanism underlying radiative-convective equilibrium: To maintain equilibrium, upward convective ascent must be matched by radiatively driven clear-sky subsidence. Thus, the vertical profile of clear-sky diabatic cooling acts as a control on the vertical development of tropical deep convection. Using observations from multiple A-train sensors, Zelinka and Hartmann (2011) found support for this mechanism by regressing monthly mean anomalies against sea surface temperature (SST). Li et al. (2012), in a similar study, point out that cloud profiles, as can obtained from active sensors, are a more appropriate indicator of the cloud detrainment level than cloud top temperature derived from passive sensors, and would provide better quantification of the SST–cloud top relationships.

Other studies have looked in observations for the trends in tropical cloud heights predicted by models. Norris et al. (2016) analyzed the 30-year (1983–2009) ISCCP and PATMOS-x datasets, both based on observations from a series of operational geostationary and polar orbiting satellites. They found it necessary to remove systematic biases due to satellite orbit drifts and sensor calibration differences, with the result of also removing global-mean variability and any climate signal which is geographically correlated with the biases removed (Evan et al. 2007; Norris and Evan 2015). They found changes in cloud height consistent with rising cloud altitude, but uncertainties in the datasets used precluded a quantitative estimation of the magnitude. Analysis of the two different cloud datasets—derived from similar sets of satellites, although using different algorithms—show significantly different patterns of changing cloud vertical occurrence, illustrating the uncertainties involved in using the existing long-term passive cloud climatologies for trend analysis.

Attention has focused recently on a more fundamental problem in deriving cloud feedbacks from observed trends. It is known that the SST change expected from rising CO_2 levels is more uniform than that from natural variability. Recent work shows that climate feedbacks are not constant but vary over the historical period (Armour et al. 2013; Xie et al. 2016; Gregory and Andrews 2016; Rugenstein et al. 2016) and that this comes, in part, from the dependence of cloud feedbacks on patterns of surface temperature change. In particular, Zhou et al. (2016) find the spatial patterns of surface warming have varied in a way that cloud feedbacks over recent decades are significantly more negative than long-term feedbacks. Su and Jiang (2013) provide an illustration of this, using observations from CALIPSO and CloudSat to show that El Nino events with different patterns of SST change can produce nearly opposite responses in the vertically resolved tropical cloud occurrence and cloud water content. They further diagnose that a significant part of the cloud response to SST changes is mediated by changes in large-scale circulation. There are also longer-term modes of climate variability operating on decadal and multidecadal time scales such as the Atlantic multidecadal oscillation (Enfield et al. 2001) and the interdecadal Pacific oscillation (Folland et al. 1999). These modes are associated with patterns of SST change distinct from those of both ENSO and global warming, further complicating the task of relating cloud feedbacks estimated from current observations to long-term feedbacks.

Given the difficulty of deriving constraints on long-term cloud feedbacks from observed trends, there has been a recent focus on relating the observed variability of shallow clouds—on time scales ranging from weekly to interannual—to controlling factors such as SST and the strength of the marine boundary layer inversion. In the so-called emergent constraints approach (Klein and Hall 2015), observed relationships between clouds and their controlling factors are used to constrain long-term feedbacks from models under the assumption there is a relationship between simulated present day and centennial scale variabilities. In one recent paper, Zhai et al. (2015) use observations from AIRS and CALIPSO–CloudSat to compute the sensitivity of marine low-cloud cover to changes in SST in regions of large-scale subsidence. They find the sensitivities at interannual and centennial scales are correlated in models. In a comparison of observed and modeled seasonal sensitivities, the models with seasonal sensitivities consistent with observations were found to have high ECS, whereas models inconsistent with the observations tended to have low ECS. Marine low-cloud cover and shallow cloud feedbacks are not uniquely determined by SST however. Some models showed a strong link between low cloud cover and SST, but in other models the link was much weaker, indicating the importance of other controlling factors such as tropical inversion strength, lower tropospheric stability, water vapor, and variations in atmospheric circulations. Thus, there is still a need to confirm these results with long-term observations.

Qu et al. (2015) took a different approach to deriving long-term constraints from observed interannual variability. They used observations to estimate the sensitivity of shallow clouds to their controlling factors and then applied these sensitivities to the changes in the large-scale environmental simulated in future climates. The advantage of this approach is it does not rely on climate model simulations of clouds, but instead uses the more trustworthy predictions of how the large-scale controlling factors will change.

A number of other recent studies use one or both of these short-term variability approaches (Brient and Schneider 2016; Myers and Norris 2016; McCoy et al. 2017). Uncertainties in feedbacks estimated from these studies are still relatively large, but a meta-analysis by Klein et al. (2017, this issue) is able to derive a useful constraint on global models which indicates negative and near-zero tropical cloud feedbacks are unlikely. Even if these approaches are able to provide constraints on model feedbacks, however, they don't provide a straightforward path to improving the representation of cloud processes in models. Observational studies are still required to identify and constrain the mechanisms responsible for cloud feedbacks.

3 Observing Clouds with Active Sensors

While research over the last decade has greatly increased our understanding of cloud feedbacks, uncertainties in their magnitude have not been significantly reduced. Given the variety of approaches which have been explored, what are the prospects for reducing the current uncertainties and how can the (relatively new) availability of observations from active sensors contribute?

Satellite observations over multiple decades are necessary to directly detect the emergence of forced changes and separate them from natural variability. A National Institute of Standards and Technology (NIST) report (Ohring 2004) has identified stability and accuracy requirements on measurements used to characterize climate change trends. Requirements on cloud measurements were tied to current abilities to monitor solar irradiance and outgoing TOA LW fluxes to within a few tenths of a W/m^2 per decade. Climate monitoring requirements identified for clouds, necessary to constrain cloud feedbacks, are summarized in Table 1. Requirements on the accuracy of global-mean cloud cover and global-mean cloud top height were set at 1% and 150 m, respectively, with requirements on measurement stability of 0.3%/decade and 30 m/decade. These requirements call for observations with exceptional stability, and the calibration uncertainties of the satellite radiometers in current use significantly increase the time to detect cloud trends relative to a perfect instrument (Wielicki et al. 2013; Shea et al. 2017).

CALIPSO (Winker et al. 2010) and CloudSat (Stephens et al. 2002), launched together in 2006, have provided our first experience with active profiling of clouds from space. These instruments have now acquired more than 11 years of global cloud profiles over a period which includes three ENSO cycles—long enough to sufficiently characterize the mean state of clouds in the early twenty-first century. Lidar offers the necessary accuracy, sensitivity, and long-term stability to monitor the small trends in cloud cover and altitude expected under climate change and provide constraints on cloud radiative feedbacks.

Passive cloud retrievals are based on relatively simple forward models—involving assumptions on atmospheric vertical structure—and large errors and ambiguities can ensue when these assumptions are violated in the real atmosphere (Stephens and Kummerow 2007). Whereas passive retrievals are limited to retrieval of a single effective cloud height, lidar provides a profile of the vertical distribution of cloud—including multilayer clouds—down to the altitude where the return signal is extinguished by cloud attenuation. Lidar cloud altitude is directly measured from the laser pulse time-of-flight (Winker et al. 2007). Measurement of the time delay is inherently accurate, and independent of radiometric calibration. Knowledge of the distance from the satellite to the ocean surface provides an additional constraint. The satellite orbit altitude is determined to within 30 m using the

Table 1 NIST requirements for climate-accuracy cloud observations

Parameter	Accuracy	Stability
Cloud top temperature	1 K	0.2 K/decade
Cloud top cover	1%	0.3%/decade
Cloud top height	150 m	30 m/decade
Cloud base height	500 m	100 m/decade

global positioning system (GPS), so that the range from satellite to the ocean surface is known independently of laser timing. Any potential timing drifts can be identified by an apparent change in the time delay from emission of the laser pulse to detection of the laser return from the ocean surface. GPS orbit determination allows one to constrain errors in lidar altitudes due to a potential instrument clock drift to about 1 m absolute long-term accuracy.

Clouds are detected in the lidar backscatter profile via the contrast between returns from clouds and the molecular atmosphere. As this is a relative measurement, cloud detection is largely independent of potential calibration errors. For those measurements which do depend on radiometric calibration, lidar returns from the mid-stratosphere are referenced to molecular density profiles computed from global re-analysis products—essentially using the mid-stratosphere as a calibration target (Powell et al. 2009). Radiometric calibration of Cloud-Aerosol Lidar with Orthogonal Polarization (CALIOP) is currently performed using lidar returns between 35 and 40 km altitude, where aerosol contributions are negligible. Experience with CALIPSO has shown these calibrations have excellent long-term stability, as illustrated in Fig. 1. The red curve shows a time series of CALIOP 532 nm attenuated backscatter (calibrated profile data) integrated from 24 to 40 km, where the lidar backscatter is dominated by molecular scattering, and averaged over 50°S–50°N. The blue dashed curve is the normalized molecular number density from the NASA Global Modeling and Assimilation Office Forward Processing for Instrument Teams (GMAO-FP-IT) analysis product over the same altitude range, interpolated to the CALIPSO ground track. The black curve shows the time history of 532-nm laser pulse energy over the mission. Pulse energy has varied over time due to the loss of pump diodes, adjustments to the laser, and a switch between primary and backup lasers in March 2009. The calibration scheme is able to accurately correct the profile data for these variations in pulse energy, producing a stable long-term record.

With high sensitivity and high-resolution vertical profiling capability, lidar provides the most rigorous observations of cloud fraction and cloud top height. Cloud profiling radar provides complementary information on the vertical distribution of optically thick clouds. CALIOP has been used extensively to identify and characterize biases in passive cloud height retrievals due to broken or multilayer clouds, or when clouds lie underneath temperature inversions (Holz et al. 2008; di Michele et al. 2013). In particular, CALIOP and CloudSat together have been used to illustrate the biases in the cloud top pressure-optical

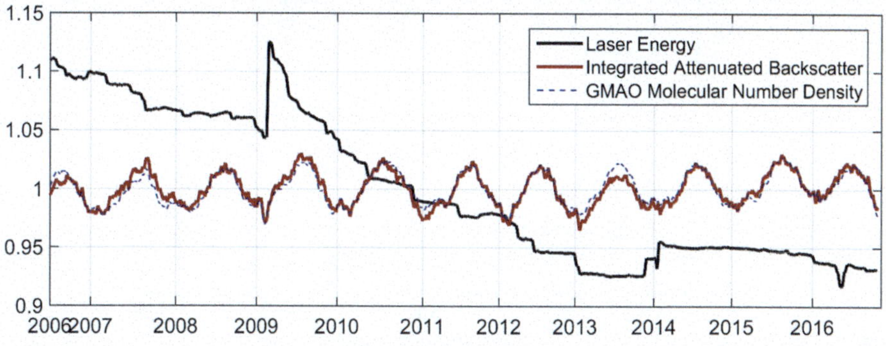

Fig. 1 Normalized trends of laser total pulse energy, mid-stratospheric 532 nm attenuated backscatter signals integrated between 25 and 40 km altitude, and molecular number density averaged over the same altitude range

depth (CP-τ) diagrams derived from passive sensors due to ambiguities in passive cloud height retrievals (Marchand et al. 2010; Mace and Wrenn 2013). Relative to using passive sensors alone, Stephens et al. (2017) illustrate the improved ability to partition CRE by cloud type using active and passive sensors together, primarily due to the ability of active sensors to identify and characterize multilayer cloud situations.

These characteristics of lidar result in a highly stable observing system which is referenced to fundamental benchmarks, so that overlap between successive lidar missions is not required to transfer calibrations. For nadir-viewing instruments such as lidar, sampling can be an important limitation, however. But even nadir-viewing measurements can provide accurate sampling on the space and time scales relevant to climate. Uncertainties in sampling cloud cover using a nadir-viewing sensor can be estimated using an approach from Key (1993), who develops a simple expression for the variance of an estimate of 2D cloud cover based on observations along one-dimensional transects. The method assumes a random, isotropic field of circular clouds. The variance of estimated cloud cover, p', is a function of the number and length of the transects and the autocorrelation scale of cloud cover:

$$\mathrm{var}(p') = 2p(1-p)\frac{1-1/aL}{aNL} \quad (2)$$

where N is the number of transects, L is the length of the transects, and a describes the correlation scale such that aN is the number of independent samples. The uncertainty in estimated cloud cover decreases linearly as aNL increases. Assuming mean cloud cover of 50% ($p = 0.5$, which maximizes the variance) and using a cloud autocorrelation scale of 50 km, Fig. 2 shows how uncertainties in cloud cover estimates decrease as space–time sampling increases. Symbols indicate a few relevant space–time scales. This theoretical analysis predicts a RMS sampling uncertainty in global monthly mean cloud cover of 0.1%, well below the National Institute of Standards and Technology (NIST) requirement, and 1% for monthly zonal mean cloud cover with 10° latitudinal resolution.

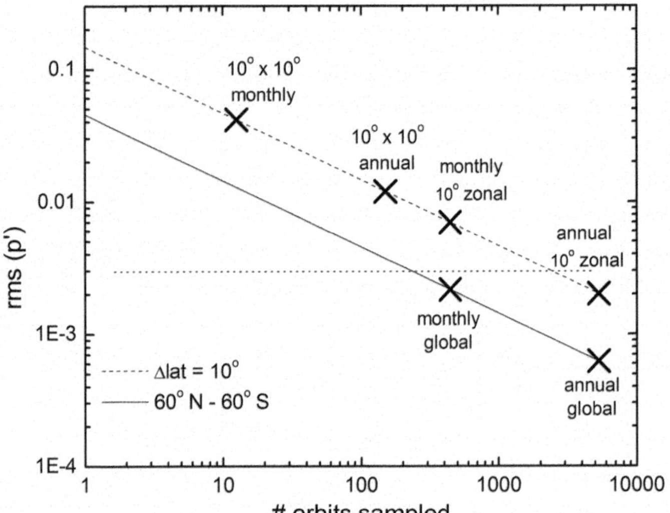

Fig. 2 RMS uncertainty in estimated cloud fraction, p', for different space–time averaging scales assuming mean cloud fraction of 50%

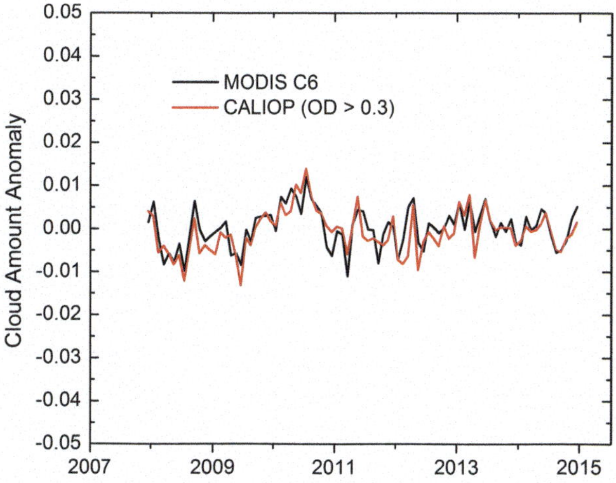

Fig. 3 Cloud amount anomalies, 60°S–60°N, from MODIS Collection 6 full-swath observations and thresholded CALIOP nadir observations (optical depth > 0.3 only)

As a check on this theoretical estimate, Fig. 3 shows the anomalies of monthly mean cloud cover between 60°N and 60°S from nadir-viewing CALIOP observations and from full-swath Moderate Resolution Imaging Spectroradiometer (MODIS) Collection 6 data. CALIOP cloud amounts were computed using only cloud layers with optical depths greater than 0.3, in an attempt to adjust for optically thin clouds below the MODIS detection threshold. It can be seen the anomalies track closely between the two instruments with an RMS cloud fraction difference of 0.0031 and a trend which is not statistically different from zero. This result is reasonably consistent with the theoretical prediction shown in Fig. 2 (monthly global ~ 0.2%), in spite of the rather simple correction for sensitivity differences between CALIOP and MODIS. Similar behavior is seen in comparing anomalies of tropical cloud amount. These results support the ability of nadir-only observations to provide sufficient sampling to meet or exceed climate-accuracy requirements at monthly global scales and to provide sufficiently accurate estimates at monthly zonal scales.

Additionally, CALIOP has one advantage over MODIS related to sampling. Cloud detections from MODIS and many other satellite sensors rely on observations using 1 km pixels or larger. Many clouds smaller than a kilometer are found in trade cumulus regions, and it has been shown that this can lead to overestimates of cloud fraction relative to sensors having smaller footprints (Wielicki and Parker 1992; Zhao and Di Girolamo 2006). Moreover, this overestimate is sensitive to the spatial distribution of clouds so that it may be difficult to separate changes in cloud fraction from changes in the spatial organization of trade cumulus (Zhao and Di Girolamo 2006). The lidar footprint of CALIOP (about 100 m diameter) is much smaller than the 1 km pixels used for MODIS cloud detection and greatly reduces the occurrence of broken cloud within a footprint.

Another limitation of lidar is attenuation; lidar signals do not penetrate through optically thick clouds. A typical rule of thumb is that usable lidar signals penetrate to an optical depth of about 3. Because of the 100-m diameter footprint of CALIOP, relatively large for lidar, small-angle forward scattering in ice clouds provides enhanced penetration to optical depths of about 5. Figure 4 compares annual average cloud cover of all ice clouds (42%) and of ice-topped clouds which are opaque to the lidar (9.1%). The relatively rare opaque

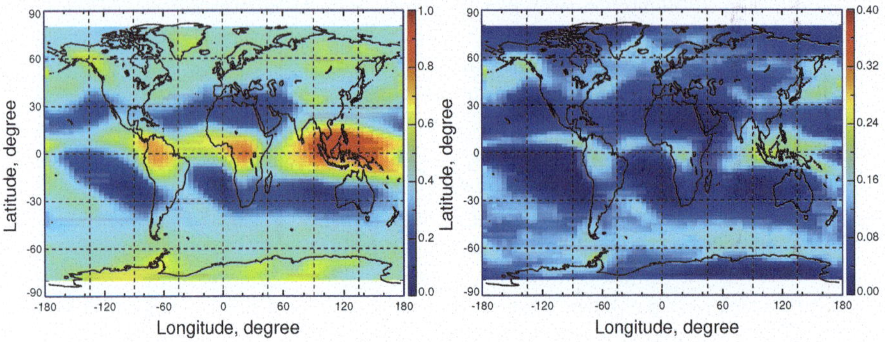

Fig. 4 Annual mean coverage during 2008 of cirrus and ice-topped clouds (left panel) and opaque ice-topped clouds (right panel)

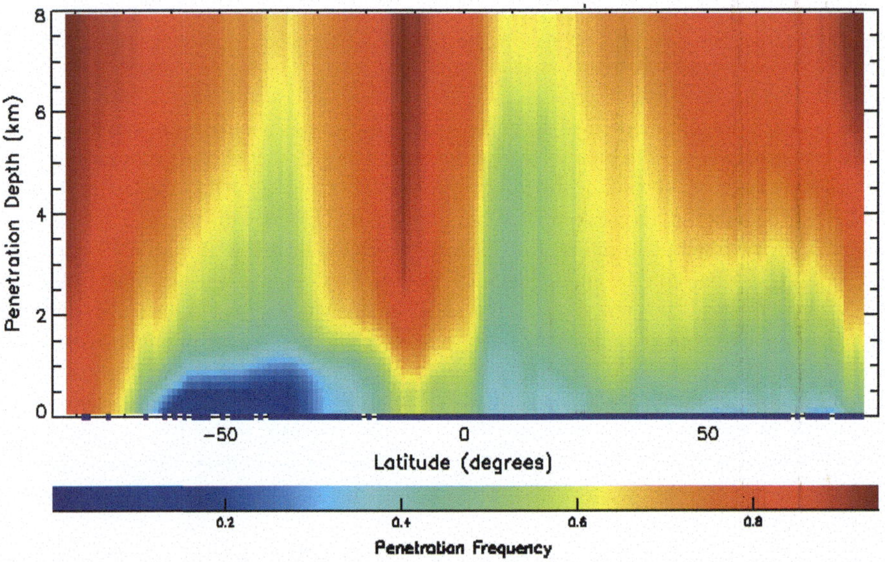

Fig. 5 Zonal mean CALIOP frequency of penetration in cloudy columns, from TOA down to altitude *y*, for July 2009 based on single-shot 532 nm data. In the vicinity of the InterTropical Convergence Zone (ITCZ), about 20% of single shots are fully attenuated above 8 km

ice clouds are found mostly at mid-to-high latitudes and in regions of tropical deep convection. As a consequence, warm clouds in subsidence regions are well observed by satellite lidar. Figure 5 presents another view of lidar attenuation, showing zonal mean penetration statistics from CALIOP. The lidar signal is seen to frequently penetrate into the lower troposphere, essentially penetrating most ice cloud and being attenuated within most water clouds.

4 Constraints on Cloud Feedbacks from Lidar

As discussed above, cloud feedbacks result from the small but radiatively important changes in cloud fraction, height, optical depth, or water phase in response to rising greenhouse gas concentrations. These parameters have large natural variability making the detection of forced trends difficult. The small changes in these properties which are driven by warming will only emerge from the noise of natural variability on multidecade time scales. This places stringent requirements on the long-term accuracy and stability of sensors used to detect change. Rigorous analysis shows the time required for a perfect instrument to detect a trend of magnitude $|\omega_o|$ with probability of 90% is:

$$n = \left[\frac{3.3\sigma_\varepsilon}{|\omega_o|(1-\phi)}\right]^{2/3} \tag{3}$$

where n is the number of years of observations required, σ_ε^2 is the variance of a white noise process representing the natural variability of the parameter being trended, and ϕ is the autocorrelation (Weatherhead et al. 1998). For a given value of ϕ it can be seen that the trend detection time increases as $[\sigma_\varepsilon/|\omega_o|]^{2/3}$. Given the long time-scales required to detect climate trends, measurement accuracy and long-term stability are more important than the precision of individual measurements, and calibration uncertainties can significantly extend the trend detection time.

Global-mean surface air temperature is expected to increase at about 0.2 K/decade over the next few decades. Wielicki et al. (2013) show that the trend uncertainty for an infrared radiometer with perfect calibration reaches 0.2 K/decade in about 12 years, but if an instrument with calibration accuracy similar to that of IASI or AIRS was used, it would take more than 20 years to reach the same trend uncertainty. Additional sources of bias such as orbit drifts or changes in channel spectral widths between instruments extend the detection time even further.

In addition to problems with instrument calibrations, passive retrievals of cloud height and optical depth are under-constrained. The forward models of the atmosphere which form the basis of cloud retrievals are highly simplified and often involve poorly justified assumptions (see discussions by Stephens and Kummerow (2007) and Pincus et al. (2012) for example). The presence of just a single cloud layer is one of the most common assumptions of passive cloud retrievals, whereas multilayer cloud occurrence ranges from 24 to 40%, depending on the definition of the gap separating single- and multilayer clouds, and as high as 60% in certain parts of the tropics (Mace et al. 2009; Matus and L'Ecuyer 2017). The most common situation is high ice cloud located over low warm cloud. Retrieval of an effective single-layer cloud height in the presence of multiple cloud layers can result in cloud height errors of as much as several kilometers (Holz et al. 2008), causing high clouds overlying low clouds to be reported as mid-level clouds. Other biases are found in the retrieval of broken cloud scenes or nighttime retrievals of low clouds where there is limited thermal contrast with the surface.

For all these reasons, evaluations of cloud vertical distributions derived from ISCCP show significant differences with CALIPSO–CloudSat (Rossow and Zhang 2010; Tselioudis et al. 2013). Comparison of joint histograms of cloud top pressure and optical depth (CP-τ diagrams) from ISCCP and from CALIPSO–CloudSat shows that ISCCP retrievals tend to place high clouds too low and low clouds too high (Mace and Wrenn 2013). While the ISCCP dataset provides our best long-term (30 + years) cloud observational dataset, these ambiguities and biases in the retrieved vertical distribution of clouds impact our ability to interpret the observed distribution of clouds. As an example, combining cloud profiles from CALIOP and CloudSat with other A-train observations have significantly improved previous estimates of the surface radiation budget due to improved estimates of cloud base height from the active sensors (Kato et al. 2011; Stephens et al. 2012).

To understand and model cloud feedbacks, we must be able to link changes in cloud radiative effects to the types of cloud which are changing and active sensors are the best tools we have for doing this. Evaluation of errors in CMIP5 clouds against CALIPSO–CloudSat observations has shown that errors in cloud simulation are due primarily to the cloud parameterizations used in the models rather than errors in simulating the large-scale atmospheric state (Su et al. 2013). Caldwell et al. (2013) reach a similar conclusion based on a number of recent General Circulation Model (GCM) and Large Eddy Simulation (LES) studies. Therefore, in addition to constraints on cloud radiative feedbacks, we also need observational constraints on the processes driving the changes and active sensors provide unique capabilities to do this.

Much recent work on cloud feedbacks has focused on deep convection in the tropics and clouds in regions of moderate to strong subsidence. In the following sections we focus on how active sensors can help in these specific areas of interest.

4.1 Tropical Convective Regions

While the tropical LW cloud feedback is robustly positive in climate models, it varies substantially from weakly positive (0.2) to more than 1 $W/m^2/K$ (Zelinka and Hartmann 2010; Tomassini et al. 2013) and the predicted change in cloud height over the twenty-first century varies by a factor of two, from about 600 to 1100 m (Zelinka and Hartmann 2010). A decrease in tropical anvil coverage tends to accompany the cloud rise, and the changes in high cloud amount represent the primary source of diversity in the LW feedback. A "stability-iris" mechanism, which could be responsible for decreasing anvil coverage, has recently been proposed (Bony et al. 2016), but the net radiative effect of the changes in anvil coverage are ambiguous. For tropical high clouds, LW and SW CRE tend to compensate but the net balance depends on optical depth, which could change as part of changes in anvil coverage. Further, decreasing anvil coverage might induce a change in lower-level clouds via changes in downwelling radiation. Whether anvil clouds exhibit a stability-iris effect will be an important observational test of climate models and of our understanding of tropical feedback processes.

Evaluation of the predicted ascent of tropical clouds using passive observations is limited by (1) systematic errors in the long-term cloud record due to drifts in calibration and orbit drifts, and (2) ambiguities in cloud height–optical depth relationships (Mace and Wrenn 2013). Studies based on passive satellite observations have been used to confirm the underlying mechanism and that the sign of high cloud feedback is positive, but ambiguities in passive cloud datasets are too large to provide useful constraints on the magnitude of the feedback (Zelinka and Hartmann 2011; Marvel et al. 2015; Li et al. 2012; Norris et al. 2016). Shea et al. (2017) present a trend detection analysis for the global-mean difference between effective cloud top temperature and surface temperature, finding that the trend simulated by CMIP5 models could be constrained with about 12 years of observations from a perfect radiometer. Using an instrument with calibration accuracy similar to that of MODIS or Visible Infrared Imaging Radiometer Suite (VIIRS), however, requires 40–50 years of observations.

Rather than retrieving only an effective cloud height, lidar produces vertically resolved profiles of optically thin clouds and the upper portions of optically thick clouds; visible optical depth and infrared emissivity are related by:

$$\epsilon = 1 - e^{-k\tau} \tag{4}$$

where τ is the visible optical depth and k is the ratio of visible optical depth to 12 μm absorption optical depth. The value of k depends on particle size and has a weak dependence on crystal habit but for typical conditions has a value close to 1/2 (Garnier et al. 2015). As mentioned above, small-angle forward scattering in ice clouds enhances the penetration of a satellite lidar, so that CALIOP profiles the upper part of deep convective clouds to an optical depth of about 5. This corresponds to an emissivity of about 0.9 and is the part of the cloud which is relevant to radiative interactions with the top of the atmosphere—which can extend over depths of several kilometers in dissipating tropical convective systems.

All satellite sensors have limitations in characterizing clouds, and the various climatologies derived from them reflect these limitations (Stubenrauch et al. 2012). In recent years "satellite simulators" have been developed which attempt to sample model output in a way which is consistent with the limitations of the various satellite sensors which provide cloud observations (Webb et al. 2001; Bodas-Salcedo et al. 2011). These simulators provide a more consistent way of comparing observed clouds with model simulations. In particular, the "COSP/lidar simulator" (Chepfer et al. 2008, 2010) simulates the sampling limitations of the CALIPSO lidar due to nadir-only viewing and attenuation of the signal in dense clouds, as described in the previous section.

Chepfer et al. (2014) analyzed climate model simulations to show the predicted cloud rise is manifested as an upward shift in the profile of cloud occurrence. This ascending cloud profile is expected to be the first signature of the response of clouds to warming to emerge from the noise of natural variability. Use of COSP/lidar simulator output showed the signal of cloud rise can be robustly detected using satellite lidar observations, and observation of the upward shift in cloud profile is a more robust diagnostic than changes in effective cloud top heights from passive sensors (Li et al. 2012).

As a follow-on to Chepfer et al. (2014), the fraction of opaque cloud at an altitude of 6 km was defined as a parameter related to the ascent of tropical deep convection. While not an optimum metric of LW feedback, this was the best option from the available simulator output saved from CMIP5 experiments. COSP/lidar outputs from HadGEM2-A Control, and + 4 K AMIP experiments were used to determine the change in this metric over 100 years. Natural variability of the metric was derived from 7 years of CALIOP observations. While long enough to capture ENSO, the 7-year record is too short to capture longer-term modes of variability and so somewhat underestimates natural variability. Following the approach of Loeb et al. (2009), a 100-year time series of synthetic observations was constructed by adding zero-mean random fluctuations based on observed variability and autocorrelations to the linear trend estimated from model COSP/lidar output. A 200-member ensemble of 100-year time series of the metric, averaged over the tropics, was then produced, and trends were computed for each ensemble member by applying linear regression for all months between 2008 and 2108.

Results averaged over tropical ascent regions are shown in Fig. 6, illustrating the detectability of long-term cloud trends. Starting in 2008, the blue curve in Fig. 6 shows the mean trend in simulated lidar observations of opaque clouds computed from one ensemble member. Shading indicates the probability envelopes of trends computed from the entire

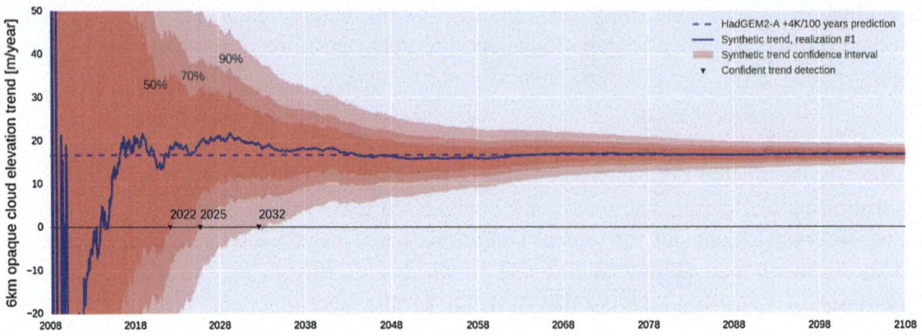

Fig. 6 Slope of the linear regression as a function of lidar observation record length from the HadGEM-2 CALIPSO simulator output. Shading shows 50, 70, and 90% confidence envelopes

ensemble. The results show the current CALIPSO record is too short to constrain even the sign of the LW feedback due to tropical ascent, but extending the record another 15 years to 2032 would provide a highly confident constraint on the sign of the feedback and a significant constraint on the magnitude. The lidar observations would of course also provide constraints on changes in coverage of high clouds.

Because of the inherent high accuracy of lidar cloud altitudes, measurements from one lidar instrument can be related to that of another without having measurement overlap. Thus, gaps between CALIPSO and future lidar missions do not represent a fundamental impediment to constructing a consistent long-term data record. The EarthCARE mission (Illingworth et al. 2015), currently scheduled for launch in 2019, will probably not last long enough to provide a significant constraint on tropical cloud ascent. Cloud heights from a future lidar system operating in the late 2020s, however, could be related to those from CALIPSO regardless of whether there are additional lidar missions in the interim period. Observations from such a future lidar mission, combined with the existing record from CALIPSO, would then provide a baseline of more than 20 years, sufficient to detect forced trends in cloud top heights as discussed in Chepfer et al. (2014) or shown in Fig. 6.

4.2 Low-Latitude Subsidence Regions

The dominant uncertainty in the cloud feedbacks simulated by climate models is associated with shallow marine clouds—stratus, stratocumulus, and trade cumulus (Bony et al. 2006). Due to their low altitude, these clouds drive SW feedbacks more strongly than LW. Shallow marine clouds are highly sensitive to changes in their environment and so may potentially undergo significant changes as SST increases and the tropical overturning circulation weakens. Most global climate models predict a positive cloud radiative feedback from shallow marine clouds, arising from a small but radiatively significant decrease in cloud fraction, but there is significant uncertainty in the magnitude and some models predict a near-zero or somewhat negative shallow cloud feedback. Understanding the physical mechanisms behind the diversity in shallow cloud feedbacks and improving their representation in models is one of the current research priorities.

The radiative properties of these shallow clouds are the result of a complex interplay of local processes (radiative cooling, entrainment, turbulent mixing, precipitation, etc.) and interactions with the large-scale environment, taking place over a range of space and time scales (Wood 2012; Bretherton 2015). The diversity of SW cloud feedback estimates from global models is a result of the difficulty of adequately representing these processes and their interactions. Computational limitations prevent models from explicitly resolving all these processes, and they must necessarily rely on parameterizations to a greater or lesser extent. While simulations from high-resolution models tend to exhibit more consistency than global models, even large eddy simulation (LES) models still rely on parameterized cloud microphysics. More importantly, LES studies tend to be run under simplified configurations on small spatial domains, cannot realistically represent interactions with the large-scale environment, and so are limited in their ability to estimate long-term feedbacks (Vial et al. 2017).

While LES models have limitations, they have been used extensively over the last 20 years or so, along with simpler mixed-layer models, to gain insight into the mechanisms driving shallow clouds and their responses to climate change. While the dominant control on shallow clouds in the present climate is the strength of the marine inversion, LES studies indicate an increasing moisture gradient at the marine inversion with climate warming will enhance turbulent entrainment-driven drying of the marine boundary layer, leading to thinner clouds and reduced cloud cover (Bretherton 2015). Reduced radiative cooling at cloud top from increasing greenhouse gas concentrations can reinforce this cloud thinning. But this cloud thinning would be partially compensated by the expected increased temperature gradient at the marine inversion, promoting thicker clouds and increased cloud cover. Similarly, weakening of the tropical overturning circulation could induce a compensating effect via reduced subsidence, deepening of the marine boundary layer, and thicker clouds (Bretherton 2015). Overall, LES studies point to a cloud thinning in response to climate warming, representing a positive cloud feedback. A degree of consistency in these predictions has been found between simulations from different high-resolution models (Blossey et al. 2013), but LES studies are inconclusive regarding the magnitude of the shallow cloud feedback.

Given the much larger areal coverage of the trade regime relative to marine stratus, global net shallow cloud feedbacks may be driven more by the response of trade cumulus than by clouds in regions of strong subsidence. Recent research has focused on the potential role of shallow convective mixing in controlling low-cloud amount (Sherwood et al. 2014). Vial et al. (2016) found that the strength of low-level cloud feedback in a global model depended on the closure scheme adopted for the convective parameterization. More generally, the feedback strength predicted by the model was found to depend sensitively on parameterization assumptions and on subtle couplings between different boundary layer processes. Recent results from LES modeling on large domains ($\sim$ 50 km) also point to a significant role for mesoscale circulations in regulating properties of the cloud field in trade cumulus regimes (Seifert et al. 2015). These circulations occur on scales typically too small to resolve in global models and too large for standard LES domains, and are just beginning to be explored as modeling capabilities expand to better capture scale interactions.

Recent high-resolution modeling studies have shown that low-cloud feedbacks arise from a number of competing mixing and stabilization processes and that cloud vertical structure may change in response (Blossey et al. 2013; Bretherton 2015). Brient et al. (2015) find that boundary layer cloud fraction in regions of weak subsidence is primarily driven by two opposing mechanisms: shallow convective mixing which dries the lower atmosphere—reducing cloud fraction—and turbulent moistening of the boundary layer, which tends to stabilize and increase cloud fraction. They suggest that long-term changes in cloud vertical structure are associated with the processes responsible for SW shallow cloud feedbacks and show the sign of the net balance between competing stabilization and mixing processes leads to either increasing or decreasing cloud top heights. Therefore, watching how the vertical structure of shallow clouds responds to warming will provide insights into the mechanisms responsible for shallow cloud feedbacks in the real world.

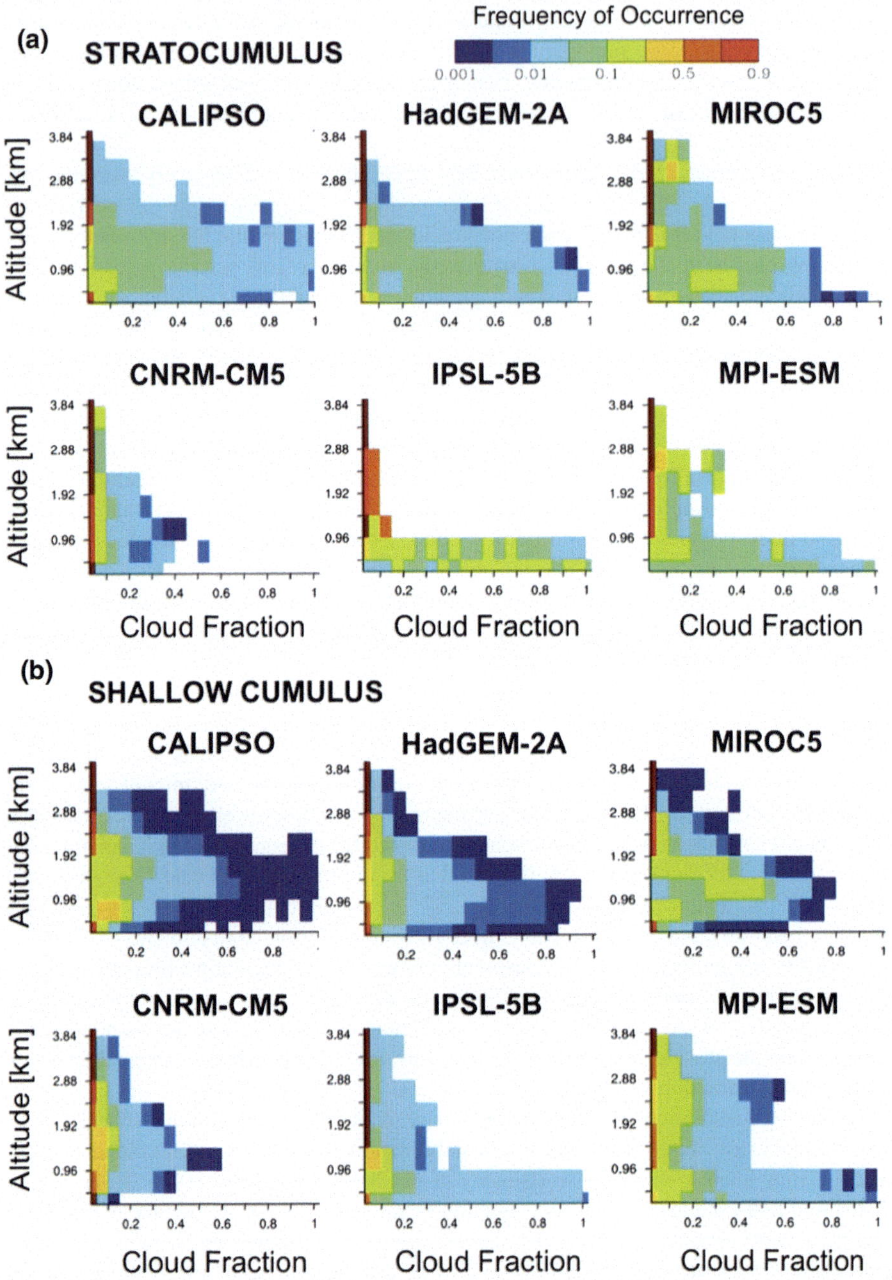

Fig. 7 Height-resolved frequency of occurrence of cloud layers for non-overlapped low-level cloud conditions in subtropical stratocumulus and shallow cumulus regimes as observed by CALIOP and as simulated by five CMIP5 models. Adapted from Nam et al. 2012

TOA radiances provide only a weak test of how well clouds are represented in models. Lidar provides the highest accuracy and the best vertical resolution for evaluating model representations of cloud fraction and vertical development in the shallow marine boundary layer. Nuijens et al. (2015) used ground-based lidar observations from the Barbados Cloud Observatory (Stevens et al. 2016b) to evaluate trade cumulus simulated by CMIP5 models, finding the models simulated a reasonable representation of mean cloudiness in the trade-wind layer but didn't reproduce the variability in cloudiness observed on short time-scales. In Nam et al. (2012), observations from CALIOP were used to discriminate between CMIP5 simulations of shallow marine clouds in the present climate (Fig. 7). Their evaluation is based on the altitude-resolved cloud fraction of non-overlapped low clouds from CALIOP observations, aggregated over global stratocumulus regimes (Fig. 7a) and global shallow cumulus regimes (Fig. 7b). It can be seen that the models tend to underestimate low-cloud cover and often predict stratocumulus-type clouds in regimes where shallow cumulus cloud types should prevail. Other studies have also noted that the variability of low-cloud cover in models and the relation to large-scale conditions correlate poorly with observations (Qu et al. 2015; Myers and Norris 2015). Further, in these studies the models in better agreement with observations tend to have higher ECS and/or larger cloud feedback.

Using the same method as described above for tropical ascent regions (Fig. 6), a trend analysis based on time series of synthetic observations was performed for tropical subsidence regions. For subsidence regions, the metric which was simulated was "Low Cloud Volume," computed as the sum of the vertically integrated cloudy sections of lidar profiles between the ocean surface and 4 km. A change in Low Cloud Volume would be produced by either a change in boundary layer cloud coverage or a change in the boundary layer cloud vertical extent. Simulated trends of this diagnostic were constructed from two CMIP5 models in a similar way as for Fig. 6, again using output from the COSP/lidar simulator. Figure 8 shows the long-term trend of this diagnostic derived from COSP/lidar simulator outputs. The SW cloud feedbacks predicted by these two models are significantly different from each other and trends of the low-cloud diagnostic start to separate in the early 2020s. By the late 2020s the trends separate with some confidence and would provide a test of model performance.

The above discussion points to the need to better understand shallow cloud processes and improve model parameterizations. Satellite retrievals of cloud properties (cloud height,

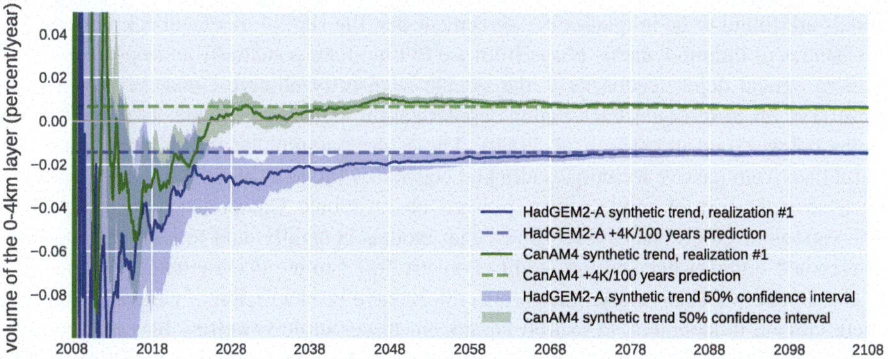

Fig. 8 Same as Fig. 6 but for Low Cloud Volume, comparing CALIPSO simulator output from the HadGEM2 and CanAM4 models

albedo, phase, etc.) can be used to gain insight into the processes responsible for cloud responses to environmental changes. Satellite lidar is a relatively new tool to observe cloud properties: In addition to cloud top height and altitude-resolved cloud fraction, lidar can detect the height of the trade inversion in clear-sky conditions when there is a gradient in aerosol concentration across the inversion. Direct observation of the small predicted changes in height of the marine inversion (Blossey et al. 2013) could provide insight into the processes underlying observed changes in shallow clouds. In addition to lidar, passive radiometers and cloud profiling radar are needed to provide complementary measurements of liquid water path, droplet size, and precipitation—all necessary to identify and characterize the processes responsible for the changes in cloud radiation.

One critical parameter which is not currently retrieved is the geometric thickness of shallow clouds—particularly stratiform boundary layer clouds—which would provide a valuable constraint on cloud processes (Wood 2007). A cloud thickness retrieval would also provide an observational test of cloud adiabaticity. For an adiabatic cloud, a relation between cloud optical depth, cloud droplet number concentration (CDNC) and cloud thickness can be derived from theory. Cloud optical depth and CDNC are currently retrieved from imagers such as MODIS. Addition of an independently retrieved cloud thickness would give an observational test of the adiabaticity relationship. Cloud thickness combined with lidar cloud top heights would give cloud base height and an estimate of the lifting condensation level, also providing insight into the thermodynamic state of the marine boundary layer (Stevens 2007). Retrieval of cloud thickness from airborne multiple-field-of-view lidar (Cahalan et al. 2005) has been demonstrated and the feasibility of implementing this capability in a future satellite lidar should be pursued.

4.3 Middle and High Latitudes

This paper has focused on cloud feedbacks from low-latitude clouds, but there has also been interest in shortwave optical depth feedbacks at mid- and high latitudes. Climate models are consistent in predicting that extratropical clouds will become optically thicker and brighter due to a combination of increases in total water path and changes of thermodynamic phase from ice to water. A number of recent studies have looked for support for these predictions in the satellite record, primarily relying on passive observations from the ISCCP, MODIS and CERES datasets (Ceppi et al. 2016; Gordon and Klein 2014; Terai et al. 2016). While observational uncertainties and disagreements between datasets have prevented quantitative estimates of the cloud optical depth feedback, observations and models are found to be in qualitative agreement that the sign of this feedback is negative.

Changes of thermodynamic phase from ice to liquid are potentially an important driver of these optical depth feedbacks. Lidar is able to directly observe cloud thermodynamic phase (via range-resolved backscatter depolarization signatures) in the first few optical depths below cloud top (Hu et al. 2009). These observations are far more accurate and useful than from passive techniques (Jin and Nasiri 2014; Cho et al. 2009) and so provide a much stronger test of models than passive observations. Comparisons with CALIOP observations of cloud phase have shown that models generally tend to simulate too little supercooled liquid water and too readily convert liquid to ice (Cesana et al. 2015).

The Southern Ocean mid-latitude storm tracks have been a region of particular interest, where climate models tend to exhibit significant biases in downwelling SW surface radiation. These SW radiation biases have been linked to the representation of shallow marine clouds in the cold sector of extratropical storms. CALIOP observations have shown a lack of supercooled water in these regions in models, due to biases in the partitioning of cloud water between liquid and ice (Bodas-Salcedo et al. 2014, 2016; Forbes et al. 2016). These biases in how models partition liquid and ice also have implications for model estimates of climate sensitivity (Tan et al. 2016). Continuation of lidar cloud phase observations is essential for monitoring trends in cloud phase driven by climate warming.

4.4 Aerosols as a Cloud Controlling Factor

Cloud cover or albedo might also change in response to changing aerosol concentrations. While aerosols are not thought to play a significant role in cloud feedbacks, simple conceptual models (Twomey 1977; Albrecht 1989) make clear the potential of aerosol microphysical and radiative effects to influence cloud properties. In the real world, however, numerous other microphysical and macrophysical processes likely act to reduce the response of cloud to changing aerosols predicted by these simple conceptual models (Stevens and Feingold 2009; Seifert et al. 2015). Influences of aerosol on cloud also depend on the vertical location of the aerosol relative to clouds (Amiri-Farahani et al. 2017), which can change even the sign of the effect. Given these uncertainties, the magnitude of long-term trends in cloud properties due to aerosol influences is highly uncertain.

If the changes in aerosol concentrations are due to changes in anthropogenic emissions this would represent an aerosol indirect forcing, whose effects might be mistakenly attributed to cloud feedbacks (Gettelman et al. 2016). If emissions of natural aerosols change in response to warming surface temperatures and then alter cloud properties, this could be thought of as a cloud-mediated aerosol feedback (Carslaw et al. 2010). To control for these potential aerosol-induced effects on clouds requires, at a minimum, vertically resolved monitoring of trends in aerosol loading. With regard to the long-term monitoring of aerosols, passive aerosol retrievals are very sensitive to calibration drifts and can be biased by a variety of cloud-induced artifacts including cloud masking errors and side-scattering of sunlight from clouds into cloud-free columns (Várnai and Marshak 2009). If cloud fraction changes over time, these cloud-induced aerosol retrieval artifacts could produce spurious trends in aerosol, leading to mistaken attribution of some of the observed cloud changes to aerosol indirect effects. Lidar has more stable long-term calibration, is less susceptible to cloud-induced artifacts and would provide a higher confidence indication of long-term aerosol change, in addition to providing a vertically resolved measurement rather than just a column integral. Satellite lidar is also able to retrieve cloud top extinction and cloud droplet number concentration (Zeng et al. 2014). Retrieval of droplet size at cloud top is likely also possible, though still under development at this point. These retrievals can be performed with even a simple elastic backscatter lidar (such as CALIOP) but satellite High Spectral Resolution Lidar (HSRL, She et al. 1992; Hair et al. 2008) would provide the additional advantage of direct measurement of aerosol extinction.

5 Meeting the Observational Challenge

The availability of colocated observations from cloud lidar and cloud profiling radar in the A-train have spurred substantial advances in our understanding of cloud-climate processes. The existing 11-year record from CALIOP and the A-train represents an initial baseline sufficient to characterize the current climate state (Chung et al. 2012) but is not yet long enough to fully capture interannual variability or to detect climate trends. Continued advancement of our understanding of cloud processes and cloud feedbacks depends on our ability to continue the types of measurements that CALIPSO, CloudSat and the A-train have provided. Observing the small forced trends in cloud height predicted by models and proper attribution of changes in CRE to cloud type requires cloud profiling from radar and lidar. Sustained observation by a small suite of core instruments is necessary to advance beyond the current state of understanding achieved from the A-train and globally monitor the evolution of clouds as they begin to move outside the envelope of current variability.

In addition to monitoring changes in clouds, we need to understand and derive observational constraints on the processes responsible for those changes. Ongoing observations are needed to understand and constrain physical mechanisms responsible for cloud-climate-feedbacks which are now being proposed (e.g., Hartmann and Larson 2002; Sherwood et al. 2014; Bony et al. 2016). The ability of lidar to accurately observe changes in the vertical distribution and coverage of clouds is part of the essential capabilities required.

Table 2 outlines a set of continued satellite observations to address the issues which have been discussed above. A minimum set would be lidar plus cloud profiling radar and a passive radiometer covering key channels in the visible and thermal infrared. This assumes that crucial monitoring of fundamental integral constraints on the overall energy budget—such as TOA broadband radiative fluxes and sea surface altimetry for ocean heat content–are available from sensors flying on other platforms. These measurements of integral constraints would not need to be colocated with the cloud sensors. Together, these would constitute the long-term observing system necessary to understand, quantify, and constrain cloud feedbacks.

With the A-train era soon drawing to a close, follow-on lidar missions are necessary to extend the initial data record established by CALIOP. Global active profiling will be continued by the ESA EarthCARE mission (Illingworth et al. 2015), currently scheduled to launch in 2019. There will likely be a gap between CALIPSO and EarthCARE and, given the length of time required to develop new satellite missions, it may be already too late to

Table 2 Summary of observation requirements for cloud feedbacks and colocation needs

Science objective	Geophysical variable	Sensors	Spatially matched?
LW cloud feedbacks			
	TOA LW radiative flux	Broadband flux radiometer	No
	Cloud profiles	Lidar and W-band radar	Yes
	Cloud emissivity	Thermal IR radiometer, lidar	Yes
	Water vapor and temperature profiles	AIRS/AMSR, GPS-Radio Occultation	Yes
SW cloud feedbacks			
	TOA SW radiative flux	Broadband flux radiometer	No
	Light precipitation	W-band radar	Yes
	Cloud albedo, OD, Re	Visible radiometer	Yes
	Cloud top properties	Visible radiometer, lidar	Yes
	Liquid water path	Visible and microwave radiometers	Yes
	Cloud top height, thickness, phase	Lidar	Yes
Aerosol–cloud interactions (in addition to measurements above)			
	Aerosol near cloud base	Lidar (backscatter or HSRL)	Yes

avoid a gap between EarthCARE and a follow-on mission. However, if future lidar instruments are designed with repeatability in mind—with instrument characteristics and orbits designed to provide consistency with the existing CALIOP record—then mission overlap is not strictly required due to the inherent accuracy and stability of the primary cloud measurements, as discussed above. However, as of this writing, there are no plans within ESA or the national space agencies to continue—much less improve—this vital observational record after EarthCARE.

The focus of this paper has been on the use of active sensors to characterize the response of clouds to a warming climate. A decade of observation by satellite-borne active sensors has transformed our understanding of the vertical distribution of condensate, especially cloud ice, but these active measurements are equally important to understanding the coupling of clouds to circulation. The sensitivity of cloud feedbacks to patterns of SST change, discussed above, demonstrates the important influence of large-scale dynamics on the cloud response to warming. Experience with the A-train has demonstrated that cloud profiling from active sensors is essential to describing the vertical distribution of radiative and latent heating (Haynes et al. 2013), which are the primary ways that clouds interact with the large-scale circulation. Sustained observations of atmospheric heating derived from cloud profiling would help our understanding of processes responsible for the large-scale structure of atmospheric circulations systems, such as the jets, storm tracks and rain bands. Continuing active measurements beyond the A-train and EarthCARE is essential to our ability to understand regional climate change and to develop adaptive strategies (Bony et al. 2015).

Acknowledgements The authors acknowledge Sandrine Bony, Steve Klein, Robert Pincus, Bjorn Stevens, and Rob Wood for valuable comments and technical discussions. We acknowledge the assistance of Jason Tackett, who generated Fig. 1. This paper additionally benefited from discussions at the Workshop on "Shallow clouds and water vapor, circulation and climate sensitivity" at the International Space Science Institute (ISSI) and we would like to thank the two reviewers for valuable comments and suggestions, which allowed us to significantly improve this paper. This work has been supported by NASA and CNES.

References

Albrecht BA (1989) Aerosols, cloud microphysics and fractional cloudiness. Science 245:1227–1230

Amiri-Farahani A, Allen RJ, Neubauer D, Lohmann U (2017) Impact of Saharan dust on North Atlantic marine stratocumulus clouds: importance of the semidirect effect. Atmos Chem Phys 17:6305–6322. https://doi.org/10.5194/acp-17-6305-2017

Armour KC, Bitz CM, Roe GH (2013) Time-varying climate sensitivity from regional feedbacks. J Climate 26:4518–4534. https://doi.org/10.1175/JCLI-D-12-00544.1

Blossey PN et al (2013) Sensitivity of marine low clouds to an idealized climate change: the CGILS LES intercomparison. JAMES 5:234–258. https://doi.org/10.1002/jame.20025

Bodas-Salcedo A, Webb MJ, Bony S, Chepfer H, Dufresne J-L, Klein SA, Zhang Y, Marchand R, Haynes JM, Pincus R, John VO (2011) COSP: satellite simulation software for model assessment. Bull Am Meteorol Soc 92:1023–1043. https://doi.org/10.1175/2011BAMS2856.1

Bodas-Salcedo A, Williams KD, Ringer MA, Beau I, Cole JNS, Dufresne J-L, Koshiro T, Stevens B, Wang Z, Yokohata T (2014) Origins of the solar radiation biases over the Southern Ocean in CFMIP2 models. J Climate 27:41–56. https://doi.org/10.1175/JCLI-D-13-00169.1

Bodas-Salcedo A, Hill PG, Furtado K, Williams KD, Field PR, Manners JC, Hyder P, Kato S (2016) Large contribution of supercooled liquid clouds to the solar radiation budget of the Southern Ocean. J Climate 29:4213–4228. https://doi.org/10.1175/JCLI-D-15-0564.1

Bony S, Dufresne J-L (2005) Marine boundary layer clouds at the heart of tropical cloud feedback uncertainties in climate models. Geophys Res Lett 32:L20806. https://doi.org/10.1029/2005GL023851

Bony S et al (2006) How well do we understand and evaluate climate change feedback processes? J Climate 10:3445–3482

Bony S, Stevens B, Frierson D et al (2015) Clouds, circulation, and climate sensitivity. Nat Geosci 8:261–268. https://doi.org/10.1038/ngeo2398

Bony S, Stevens B, Coppin D, Becker T, Reed K, Voigt A, Medeiros B (2016) Thermodynamic control of anvil-cloud amount. Proc Nat Acad Sci. https://doi.org/10.1073/pnas.1601472113

Bretherton CS (2015) Insights into low-latitude cloud feedbacks from high-resolution models. Philos Trans R Soc A 373:20140415. https://doi.org/10.1098/rsta.2014.0415

Brient F, Schneider T (2016) Constraints on climate sensitivity from space-based measurements of low-cloud reflection. J Climate 29:5821–5834. https://doi.org/10.1175/JCLI-D-15-00897.1

Brient F, Schneider T, Tan Z, Bony S, Qu X, Hall A (2015) Shallowness of tropical low clouds as a predictor of climate models response to warming. Clim Dyn 47:433–449

Cahalan RF, McGill M, Kolasinski J, Varnai T, Yetzer K (2005) THOR: cloud thickness from off-beam lidar returns. J Atmos Ocean Technol 22:605–627

Caldwell PM, Zhang Y, Klein SA (2013) CMIP3 subtropical stratocumulus cloud feedback interpreted through a mixed-layer model. J Climate 26:1607–1623. https://doi.org/10.1175/JCLI-D-12-00188.1

Carslaw KS, Boucher O, Spracklen DV, Mann GW, Rae JGL, Woodward S, Kulmala M (2010) A review of natural aerosol interactions and feedbacks within the Earth system. Atmos Chem Phys 10:1701–1737

Ceppi P, McCoy DT, Hartmann DL (2016) Observational evidence for a negative shortwave cloud feedback in middle to high latitudes. Geophys Res Lett 43:1331–1339

Cesana G, Waliser DE, Jiang X, Li J-LF (2015) Multi-model evaluation of cloud phase transition using satellite and reanalysis data. J Geophys Res 120:7871–7892. https://doi.org/10.1002/2014JD022932

Charney J, Arakawa A, Baker D et al (1979) Carbon dioxide and climate: a scientific assessment. National Research Council, Washington, DC, p 22

Chepfer H, Bony S, Winker DM, Chiriaco M, Dufresne J-L, Seze G (2008) Use of CALIPSO lidar observations to evaluate the cloudiness simulated by a climate model. Geophys Res Lett 35:L15704. https://doi.org/10.1029/2008GL034207

Chepfer H, Bony S, Winker D, Cesana G, Dufresne JL, Minnis P, Stubenrauch CJ, Zeng S (2010) The GCM oriented CALIPSO cloud product (CALIPSO-GOCCP). J Geophys Res 115:D00H16. https://doi.org/10.1029/2009JD012251

Chepfer H, Noel V, Winker D, Chiriaco M (2014) Where and when will we observe cloud changes due to climate warming? Geophys Res Lett 41:8387–8395. https://doi.org/10.1002/2014GL061792

Cho H-M, Nasiri SL, Yang P (2009) Application of CALIOP measurements to the evaluation of cloud phase derived from MODIS infrared channels. J Appl Meteorol Climatol 48:2169–2180. https://doi.org/10.1175/2009JAMC2238.1

Chung E-S, Soden BJ, Clement AC (2012) Diagnosing climate feedbacks in coupled ocean–atmosphere models. Surv Geophys 33:733–744. https://doi.org/10.1007/s10712-012-9187-x

Cooke R, Wielicki BA, Young DF, Mlynczak MG (2013) Value of information for climate observing systems. Environ Syst Decis 34:98–109

di Michele S, McNally T, Bauer P, Genkova I (2013) Quality assessment of cloud-top height estimates from satellite IR radiances using the CALIPSO lidar. IEEE Trans Geosci Remote Sens 51:2454–2464. https://doi.org/10.1109/TGRS.2012.2210721

Enfield DB, Mestaz-Nunez AM, Trimble PJ (2001) The Atlantic multidecadal oscillation and its relation to rainfall and river flows in the continental US. Geophys Res Lett 28:2077–2080

Evan AT, Heidinger AK, Vimont DJ (2007) Arguments against a physical long-term trend in global ISCCP cloud amounts. Geophys Res Lett 34:L04701. https://doi.org/10.1029/2006GL028083

Folland CK, Parker DE, Colman A (1999) Large scale modes of ocean surface temperature since the late nineteenth century. In: Navarra A (ed) Beyond El Nino: decadal and interdecadal climate variability. Springer, New York, pp 73–102

Forbes R, Geer A, Lonitz K, Ahlgrimm M (2016) Reducing systematic errors in cold-air outbreaks. ECMWF Newsl 146:17–22

Forster PM (2016) Inference of climate sensitivity from analysis of Earth's energy budget. Ann Rev Earth Planet Sci 44:85–106

Garnier A, Pelon J, Vaughan MA, Winker DM, Trepte CR, Dubuisson P (2015) Lidar multiple scattering factors inferred from CALIPSO lidar and IIR retrievals of semi-transparent cirrus cloud optical depths over oceans. Atmos Meas Technol 8:2759–2774. https://doi.org/10.5194/amt-8-2759-2015

Gettelman A, Lin L, Medeiros B, Olson J (2016) Climate feedback variance and the interaction of aerosol forcing and feedbacks. J Climate. https://doi.org/10.1175/JCLI-D-16-0151.1

Gordon ND, Klein SA (2014) Low-cloud optical depth feedback in climate models. J Geophys Res 119:6052–6065. https://doi.org/10.1002/2013JD021052

Gregory JM, Andrews T (2016) Variation in climate sensitivity and feedback parameters during the historical period. Geophys Res Lett 43:3911–3920. https://doi.org/10.1002/2016GL068406

Hair JW, Hostetler CA, Cook AL, Harper DB, Ferrare RA, Mack TL, Welch W, Izquierdo LR, Hovis FE (2008) Airborne high spectral resolution lidar for profiling aerosol optical properties. Appl Opt 47:6734–6753

Hartmann DL, Larson K (2002) An important constraint on cloud-climate feedback. Geophys Res Lett 20:1951–1954

Haynes JM, Vonder Haar TH, L'Ecuyer T, Henderson D (2013) Radiative heating characteristics of Earth's cloudy atmosphere from vertically resolved active sensors. Geophys Res Lett 40:624–630. https://doi.org/10.1002/grl.50145

Holz R, Ackerman S, Nagle F, Frey R, Dutcher S, Kuehn R, Vaughan M, Baum B (2008) Global MODIS cloud detection and height evaluation using CALIOP. J Geophys Res 113:D00A19. https://doi.org/10.1029/2008JD009837

Hu Y, Winker D, Vaughan M et al (2009) CALIPSO/CALIOP cloud phase discrimination algorithm. J Atmos Ocean Tech 26:2293–2309. https://doi.org/10.1175/2009JTECHA1280.1

Illingworth AJ, Barker HW, Beljaars A et al (2015) The EarthCARE satellite. Bull Am Meteorol Soc 96:1311–1332. https://doi.org/10.1175/BAMS-D-12-00227.1

IPCC (2013) Summary for policy makers. In: Stocker TF, Qin D, Plattner G-K, Tignor M, Allen SK, Boschung J, Nauels A, Xia Y, Bex V, Midgley PM (eds) Climate change 2013: the physical science basis. Cambridge University Press, Cambridge

Jin H, Nasiri SL (2014) Evaluation of AIRS cloud thermodynamic phase determination with CALIPSO. J Appl Meteorol Climatol 53:1012–1027. https://doi.org/10.1175/JAMC-D-13-0137.1

Kato S, Rose FG, Mack SS, Miller WF et al (2011) Improvements of top-of-atmosphere and surface irradiance computations with CALIPSO-, CloudSat-, and MODIS-derived cloud and aerosol properties. J Geophys Res 116:D19209. https://doi.org/10.1029/2011JD016050

Key JR (1993) Estimating the area fraction of geophysical fields from measurements along a transect. IEEE Trans Geosci Remote Sci 31:1099–1102

Klein SA, Hall A (2015) Emergent constraints for cloud feedbacks. Curr Clim Change Rep 1:276–287. https://doi.org/10.1007/s40641-015-0027-1

Klein SA, Hall A, Norris JR, Pincus R (2017) Low-cloud feedbacks from cloud-controlling factors: a review. Surv Geophys. https://doi.org/10.1007/s10712-017-9433-3

Knutti R, Hegerl GC (2008) The equilibrium climate sensitivity of the Earth's temperature to radiation changes. Nat Geosci 1:735–743

L'Ecuyer TS, Jiang JH (2010) Touring the atmosphere aboard the A-Train. Phys Today 63(10):36–41

Lewis N, Curry JA (2015) The implications for climate sensitivity of AR5 forcing and heat uptake estimates. Clim Dyn 45:1009–1023. https://doi.org/10.1007/s00382-014-2342-y

Li Y, Yang P, North GR, Dessler A (2012) Test of the fixed anvil temperature hypothesis. J Atmos Sci 69:2317–2328. https://doi.org/10.1175/JAS-D-11-0158.1

Loeb NG, Wielicki BA, Wong T, Parker PA (2009) Impact of data gaps on satellite broadband radiation records. J Geophys Res 114:D11109. https://doi.org/10.1029/2008JD011183

Mace GG, Wrenn FJ (2013) Evaluation of the hydrometeor layers in the East and West Pacific within ISCCP cloud-top pressure-optical depth bins using merged CloudSat and CALIPSO data. J Climate 26:9429–9444. https://doi.org/10.1175/JCLI-D-12-00207.1

Mace G, Zhang Q, Vaughan M, Marchand R, Stephens G, Trepte C, Winker D (2009) A description of hydrometeor layer occurrence statistics derived from the first year of merged Cloudsat and CALIPSO data. J Geophys Res 114:D00A26. https://doi.org/10.1029/2007JD009755

Marchand R, Ackerman T, Smyth M, Rossow WB (2010) A review of cloud top height and optical depth histograms from MISR, ISCCP, and MODIS. J Geophys Res 115:D16206. https://doi.org/10.1029/2009JD013422

Marvel K, Zelinka M, Klein SA, Bonfils C, Caldwell P, Doutriaux C, Santer BD, Taylor KE (2015) External influences on modeled and observed cloud trends. J Climate 28:4820–4840. https://doi.org/10.1175/JCLI-D-14-00734.1

Matus AV, L'Ecuyer TS (2017) The role of cloud phase in Earth's radiation budget. J Geophys Res Atmos 122:2559–2578. https://doi.org/10.1002/2016JD025951

McCoy DT, Eastman R, Hartmann DL, Wood R (2017) The change in low-cloud cover in a warmed climate inferred from AIRS, MODIS, and ECMWF-interim analyses. J Climate 30:3609–3620. https://doi.org/10.1175/JCLI-D-15-0734.1

Myers TA, Norris JR (2015) On the relationships between subtropical clouds and meteorology in observations and CMIP3 and CMIP5 models. J Climate 28:2945–2967. https://doi.org/10.1175/JCLI-D-14-00475.s1

Myers TA, Norris JR (2016) Reducing the uncertainty in subtropical cloud feedback. Geophys Res Lett 43:2144–2148. https://doi.org/10.1002/2015GL067416

Myhre G, Myhre A, Stordal F (2001) Historical time evolution of total radiative forcing. Atmos Environ 35:2361–2373

Nam CCW, Bony S, Dufresne J-L, Chepfer H (2012) The 'too few, too bright' tropical low-cloud problem in CMIP5 models. Geophys Res Lett 39:L21801. https://doi.org/10.1029/2012GL053421

Neubersch D, Held H, Otto A (2014) Operationalizing climate targets under learning: an application of cost-risk analysis. Clim Change 126:305–318. https://doi.org/10.1007/s10584-014-1223-z

Norris JR, Evan AT (2015) Empirical removal of artifacts from the ISCCP and PATMOS-x satellite cloud records. J Atmos Ocean Technol 32:691–702

Norris JR, Allen RJ, Evan AT, Zelinka MD, O'Dell CW, Klein S (2016) Evidence for climate change in the satellite cloud record. Nature 536:72–75. https://doi.org/10.1038/nature18273

Nuijens L, Medeiros B, Sandu I, Ahlgrimm M (2015) The behavior of trade-wind cloudiness in observations and models: the major cloud components and their variability. JAMES 7:600–616. https://doi.org/10.1002/2014MS000390

Ohring G (2004) Satellite instrument calibration for measuring global climate change. In: Ohring G, Wielicki B, Spencer R, Emery B, Datla R (eds) NISTIR 7047

Otto AF, Otto O Boucher et al (2013) Energy budget constraints on climate response. Nat Geosci 6:415–416

Pincus R, Platnick S, Ackerman SA, Hemler RS, Hofmann RJP (2012) Reconciling simulated and observed view of clouds: MODIS, ISCCP, and the limits of instrument simulators. J Climate 25:4699–4720. https://doi.org/10.1175/JCLI-D-11-00267.1

Powell KA, Hostetler CA, Liu Z, Vaughan MA, Kuehn RE, Hunt WH, Lee K, Trepte CR, Rogers RR, Young SA, Winker DM (2009) CALIPSO lidar calibration algorithms: part I—nighttime 532 nm parallel channel and 532 nm perpendicular channel. J Atmos Ocean Technol 26:2015–2033. https://doi.org/10.1175/2009-JTECHA1242.1

Qu X, Hall A, Klein SA, DeAngelis AM (2015) Positive tropical marine low-cloud cover feedback inferred from cloud-controlling factors. Geophys Res Lett 42:7767–7775. https://doi.org/10.1002/2015GL065627

Rossow WB, Zhang Y (2010) Evaluation of a statistical model of cloud vertical structure using combined CloudSat and CALIPSO cloud layer profiles. J Clim 23:6641–6653. https://doi.org/10.1175/2010JCLI3734.1

Rugenstein MAA, Caldeira K, Knutti R (2016) Dependence of global radiative feedbacks on evolving patterns of surface heat fluxes. Geophys Res Lett 43:9877–9885. https://doi.org/10.1002/2016GL070907

Seifert A, Heus T, Pincus R, Stevens B (2015) Large-eddy simulation of the transient and near-equilibrium behavior of precipitating shallow convection. JAMES 7:1918–1937. https://doi.org/10.1002/2015MS000489

She C, Alvarez RJ II, Caldwell LM, Krueger DA (1992) High-spectral-resolution Rayleigh–Mie lidar measurement of aerosol and atmospheric profiles. Opt Lett 17:541–543

Shea YL, Wielicki BA, Sun-Mack S, Minnis P (2017) Quantifying the dependence of satellite cloud retrievals on instrument uncertainty. J Clim. https://doi.org/10.1175/JCLI-D-16-0429.1

Sherwood SC, Bony S, Dufresne JL (2014) Spread in model climate sensitivity traced to atmospheric convective mixing. Nature 505:37–42. https://doi.org/10.1038/nature12829

Soden BJ, Held IM (2006) An assessment of climate feedbacks in coupled ocean–atmosphere models. J Climate 19:3354–3360

Soden BJ, Held IM, Colman R, Shell KM, Kiehl JT, Shields CA (2008) Quantifying climate feedbacks using radiative kernels. J Climate 21:3504–3520. https://doi.org/10.1175/2007JCLI2110.1

Stephens GL, Kummerow CD (2007) The remote sensing of clouds and precipitation from space: a review. J Atmos Sci 64:3742–3765. https://doi.org/10.1175/2006JAS2375.1

Stephens GL, Vane DG, Boain RJ et al (2002) The CloudSat mission and the A-Train. Bull Am Meteorol Soc 83:1771–1790. https://doi.org/10.1175/BAMS-83-12-1771

Stephens GL, Wild M, Stackhouse PW, L'Ecuyer T, Kato S, Henderson DS (2012) The global character of the flux of downward longwave radiation. J Clim 25(7):2329–2340. https://doi.org/10.1175/JCLI-D-11-00262.1
Stephens GL, Winker DM, Pelon J et al (2017) CloudSat and CALIPSO within the A-Train: ten years of actively observing the Earth system. Bull Am Meteorol Soc. https://doi.org/10.1175/BAMS-D-16-0324.1
Stevens B (2007) On the growth of layers of nonprecipitating cumulus convection. J Atmos Sci 64:2916–2931. https://doi.org/10.1175/JAS3983.1
Stevens B, Feingold G (2009) Untangling aerosol effects on clouds and precipitation in a buffered system. Nature 461:607–613. https://doi.org/10.1038/nature08281
Stevens B, Sherwood SC, Bony S, Webb MJ (2016a) Prospects for narrowing bounds on Earth's equilibrium climate sensitivity. Earth's Future. https://doi.org/10.1002/11016EF000376
Stevens B, Farrell D, Hirsch L, Jansen F, Nuijens L, Serikov I, Brügmann B, Forde M, Linne H, Lonitz K, Prospero JM (2016b) The Barbados Cloud Observatory: anchoring investigations of clouds and circulation on the edge of the ITCZ. Bull Am Meteorol Soc 97:787–801. https://doi.org/10.1175/BAMS-D-14-00247.1
Stubenrauch C, Rossow W, Kinne S (2012) Assessment of global cloud data sets from satellites. WCRP Report No. 23
Su H, Jiang JH (2013) Tropical clouds and circulation changes during the 2006–07 and 2009–10 El Niños. J Climate 26:399–413. https://doi.org/10.1175/JCLI-D-12-00152.1
Su H, Jiang JH, Zhai C, Perun V, Shen JT, Del Genio AD, Nazarenko LS, Donner LJ, Horowitz LW, Seman CJ, Morcrette CJ, Petch J, Ringer MA, Cole J, dos Santos Mesquita M, Iversen T, Kristjansson JE, Gettelman A, Rotstayn LD, Jeffrey SJ, Dufresne J-L, Watanabe M, Kawai H, Koshiro T, Wu T, Volodin EM, L'Ecuyer T, Teixeira J, Stephens GL (2013) Diagnosis of regime-dependent cloud simulation errors in CMIP5 models using 'A-Train' satellite observations and reanalysis data. J Geophys Res 118:2762–2780. https://doi.org/10.1029/2012JD018575
Su H, Jiang JH, Zhai C, Shen TJ, Neelin JD, Stephens GL, Yung YL (2014) Weakening and strengthening structures in the Hadley circulation change under global warming and implications for cloud response and climate sensitivity. J Geophys Res 119:5787–5805. https://doi.org/10.1002/2014JD021642
Tan I, Storelvmo T, Zelinka MD (2016) Observational constraints on mixed-phase clouds imply higher climate sensitivity. Science 352:224–227. https://doi.org/10.1126/science.aad5300
Terai CR, Klein SA, Zelinka MD (2016) Constraining the low-cloud optical depth feedback at middle and high latitudes using satellite observations. J Geophys Res 121:9696–9716. https://doi.org/10.1002/2016JD025233
Tomassini L et al (2013) The respective roles of surface temperature driven feedbacks and tropospheric adjustment to CO_2 in CMIP5 transient climate simulations. Clim Dyn. https://doi.org/10.1007/s00382-013-1682-3
Tselioudis G, Rossow W, Zhang Y, Konsta D (2013) Global weather states and their properties from passive and active satellite cloud retrievals. J Clim 26:7734–7746. https://doi.org/10.1175/JCLI-D-13-00024.1
Twomey S (1977) The influence of pollution on the shortwave albedo of clouds. J Atmos Sci 34:1149–1152
Várnai T, Marshak A (2009) MODIS observations of enhanced clear sky reflectance near clouds. Geophys Res Lett 36:L06807
Vial J, Dufresne J-L, Bony S (2013) On the interpretation of inter-model spread in CMIP5 climate sensitivity estimates. Clim Dyn 41:3339–3362. https://doi.org/10.1007/s00382-013-1725-9
Vial J, Bony S, Dufresne J-L, Roehrig R (2016) Coupling between lower-tropospheric convective mixing and low-level clouds: physical mechanisms and dependence on convection scheme. JAMES 8:1892–1911. https://doi.org/10.1002/2016MS000740
Vial J, Bony S, Stevens B, Vogel R (2017) Mechanisms and model diversity of trade-wind shallow cumulus cloud feedbacks: a review. Surv Geophys. https://doi.org/10.1007/s10712-017-9418-2
Weatherhead EC, Reinsel GC, Tiao GC et al (1998) Factors affecting the detection of trends: statistical considerations and applications to environmental data. J Geophys Res 103:17149–17161
Webb M, Senior C, Bony S, Morcrette J-J (2001) Combining ERBE and ISCCP data to assess clouds in the Hadley Centre, ECMWF, and LMD atmospheric climate models. Clim Dyn 17:905–922
Webb MJ, Lambert FH, Gregory JM (2013) Origins of differences in climate sensitivity, forcing, and feedback in climate models. Clim Dyn 40:677–707. https://doi.org/10.1007/s00382-012-1336-x
Wielicki BA, Parker L (1992) On the determination of cloud cover from satellite sensors: the effect of sensor spatial resolution. J Geophys Res 97:12799–12823
Wielicki BA, Young DF, Mlynczak MG et al (2013) Achieving climate change absolute accuracy in orbit. Bull Am Meteorol Soc 94:1520–1539. https://doi.org/10.1175/BAMS-D-12-00149.1

Winker DM, Hunt WH, McGill MJ (2007) Initial performance assessment of CALIOP. Geophys Res Lett 34:L19803. https://doi.org/10.1029/2007GL030135

Winker DM, Pelon J, Coakley JA Jr, Ackerman SA, Charlson RJ, Colarco PR, Flamant P, Fu Q, Hoff R, Kittaka C, Kubar TL, LeTreut H, McCormick MP, Megie G, Poole L, Powell K, Trepte C, Vaughan MA, Wielicki BA (2010) The CALIPSO mission: a global 3D view of aerosols and clouds. Bull Am Meteorol Soc 91:1211–1229. https://doi.org/10.1175/2010BAMS3009.1

Wood R (2007) Cancellation of aerosol indirect effects in marine stratocumulus through cloud thinning. J Atmos Sci 64:2657–2669

Wood R (2012) Stratocumulus clouds. Mon Weather Rev 140:2373–2423. https://doi.org/10.1175/MWR-D-11-00121.1

Xie S-P, Kosaka Y, Okumura YM (2016) Distinct energy budgets for anthropogenic and natural changes during global warming hiatus. Nat Geosci 9:29–34

Zelinka MD, Hartmann DL (2010) Why is longwave cloud feedback positive? J Geophys Res 115:D16117. https://doi.org/10.1029/2010JD013817

Zelinka MD, Hartmann DL (2011) The observed sensitivity of high clouds to mean surface temperature anomalies in the tropics. J Geophys Res 116:D23103. https://doi.org/10.1029/2011JD016459

Zelinka MD, Zhou C, Klein SA (2016) Insights from a refined decomposition of cloud feedbacks. Geophys Res Lett 43:9259–9269. https://doi.org/10.1002/2016GL069917

Zeng S, Riedi J, Trepte CR, Winker DM, Hu Y-X (2014) Study of global droplet number concentration with A-Train satellites. Atmos Chem Phys 14:7125–7134. https://doi.org/10.5194/acp-14-7125-2014

Zhai C, Jiang JH, Su H (2015) Long-term cloud change imprinted in seasonal cloud variation: more evidence of high climate sensitivity. Geophys Res Lett 42:8729–8737. https://doi.org/10.1002/2015GL065911

Zhao G, Di Girolamo L (2006) Cloud fraction errors for trade wind cumuli from EOS-Terra instruments. Geophys Res Lett 33:L20802. https://doi.org/10.1029/2006GL027088

Zhou C, Zelinka MD, Klein SA (2016) Impact of decadal cloud variations on the Earth's energy budget. Nat Geosci 9:871–875. https://doi.org/10.1038/NGEO2828

Surv Geophys (2017) 38:1509–1528
https://doi.org/10.1007/s10712-017-9442-2

Shallow Circulations: Relevance and Strategies for Satellite Observation

Gilles Bellon[1] · **Oliver Reitebuch**[2] · **Ann Kristin Naumann**[3]

Received: 20 February 2017 / Accepted: 19 October 2017 / Published online: 1 November 2017

Abstract Shallow circulations are central to many tropical cloud systems. We investigate the potential of existing and upcoming data to document these circulations. Different methods to observe or constrain atmospheric circulations rely on satellite-borne instruments. Direct observations of the wind are currently possible at the ocean surface or using tracer patterns. Satellite-borne wind lidar will soon be available, with a much better coverage and accuracy. Meanwhile, circulations can be constrained using satellite observations of atmospheric diabatic heating. We evaluate the commonalities and discrepancies of these estimates together with reanalysis in systems that include shallow circulations. It appears that existing datasets are in qualitative agreement, but that they still differ too much to provide robust evaluation criteria for general circulation models. This state of affairs highlights the potential of satellite-borne wind lidar and of further work on current satellite retrievals.

Keywords Shallow circulations · Winds · Diabatic heating

1 Introduction

Since the discovery of a meridional shallow circulation in the eastern Pacific (Trenberth et al. 2000; Zhang et al. 2004), meridional, large-scale, shallow circulations have been shown to occur in very different regions of the tropics, over land and over ocean (Zhang et al. 2008). These large-scale shallow circulations are regional and seasonal specificities

✉ Gilles Bellon
gilles.bellon@auckland.ac.nz

[1] Department of Physics, University of Auckland, 38 Princes Street, Auckland, New Zealand

[2] Deutsches Zentrum für Luft- und Raumfahrt (DLR), Institut für Physik der Atmosphäre, Oberpfaffenhofen, Münchener Strasse 20, 82234 Wessling, Germany

[3] Max Planck Institute for Meteorology, Bundesstrasse 53, D-20146 Hamburg, Germany

of the Hadley circulation; as an example, Fig. 1 shows the climatological July circulation over the eastern Pacific ocean (90°W–120°W) as described by the European Centre for Medium-range Weather Forecasts (ECMWF) Reanalysis ERA-Interim. The boundary-layer flow converges into the Intertropical Convergence Zone (ITCZ) around 10°N, and a return cross-equatorial flow can be seen just above the boundary layer between 3 and 7 km of altitude; above it, a cross-equatorial southerly flow can be seen converging into the ITCZ up to 9 km of altitude, below the upper-tropospheric return flow. The shallow circulation extends over most of the winter Hadley cell (down to 25°S), and the climatological intensity of the associated meridional wind is about 0.5–1 m s^{-1} as seen from the return flow in the lower free troposphere. Some in situ observations have measured winds that reached 6 m s^{-1} (Zhang et al. 2004).

These circulations have recently attracted a lot of interest because they modulate the moisture and energy transports associated with the Hadley circulation (Zhang et al. 2004, 2008). Compared to tropospheric-deep overturning circulations, shallow circulations transport more moisture and less dry static energy away from the ascending branch, because lower-tropospheric air is moister than upper-tropospheric air, and because its potential energy content is lower than the upper-tropospheric value, resulting in lower dry static energy despite the larger thermal energy. Since humidity in subsiding regions is exclusively determined by transport (Pierrehumbert and Roca 1998), shallow circulations have the potential to considerably alter the humidity of the subsiding lower troposphere and control radiative cooling there. Because of their influence on the thermodynamic state of the lower troposphere in subsiding regions, shallow circulations also have the potential to influence cloudiness in the boundary layer of these regions: low clouds have been shown to be sensitive to the lower-tropospheric stability and humidity (Klein and Hartmann 1993; Norris 1998), with consequences for low-cloud feedbacks (Rieck et al. 2012). In particular, shallow circulations are suggested to contribute significantly to the spread in cloud feedbacks in climate-change projections by general circulation models (GCMs) (Sherwood et al. 2014).

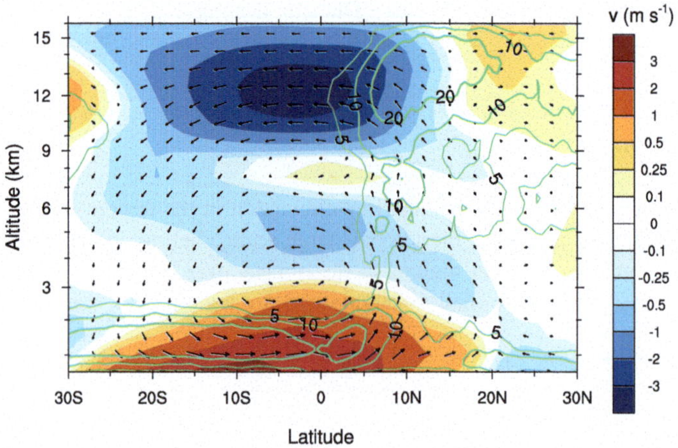

Fig. 1 July climatology of meridional wind (shadings), meridional and vertical velocity vectors, and cloud fraction (green contours, in %), for the eastern Pacific region (90°W–120°W). Data from the reanalysis ERA-Interim for winds and from the GCM-Oriented Cloud-Aerosol Lidar and Infrared Pathfinder Satellite Observation (CALIPSO) Cloud Product for cloud fraction

Shallow circulations have been observed in the monsoons of West Africa (Hagos and Zhang 2010) and Australia (Nie et al. 2010), in which they play an important role in determining the off-equatorial location of the monsoon convergence zone. They are also observed in the onsets of the West African and Indian monsoons and thought to be important for the poleward migration of rainfall during these onsets (Nguyen et al. 2011; Thorncroft et al. 2011; Parker et al. 2016). These large-scale shallow circulations result from a reversal of the geopotential gradients close to the top of the boundary layer due to surface forcing (Nolan et al. 2007) or radiative cooling in the subsiding regions (Nishant et al. 2016), although momentum transport has been suggested to constrain them (Dixit and Srinivasan 2016).

Shallow circulations are also part of the Madden–Julian Oscillation (MJO), the main mode of intraseasonal variability in the tropics (Madden and Julian 1971; Zhang 2005). The circulation associated with the MJO involves troposphere-deep ascent colocated with the eastward-propagating convective envelope characteristic of the MJO, but also a shallow circulation ahead of the convective disturbance, during the transition from suppressed to active phase. The associated moisture transport contributes to the moistening of the lower free troposphere that is instrumental to the gradual transition from shallow to deep convection and the propagation of the MJO (Kemball-Cook and Weare 2001; Zermeño-Díaz et al. 2015). This shallow circulation results from shallow heating (Zhang and Hagos 2009), which has been identified ahead of the MJO deep-convective disturbances during field campaigns (Lin and Johnson 1996; Johnson et al. 2015), and gives the overall diabatic heating associated with the MJO a slanted pattern on the eastern side of the deep-convective disturbance (Kiladis et al. 2005).

Smaller, mesoscale shallow circulations have recently drawn additional interest: cloud-modeling simulations of radiative–convective equilibrium (RCE) suggest that they have a role in self-aggregation of deep convection, a phenomenon that causes deep-convective clouds to horizontally aggregate, leaving the rest of the domain in convectively suppressed conditions. Because shallow circulations can export less energy in their upper branch than deep circulations, they can result in a net import of moist static energy (MSE) from their subsiding branch to their ascending branch; as a result, shallow circulations can act as a positive feedback on the MSE difference between convective and subsiding regions, reinforcing the contrast between convective and non-convective regions characteristic of the self-aggregated RCE (Bretherton et al. 2005; Muller and Held 2012; Wing and Emanuel 2013; Muller and Bony 2015; Naumann et al. 2017). Figure 2 shows an example of shallow circulation in the early stage of convective aggregation. A dry, cold patch is developing, with a horizontal scale of 100 km. Boundary-layer winds diverge from this patch and compensate the convergence that occurs just above the boundary layer around 1 km of altitude, with wind speeds up to 1 m s^{-1}.

Although shallow circulations are documented in reanalyses and field-campaign measurements, there is still a lot of uncertainty in the intensity and exact altitudes of the inflows and outflows and their variability. Better observing these circulations is required to understand their mechanisms, their roles in the water and energy budget of the ITCZ and the monsoons, their role in controlling cloudiness in the lower troposphere and the associated climate feedbacks. These shallow circulations are also hypothesized to participate in the propagation of convergence zones at the seasonal scale (in monsoons) and at the intraseasonal scale (in the MJO). Better observing these circulations would permit the testing of these hypotheses. Finally, as shallow circulations are a prime candidate to explain the self-aggregation of deep convection in cloud-resolving simulations, high-resolution observations of wind would determine whether shallow circulations are associated

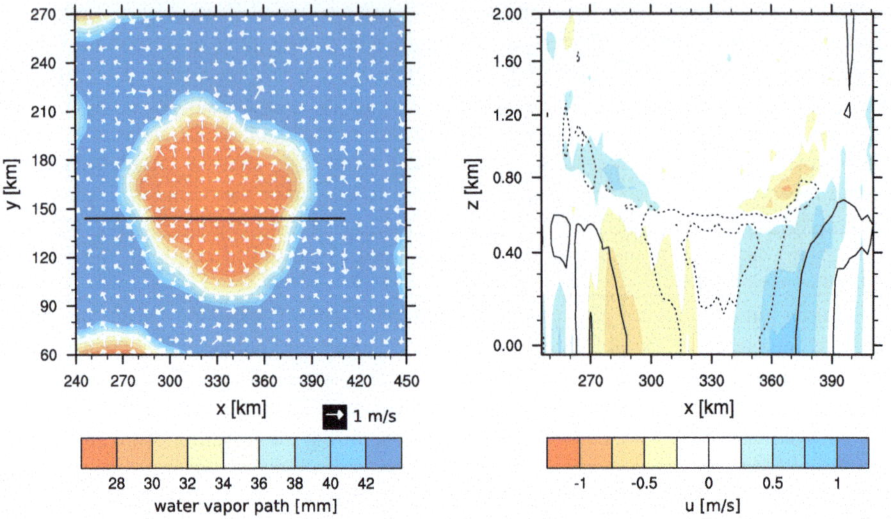

Fig. 2 Boundary-layer structure of an expanding dry region at the onset of convective aggregation from an RCE simulation with 3-km grid spacing. (left) Water-vapor path and surface wind in a subset of the domain. (right) Cross section as indicated in the left panel showing the x-component of the wind vector (color) and the virtual potential temperature deviation from the horizontal mean (lines). Contour intervals are at 0.1 K with positive values indicated by solid lines and negative values by dashed lines. The data correspond to the simulation named U50 in Hohenegger and Stevens (2016)

with the formation of observed cloud clusters, superclusters and possibly intraseasonal disturbances.

Currently, satellite retrievals of winds are limited to ocean surface winds measured by scatterometers and altimeters, and atmospheric motion vectors (AMV) derived from the motion of detectable entities transported by the wind (clouds, water-vapor patterns). Section 2 details these observations and their potential. As a result of the scarcity of direct wind observations, most observational studies of atmospheric circulation rely on reanalysis because they are the only available large-scale datasets of winds. Despite the observational constraints from wind observations at the surface and radiosondes, much uncertainty remains in the reanalyzed winds, as the study of shallow meridional circulations illustrates (Zhang et al. 2008). Now, lidar technology on satellite platforms opens new perspectives for direct wind measurements over the entire depth of the atmosphere; these exciting perspectives are detailed in Sect. 2. As the available direct observations of wind will remain scarce until the promises of satellite-borne wind lidar become reality, researchers have turned to observations of diabatic heating, which is an indirect constraint on winds, to better understand atmospheric circulations and to evaluate GCMs, with limited success as explained in detail in Sect. 3.

2 Direct Satellite Estimates of Winds

2.1 Existing Wind Observations from Satellites

Direct observations of wind profiles of the atmospheric circulation over the tropical oceans are scarce. Sea surface winds over the oceans are available from microwave scatterometers on polar-orbiting satellites (Sampe and Xie 2007). Here an equivalent 10-m wind is derived by measuring microwave backscatter from capillary waves at the ocean surface, which is proportional to the surface roughness and thus the wind speed. Microwave measurements from ERS-1/-2 and QuikSCAT have been used to study the oceanic shallow overturning circulation in the Indian ocean (Lee 2004; Schott et al. 2002) and the tropical south Pacific (Zilberman et al. 2013). Currently the instruments ASCAT (Advanced Scatterometer) on the satellites Metop-A and Metop-B are in operation and provide sea surface winds with a horizontal resolution of 25 km within two 500-km-wide swaths (Klaes et al. 2007). In addition the magnitude (but not direction) of the ocean surface wind speed can be derived from nadir-pointing radar altimeters (Abdalla 2012; Lillibridge et al. 2014), which are used to measure significant wave heights.

Atmospheric winds can also be derived from the movement of tracers, e.g., clouds or water-vapor structures using consecutive images from geostationary satellites (Menzel 2001; Velden et al. 2005). Those AMVs are only available from cloud layers or from the mid-troposphere in case of water vapor and are prone to significant height assignment errors. In addition it must be assumed in the retrieval of AMVs that the atmospheric tracers move with the mean wind, which is not the case for convective clouds bounded to a certain location.

2.2 Potential of Airborne Wind Lidar

Wind observations with lidar (light detection and ranging) instruments have been used to study atmospheric flow on a large variety of scales: from atmospheric turbulence to aircraft wake vortices or flows around wind turbines to mesoscale atmospheric flows, e.g., within mountain valleys or land-sea breeze circulation, to large-scale flows like jet streams or typhoon outflows (Reitebuch 2012b). Different types of wind lidars allow measurements at a wide range of altitudes, from the atmospheric boundary layer with high aerosol content to the upper troposphere, stratosphere and even mesosphere with molecules as backscattering targets. Optical remote sensing with lidar is a very powerful method for measuring the atmospheric wind vector. It is possible to measure the horizontal wind vector as well as the vertical wind component with different viewing geometries. Wind lidars are operated from ground, balloons, ships or aircraft, and will be deployed on satellites in the future. While first demonstrations of airborne wind-lidar technologies were initiated in the 1980s, the first airborne wind lidar for atmospheric research capable of measuring vertical profiles of horizontal wind vector was deployed on the Deutsches Zentrum für Luft-und Raumfahrt (DLR) Falcon aircraft in 1999 (Reitebuch et al. 2001, 2003). Numerous field experiments with different types of airborne wind lidars showed the potential of this new technology for studying mesoscale flows (e.g., land-sea breeze, Mistral wind, mountain circulation, West African monsoon), transport of aerosol and volcanic ash, targeted observations as input to numerical weather prediction (NWP) models, or characterization of gravity waves (Reitebuch 2012b). Recently the first airborne wind-lidar observations were performed above the tropical Atlantic during the Saharan Aerosol Long-range Transport and Aerosol-Cloud-Interaction Experiment (SALTRACE) field campaign in June/July 2013. SALTRACE was planned as a closure experiment to investigate the Saharan dust long-range transport between Africa and the Caribbean, with a focus on the dust aging and deposition processes and the characterization of its optical properties. Airborne wind-lidar observations allow sampling of the horizontal wind vector using conical scanning techniques (as for Doppler radars) with a vertical resolution of 100 m and horizontally of 4–8 km from the ground up to the flight level of the aircraft (Reitebuch 2012b). Precisions of better than 1 m s^{-1} and systematic differences below 0.1 m s^{-1} were achieved using comparisons with dropsondes (Chouza et al. 2016). Thus the shallow circulation patterns could be resolved with 100 m vertically and several km horizontally using wind lidars on aircrafts. This is demonstrated in Chouza et al. (2016) with wind profile observations in the trade wind zone of 1–2 km vertical extent in the tropical Atlantic north of the ITCZ during a flight from Cape Verde Island to Brazil [see Fig. 9 of Chouza et al. (2016)]. Another case from SALTRACE shows a complex interaction of different circulation patterns obtained during a flight from Cape Verde Islands toward Senegal, West Africa, shown in Fig. 3. The Saharan air layer with high amount of dust aerosol up to 6 km (Fig. 3, top panel) is transported toward the Atlantic ocean by the African Easterly Jet, with easterly winds above 20 m s^{-1} (see red colors in the middle panel of Fig. 3). A very shallow layer of northeasterly trade winds with an upper altitude of less than 1 km was observed close to Cape Verde (red colors in the bottom panel of Fig. 3), while a sea breeze toward the continent was observed around the West African coast (dark blue in the bottom panel of Fig. 3). Both the trade winds and the sea breeze are associated with low wind speeds of a few m/s and limited to a shallow layer close to the ground, which demonstrates the potential of airborne wind-lidar observations to resolve these shallow circulation patterns.

Vertical wind measurements were recently achieved with vertical resolutions of 100 m and horizontal resolutions of even 200 m to study island-induced gravity waves (Chouza

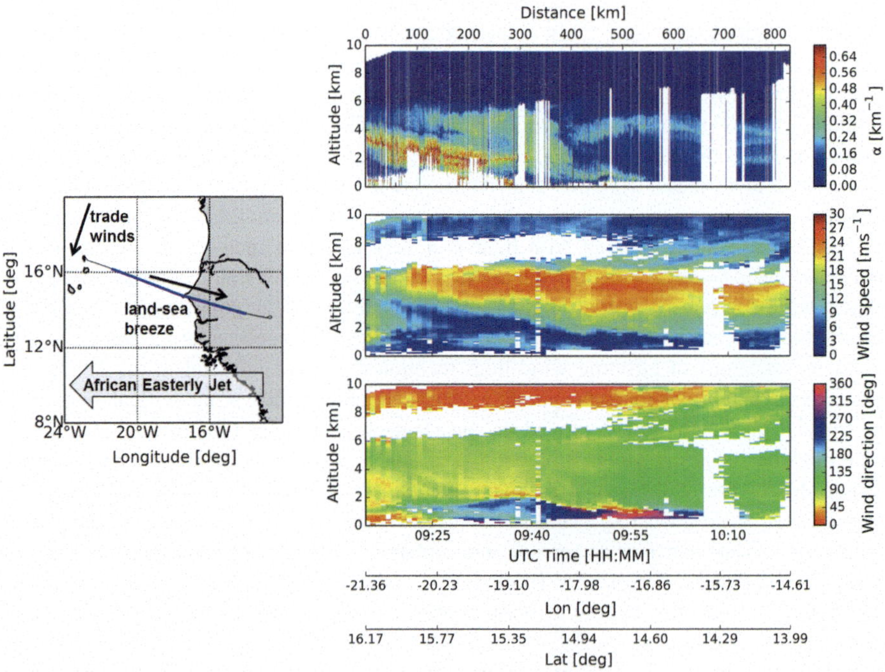

Fig. 3 Left: Flight track of the DLR Falcon aircraft on June 12, 2013, from Cape Verde Islands toward Senegal in West Africa (black and overlaid blue line of measurement locations) and the observed circulation patterns of the trade winds, land-sea breeze and African Easterly Jet. Right: airborne lidar observations of aerosol extinction (top), horizontal wind speed (middle) and direction (bottom, in degrees clockwise from south); regions were no atmospheric signal is available (e.g., below clouds, low lidar signal) are colored white; the West African coast is located at a flight distance of about 450 km [figure adapted from Chouza et al. (2016)]

et al. 2016) with estimated random errors of below 0.2 m s^{-1} and systematic errors below 0.05 m s^{-1}. Thus, the subsidence related to shallow circulations could be quantified using both the divergence of the horizontal flow from horizontal wind vector observations using a circular or rectangular flight patterns of 100 km horizontal scale and by vertical wind observations. As the vertical wind random error of 0.2 m s^{-1} refers to a horizontal resolution of 200 m, this error will be reduced by a factor of $\sqrt{100/0.2} \approx 22$ when averaging over 100 km.

2.3 Upcoming Wind-Lidar Mission Aeolus and Perspectives

After the first technology demonstration for a spaceborne lidar during a space shuttle mission (Lidar In-space Technology Experiment, LITE) during 2 weeks in September 1994, the first lidar was deployed on the satellite mission Ice, Cloud, and land Elevation Satellite (ICEsat) in 2003 mainly to monitor the ice sheet elevation and ice mass balance in addition to capturing atmospheric aerosol and cloud layer heights. Currently two lidars for the characterization of cloud and aerosol optical properties are in orbit: the lidar on-board the NASA-CNES satellite Cloud-Aerosol Lidar and Infrared Pathfinder Satellite Observation (CALIPSO, launched in 2006) and the Cloud-Aerosol Transport System (CATS, launched in 2015) lidar on the International Space Station (ISS). Aerosol lidars need to measure the backscattered signal intensity only, but wind lidars additionally need to measure the spectral properties of the backscattered signal with accuracy on the order of 10^{-8}. Thus the laser and optical receiver technology of a wind lidar is far more challenging; compared to an aerosol lidar, more powerful laser transmitters and larger telescopes are required.

Although global observations of wind profiles are recognized as the number one priority for global numerical weather prediction (NWP), the wind profile is still poorly observed globally because most satellite sounding instruments provide temperature rather than wind information. A space-borne wind lidar is considered to be the only candidate to provide the global coverage of wind profile observations throughout the troposphere and lower stratosphere.

In the near-future global wind profile observations will be available from the Aeolus mission of the European Space Agency, ESA (Stoffelen et al. 2005; European Space Agency 2008; Reitebuch 2012a). This mission, which is planned for launch in 2018, will carry the first wind lidar (light detection and ranging) in space. On a polar-orbiting satellite (Fig. 4, left panel), Aeolus will sense one component of the horizontal wind vector—mainly in the zonal direction—from ground up to 20–30 km with a vertical resolution of 250–2000 m depending on altitude. The vertical resolution of the instrument can be commanded and varied along the orbit, which allows sampling of the lowest altitudes in the boundary layer with the highest resolution of 250 m for specific geographical regions of interest, e.g., the shallow circulation regions. The horizontal along-track resolution of the wind profiles will be 90 km to achieve the required precision of 1–3 m s^{-1} (depending on altitude). Nevertheless, higher-resolution measurements of 3 km along track are available for on-ground processing and could be used for cases with high signals, e.g., from clouds or aerosol layers. The horizontal coverage of the Aeolus observations, which samples essentially a vertical curtain through the atmosphere, is visualized in Fig. 4 (right panel) for 1 week over the Atlantic. These measurement locations are periodically sampled every week, as this period corresponds to the orbit repeat cycle of Aeolus. The Aeolus observations will be assimilated by the European Center for Medium-range Weather

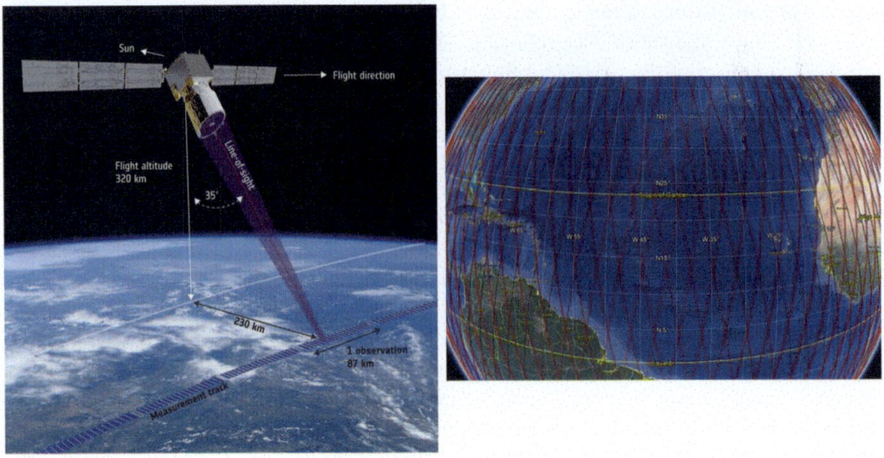

Fig. 4 Left: Schematic view of Aeolus measurement geometry with a polar-orbiting satellite at 320 km altitude and a telescope pointing direction 35°off-nadir and perpendicular to the satellite flight direction, which separates satellite (white) and measurement track (purple stripes indicating raw data resolution) by 230 km; line-of-sight wind profiles are obtained after on-ground averaging over 87 km to one observation (Fig. copyrights ESA/ATG medialab); right: Aeolus satellite tracks (red) over the Atlantic during an orbit repeat cycle of 1 week; consecutive Aeolus satellite tracks are separated by about 2500 km at the equator, resulting from the duration of 91 min for one orbit

Forecasts (ECMWF) and other meteorological services, and they are expected to have a large positive impact on analysis and forecast quality, especially in the tropics (Zagar 2004; Zagar et al. 2008), where the wind field cannot be derived from mass observations (e.g., temperature from passive sounders) due to the missing geostrophic balance.

In order to keep the instrument concept simple for Aeolus, only one component of the horizontal wind vector will be sensed using one telescope with a fixed pointing direction. By itself the observations from Aeolus will provide mainly the zonal-wind component in the tropics. Aeolus's goals are (1) to demonstrate the wind-lidar technology on a satellite platform and (2) to constrain the atmospheric circulation in the analysis through data assimilation. A significant positive impact is expected from Aeolus as shown by Observing System Simulation Experiments (OSSE) (Stoffelen et al. 2006) and ensemble data assimilation experiments at ECMWF (Tan et al. 2007). Studies of the atmospheric circulations will be able to rely on a better reanalysis, with assimilated Aeolus observations, which allow to better constrain the atmospheric flow and momentum budget in the NWP model used for the reanalysis. Future satellite wind-lidar concepts for observing both components of the horizontal wind vector by using two telescopes with different line-of-sight (LOS) pointing directions are studied (Baker et al. 2014) and proposed for deployment on the ISS (Tucker et al. 2016).

Future studies of shallow circulations will strongly benefit from the regular availability of wind profile observations from the Aeolus mission, which is complemented by the sea surface wind speed from scatterometers. In particular, due to the polar orbit and the perpendicular LOS wind direction, Aeolus is sensing mainly the zonal-wind component, especially in the tropics, which can be used to detect the low-level zonal wind associated with the MJO. Airborne wind-lidar observations during limited campaign periods will ideally be suited to provide higher resolved horizontal wind vector and even vertical wind observations for characterizing the dynamical properties of shallow circulation regimes for case studies.

3 Constraining Circulation with Diabatic Heating

Considering the scarcity of direct existing observations of wind, indirect approaches aiming to constrain the circulation have been developed. One of these approaches is to estimate the diabatic heating Q_1 in the atmosphere. Indeed, if the spatiotemporal distribution of diabatic heating and initial conditions of temperature, wind and surface pressure are known, the equations of momentum, continuity and thermodynamics can be integrated in time (using, for example, the dynamical core of an existing model) to deduce the time evolution of the circulation. Troposphere-deep heating creates pressure gradients that force deep circulation with ascent in the region of heating; shallow circulations can be created by shallow heating that causes shallow ascent as in the case of observed circulations in the ITCZ and the MJO, or by shallow cooling that cause shallow descent as in the case of simulated radiative cold pools leading to self-aggregation (Coppin and Bony 2015). The details of the profile of diabatic heating have been shown to be crucial to the dynamical response both for the climatological mean (Hartmann et al. 1984; Bellon and Sobel 2010; Nishant et al. 2016) and for tropical variability (Li et al. 2009; Lappen and Schumacher 2012).

In this section, we will evaluate the usefulness of existing diabatic-heating estimates from a modeler's point of view. Are there diabatic-heating features which are found in all estimates and therefore robust enough to be used to evaluate GCMs?

3.1 Current Estimates of Diabatic Heating

Satellite estimates The atmospheric diabatic heating results essentially from two processes: phase changes and radiation. In most precipitating clouds, latent heat release is the dominant process, but radiative cooling is dominant in stratocumulus and clear sky (Bellon and Bony 2017), and it can account for about 15–20% of heating anomalies in the MJO (Ciesielski et al. 2017). At times, mixing by small-scale circulations is considered as diabatic heating for the large-scale circulation, but it appears to be one to two orders of magnitude smaller than the other terms, except in boundary-layer clouds. Here, we review the existing satellite estimates of latent and radiative heating.

With the advent of satellite-based precipitation radar (PR) on the Tropical Rainfall Measuring Mission (TRMM) in 1998 (Simpson et al. 1996), the estimates of precipitating water mixing ratio and precipitation fluxes have opened new perspectives to quantify latent heating Q_c, first over the tropics with TRMM (1998–2015) and now almost globally with GPM launched in 2014 (Smith et al. 2007). Various approaches have been used to reconstitute the profiles of latent heating from precipitation fluxes (see (Tao et al. 2006) and (Hagos et al. 2010) for reviews, but also other articles of the *Journal of Climate's special* collection on **TRMM Diabatic Heating**). Many of them rely on tabulated reference heating profiles produced by a cloud-resolving model (CRM). The main algorithms are:

- The convective-stratiform heating (CSH) algorithm (Tao et al. 1993, 2000) uses surface rainfall rate and its stratiform fraction in combination with reference rain-normalized heating profiles from CRM tabulated according to rain type (convective or stratiform) and region of the atmosphere (lower troposphere, oceanic and continental);

- The spectral latent heating (SLH) algorithm (Shige et al. 2004, 2007, 2008) divides the simulated precipitation into subgroups according to the PR echo-top heights for convective rain and to the melting-level precipitation rate for deep stratiform rain, and uses a set of spectral look-up tables independent of location;
- The TRAIN algorithm (Grecu et al. 2009) uses an even larger database of CRM profiles that include their dependence on the brightness temperature, and the observed brightness temperature from the TRMM Microwave Imager (TMI) is exploited, capitalizing on previous satellite estimates from microwave sensors such as the Goddard profiling (GPROF) heating algorithm (Olson et al. 1999) and the hydrometeor heating (HH) algorithm (Yang and Smith 1999a, b).
- The precipitation radar heating (PRH) algorithm (Kodama et al. 2009) does not rely on CRM inputs. It estimates the growth evaporation of raindrops during their vertical displacement, by comparing the vertical flux of precipitation water with the motion of precipitating hydrometeors and diagnosing the phase changes from the difference; to do so, it utilizes tabulated vertical profiles of vertical velocity.

Multiple satellite-based sensors of increasing sophistication have allowed increasingly precise measurements of radiative heating rates Q_r:

- The Hydrologic Cycle and Earth Radiation Budget (HERB) algorithm (L'Ecuyer and Stephens 2003, 2007; L'Ecuyer and McGarragh 2010) combines measurements from the Visible and Infrared Scanner (VIRS) on TRMM and TMI with a radiative transfer model to provide the spatiotemporal distribution of radiative heating. The radiative transfer model in HERB uses hydrometeors information from GPROF, aerosol datasets (initially the Global Aerosol Climatology Project, now the Spectral Radiation-Transport Model for Aerosol Species) and temperature and humidity profiles from reanalysis as inputs.
- Recent developments provide radiative heating rates based on observations from the A-Train constellation of satellites. The latest version is the 2B-FLXHR-LIDAR dataset (Henderson et al. 2013); it utilizes CALIPSO lidar observations together with a radiative transfer model that uses hydrometeor properties from CloudSat radar observations and from the Moderate Resolution Imaging Spectroradiometer (MODIS), as well as humidity and temperature profiles from reanalysis.

Reanalysis Reanalyses are produced by general circulation models coupled with a data assimilation system to constrain the models to follow all available observations as closely as possible. They provide global fields of meteorological data. Diabatic heating in reanalyses is the sum of the heating outputs from the parameterizations of diabatic subgrid processes (convection, clouds, radiation, mixing). It therefore depends on the simplifications and assumptions made in the parameterization schemes of these processes that are included in the reanalysis's general circulation model, as well as the reanalyzed large-scale variables that these parameterizations use as input.

Three global, state-of-the-art reanalyses are used here:

- The European Centre for Medium-Range Weather Forecasts (ECMWF) Interim Re-Analysis ERA-Interim (Dee et al. 2011);
- The Climate Forecast System Reanalysis CFSR (Saha et al. 2010) from the National Centers for Environmental Prediction;

- The Modern Era Retrospective Analysis for Research and Applications (MERRA) (Rienecker et al. 2011) from the National Aeronautics and Space Administration.

3.2 Evaluation of Existing Estimates

Diabatic heating in the eastern Pacific shallow circulation The climate in the eastern tropical Pacific has proven difficult to simulate with GCMs, with most GCMs simulating spurious precipitation south of the equator, the so-called double-ITCZ bias (Li and Xie 2014; Oueslati and Bellon 2015). Understanding the circulation in this region is therefore of particular interest in order to better constrain new model developments. As an example, we investigate the case of boreal summer in the tropical eastern Pacific (90°W–120°W). Figure 1 shows the July climatological circulation as described by the ERA-Interim reanalysis, with a prominent shallow circulation.

Figure 5 shows the associated diabatic heating as estimated by reanalyses (Fig. 5a, b) and from TRMM using two different algorithms (SLH and CSH) for latent heating and HERB for radiative heating (Fig. 5c, d). The overall patterns of diabatic heating are similar in these estimates: troposphere-deep heating in the ITCZ around 10°N extending northward in the mid-troposphere, heating within the boundary layer south of the ITCZ and intense cooling at its top due to stratocumulus-top radiative cooling (Lilly and Schubert 1980; Bellon and Geoffroy 2016) and more moderate cooling in the clear-sky free troposphere

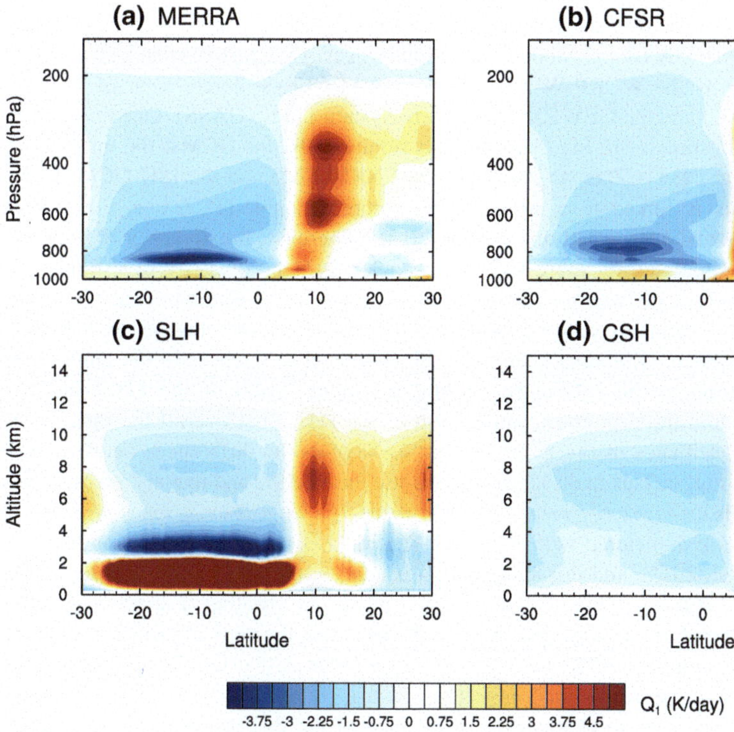

Fig. 5 July climatology of diabatic heating in the eastern tropical Pacific region (90°W–120°W) estimated by two reanalysis (**a**, **b**) and by two different algorithms using TRMM data for latent heating and HERB for radiative heating (**c**, **d**)

above. But in terms of the details of the heating vertical profiles, there are significant differences between the estimates. First, the number of heating maxima on the vertical within the ITCZ varies from one in the mid-troposphere (in CSH estimates) to four (two in the boundary layer, one in the lower free troposphere and one in the upper troposphere, in MERRA estimates). Second, the extent, altitude and intensity of the northward tail of the mid-tropospheric northward extension of the ITCZ heating are significantly different from one estimate to the other. Last, while the heating–cooling dipole in the boundary layer south of the ITCZ is similar in the two reanalyses, with heating up to 900 hPa and cooling above that decreases progressively, this pattern is very different in and in between the two TRMM estimates: this dipole is very weak in CSH estimates, with barely any heating close to the surface and weak cooling above; it is very large in the SLH estimates, with very intense heating up to 2 km and very intense cooling above. The large heating within the boundary layer was documented in regions of subsidence in a previous study of SLH latent heating (Takayabu et al. 2010). In the presence of a shallow circulation, we would expect shallow heating and we do see such heating in SLH, MERRA and CFSR, but not in CSH estimates. The detailed characteristics of the shallow circulation depend on the location and intensity of the low-level heating maximum, which are significantly different between the reanalyses and SLH. These results are in agreement with previous studies of diabatic heating and circulation in this region that showed that there is limited consensus between estimates of diabatic-heating profiles in this region (Ling and Zhang 2013; Yokoyama and Takayabu 2012; Huaman and Takahashi 2016).

Figure 6 shows the corresponding figures for the 20-year climate simulations of four GCMs that participated to the intercomparison project “Vertical Structure and Diabatic Processes of the MJO: A Global Model Evaluation Project” (MJO-MIP) (Petch et al. 2011) organized jointly by the WCRP-WWRP/THORPEX MJO Task Force, Year of Tropical Convection (YoTC) and the GEWEX Atmosphere System Study (GASS). One feature that is common to all estimates in Fig. 5 is not well simulated by the GCMs: the intensity of diabatic heating in the ITCZ is very weak in the CNRM-CM5 and the ISU-GCM. In the CNRM-CM5, this is linked to a strong boreal-summer double-ITCZ bias that damps convection in the ITCZ (Oueslati and Bellon 2013). Otherwise, the diversity of diabatic-heating patterns simulated by these GCMs is smaller than the diversity of estimates in Fig. 5: all GCMs simulate two maxima of diabatic heating on the vertical in the ITCZ and, south of the ITCZ, heating up to 900 hPa and cooling above, which decreases with altitude in the free troposphere. Of course, there are still significant differences in the altitudes of the heating maxima in the ITCZ, the cooling magnitude south of the ITCZ and the extension of the mid-tropospheric heating north of the ITCZ, but overall the differences between heating estimates in Fig. 5 are at least as large as the differences between simulated heating in Fig. 6. This makes it difficult to use the satellite and reanalyzed estimates to evaluate GCMs.

Diabatic heating associated with the MJO The MJO is unsatisfactorily simulated by GCMs (Hung et al. 2013), and this shortcoming is attributed to the inability of the GCMs to simulate a realistic spatiotemporal distribution of diabatic heating (Lappen and Schumacher 2012). One of the leading specific hypotheses is that the vertical profile of diabatic heating is poorly simulated. (The other leading hypothesis is that the GCMs cannot represent the spatial organization of convective clouds on the horizontal.) The MJO is therefore a good test case to evaluate the usefulness of the diabatic-heating approach to constrain tropical circulations and in particular the shallow circulation that propagates ahead of the convective disturbance.

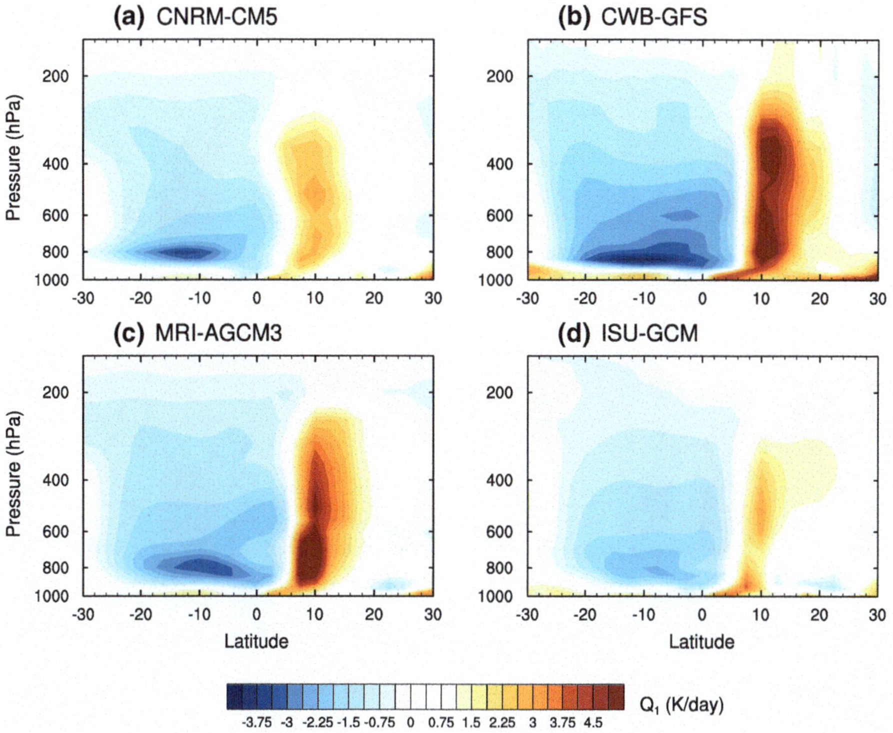

Fig. 6 July climatology of diabatic heating in the eastern tropical Pacific region (90°W–120°W) simulated by four GCMs

The MJO propagation can be divided into 8 phases (Wheeler and Hendon 2004): Phase 1 corresponds to decreased convection over the Indo-Pacific region, Phases 2 and 3 to the initiation and intensification of an MJO event in the Indian ocean, Phases 4 and 5 to the propagation of this MJO event over the Maritime Continent, Phases 6 and 7 to the propagation and decay of this MJO event over the Pacific ocean, and Phase 8 to a residual convective perturbation in the Pacific with the establishment of a dry spell over the maritime continent.

Figure 7 [adapted from Jiang et al. (2011)] shows the vertical profile of diabatic heating and the precipitation anomaly as a function of the phase of the MJO in the western Pacific. The diabatic heating displayed in Fig. 7a–c is from reanalysis datasets described in the previous section, while in Fig. 7d–f it is the sum of the latent heating estimated by three of the latent heat algorithms presented in the previous section (whose names are used to label the panels) and the radiative heating estimated by the algorithm HERB. The different datasets agree on the main feature of the diabatic structure of the MJO: the precipitation anomaly and diabatic heating are positive during phases 4–6, negative during phases 8–2, with maximum diabatic heating/cooling in the mid-troposphere typical of convective rainfall. The pattern of diabatic heating is slightly slanted with heating during phase 3 in the lower troposphere, as expected since this feature is associated with the shallow circulation ahead of the main convective disturbance (Kiladis et al. 2005) and cooling during phase 6. But the different datasets disagree on the exact altitude of maximum heating and the amplitude of the maximum heating, with slightly lower maxima in CFSR and CSH, and

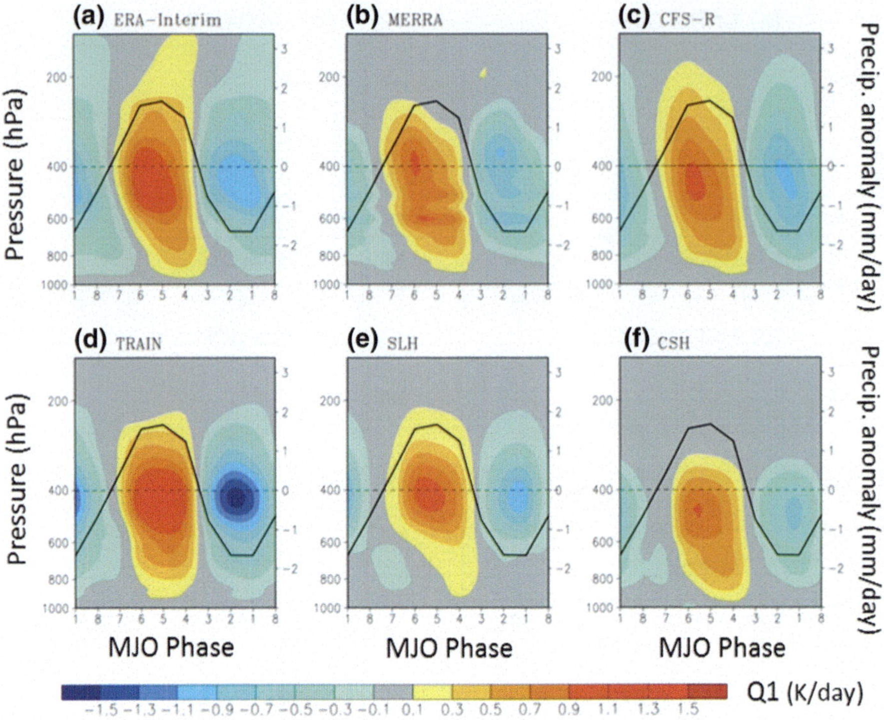

Fig. 7 Profiles of diabatic-heating rate (shadings) and precipitation anomaly (black lines) in the different phases of the MJO over the equatorial western Pacific (150°E–160°E, 10°S–10°N), estimated by three reanalysis (**a–c**) and three different algorithms using TRMM data for latent heating and HERB for radiative heating (**d–f**). Adapted from (Jiang et al. 2011) © American Meteorological Society. Used with permission

a second maximum heating at 600 hPa in MERRA. The slant of the pattern also varies depending on the dataset, suggesting that the corresponding shallow circulation differs. Also, CSH provides an estimate of heating that does not extend on the vertical as much as the other estimates. The interalgorithm spread in satellite estimates is roughly similar to the interreanalysis spread.

Figure 8 shows the corresponding plots for the same four simulations as above. Two of these models (CNRM-CM5 and MRI-AGCM3) have shown decent skills in simulating the MJO, while the two others (CWB-GFS and ISU-GCM) struggle to simulate realistic intraseasonal perturbations (Jiang et al. 2015). Figure 8b suggests that CWB-GFS simulates MJO events that are less intense and less spatially developed than the observed events, while Fig. 8d suggests that the intraseasonal variability simulated by the ISU-GCM is very different from the observed variability, to the point that the analysis in phases following (Wheeler and Hendon 2004) does not make much sense. The profiles of MJO diabatic heating simulated by the models that do simulate an MJO, even a weak one as CWB-GFS, are similar to the satellite estimates, with a maximum in the mid-troposphere at or slightly below 400 hPa and an amplitude roughly proportional to the precipitation anomaly, and shallow heating during phase 3. Intermodel spread of these profiles is similar to the interalgorithm spread of the satellite estimates and the interreanalysis spread illustrated by Fig. 7. Even in the case of the ISU-GCM that does not simulate a proper MJO,

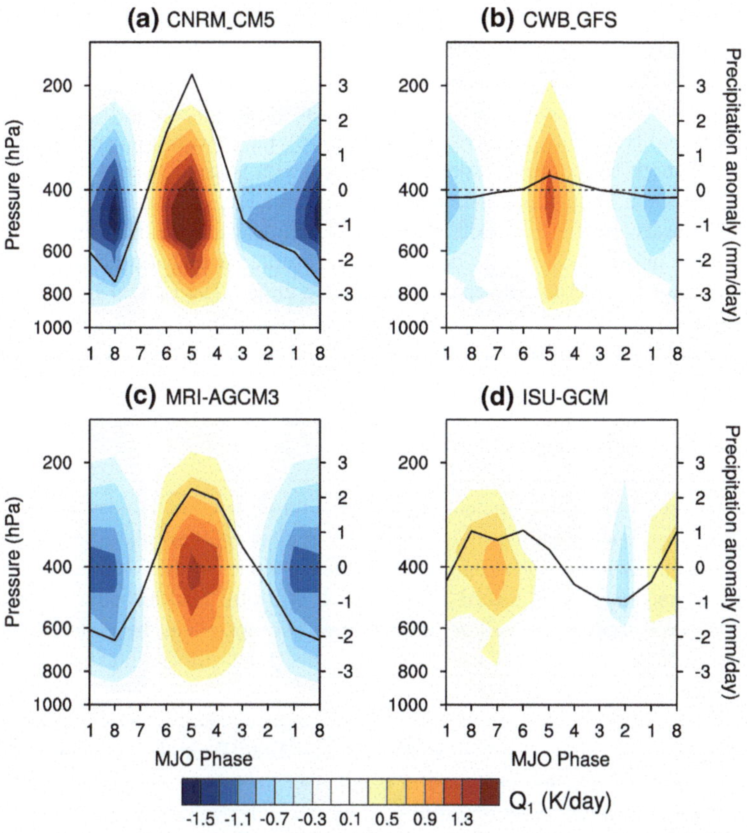

Fig. 8 Profiles of diabatic-heating rate (shadings) and precipitation anomaly (black line) in the different phases of the MJO over the equatorial western Pacific (150°E–160°E, 10°S–10°N), simulated by four GCMs

the profile of diabatic heating exhibits the same general characteristics. Indeed, an analysis of the vertical profiles of diabatic heating in terms of typical profiles associated with different types of rainfall (convective, shallow and stratiform) might shed some light on intermodel differences. But it is clear from Fig. 7 that the same analysis on estimates will also reveal significant interalgorithm and interreanalysis spread. At present, the interdataset spread of diabatic-heating profiles estimated from satellite observations (TRMM) or by reanalysis systems is significant, and this absence of consensus limits the usefulness of these estimates to evaluate GCMs when it comes to their ability to simulate the MJO.

At this point, it appears that the satellite estimates of diabatic heating exhibit too much dependency on the choice of algorithm to provide useful constraints on GCMs, even when it comes to troposphere-deep circulations such as the main pattern associated with the MJO. The shallow diabatic heating associated with shallow circulations which involves the latent and radiative heating in low clouds, is particularly difficult to observe from satellites because screening by clouds aloft is frequent and because by design the TRMM precipitation radar is unable to detect small hydrometeors in light drizzle, frequent in shallow clouds; as a result the satellite estimates of low-level heating are particularly algorithm-dependent, as illustrated in Figs. 5c, d and 7d–f for the examples of the eastern Pacific circulation and the MJO. Ongoing work on algorithms and the exploitation of new satellites will improve this situation. In particular CloudSat's cloud-water radar is able to detect drizzle and cloud hydrometeors, particularly significant in low clouds and therefore for shallow circulations; its data can provide better constraints on the structure of clouds and light rain and work is underway to estimate diabatic heating from this data. Also, the recently launched Himawari 8 and 9 satellites will provide observations of the Indo-Pacific region at very high temporal and spatial resolution with passive sensors, complementing existing cloud data. Despite sophisticated data assimilation systems and models, current reanalysis datasets exhibit as much spread as satellite estimates and therefore do not provide robust information on the details of the profile of diabatic heating in the east Pacific ITCZ or in the MJO convective disturbance either. The current limitations of the diabatic-heating approach to constrain circulations are certainly an incentive to maintain and develop alternative strategies such as the development of satellite-based wind lidars.

4 Conclusion and Perspectives

The observation of winds over the entire globe and throughout the atmosphere would lead to a more detailed understanding of atmospheric circulations at all scales and in particular shallow circulations in the Intertropical Convergence Zone, in the Madden–Julian Oscillation and around radiatively forced cold pools. It will also lead to better constraints on climate and NWP models, as well as on reanalysis. To document the global atmospheric circulation, a global coverage is necessary, and this points to satellites as the optimum platforms of observations.

As shown in the previous sections, the observation of winds from satellites is subject to a number of challenges, but also with significant opportunities in the future. Currently and in the foreseeable future, this observation is expected to pursue the two approaches presented above: direct observation and indirect constraint through the estimation of diabatic heating. The first approach is currently limited to the ocean surface or to the presence of detectable cloud and water-vapor structures advected by the wind, but the advent of satellite-borne wind lidar opens far-reaching perspectives. Nevertheless, the first wind-lidar mission (and the only one programmed so far), Aeolus, will measure only one component of the wind vector and will provide only a partial documentation of atmospheric motions. On the other hand, the indirect method is faced with significant methodological uncertainties, with, so far, no independent dataset to discriminate between the different algorithms used to estimate diabatic-heating profiles from precipitation radar. But existing satellites such as CloudSat could provide this independent data and field campaigns could help constrain these methods.

Reanalyses assimilate the existing observations used in both approaches and will assimilate the upcoming wind-lidar observations. By doing so, they take advantage of both direct and indirect estimates of atmospheric circulation. With the assimilation of Aeolus data, reanalyses will probably become some of the best datasets to study atmospheric circulations in the foreseeable future.

Acknowledgements This paper arises from the International Space Science Institute (ISSI) Workshop on Shallow clouds and water vapor, circulation and climate sensitivity. G. B. acknowledges the support of the Pacific Fund grant Pluvar and the support of H. Glavish. A. K. N. was supported by the Hans-Ertel Centre for Weather Research. This research network of universities, research institutes and the Deutscher Wetterdienst is funded by the Federal Ministry of Transport and Digital Infrastructure (BMVI). Thanks are extended to Yi Song for her help with the CSH data.

References

Abdalla S (2012) Ku-band radar altimeter surface wind speed algorithm. Mar Geod 35(supplement 1):276–298

Baker WE, Atlas R, Cardinali C, Clement A, Emmitt GD, Gentry BM, Hardesty RM, Källén E, Kavaya MJ, Langland R, Ma Z, Masutani M, McCarty W, Pierce RB, Pu Z, Riishojgaard LP, Ryan J, Tucker S, Weissmann M, Yoe JG (2014) Lidar-measured wind profiles: the missing link in the global observing system. Bull Am Meteorol Soc 95(4):543–564

Bellon G, Bony S (2017) Tropical and subtropical cloud systems. In: Bony S, Jakob C, Siebesma AP, Stevens B (eds) Clouds and climate, Chap 9. Cambridge University Press, Cambridge

Bellon G, Geoffroy O (2016) Stratocumulus radiative effect, multiple equilibria of the well mixed boundary layer and transition to shallow convection. Q J R Meteorol Soc 142:1685–1696

Bellon G, Sobel AH (2010) Multiple equilibria of the Hadley circulation in an intermediate-complexity axisymmetric model. J Clim 23(7):1760–1778

Bretherton CS, Blossey PN, Khairoutdinov M (2005) An energy-balance analysis of deep convective self-aggregation above uniform SST. J Atmos Sci 62:4273–4292

Chouza F, Reitebuch O, Benedetti A, Weinzierl B (2016) Saharan dust long-range transport across the Atlantic studied by an airborne Doppler wind lidar and the MACC model. Atmos Chem Phys 16:11581–11600

Chouza F, Reitebuch O, Jähn M, Rahm S, Weinzierl B (2016) Vertical wind retrieved by airborne lidar and analysis of island induced gravity waves in combination with numerical models and in-situ particle measurements. Atmos Chem Phys 16:4675–4692

Ciesielski PE, Johnson RH, Jiang X, Zhang Y, Xie S (2017) Relationships between radiation, clouds, and convection during DYNAMO. Geophys Res Atmos 122:2529–2548

Coppin D, Bony S (2015) Physical mechanisms controlling the initiation of convective self-aggregation in a General Circulation Model. J Adv Model Earth Syst 7(4):2060–2078

Dee DP et al (2011) The ERA-Interim reanalysis: configuration and performance of the data assimilation system. Q J R Meteorol Soc 137:553–597

Dixit V, Srinivasan J (2016) The momentum constraints on the shallow meridional circulation associated with the marine ITCZ. Meteorol Atmos Phys. https://doi.org/10.1007/s00703-016-0489-2

European Space Agency (2008) ADM-Aeolus Science Report, ESA SP-1311

Grecu M, Olson WS, Shie C-L, L'Ecuyer TL, Tao WK (2009) Combining satellite microwave radiometer and radar observations to estimate atmospheric heating profiles. J Clim 22:6356–6376

Hagos S, Zhang C (2010) Diabatic heating, divergent circulation and moisture transport in the African monsoon system. Q J R Meteorol Soc 136:411–425

Hagos S, Zhang C, Tao WK, Lang S, Takayabu YN, Shige S, Katsumata M, Olson B, L'Ecuyer T (2010) Estimates of tropical diabatic heating profiles: commonalities and uncertainties. J Clim 23(3):542–558

Hartmann DL, Hendon HH, Houze RA Jr (1984) Some implications of the mesoscale circulations in tropical cloud clusters for large-scale dynamics and climate. J Atmos Sci 41(1):113–121

Henderson DS, L'Ecuyer TS, Stephens GL, Partain P, Sekiguchi M (2013) A multisensor perspective on the radiative impacts of clouds and aerosols. J Appl Meteorol Climatol 52(4):853–871

Hohenegger C, Stevens B (2016) Coupled radiative convective equilibrium simulations with explicit and parameterized convection. J Adv Model Earth Syst 8(3):1468–1482

Huaman L, Takahashi K (2016) The vertical structure of the eastern Pacific ITCZs and associated circulation using the TRMM Precipitation Radar and in situ data. Geophys Res Lett 43(15):8230–8239

Hung MP, Lin JL, Wang W, Kim D, Shinoda T, Weaver SJ (2013) MJO and convectively coupled equatorial waves simulated by CMIP5 climate models. J Clim 26(17):6185–6214

Jiang X, Waliser DE, Olson WS, Tao WK, L'Ecuyer TS, Li KF, Yung YL, Shige S, Lang S, Takayabu YN (2011) Vertical diabatic heating structure of the MJO: intercomparison between recent reanalyses and TRMM estimates. Mon Weather Rev 139(10):3208–3223

Jiang X, Waliser DE, Xavier PK, Petch J, Klingaman NP, Woolnough SJ, Guan B, Bellon G, Crueger T, DeMott C, Hannay C, Lin H, Hu W, Kim D, Lappen C-L, Lu M-M, Ma H-Y, Miyakawa T, Ridout JA, Schubert SD, Scinocca J, Seo K-H, Shindo E, Song X, Stan C, Tseng W-L, Wang W, Wu T, Wu X, Wyser K, Zhang GJ, Zhu H (2015) Vertical structure and physical processes of the Madden–Julian oscillation: exploring key model physics in climate simulations. J Geophys Res Atmos 120(10):4718–4748

Johnson RH, Ciesielski PE, Ruppert JH Jr, Katsumata M (2015) Sounding-based thermodynamic budgets for DYNAMO. J Atmos Sci 72:598–622

Kemball-Cook SR, Weare BC (2001) The onset of convection in the Madden–Julian oscillation. J Clim 14:780–793

Kiladis GN, Straub KH, Haertel PT (2005) Zonal and vertical structure of the Madden–Julian oscillation. J Atmos Sci 62:2790–2809
Klaes KD, Cohen M, Buhler Y, Schlüssel P, Munro R, Luntame JP, von Engelin A, Clerigh EO, Bonekamp H, Ackermann J, Schmetz J (2007) An introduction to the EUMETSAT polar system. Bull Am Meteorol Soc 88(7):1085–1096
Klein SA, Hartmann DL (1993) The seasonal cycle of low stratiform clouds. J Clim 6(8):1587–1606
Kodama Y-M, Katsumata M, Mori S, Satoh S, Hirose Y, Ueda H (2009) Climatology of warm rain and associated latent heating derived from TRMM-PR observations. J Clim 22:4908–4929
Lappen CL, Schumacher C (2012) Heating in the tropical atmosphere: what level of detail is critical for accurate MJO simulations in GCMs? Clim Dyn 39(9–10):2547–2568
L'Ecuyer TS, McGarragh G (2010) A 10-year climatology of tropical radiative heating and its vertical structure from TRMM observations. J Clim 23(3):519–541
L'Ecuyer TS, Stephens GL (2003) The tropical atmospheric energy budget from the TRMM perspective. Part I: algorithm and uncertainties. J Clim 16:1967–1985
L'Ecuyer TS, Stephens GL (2007) The tropical atmospheric energy budget from the TRMM perspective. Part II: evaluating GCM representations of the sensitivity of regional energy and water cycles to the 199899 ENSO cycle. J Clim 20:4548–4571
Lee T (2004) Decadal weakening of the shallow overturning circulation in the South Indian Ocean. Geophys Res Lett 31:L18305
Li G, Xie SP (2014) Tropical biases in CMIP5 multimodel ensemble: the excessive equatorial Pacific cold tongue and double ITCZ problems. J Clim 27:1765–1780
Li C, Jia X, Ling J, Zhou W, Zhang C (2009) Sensitivity of MJO simulations to diabatic heating profiles. Clim Dyn 32(2–3):167–187
Lillibridge J, Scharroo R, Abdalla S, Vandemark D (2014) One- and two-dimensional wind speed models for Ka-band altimetry. J Atmos Ocean Technol 31:630–638
Lilly DK, Schubert WH (1980) The effects of radiative cooling in a cloud-topped mixed layer. J Atmos Sci 37:482–487
Lin X, Johnson RH (1996) Kinematic and thermodynamic characteristics of the flow over the western Pacific warm pool during TOGA COARE. J Atmos Sci 53:695–715
Ling J, Zhang C (2013) Diabatic heating profiles in recent global reanalyses. J Clim 26:3307–3325
Madden RA, Julian PR (1971) Detection of a 40–50 day oscillation in the zonal wind in the tropical Pacific. J Atmos Sci 28(5):702–708
Menzel PW (2001) Cloud tracking with satellite imagery: from the pioneering work of Ted Fujita to the present. Bull Am Meteorol Soc 82(1):33–47
Muller C, Bony S (2015) What favors convective aggregation and why? Geophys Res Lett 42:5626–5634
Muller CJ, Held IM (2012) Detailed investigation of the self-aggregation of convection in cloud-resolving simulations. J Atmos Sci 69:2551–2565
Naumann AK, Stevens B, Hohenegger C, Mellado JP (2017) A conceptual model of a shallow circulation induced by low-level radiative cooling. J Atmos Sci 74:3129–3144
Nguyen H, Thorncroft CD, Zhang C (2011) Guinean coastal rainfall of the West African Monsoon. Q J R Meteorol Soc 137:1828–1840
Nie J, Boos WR, Kuang Z (2010) Observational evaluation of a convective quasi-equilibrium view of monsoons. J Clim 23:4416–4428
Nishant N, Sherwood SC, Geoffroy O (2016) Radiative driving of shallow return flows from the ITCZ. J Adv Model Earth Syst 8:831–842
Nolan DS, Zhang C, Chen SH (2007) Dynamics of the shallow meridional circulation around intertropical convergence zones. J Atmos Sci 64(7):2262–2285
Norris JR (1998) Low cloud type over the ocean from surface observations. Part I: relationship to surface meteorology and the vertical distribution of temperature and moisture. J Clim 11(3):369–382
Olson WS, Kummerow CD, Hong Y, Tao WK (1999) Atmospheric latent heating distributions in the tropics derived from satellite passive microwave radiometer measurements. J Appl Meteorol 38:633–664
Oueslati B, Bellon G (2013) Convective entrainment and large-scale organization of tropical precipitation: sensitivity of the CNRM-CM5 hierarchy of models. J Clim 26(9):2931–2946
Oueslati B, Bellon G (2015) The double ITCZ bias in CMIP5 models: interaction between SST, large-scale circulation and precipitation. Clim Dyn 44:585–607
Parker DJ, Willetts P, Birch C, Turner AG, Marsham JH, Taylor CM, Kolusu S, Martin GM (2016) The interaction of moist convection and mid level dry air in the advance of the onset of the Indian monsoon. Q J R Meteorol Soc 142:2256–2272
Petch J, Waliser D, Jiang X, Xavier PK, Woolnough S (2011) A global model intercomparison of the physical processes associated with the Madden–Julian oscillation. GEWEX News 21(3):3–5

Pierrehumbert RT, Roca R (1998) Evidence for control of Atlantic subtropical humidity by large scale advection. Geophys Res Lett 25(24):4537–4540

Reitebuch O (2012a) The space-borne wind lidar mission ADM-Aeolus. In: Schumann U (ed) Atmospheric physics background. Methods, trends. Springer series on research topics in aerospace. Springer, Berlin, pp 815–827

Reitebuch O (2012b) Wind lidar for atmospheric research. In: Schumann U (ed) Atmospheric physics background. Methods, trends. Springer series on research topics in aerospace. Springer, Berlin, pp 487–507

Reitebuch O, Werner C, Leike I, Delville P, Flamant PH, Cress A, Engelbart D (2001) Experimental validation of wind profiling performed by the airborne 10 μm-Heterodyne Doppler Lidar WIND. J Atmos Ocean Technol 18:1331–1344

Reitebuch O, Volkert H, Werner C, Dabas A, Delville P, Drobinski P, Flamant PH, Richard E (2003) Determination of air flow across the Alpine ridge by a combination of airborne Doppler lidar, routine radio-sounding and numerical simulation. Q J R Meteorol Soc 129:715–728

Rieck M, Nuijens L, Stevens B (2012) Marine boundary layer cloud feedbacks in a constant relative humidity atmosphere. J Atmos Sci 69(8):2538–2550

Rienecker MM et al (2011) MERRA: NASAs modern-era retrospective analysis for research and applications. J Clim 24:3624–3648

Saha S et al (2010) The NCEP climate forecast system reanalysis. Bull Am Meteorol Soc 91:1015–1057

Sampe T, Xie SP (2007) Mapping high sea winds from space a global climatology. Bull Am Meteorol Soc 88(12):1965–1978

Schott FA, Dengler M, Schoenefeldt R (2002) The shallow overturning circulation of the Indian Ocean. Prog Oceanogr 53:57–103

Sherwood SC, Bony S, Dufresne J-L (2014) Spread in model climate sensitivity traced to atmospheric convective mixing. Nature 505(7481):37–42

Shige S, Takayabu YN, Tao W-K, Johnson DE (2004) Spectral retrieval of latent heating profiles from TRMM PR data. Part I: development of a model-based algorithm. J Appl Meteorol 43:1095–1113

Shige S, Takayabu YN, Tao W-K, Shie C-L (2007) Spectral retrieval of latent heating profiles from TRMM PR data. Part II: algorithm improvement and heating estimates over tropical ocean regions. J Appl Meteorol Climatol 46:1098–1124

Shige S, Takayabu YN, Tao W-K (2008) Spectral retrieval of latent heating profiles from TRMM PR data. Part III: estimating apparent moisture sink profiles over tropical oceans. J Appl Meteorol Climatol 47:620–640

Simpson J, Kummerow C, Tao W-K, Adler RF (1996) On the tropical rainfall measuring mission (TRMM). Meteorol Atmos Phys 60:19–36

Smith EA, Asrar G, Furuhama Y, Ginati A, Mugnai A, Nakamura K, Entin JK et al (2007) International global precipitation measurement (GPM) program and mission: an overview. In: Levizzani V, Bauer P, Turk FJ (eds) Measuring precipitation from space. Springer, Dordrecht, pp 611–653

Stoffelen A, Pailleux J, Källén E, Vaughan JM, Isaksen LP, Flamant P, Wergen W, Andersson E, Schyberg H, Culoma A, Meynart R, Endemann M, Ingmann P (2005) The atmospheric dynamics mission for global wind field measurement. Bull Am Meteorol Soc 86(1):73–87

Stoffelen A, Marseille GJ, Bouttier F, Vasiljevic V, de Haan S, Cardinali C (2006) ADM-Aeolus Doppler wind lidar observation system simulation experiment. Q J R Meteorol Soc 132:1927–1947

Takayabu YN, Shige S, Tao WK, Hirota N (2010) Shallow and deep latent heating modes over tropical oceans observed with TRMM PR spectral latent heating data. J Clim 23:2030–2046

Tan D, Andersson E, Fisher M, Isaksen L (2007) Observing-system impact assessment using a data assimilation ensemble technique: application to the ADM-Aeolus wind profiling mission. Q J R Meteorol Soc 133:381–390

Tao W-K, Lang S, Simpson J, Adler R (1993) Retrieval algorithms for estimating the vertical profiles of latent heat release: their applications for TRMM. J Meteorol Soc Jpn 71:685–700

Tao W-K, Lang S, Simpson J, Olson WS, Johnson D, Ferrier B, Kummerow C, Adler R (2000) Vertical profiles of latent heat release and their retrieval in TOGA-COARE convective systems using a cloud resolving model. SSM/I and radar data. J Meteorol Soc Jpn 78:333–355

Tao W-K, Smith EA, Adler RF, Haddad ZS (2006) Retrieval of latent heating from TRMM measurements. Bull Am Meteorol Soc 87(11):1555–1571

Thorncroft CD, Nguyen H, Zhang C, Peyrillé P (2011) Annual cycle of the West African monsoon: regional circulations and associated water vapour transport. Q J R Meteorol Soc 137:129–147

Trenberth KE, Stepaniak DP, Caron JM (2000) The global monsoon as seen through the divergent atmospheric circulation. J Clim 13(22):3969–3993

Tucker S, Weimer C, Hardesty RM (2016) The Athena-OAWL Doppler wind lidar mission. In: EPJ web of conferences 27th international laser radar conference, vol 119. 01002
Velden C, Daniels J, Stettner D, Santek D, Key J, Dunion J, Holmlund K, Dengel G, Breskey W, Menzel P (2005) Recent innovations in deriving tropospheric winds from meteorological satellites. Bull Am Meteorol Soc 86(2):205–223
Wheeler MC, Hendon HH (2004) An all-season real-time multivariate MJO index: development of an index for monitoring and prediction. Mon Weather Rev 132(8):1917–1932
Wing AA, Emanuel KA (2013) Physical mechanisms controlling self-aggregation of convection in idealized numerical modeling simulations. J Adv Model Earth Syst 6:59–74
Yang S, Smith EA (1999a) Four-dimensional structure of monthly latent heating derived from SSM/I satellite measurements. J Clim 12:1016–1037
Yang S, Smith EA (1999b) Moisture budget analysis of TOGA COARE area using SSM/I-retrieved latent heating and large-scale Q2 estimates. J Atmos Ocean Technol 16:633–655
Yokoyama C, Takayabu YN (2012) Relationships between rain characteristics and environment. Part II: atmospheric disturbances associated with shallow convection over the eastern tropical Pacific. Mon Weather Rev 140:2841–2859
Zagar N (2004) Assimilation of equatorial waves by line of sight wind observations. J Atmos Sci 61:1877–1893
Zagar N, Stoffelen A, Marseille GJ, Accadia C, Schlüssel P (2008) Impact assessment of simulated Doppler wind lidars with a multivariate variational assimilation in the tropics. Mon Weather Rev 136:2443–2460
Zermeño-Díaz DM, Zhang C, Kollias P, Kalesse H (2015) The role of shallow cloud moistening in MJO and non-MJO convective events over the ARM Manus site. J Atmos Sci 72:4797–4820
Zhang C (2005) Madden–Julian oscillation. Rev Geophys 43(2):1–36
Zhang C, Hagos SM (2009) Bi-modal structure and variability of large-scale diabatic heating in the tropics. J Atmos Sci 66:3621–3640
Zhang C, McGauley M, Bond NA (2004) Shallow meridional circulation in the tropical eastern Pacific. J Clim 17(1):133–139
Zhang C, Nolan DS, Thorncroft CD, Nguyen H (2008) Shallow meridional circulations in the tropical atmosphere. J Clim 21(14):3453–3470
Zilberman NV, Roemmich DH, Gille ST (2013) The mean and the time variability of the shallow meridional overturning circulation in the tropical South Pacific Ocean. J Clim 26:4069–4087

Surv Geophys (2017) 38:1529–1568
https://doi.org/10.1007/s10712-017-9428-0

EUREC⁴A: A Field Campaign to Elucidate the Couplings Between Clouds, Convection and Circulation

Sandrine Bony[1] · Bjorn Stevens[2] · Felix Ament[4] · Sebastien Bigorre[5] · Patrick Chazette[3] · Susanne Crewell[6] · Julien Delanoë[7] · Kerry Emanuel[8] · David Farrell[9] · Cyrille Flamant[7] · Silke Gross[10] · Lutz Hirsch[2] · Johannes Karstensen[11] · Bernhard Mayer[12] · Louise Nuijens[13] · James H. Ruppert Jr.[2] · Irina Sandu[14] · Pier Siebesma[16] · Sabrina Speich[15] · Frédéric Szczap[17] · Julien Totems[3] · Raphaela Vogel[2] · Manfred Wendisch[18] · Martin Wirth[10]

Received: 14 January 2017 / Accepted: 14 September 2017 / Published online: 27 September 2017

Abstract Trade-wind cumuli constitute the cloud type with the highest frequency of occurrence on Earth, and it has been shown that their sensitivity to changing environmental conditions will critically influence the magnitude and pace of future global warming. Research over the last decade has pointed out the importance of the interplay between clouds, convection and circulation in controling this sensitivity. Numerical models represent this interplay in diverse ways, which translates into different responses of trade-cumuli to climate perturbations. Climate models predict that the area covered by shallow

✉ Sandrine Bony
bony@lmd.jussieu.fr

1 LMD/IPSL, CNRS, Sorbonne Université, UPMC, 4 Place Jussieu, 75252 Paris, France

2 Max Planck Institute for Meteorology, Bundesstr. 53, 20146 Hamburg, Germany

3 LSCE/IPSL, CNRS-CEA-UVSQ, CEA Saclay, 91191 Gif sur Yvette, France

4 University of Hamburg, Bundesstrasse 55, 20146 Hamburg, Germany

5 Woods Hole Oceanographic Institution, 266 Woods Hole Rd, Woods Hole, MA 02543, USA

6 University of Cologne, Albertus-Magnus-Platz, 50923 Cologne, Germany

7 LATMOS/IPSL, CNRS-UPMC-UVSQ, 11 Boulevard D'Alembert, 78280 Guyancourt, France

8 Massachusetts Institute of Technology, 77 Massachussetts Avenue, Cambridge, MA 02139, USA

9 Caribbean Institute for Meteorology and Hydrology, P.O. Box 130, Bridgetown, Barbados

10 German Aerospace Center, Múnchener Str. 20, 82234 Oberpfaffenhofen-Wessling, Germany

11 GEOMAR Helmholtz Centre for Ocean Research, Duesternbrooker Weg 20, 24105 Kiel, Germany

12 Ludwig-Maximilians University of Munich, Theresienstrasse 37, 80333 Munich, Germany

13 Delft University of Technology, P.O. Box 5048, 2600 GA Delft, The Netherlands

14 ECMWF, Shinfield Park, Reading RG2 9AX, UK

15 LMD/IPSL, Ecole Normale Supérieure, 24 rue Lhomond, 75231 Paris, France

cumuli at cloud base is very sensitive to changes in environmental conditions, while process models suggest the opposite. To understand and resolve this contradiction, we propose to organize a field campaign aimed at quantifying the physical properties of trade-cumuli (e.g., cloud fraction and water content) as a function of the large-scale environment. Beyond a better understanding of clouds-circulation coupling processes, the campaign will provide a reference data set that may be used as a benchmark for advancing the modelling and the satellite remote sensing of clouds and circulation. It will also be an opportunity for complementary investigations such as evaluating model convective parameterizations or studying the role of ocean mesoscale eddies in air–sea interactions and convective organization.

Keywords Trade-wind cumulus · Shallow convection · Cloud feedback · Atmospheric circulation · Field campaign

1 Introduction

Of all the clouds that populate the Earth's atmosphere, trade-cumuli count among the most fascinating expressions of the interplay between clouds and circulations. These broken shallow clouds form within the lowest kilometres of the atmosphere, influenced at their base by small-scale turbulent motions of the warm, moist surface layer, and at their top by the large-scale sinking motions of the warm and dry overlying free troposphere. If many of these clouds do not rise by more than a few hundred metres above their base, some reach higher levels (e.g Nuijens et al. 2014), detrain and help sustain (evaporatively and radiatively) the trade-wind temperature-inversion layer higher up. Trade-cumuli warm the layer in which they form through condensation, but cool the subcloud layer and the trade inversion through the evaporation of falling raindrops and detrained droplets. In addition, the emission of infrared radiation to space produces an efficient cooling of the lower atmosphere in which clouds form. This radiative cooling contributes to generate shallow mesoscale circulations (Naumann et al. 2017) which, depending on local conditions and remote convective activity, can organize either randomly or into streets, arcs or circles of cloud clusters. In certain conditions, these mesoscale circulations can also trigger remotely the aggregation of deep convection (Muller and Held 2012).

This coupling between shallow clouds and circulation greatly matters for climate sensitivity. Trade-cumuli are so ubiquitous over tropical oceans that their radiative properties substantially influence the Earth's radiation budget. Their response to global warming is thus critical for global-mean cloud feedbacks, and actually it is their differing response to warming that explains most of the spread of climate sensitivity across climate models (Bony et al. 2004; Bony and Dufresne 2005; Webb et al. 2006; Medeiros et al. 2008; Vial et al. 2013; Boucher et al. 2013; Medeiros et al. 2015). Model diversity in the strength of the vertical mixing of water vapour within the first few kilometres above the ocean surface (in association with both convective and large-scale circulations) is thought to explain half of the variance in climate sensitivity estimates across models (Sherwood et al. 2014): the

[16] Delft University of Technology and Royal Netherlands Meteorological Institute, De Bilt, Netherlands

[17] Laboratoire de Météorologie Physique, UMR6016, CNRS, Aubière, France

[18] University of Leipzig, Stephanstr. 3, 04103 Leipzig, Germany

lower-tropospheric mixing dehydrates the cloud layer near its base at an increasing rate as the climate warms and this rate scales with the mixing strength in the current climate (Sherwood et al. (2014); Gettelman et al. (2012); Tomassini et al. (2015); Brient et al. (2016); Stevens et al. (2016); Vial et al. (2016), Fig. 1). There is increasing evidence that the diversity of the modelled response to warming reflects model diversity in how this coupling between convective mixing, surface turbulent fluxes, and low-cloud radiative effects is represented in regimes of large-scale subsidence, and that it can be partly related to the numerical representation (or parameterization) of convection (Webb et al. 2015; Vial et al. 2016). However, so far it has not been possible to constrain this coupling observationally due to a lack of appropriate measurements.

On the contrary, in large-eddy simulations (LES) and in observations the cloud-base fraction of trade-wind cumuli appears to be much more resilient to changes in environmental conditions than in climate models, both in the current (Nuijens et al. 2014, 2015a) and projected warmer climate (Rieck et al. 2012; Bretherton 2015). Interpreting these results remains difficult. For the observations, in the past it has not been possible to link cloud amount to the large-scale circulation in which the clouds form. The cloud amount predicted by LES, though often resilient to changes in thermodynamic conditions, is known to be sensitive to various aspects of the simulation such as resolution, microphysics, numerics or domain size (Vial et al. 2017, and references therein). Theoretically, the apparent resilience of cloud-base cloud fraction has been interpreted as the consequence of a "cumulus-valve mechanism" whereby clouds act as a valve which helps maintain the top of the subcloud layer close to the lifting condensation level and thus regulate the area covered by cumulus updrafts at cloud base (Albrecht et al. 1979; Neggers et al. 2006; Stevens 2006; Nuijens et al. 2015a). However, this idea has not been tested observationally. Moreover, recent studies running large-eddy simulations over large domains question this idea of cloud-base resilience, as they show that changes in the mesoscale organization of shallow cumuli can significantly influence the cloud fraction (Seifert and Heus 2013; Vogel et al. 2016; Vial et al. 2017). It is thus paramount to assess the ability of LES to predict the cloud cover and its dependence on the organization of convection and on environmental conditions.

The discussion above illustrates how the science has matured to the point where it is now possible to identify a few key hypotheses or questions that, if tested or answered, would enable a step improvement in understanding of the interplay between clouds,

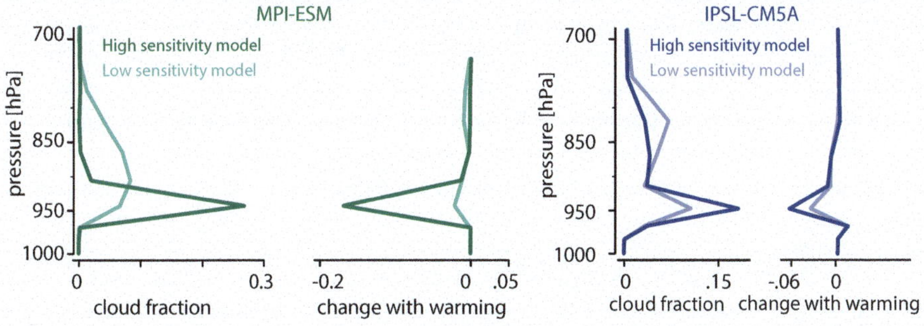

Fig. 1 Vertical profiles of the low-cloud fraction, and of its response to global warming, predicted by two general circulation models (MPI and IPSL) in the trade-wind cumulus regime. For each model, results are shown for two versions differing only by their representation of lower-tropospheric mixing (after Stevens et al. 2016; Vial et al. 2016)

convection and circulation, and their role in climate change: How strong is the convective mixing in regimes of shallow cumulus and how much does it couple to surface turbulent fluxes, radiative effects and water vapour? Is the cloud-base fraction of trade-wind cumuli insensitive to variations in convective mixing and large-scale circulations? Does the cumulus mass flux act as a valve to restrict the turbulent boundary layer from growing appreciably beyond its lifting condensation level? Do the statistics of shallow convection depend on the form of spatial organization?

Improved observations are also necessary to help advance space-based remote sensing. The trade-wind regions are often characterized by a strongly layered vertical structure, and by warm, small, and thin broken clouds. Current satellite observations are inadequate to detect sharp vertical gradients of water vapour (Chazette et al. 2014; Asrar et al. 2015, Stevens et al. this volume), and the detection of shallow clouds from space remains difficult. Biases in cloud detection leads to significant discrepancies among the various satellite estimates of the trade-wind cloud fraction (Stubenrauch et al. 2013) and are detrimental to the quality of other satellite retrievals such as those of the cloud water path (Horváth and Gentemann 2007), precipitation and cloud microphysical properties. In these conditions, in situ observations are not only critical to investigate the physics of trade-wind clouds, but also to test—and eventually improve—the instruments and algorithms of remote sensing that are used to observe the Earth's atmosphere and surface from space.

Past field campaigns in regions of shallow cumulus such as the *Atlantic Expedition* in September to October 1965 (Augstein et al. 1973), the *Atlantic Tradewind EXperiment* in February 1969 (ATEX, Augstein et al. 1974), the *Barbados Oceanographic and Meteorological Experiment* from May to July 1969 (BOMEX, Holland 1970) or the *Puerto-Rico Experiment* in December 1972 (LeMone and Pennell 1976), did focus on the environment of clouds, on vertical transports of water, heat and momentum in the trade-wind boundary layer, and also included attempts to measure the large-scale vertical motion in the atmosphere. However, the microphysical and macrophysical properties of the shallow cumuli were not characterized. Because these campaigns took place at the dawn of the satellite era, no observations from space could help fill the gap. In June 1992, the *Atlantic Stratocumulus Transition Experiment* (ASTEX, Albrecht et al. 1995) was conducted off North Africa, in the area of Azores and Madeira Islands, to address issues related to the stratocumulus to trade-cumulus transition and cloud-mode selection. Satellites and upper-level aircraft provided a description of large-scale cloud features, and instrumented aircraft flying in the boundary layer and surface-based remote sensing systems described the mean, turbulence, and mesoscale variability in microphysical properties of boundary-layer clouds. Attempts were also made to infer large-scale divergence in the boundary layer using lagrangian balloons and an array of three rawinsonde stations (Ciesielski et al. 1999). In 2005, the *Rain in shallow cumulus over the ocean* (RICO, Rauber et al. 2007) campaign which took place off the Caribbean islands of Antigua and Barbuda did focus on cloud microphysical properties and pointed out the importance and diversity of mesoscale organizations, but the large-scale dynamical environment and the interplay between cloud macrophysical properties (e.g., the low-level cloud fraction) and their environment were not characterized.

The establishment of the Barbados Cloud Observatory (BCO) in 2010, and the two *Next-Generation Aircraft Remote Sensing for Validation Studies* airborne field campaigns (NARVAL and NARVAL2) held in December 2013 and August 2016 have since created an observational foothold to better understand the coupling between clouds and their environment (Stevens et al. 2016). The BCO provides long-term context for intensive observations, and when combined with the NARVAL measurements, helps advance and test new approaches for bridging the gap between measurements of cloud macro-structure and the large-scale environment.

A new field campaign, EUREC^{4}A (*Elucidating the role of clouds-circulation coupling in climate*), has been designed to take advantage of and extend these advances. Anchored at the BCO, it will measure clouds in the winter trades of the North Atlantic, windward of Barbados, in early 2020. EUREC^{4}A will have two primary objectives:

- To quantify macrophysical properties of trade-wind cumuli as a function of the large-scale environment, and
- To provide a reference data set that may be used as a benchmark for the modelling and the satellite observation of shallow clouds and circulation.

To address these objectives EUREC^{4}A will provide, for the first time, simultaneous measurement of cloud macrophysical properties (cloud fraction, vertical extent and cloud-size distributions), cloud radiative properties (large-scale albedo, broadband solar and terrestrial net fluxes and derived quantities such as radiative divergence and heating/cooling rates), convective activity (cloud-base mass flux, mesoscale organization), and the large-scale environment in which clouds and convection are embedded (large-scale vertical motion, thermodynamic stratification, surface properties, turbulent and radiative sources or sinks of energy).

In Sect. 2, we present an overview of the experimental strategy for the EUREC^{4}A field campaign. In Sect. 3, we discuss the premises which are at the basis of this strategy, namely the possibility to measure cloud profiles (especially cloud amount at cloud base), convective mass flux and large-scale vertical velocity, as only this can connect the macrophysical properties of clouds to the environment. Then, we discuss how the results from the campaign could be used to build a reference data set for evaluating process and climate models, and for assessing retrievals from space-borne observations (Sect. 4). Beyond the study of clouds-circulation interactions, EUREC^{4}A will be an opportunity for complementary scientific investigations. We describe some of these possibilities as of EUREC^{4}A^{++} in Sect. 5. A brief conclusion is presented in Sect. 6.

2 Overview of the EUREC^{4}A Experimental Strategy

The core objective of the EUREC^{4}A field campaign is to elucidate how the macrophysical properties of trade-cumuli depend on the dynamic and thermodynamic properties of the environment in which the clouds form. More specifically, EUREC^{4}A aims to answer the following questions:

- What controls the convective mass flux, mesoscale organization, and depth of shallow clouds?
- How does the cumulus cloud amount in the trade-wind boundary layer vary with turbulence, convective mixing and large-scale circulations, and what impact does this variation have on the atmospheric radiation field?

The EUREC^{4}A field campaign will take place in the lower Atlantic trades, over the ocean east of Barbados (13 °N, 59 °W) from 20 January to 20 February 2020. Several reasons motivate the choice of this specific location. First, shallow cumuli are prominent in this area, especially during winter (Norris 1998; Nuijens et al. 2014). Second, the cloudiness in the vicinity of Barbados is representative of clouds across the whole trade-wind regions of the tropical ocean, both in models and in observations (Medeiros and

Nuijens 2016). Finally, it anchors the measurements to the extensively instrumented Barbados Cloud Observatory which has been monitoring clouds continuously since 2010, and allows it to benefit from the legacy of the NARVAL series of flight campaigns organized in the area in the last few years (Stevens et al. 2016).

2.1 Aircraft Measurements

The primary motivation for EUREC^{4}A is the need to characterize simultaneously the trade-cumulus field and the dynamic and thermodynamic environment in which it forms. For this purpose, the core of the EUREC^{4}A field campaign will be the deployment of two research aircraft (Fig. 2): The French ATR-42 operated by the Service des Avions Français Instrumentés pour la Recherche en Environnement (SAFIRE), which will fly in the lower-troposphere with a payload of up to two tons and will be equipped with both remote sensing instrumentation and a suite of in situ sensors (Table 1), and the German HALO (High Altitude and Long Range Research Aircraft) operated by the Deutsches Zentrum für Luft- and Raumfahrt (DLR), which has a payload of up to three tons, a range of up to 8000 km, and a ceiling of up to 15 km, an advanced instrumentation (Table 2) and the ability to launch dropsondes (see for instance Wendisch et al. 2016). In addition to these aircraft, we will use the Barbados Cloud Observatory (BCO), buoys, drifters plus several research vessels deployed in the area and equipped with radiosondes and additional remote sensing instruments to complement surface, atmospheric and ocean measurements (Sect. 2.2).

HALO will fly large circle patterns (45–50 min, corresponding to a circumference of about 500 km) at 9 km altitude (flight level 300, FL300), and will densely distribute dropsondes around the circles. The dropsondes will characterize the vertical thermodynamic structure of the trade-wind atmosphere and will make it possible to infer the vertical profile of large-scale divergence over the area (Sect. 3.1). The advanced remote sensing instrumentation on HALO will characterize the cloud field and its environment (water vapour, hydrometeors, cloud particle phase, cloud vertical structure, cloud albedo, etc).

Simultaneously, the ATR-42 will characterize the shallow cumulus field and boundary-layer properties within the area through a series of low-level legs, flown primarily near the cloud-base level (~ 1 km), with additional legs near the trade inversion level (~ 2 km), and (by flying at the lowest safe flight level) near the sea surface. Sideways-looking lidar and radar instruments will measure the cloud fraction at the flight level (Sect. 3.2). Upward-pointing high-spectral-resolution (HSR) backscatter lidar plus a vertically pointing Doppler radar will be used to assess the boundary-layer depth and measure the vertical velocity in the aerosol-laden lower troposphere above the aircraft. Other instruments on board the aircraft will characterize cloud microphysical properties, the tri-dimensional wind field along the trajectory of the aircraft, and turbulence statistics (Table 1, Appendix 1).

The instrumentation on board both aircraft will provide a detailed characterization of the vertical distribution of water vapour, clouds and aerosol particles, and of vertical velocities within clouds (Appendix 1). Measurements of radiative fluxes at different altitudes, as well as radiative transfer calculations using observed atmospheric and cloud properties, will help infer vertical profiles of radiative solar heating and terrestrial cooling rates above, within and below the observed clouds.

The characterization of surface (Sect. 2.2) and subcloud layer properties, and of the difference between the subcloud layer and the air just above it, combined with estimates of surface turbulent fluxes, radiative cooling and large-scale mass divergence, should enable

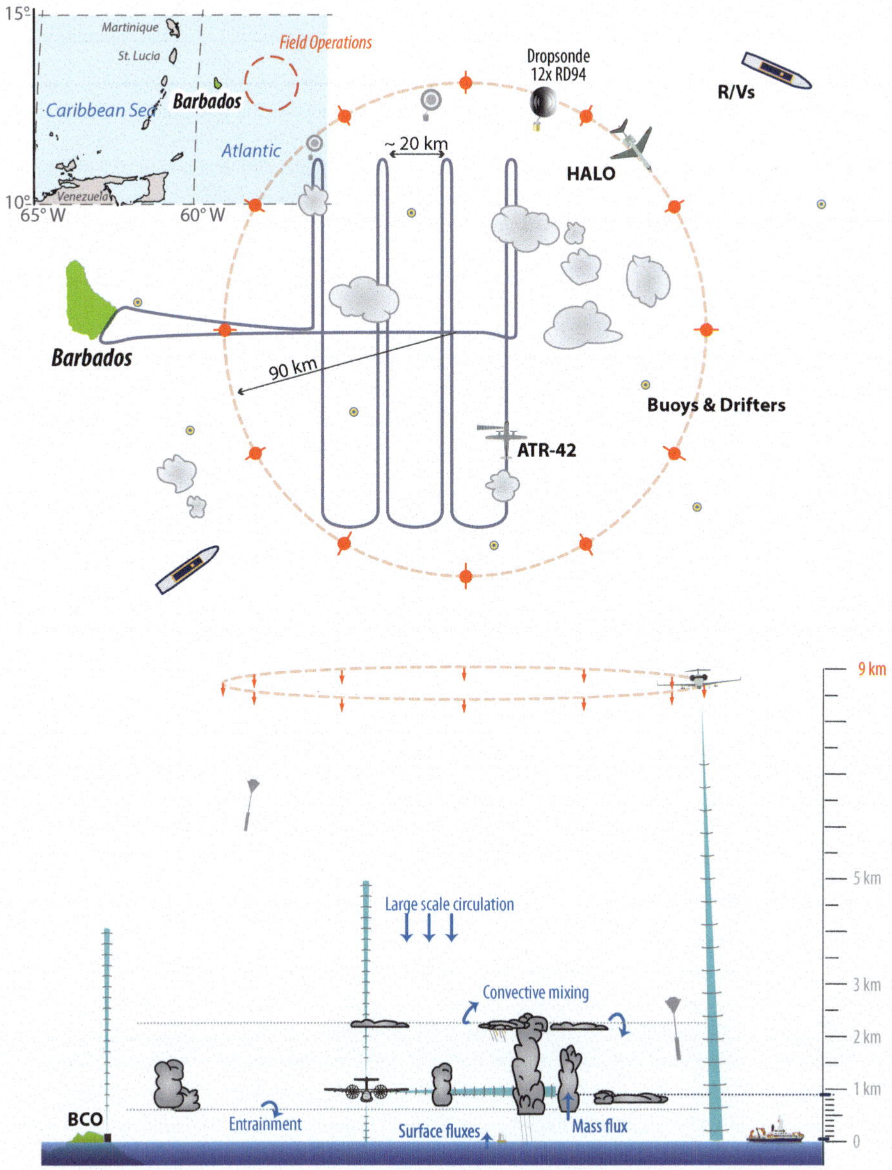

Fig. 2 Envisioned flight strategy for the EUREC^{4}A core measurements

closure of the mass and moist static energy budgets of the subcloud layer. The analysis of the mass budget will make it possible to estimate the cumulus mass flux at cloud base (Sect. 3.3). The moist static energy and water budgets will be used to verify the consistency among the different measurements and provide insight into the factors influencing shallow convective development. The moist static energy budget should also make it possible to test the boundary-layer quasi-equilibrium hypothesis (Raymond 1995), which

Table 1 Synopsis of ATR-42 instrumentation

Instrument	Brief description
Thermodynamics and turbulence	In situ water vapour, temperature, pressure and 3D wind; momentum and heat fluxes
Cloud particles	In situ liquid and total water contents; droplet size distribution (0.5–6000 μm); 2D particle imaging (25–6000 μm)
BASTA cloud radar	Bistatic 95 GHz Doppler cloud radar to be deployed in sidewards looking mode (Delanoë et al. 2016)
ALiAS Lidar	Lightweight backscatter lidar (355 nm) to be deployed in sidewards looking mode (Chazette 2016)
RASTA cloud radar	Upward- and downward-looking 95 GHz Doppler cloud radar with six antenna configuration for wind-vector retrievals (Delanoë et al. 2013)
LNG lidar	Three-wavelength (1064, 532 and 355 nm) high-spectral-resolution polarized backscatter lidar (upwards, downwards or 35° pointing) (Bruneau et al. 2015)
CLIMAT-AV	Three-channel downward-staring measurements of infrared irradiance at 8.7, 10.8, and 12.0 μm (Brogniez et al. 2003)
Pyrgeometer	Hemispheric broadband upwelling and downwelling thermal infrared radiative fluxes (Kipp and Zonen CGR4)
Pyranometer	Hemispheric broadband upwelling and downwelling solar radiative fluxes (Kipp and Zonen CMP22)

Table 2 Synopsis of HALO instrumentation

Instrument	Brief description
BAHAMAS	In situ water vapour, temperature, gust probe winds and aircraft state vector. Up- and downward shortwave and longwave broadband irradiances (in development)
HAMP cloud radar	Downward-staring polarized Doppler 36 GHz cloud radar (Mech et al. 2014)
HAMP radiometer	Downward-staring microwave radiometers with 26 channels between 22 and 183 GHz (Mech et al. 2014)
WALES	Downward-staring water vapour DIAL and backscatter HSRL lidar (Wirth et al. 2009)
SMART	Up- and downward-looking spectral (300–2200 nm) radiance and irradiance measurements (Wendisch et al. 2001; Ehrlich et al. 2008)
SpecMACS	Downward-looking hyper-spectral (400–2500 nm) line imager (Ewald et al. 2016)
Thermal imager	Downward-looking (10.8 and 12 μm) two channel line imager (in development)
Dropsondes	AVAPs system with four-channel receiver supporting Vaisala RD94 Sondes (ten channel receiver in development)

holds that the cumulus enthalpy flux out of the subcloud layer balances surface fluxes and large-scale ascent at the top of the subcloud layer.

To maximize the chance of sampling a large diversity of environmental conditions and mesoscale organizations, the campaign will consist of about 90 h of research flights (for each aircraft) over four weeks for operations out of Grantley Adams International Airport on Barbados. Ten HALO research flights from Barbados are envisioned, each with a duration of 9 h bracketing two 4-h flights of the ATR-42 (with a refuelling in between).

The zone of operations will not change much from day to day, but owing to daily meteorological variability one may hope to sample different types of cloud conditions and organizations. On any given research flight the same cloud and environmental conditions will be observed for many hours by both aircraft, making it possible to characterize these conditions in a statistically consistent and representative way.

2.2 Surface and Ship-Based Observations

In addition to the aircraft missions that will characterize clouds and their surrounding environment up to scales of O(100 km), surface and ship-based observations will be distributed over the area of and around flight operations augmenting measurements from the BCO so as to better characterize the surface and atmospheric environment of clouds on a scale of O(1000 km) and over a longer, uninterrupted time period, as well as strengthen the fidelity of the large-scale analyses.

The large-scale array of observations will be comprised of three to five stations (Fig. 3): the Barbados Cloud Observatory (BCO, Stevens et al. 2016) and a network of research vessels (RVs). Applications for ship measurement time from Germany (*Meteor* and *Maria S. Merian*), France (*Atalante*), The Netherlands (*Pelagia*), the USA and Spain are pending. The research vessels will serve as advanced surface remote sensing platforms, atmosphere and ocean sounding stations, bases for fleets of autonomous vehicles, and means of laying down an array of drifters/floats or buoys.

Radiosondes will be launched from each station to collect simultaneous measurements of profiles of air humidity, temperature, pressure and horizontal winds (derived from GPS measurements) from the surface through the lower stratosphere. The sounding data will be assimilated by weather centres, which will improve the quality of meteorological analyses for the period of the campaign and will help diagnose large-scale divergence and vertical motion over a range of scales larger than the zone of aircraft operations.

Measurements from research vessels are intended to be operated over a period of 3 to 5 weeks overlapping with the planned airborne observations. This will establish context for the aircraft missions relative to the seasonal march of the ITCZ and the evolving strength

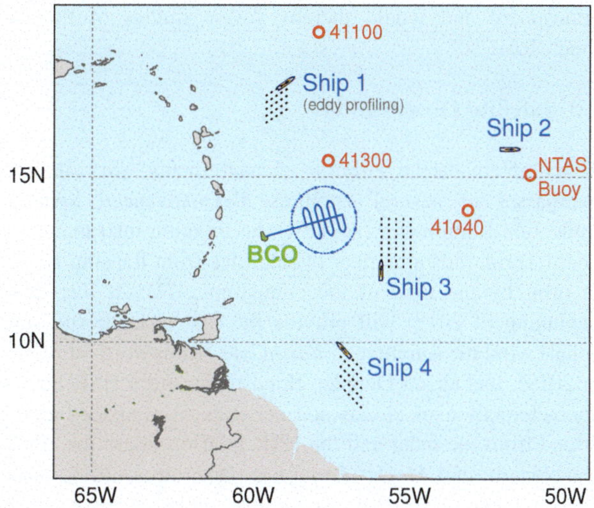

Fig. 3 Large-scale sounding array envisioned for EUREC^{4}A, comprised by the Barbados Cloud Observatory (BCO) and approximately four research vessels. Buoy stations are indicated by open circles. The ships will serve as advanced surface remote sensing platforms, atmosphere and ocean sounding stations, bases for fleets of autonomous vehicles, and means of laying down an array of drifters/floats or buoys

of overturning within the Hadley cell. Soundings will be launched with a minimum frequency of 6 per day to adequately sample the diurnal and semi-diurnal cycles, along with other subdaily variability. It will also make it possible to get thermodynamic data and large-scale divergence estimates up to the top of the atmosphere instead of up to 9 km (the flight level of HALO during EUREC^{4}A).

Besides radiosondes, in situ and remote sensing instrumentation will be installed at BCO and on-board the ships. Instruments such as lidar, radar, radiometers or ceilometers will provide additional observations of clouds, aerosols, surface turbulence and air–sea fluxes of heat and moisture, and surface and boundary-layer properties. A scanning, S-band, radar operated by the Barbados Meteorological Service can be used for research purposes in the absence of severe weather. It will help characterize the mesoscale organization of convection, and the vertical structure of the shallow cloud cover (Nuijens et al. 2009; Oue et al. 2016). The deployment of a second C-Band radar, the POLDIRAD of the Institut für Physik der Atmosphäre at the Deutschen Zentrum für Luft- und Raumfahrt, is also being considered. Shipboard deployment of drones and a HeliKite, capable of suspending an instrument of up to 100 kg at different heights in the lower 3 km of the atmosphere, can be used to characterize aerosol and cloud microphysical properties. Laser-based spectrometers could measure the isotopic composition of water and provide an additional characterization of the balance between convective drying and turbulent moistening in the boundary layer, and simple instruments such as ceilometers will aid the characterization of the vertical distribution of clouds in the observational domain.

The ships will also provide an opportunity to characterize the state of the upper ocean and more specifically the mesoscale ocean eddies which are particularly frequent east of Barbados (Sect. 5.6). Beyond their importance for the ocean transport, mesoscale ocean eddies are increasingly recognized as influencing air–sea fluxes and clouds (Chelton et al. 2004; Ferreira and Frankignoul 2008; Frenger et al. 2013; Byrne et al. 2015). This raises the question as to whether they might play a role in the organization of shallow cumuli. Oceanographic measurements of the vertical profiles of temperature, salinity, pressure, oxygen and other biogeochemical properties of the upper ocean through in situ sensors or profiling instruments, combined with the deployment of Argo profiling floats and autonomous observing platforms such as gliders or wave-gliders, would provide an unprecedented characterization of tropical mesoscale ocean eddies under a well-observed atmosphere and would thereby foster studies of their impact on air–sea interaction (Sect. 5.6).

2.3 Satellite Observations

To complement the airborne measurements, we will coordinate field operations with overpasses of several satellites: the Advanced Spaceborne Thermal Emission and Reflection Radiometer (ASTER), an imaging instrument with 15 m spatial resolution on-board Terra, plus a number of satellites from flagship space missions that we expect to be in orbit by the time of the campaign: *ADM-Aeolus* (whose launch is planned by the beginning of 2018) will provide the first space-borne vertically resolved radial (mostly zonal) wind measurements; *EarthCare* (Illingworth et al. 2015, whose launch is scheduled in 2019), includes a Doppler cloud radar and a HSR lidar which will provide a thorough characterization of clouds and aerosols from space-based products comparable to those issued from the radar and the HSR lidar on-board the ATR-42 at the same frequency and wavelength; and *Megha-Tropiques* (Roca et al. 2015, launched in 2011, the mission has been extended until 2021) measures radiative fluxes at the top of the atmosphere, the vertical distribution of relative humidity through the troposphere, and precipitation as part of the GPM (Global Precipitation Measurement) mission.

Existing satellite imagery suggests that shallow cumuli exhibit a large range of mesoscale organizations from seemingly randomly distributed cloud clusters to wind-parallel street lines or arcs (Rauber et al. 2007). An example of one form of organization, mesoscale cloud flowers, observed over the proposed EUREC^{4}A study area on 9 February 2017 is shown in Fig. 4. Space observations of the atmosphere at high spatial resolution such as derived from ASTER or other instruments such as the Geostationary Operational Environmental Satellite (GOES), Multi-angle Imaging SpectroRadiometer (MISR) or Moderate Resolution Imaging Spectroradiometer (MODIS) imagers, potentially complemented by radar observations from the surface network (Sect. 2.2), will characterize the spatial organization of clouds within the area sampled by the aircraft missions. EUREC^{4}A will be the first field study to investigate whether this organization matters for the statistical properties of the shallow cumulus field.

3 The Premises

The experimental strategy of the EUREC^{4}A campaign rests on three main premises:

- The large-scale vertical motion on scales O(100 km) can be measured using dropsondes,
- The distribution of clouds in the trade-wind boundary layer can be inferred from lidar-radar measurements, especially the cloud fraction near cloud base,
- The convective mass flux at cloud base can be inferred from the subcloud layer mass budget.

In this section, we present arguments and results from ongoing analyses that show that the first of these premises appears sound, and we discuss how the other two are currently being tested, and how the experimental strategy might be adapted based on the outcome of these tests.

3.1 Using Dropsondes to Measure the Large-Scale Vertical Motion

A main component of the large-scale mass, heat and moisture budgets is the large-scale vertical velocity ω (Yanai et al. 1973). From the equation of mass continuity, ω can be

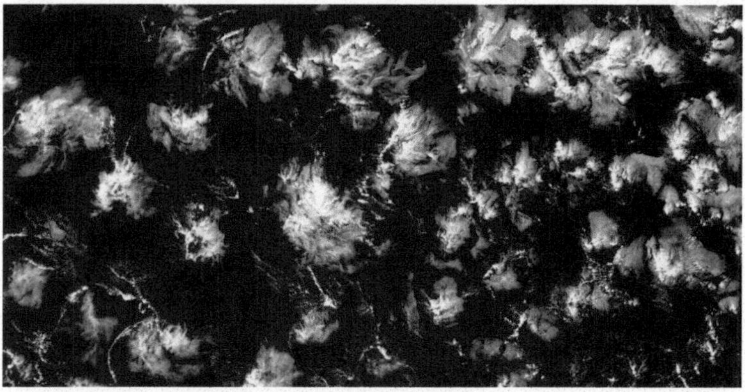

Fig. 4 Shallow cloud organization observed from MODIS on 9 February 2017 (Barbados, which is about 20 km wide, is highlighted in green). The top of cloud clusters does not exceed 3–4 km

derived from the divergence, D of the horizontal wind $\mathbf{V}_{\mathrm{H}}$ as $\omega(P) = -\int_0^P D(p)\,\mathrm{d}p$ where $D = \nabla \cdot \mathbf{V}_{\mathrm{H}}$ and P is the atmospheric pressure. D and ω are known to strongly influence the properties of the trade-cumulus boundary layer and low-level cloudiness (e.g., Albrecht et al. 1979). Our ability to measure these two quantities during EUREC^{4}A will thus critically determine the success of the campaign.

Measurements of the large-scale vertical motion on the time and space scale of individual airborne observations have long been recognized as being essential to understand how cloudiness develops and to calculate the heat and moisture budgets of the lower troposphere, but so far generally thought to be impossible. During ATEX, BOMEX, ACE-I and DYCOMS-II campaigns, attempts were made to estimate the large-scale divergence from rawinsonde sounding networks and/or aircraft data at a particular level using the "line integral" method (Holland and Rasmusson 1973; Nitta and Esbensen 1974b; Lenschow et al. 1999, 2007). This method infers D from horizontal wind measurements using:

$$D = \frac{1}{A}\oint V_n \,\mathrm{d}\ell, \tag{1}$$

where V_n is the component of the horizontal wind normal to the perimeter of measurements, and A is the area of the region enclosed by it (vorticity can be obtained similarly from the tangent component of the horizontal wind). When applied to aircraft measurements, this method requires a stationary wind field but makes no other assumption about the structure of the wind field.

An alternative method, referred to as the "regression method", has been proposed by Lenschow et al. (2007) and successfully applied to DYCOMS-II data. It assumes a particular model for the wind field, but can be more easily adopted to a wider range of sampling geometries. Lenschow et al. (2007) assumed that wind variations in longitude, latitude and time are linear for each vertical level, such that:

$$\mathbf{V}_{\mathrm{H}} = \mathbf{V}_{\mathbf{o}} + \frac{\partial \mathbf{V}_{\mathrm{H}}}{\partial x}\Delta x + \frac{\partial \mathbf{V}_{\mathrm{H}}}{\partial y}\Delta y + \frac{\partial \mathbf{V}_{\mathrm{H}}}{\partial t}\Delta t, \tag{2}$$

where $\mathbf{V}_{\mathbf{o}}$ is the mean wind velocity over the area, Δx and Δy are the eastward and northward displacements from a chosen centre point. Δt is the change in time relative to a reference, for instance the mid-point time of the sampling. An approximate solution of this overdetermined system can be found by computing the coefficients of a least squares fit to the wind field defined as (2). By measuring $\mathbf{V}_{\mathrm{H}}$ and solving (2) for its gradients, D can then be computed as: $D = \frac{\partial u}{\partial x} + \frac{\partial v}{\partial y}$.

So far, these methodologies have been applied to wind measurements from rawinsondes or flight-level estimates of winds from an aircraft gust probe. Wind measurements from GPS dropsondes (Wang et al. 2015) now offer the opportunity to measure the vertical profiles of D and ω during airborne field campaigns. However, this methodology needs to be evaluated. In particular, it has to be checked whether the divergence measured in this way would actually represent the large-scale circulation or would instead be noisy and dominated by short-term features uncharacteristic of the large-scale environment.

To answer this question, this methodology was tested during the NARVAL2 campaign, which consisted of ten research flights of HALO, and took place in the EUREC^{4}A target areas, upwind of Barbados, during August 2016. Two of the HALO flights during NARVAL2 were specifically designed as a pilot study for the proposed EUREC^{4}A divergence measurements. During the two research flights RF03 and RF06 (carried out on 12 and 19 August 2016), HALO flew horizontal circles of 45–48 min (160 km diameter) at an altitude of 9 km. Twelve dropsondes (Vaisala RD94) were released intensively along each circle, measuring the vertical profiles of pressure, temperature and humidity with an accuracy of 0.4 hPa, 0.2 °C and 2%, respectively. Equipped with a GPS receiver, the dropsondes also measured the horizontal wind speed with an accuracy of 0.1 m.s^{-1}.

To test the method, pairs of circles were flown in the same air mass (one clockwise, one counterclockwise, with the centre of the second circle slightly displaced following the mean wind relative to the first one). The idea was that if the wind field was sufficiently stationary, and the measurements by the sondes were physical, one would expect similar answers to arise between a pair of circles flown in the same airmass. Satellite imagery targeted the flights to regions of suppressed convection, with a relatively more active shallow cloud layer during the second pair of circles, with cloud tops reaching sometimes 2–3 km, than during the first pair, with cloud tops rarely exceeding 1.5 km (Bony and Stevens Measuring large-scale vertical motions with dropsondes, manuscript in preparation). As shown in Fig. 5, the vertical profiles of D and ω derived for each circle of a given pair exhibit a consistent and reproductible vertical structure over most of the troposphere. Differences between circles of a given pair are much smaller than differences from one pair to the next, where different pairs of circles were spatially dislocated. The vertical structure of D and ω measured by dropsondes in the lower troposphere (below 4 km), such as the maximum subsidence near the top of the mixed layer, is qualitatively consistent with that measured by rawinsondes or aircraft measurements during previous field campaigns in the trades (Holland and Rasmusson 1973; Nitta and Esbensen 1974b). It is also in good agreement with the vertical structure of D and ω derived over the area from ECMWF operational forecasts during periods where the horizontal wind of the forecasts is in good agreement with the dropsondes, and with storm-resolving (1 km grid) simulations

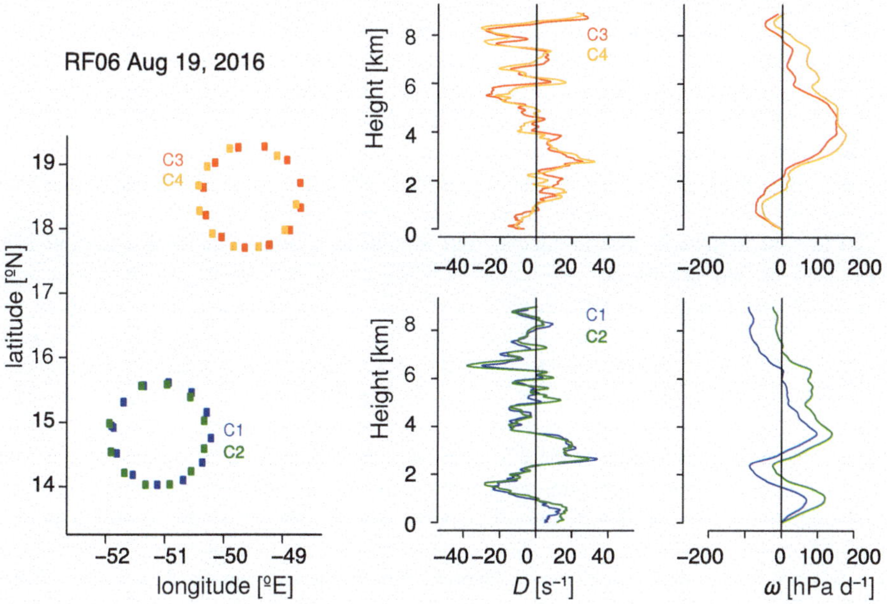

Fig. 5 (left) Research flights performed during NARVAL2 on 19 August and the vertical profiles of large-scale mass divergence D and large-scale vertical velocity ω derived from the dropsondes measurements for each circle

initialized by ECMWF analyses (not shown). It shows therefore that dropsondes can actually be used to measure the vertical profiles of D and ω on scales of O(100 km) and to discriminate the spatial heterogeneity of the environment.

Three further issues are currently being explored, also in combination with high-resolution simulations which will be used to emulate different sounding strategies: (1) The minimum number of sondes to be dropped along each circle to reach equivalent results, (2) the spatial scale over which the large-scale dynamics best correlates with the macrophysical cloud properties, and (3) the influence of vertical shear of the horizontal wind, which is much more pronounced during the winter season. Depending on the result of these investigations, the number of sondes to be dropped, as well as the size of the circular flights to be flown during EUREC^{4}A will be optimized.

3.2 Estimating the Distribution of Clouds in the Trade-Wind Boundary Layer

An additional important and novel element of the EUREC^{4}A strategy will be to measure the cloud fraction within the trade-wind boundary layer, especially around two critical levels: just above cloud base (around 1 km) and around the trade-inversion level (around 2 km).

Measurements of cloud fraction at cloud base are important for understanding what processes control its variations in the current climate and to test some of the processes involved in the climate change cloud feedbacks of climate models. However, upper-level clouds masking the field of view, they are difficult to make with downward-looking instruments. During EUREC^{4}A, we propose to use the ALiAS backscatter lidar (Chazette et al. 2007; Chazette 2016) and the Bistatic Radar System for Atmospheric Studies (BASTA) radar (Delanoë et al. 2016) on-board the ATR-42 flying just above cloud base to acquire dedicated horizontally pointing observations from the aircraft windows. At this level, LES (e.g., vanZanten et al. 2011; Vogel et al. 2016) suggest that the relative dryness of the atmosphere combined with the relatively low cloud water content (Fig. 6) should maximize the range of the lidar measurements, thereby providing useful backscatter signal over a distance of about 10 km, thus greatly enhancing the sampling volume. Beyond the mean cloud fraction, the lidar-radar measurements will help determine the spectrum of cloud sizes at cloud base, which is thought to be a crucial information for understanding the coupling between the subcloud layer and the cloud layer (Neggers 2015).

To test the approach, measurements from a field campaign which took place on 1–6 June 2017 in Ardèche (South of France, 44.4° N, 4° E) and during which a lidar was mounted horizontally on an Ultra-Light Aircraft (ULA) are being analysed. The lidar, an eye-safe 355-nm backscatter lidar similar to the ALiAS lidar (Chazette 2016) to be used during EUREC^{4}Abut five times less powerful (6 vs. 30 mJ), was pointing horizontally towards the port side of the ULA. The ULA flew a series of horizontal legs of rectangular shape within the subcloud layer and above cloud base within a field of shallow cumuli (Fig. 7). Preliminary analysis of the data suggests that above cloud base (at an altitude of about 1.2 km), the lidar signal does not saturate or vanish as soon as it encounters the first cloud edge along the line of sight, but often penetrates clouds over several hundred metres (100–200 m on average for opaque clouds) and can even go through several consecutive clouds when clouds are optically thin. Overall, this low-power lidar makes it possible to detect the presence of shallow cumuli over a distance of up to 4.5 km. Considering that the ALiAS lidar on board the ATR-42 will be five times more powerful than this one, one may hope to detect clouds over a distance of about $4.5\sqrt{5} = 10\,\mathrm{km}$ during EUREC^{4}A, thus

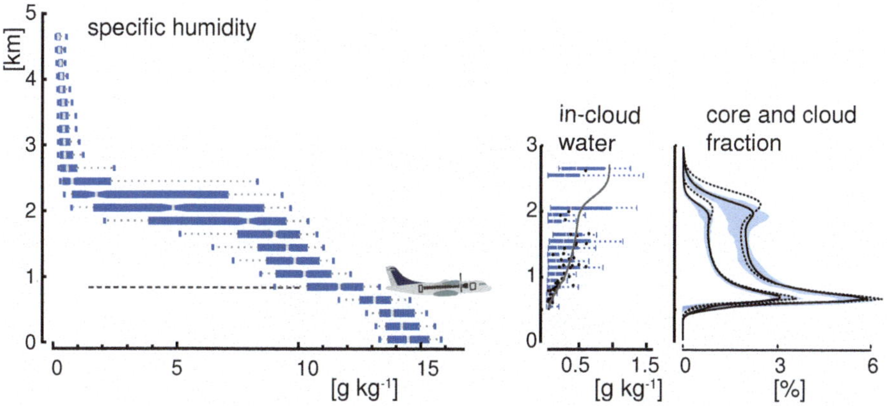

Fig. 6 Vertical profiles of water vapour mixing ratio (left) from the NARVAL 1 flights, (middle) condensed water (q_l) and (right) cloud fraction from RICO. The NARVAL 1 water vapour is derived from all sondes for which surface air temperatures exceed 25°C, as measured east of Barbados in December 2013. The distribution is described by the box plots showing range (5–95%), interquartile and median. Cloud condensate profiles are for similar conditions but during RICO and adapted from vanZanten et al. (2011). Flight date is characterized by box plots (interquartile and 5–95%) and dots (flight averaged). The line is the ensemble mean of 12 large-eddy simulations. Cloud and cloud-core fraction profiles are derived from LES simulations (adapted from vanZanten et al. 2011). Ensemble (interquartile) spread among LES simulations is given by the shading, and the mean profiles from non-precipitating simulations are shown by the thin dashed line. Approximate flight level for the cloud-base legs of the ATR-42 is also indicated

making it possible to map the cloud field in between ATR-42 legs spaced by about 20 km. The use of two gated detectors for different ranges on the lidar (one measuring near-field signals and the other far-field signals) is being considered to enhance the cloud detection range. The combination of lidar and radar measurements should further improve the restitution of the cloud mask over this distance.

The feasibility of the approach will be further tested using LES and by applying to LES outputs the McRALI (Monte Carlo Radar and LIdar) simulator of the lidar and radar instruments on board the ATR-42. McRALI is a forward Monte Carlo model (Cornet et al. 2010) enhanced to take into account light polarization, multiple scattering, high spectral resolution, Doppler effects and the three-dimensional structure of the cloudy atmosphere (Szczap et al. 2013, Alkasem et al. 2017). By diagnosing the cloud fraction that the lidar and radar would measure if they were probing an atmosphere similar to that simulated by the large-eddy model, we will assess how well the above approach can work, and/or whether the experimental strategy will have to be revised to get more accurate measurements of the cloud-base cloud fraction.

Besides the cloud base information, it will be important to determine the vertical profile of cloud fraction, especially at the top of the cloud layer. Indeed the cloud fraction near the inversion level appears to be more variable than that at cloud base (Nuijens et al. 2014), varies strongly with the intensity of the convective mass flux (e.g., Brient et al. 2016; Vial et al. 2016) and strongly influences the variations of the total cloud cover (Rodts et al. 2003). The cloud fraction near the inversion level will be estimated through different methods. First, the ATR-42 could fly around this level and measure the cloud mask through horizontal lidar/radar measurements as will be done around cloud base. However, the cloud optical thickness being much larger at this level than at cloud base, the feasibility of the

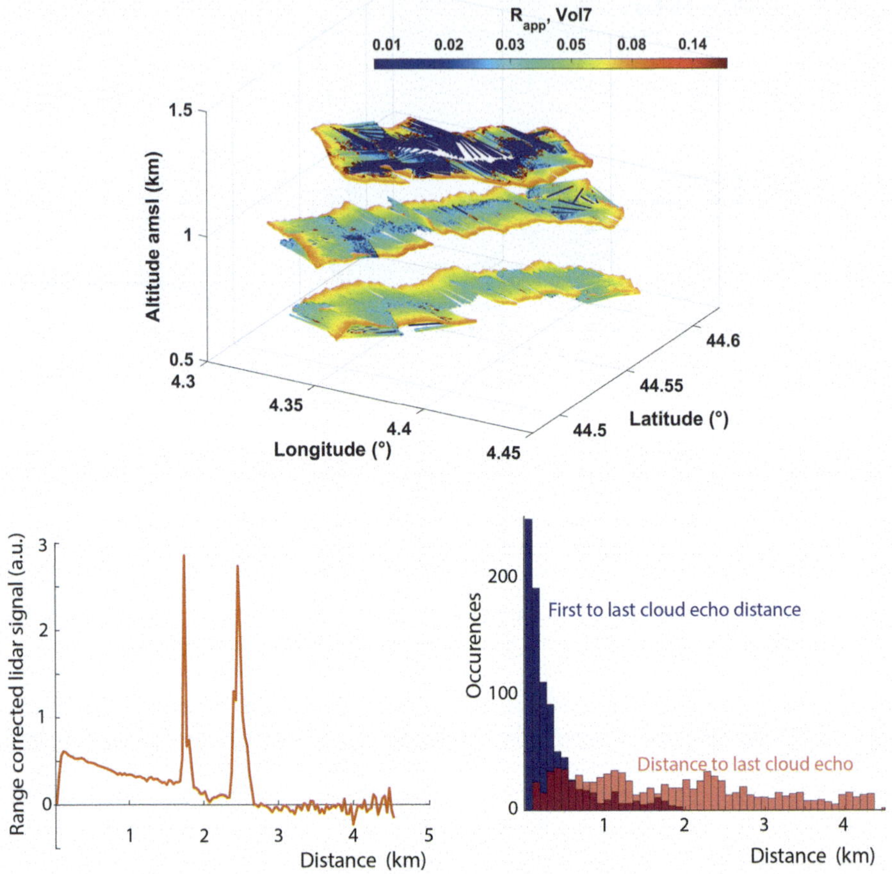

Fig. 7 (Top) Lidar backscatter ratio measured on 2 June 2017 (RF07), in the South of France from an ultra-light aircraft carrying a 355-nm horizontally pointing lidar: two rectangular legs were flown within the subcloud layer and one above the base of shallow cumuli. (Bottom left): Example of an individual lidar signal (corrected from aerosol attenuation) detecting two clouds in a row. (Bottom right): Histogram of the distance between the first and last cloud detections along each individual lidar beam and of the distance between the ULA and the last detected cloud. Note that the ALiAS lidar that will be on-board the ATR-42 will be five times more powerful than the lidar used on-board the ULA

method remains an open question at this particular level (it will be tested using simulators). Therefore, several alternative methodologies will also be considered.

One will consist in analysing the vertical distributions of the lidar backscatter signal and radar reflectivities measured from the downward-looking instruments on HALO and the upward and downward-looking instruments on the ATR-42. Another one will consist in analysing data from ground or ship-borne instruments. Ceilometers will be very useful, but a scanning radar (Oue et al. 2016) on one of the research vessels or deployed on Barbados is also being considered. Yet another approach will consist of analysing observations from the SpecMACs instrument on HALO. SpecMACs is a hyper-spectral line-imager with a field of view of about 40° which allows to map a 10 km swath with 10 m resolution, this swath being similar to the anticipated one for the sideways staring lidar on-board the ATR-42. Oxygen A-band measurements from SpecMACS can be used to measure the distance to the cloud top, as can measurements from the thermal imager which is being developed for EUREC^{4}A. Finally, satellite measurements such as those from lidar (either from CALIPSO (Winker et al. 2003) if it still operates in 2020, or from ADM-Aeolus and/or EarthCARE, that will be launched in 2017 and 2019, respectively) or high-resolution spectrometers such as ASTER, which has a 15-m horizontal resolution and many channels in the infrared, visible and near-infrared (Zhao and Di Girolamo 2007), will provide independent estimates of the vertical profile of cloud fraction.

The dynamic properties of clouds will be inferred from radar measurements (e.g., vertical velocities at cloud base will help determine whether a cloud is active or passive), and the microphysical properties will be derived from the combined analysis of radar-lidar measurements, passive radiometers and in situ measurements (Sect. 5.2). Vertically integrated cloud liquid content of shallow clouds will be measured using downward-looking radiometers flown aboard HALO. The occurrence of precipitation and mesoscale organization of precipitating shallow clouds will be characterized from the scanning precipitation (S-band) weather radar on Barbados and may be complemented by a scanning C-band research radar system.

Finally, measuring the radiative effects of clouds will be critical to assess the coupling between clouds and their large-scale environment. Vertically integrated estimates will be derived from broadband radiative fluxes measured near the surface, near the inversion level and in the upper troposphere, and vertical profiles of the radiative heating rate will be inferred from radiative transfer calculations using observed atmospheric and cloud properties.

3.3 Inferring the Convective Mass Flux at the Top of the Subcloud Layer

To test the hypothesis that lower-tropospheric mixing critically influences the trade-cumulus cloud fraction at cloud base (e. g., Rieck et al. 2012; Gettelman et al. 2012; Sherwood et al. 2014; Brient et al. 2016; Vial et al. 2016), we will need to measure the strength of the convective mixing or quantities closely related to it. Indirect measures of convective mixing, for instance in terms of the relative humidity profile or the strength of shallow overturning circulations, are straightforward to infer from the large-scale structure of humidity field, and measurements of the large-scale vertical velocity. During EUREC^{4}A a direct measure of convective mixing will also be provided in terms of the area-averaged mass flux at, or near, cloud base, M. Two independent methods will be used to estimate M: one based on direct measurements of the fractional area covered by active clouds and of the vertical velocity within them; the other based on the mass budget of the subcloud layer.

The first method is an intuitive approach and will be executed by using radar measurements from the ATR-42 flying just above cloud base. Based on vertical velocity measurements within clouds (w_{cld}), the fractional area covered by active clouds (a_{cld}) will be measured. The cloudy mass flux, just above cloud base, will then be estimated as: $M = \rho a_{\text{cld}} \cdot w_{\text{cld}}$, where ρ is the density of the air. A similar method has been applied to ground-based remote sensing measurements in the past (e.g., Kollias and Albrecht 2010; Ghate et al. 2011; Lamer et al. 2015; Ghate et al. 2016). Using an aircraft to make the same measurement greatly increases the sampling statistics, as in a given amount of time the aircraft samples many more (ten to fifteen times) cloudy updrafts. Using this method, the cloudy area (as well as properties within it) can be decomposed further into updraft and downdraft areas, or even further into different cloud parts (e.g., core or vertically coherent updraft, Fig. 8), or into a spectrum of cloud sizes (Neggers 2015). Similar and ongoing measurements of M from the cloud radar and wind lidars at the Barbados Cloud Observatory will provide additional context for these measurements, as during EUREC^{4}A it is intended to target air masses upwind of the observatory.

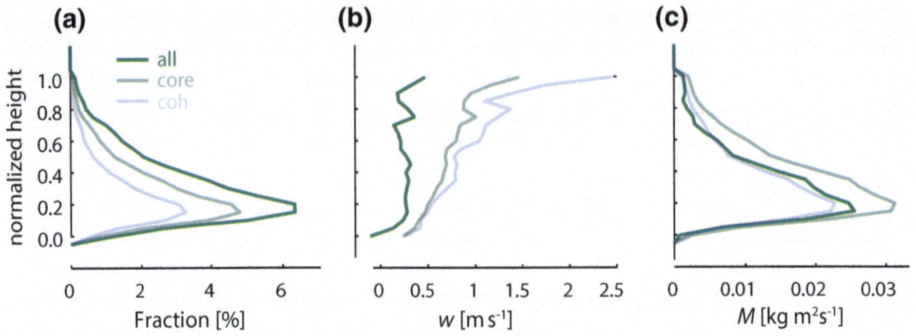

Fig. 8 Boundary-layer profiles (normalized by the maximum cloud top height and minimum cloud base height) of hourly averaged **a** cloud fraction, **b** vertical velocity and **c** mass flux for all, core and vertically coherent updraft samples collected at the island of Graciosa in the Azores. (Adapted from Ghate et al. 2011)

The second method estimates M as a residual of the subcloud layer mass budget, whereby

$$\frac{\mathrm{D}\eta}{\mathrm{D}t} = E + W - \frac{M}{\rho}. \tag{3}$$

Here η denotes the depth of the subcloud layer, E is the top entrainment velocity, W is the large-scale vertical velocity at η (which is related to ω at the same level by a coordinate tranformation) and M is the convective mass flux out of the subcloud layer. This budget is illustrated in Fig. 9.

To estimate M from Eq. (3) thus requires measurements of the other terms. W will be measured using the divergence methods discussed previously. All that remains is to estimate $\mathrm{D}\eta/\mathrm{D}t$ and E. The former, the substantial derivative of the subcloud layer depth, can be derived from a combination of soundings and downward-staring lidar aboard HALO, as this will provide both the time evolution of η and the advective contributions to the substantial derivative.

The entrainment velocity, E, will be estimated in a number of ways. One approach is to assume that the entrainment dynamics of the subcloud layer is the same as for the cloud

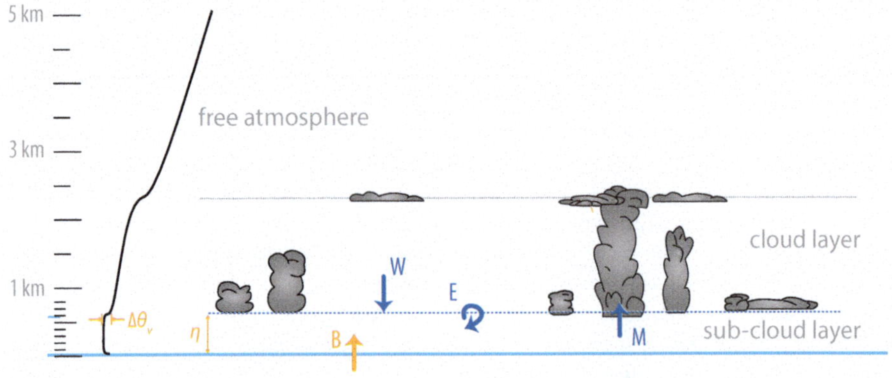

Fig. 9 Schematic representation of the subcloud layer and of the main physical processes affecting its mass budget

free convective boundary layer, as appears to be the case for LES (Siebesma and Cuijpers 1995; Siebesma et al. 2003; Stevens 2006). E can then be diagnosed with the help of the buoyancy flux closure (Lilly 1968), which states that the buoyancy (equivalently virtual potential temperature) flux at η is proportional to its flux at the surface: $(\overline{w'\theta_v'})_\eta = -A(\overline{w'\theta_v})_s$, with A a proportionality constant of about 0.4 (Naumann et al. 2017). This allows E to be estimated from the flux-jump relationship at η (Stevens 2006) as

$$E = \frac{A(\overline{w'\theta_v'})_s}{\Delta\theta_v} \tag{4}$$

where $\Delta\theta_v$, is the jump in the virtual potential temperature across η. For this calculation, $(\overline{w'\theta_v'})_s$ can be constructed from measurements of surface sensible and latent heat fluxes. To account for the thickness of the interfacial layer at η, the proportionality constant, A, the jump, $\Delta\theta_v$ and η must be estimated consistently (Garcia and Mellado 2014; Naumann et al. 2017). We propose to do so by fitting the observations of θ_v (as diagnosed by soundings and flights in the subcloud layer) with values just above cloud base (where the ATR-42 will mostly be flying) to LES in a manner consistent (as per the LES) with the chosen value of A.

A preliminary analysis of large-eddy simulations representative of typical trade-cumulus conditions (Vogel et al. 2016) shows that estimating M in this manner agrees reasonably well (within 35% for the initial calculations) with the value that is diagnosed directly from model output of cloud-core vertical velocity and cloud-core area fraction. The quantitative consistency between both estimates is sensitive to the definition of η in the LES (both the maximum gradient in total humidity or the local minimum in the vertical velocity variance close to cloud base are suitable definitions), and on how the buoyancy flux at the top of the subcloud layer relates to the surface buoyancy flux. By the time of the EUREC^{4}A field campaign, the present method will be refined by defining η such that the estimated and diagnosed mass fluxes are in closer agreement (η can be defined in several ways), and by accounting for the small temporal fluctuations in η. We will also investigate how much the method can capture the sensitivity of the mass flux to different boundary conditions such as sea surface temperature, wind speed and the large-scale divergence D, which are associated with different precipitation fluxes and different degrees of convective organization.

The entrainment velocity, E, and hence M, can also be diagnosed from tracer budgets. In principle, each independent tracer provides the basis for an independent estimate of E. As an example, the budget of the subcloud layer averaged equivalent potential temperature, $\overline{\theta_e}$, takes the form

$$\frac{D\overline{\theta_e}}{Dt} = Q_e + \frac{(\overline{w'\theta_e'})_s - E\Delta\theta_e}{\eta} \tag{5}$$

where Q_e is the radiative source/sink, and $(\overline{w'\theta_e'})_s$ the surface flux, of θ_e. Given measurements of all the other terms, thus yields E. Most of the terms can be measured using methods similar to those already discussed above. Irradiances, which are required to estimate Q_e, will be measured directly (along the near-surface and above boundary-layer legs of the ATR-42), but also estimated on the basis of radiative transfer calculations given the atmospheric state. A similar approach can, and will, be adopted for estimating E from the water budget, which will then require estimates of the precipitation rate at η and at the surface. Taken together with estimates based on Eq.(4) results in three different methods for estimating E, and hence inferring M as a residual of Eq. (3).

The comparison of the direct and budget methods for estimating M will help to assess the robustness of the estimates, especially regarding the sensitivity of the mass flux variations to changes in environmental conditions. This assessment will also test understanding of the subcloud layer budget, most importantly the extent to which M at the top of the subcloud layers as defined in Eq. (3) is related to M a short distance above cloud base and as used to parameterize cumulus convection, an equivalence that should not be taken for granted.

4 A Benchmark Data set

Previous reference observational data sets for linking clouds to circulation in the trades are those from BOMEX, ATEX and GATE. These are field studies which took place nearly a half century ago before the advent of satellite remote sensing, not to mention transformative progress in simulation science. Through a close integration with satellite remote sensing and advances in modelling, EUREC^{4}A aims to provide a reference data set for studying clouds and circulation in the trade-wind region.

4.1 A Simulation and Modelling Testbed

Large-eddy models have long been used to simulate trade-cumuli. Nowdays, they are run over increasingly larger domains, which allows shallow convection to organize into spatial patterns on the mesoscale (e.g., Seifert and Heus 2013; Vogel et al. 2016). The apparent realism of the circulations and clouds that develop often encourages their adoption as an adequate description of reality. However, LES incorporates approximations and assumptions in addition to those associated with the choice of boundary forcings for the simulation. These include the numerical methods adopted, which are known to significantly affect cloud structure and fraction (Vial et al. 2017), as well as the way in which radiative transfer, cloud microphysics and small-scale turbulent motions are parameterized. Most of the LES evaluations of cloud fields have been using observational data from ATEX, BOMEX or RICO (Stevens et al. 2001; Siebesma et al. 2003; vanZanten et al. 2011). These observations make it possible to evaluate carefully the thermodynamic structure of the boundary layer for a limited set of given large-scale forcings. However they do not answer critical questions such as: What is the typical cloud cover and what is the fraction of the cloudy air that is positively buoyant? How strong is the cumulus mass flux at cloud base? How does the cloud fraction and cloud water content vary with changes in the large-scale environment? By measuring important properties of the cumulus mass flux, the large-scale vertical velocity and the large-scale environment, the EUREC^{4}A campaign will offer opportunities to answer some of the above-mentioned questions and to critically test the fidelity of large-eddy simulations.

One process of particular interest, is the "cumulus-valve mechanism" for regulating cloud-base mass fluxes (e.g., Neggers 2015). This mechanism suggests that the mass flux is that required to maintain the cloud-base cloud fraction nearly constant. During EUREC^{4}A, as vertical velocities within clouds will be measured by radar measurements, we will evaluate to what extent it is operative, and the degree to which this indeed controls cloud base cloud fraction. Another question is to what extent mesoscale variability, which may often be the "flow-debris" of much larger-scale circulations not represented by LES, is important for determining cloudiness and its variability. For instance, the influence that cold pools or surface temperature heterogeneities associated with submesoscale processes in the ocean (Sect. 5.6), may exert on cloudiness remains an open issue.

By computing large-scale forcings (water vapour and heat large-scale advections) from EUREC^{4}A observations, it will also be possible to run single-column versions of large-scale models (Single-Column Models or SCMs). It will help us to test the model physics further, and also better understand the cloud feedbacks produced by these models. Indeed, there is ample evidence that single-column simulations of shallow cumuli can help understand low-cloud feedback processes and their dependence on process representations (e.g., Brient and Bony 2012, 2013; Zhang and Bretherton 2008; Zhang et al. 2013a; Dal Gesso et al. 2015; Brient et al. 2016). The link to observations, and the comparison between LES and SCM simulations, will allow us to investigate the relationship between the response of shallow cumuli to prescribed climate change perturbations and the realism of the simulated clouds in the present-day climate, which will help answer questions such as: How does the cloud cover depend on the strength of convective mixing? How variable is it with changes in environmental conditions? Is it possible to constrain the strength of climate change low-cloud feedbacks from present-day processes? (Vial et al., this volume). Taken all together, the new experimental methodologies being developed and deployed as part of EUREC^{4}A will provide new opportunities to provide a reference data set to inform modelling and simulation of trade-cumuli. Besides the evaluation of shallow clouds and cloud feedback processes, it will also help us evaluate the representation of physical processes in climate and weather models, including the parameterization of cumulus convection in large-scale models and the high-resolution operational forecast models used to predict weather in the trades (Sect. 5).

4.2 A Remote Sensing Testbed

Observations from field campaigns are not only fundamental to investigate the physics of trade-cumuli but also to test, and eventually improve, the instruments and algorithms of remote sensing that are used to observe the Earth from space. Beyond the evaluation of cloud retrievals from current satellites, EUREC^{4}A is expected to contribute to the evaluation of the cloud and wind retrievals from two new flagship satellite missions of the European Space Agency: ADM-Aeolus and EarthCARE, that will provide unprecedented information on clouds and circulation.

During EUREC^{4}A, the instruments on-board HALO and the ATR-42 aircraft will sample almost the full spectrum of wavelengths of atmospheric electromagnetic radiation (from the UV to the microwave), making it possible to retrieve a wide range of geophysical properties. Moreover, the UV HSR Doppler wind lidar operating at 355-nm and the 95 GHz Doppler cloud radar will provide products comparable to those obtained from the ADM-Aeolus and EarthCARE satellites. When flying underneath the satellite orbits, it will thus be possible to make direct comparisons between airborne and space-borne measurements. The in situ observations will help interpret the remote sensing in terms of geophysical variables, and the comparison between airborne and space-borne measurements will help evaluate some of the limitations of the satellite remote sensing.

The main limitations of remote sensing are due to the lack of sensitivity of the sensors (which is of particular concern when probing the lower atmosphere from space, but much less from an aircraft), the inability to exploit some measurements near the surface (e.g., the blind zone arising from the contamination of radar measurements by ground clutter or from the saturation of the lidar signal) and the poor spatial resolution of the measurements (which is particularly problematic in areas covered by small broken clouds such as shallow cumulus fields). For passive measurements, which have the best spatial coverage, a particular challenge is identifying sufficiently unique information to deconvolve the atmospheric vertical structure from signals that necessarily integrate over this structure. The synergy of in situ, airborne and space-borne measurements during EUREC^{4}A, jointly with high-resolution simulations from weather forecast models and LES simulations of the campaign area, will help quantify these different sources of uncertainty and test some of the hypotheses used in the cloud or wind retrieval algorithms. A few examples are given below.

Satellite instruments like MODIS or MetOP measure radiances in different wavelength channels, and these measurements are used to retrieve cloud droplet number, cloud phase, optical thickness, and droplet or particle size at cloud top at a spatial resolution of about 1 km. This resolution is insufficient for the observation of shallow cumuli. The specMACS instrument on-board HALO (Appendix 1), which combines hyper-spectral wavelength resolution in the visible and near-infrared wavelength range with a spatial resolution of about 10 m, will permit to observe trade-wind clouds in greater detail. Obvious products will be cloud cover and cloud-size distributions. In addition, the use of three-dimensional radiative transfer methods will permit to retrieve cloud optical thickness, droplet radius, and cloud top structure with high spatial resolution (Mayer 2009; Zinner et al. 2006) and may even be able to distinguish the cloud-core area from the optically thinner edges. The contribution of clouds to solar heating and infrared cooling rates will also be estimated from these parameters.

Another key property of clouds for which satellite retrievals remain very uncertain is the cloud liquid water path (LWP). Most of today's knowledge on the global distribution of cloud liquid water is derived from polar orbiting satellites that measure radiances in the thermal infrared and microwave spectral regions. However, given the coarse spatial resolution of these measurements (several tens of kilometres), the LWP retrievals critically depend on the estimated cloud fraction (Horváth and Gentemann 2007) and cloud vertical structure (Borg and Bennartz 2007). The synergy of the HALO and ATR-42 instrumentation will provide fine-scale information on the variability of water vapour, liquid water and cloudiness over an area of $200 \times 200\,\mathrm{km}$ which will help evaluate satellite retrievals and the validity of their underlying assumptions. By combining active (radar) and passive microwave radiometer, it may also be possible to quantify the amount of drizzle and precipitation in fields, or aggregates, of shallow cumulus. Beyond its intrinsic interest, this detection will make it possible to test the validity of the precipitation thresholds used in LWP retrievals (Wentz and Spencer 1998), and thus to improve LWP retrievals in trade-wind regions.

Active satellite remote sensing provides observations along narrow curtains aligned with the flight track. To achieve radiative closure (as envisioned with EarthCARE) or to generate precipitation fields (as done as part of the GPM mission), it is crucial to combine curtain measurements with observations from wide swath instruments. For this purpose, LES simulations are often used to get statistical information about the three-dimensional structure of the cloudy atmosphere. By providing the reference data set necessary to assess such statistics (Sect. 4.1), EUREC^{4}A will thus help assess and improve these techniques, and may also help guide future satellite measurements to, for instance, better profile lower-tropospheric water vapour.

Finally, the LNG (lidar) instrument on-board the ATR-42 will have the capability to mimic ADM-Aeolus measurements: the 355-nm HSR Doppler wind lidar of the satellite will point 35 deg from nadir (orthogonal to the ground track velocity vector to avoid contribution from the satellite velocity) to derive profiles of the horizontal wind component, and will also regularly point to nadir for calibration. LNG measurements along the satellite orbits and in the same viewing direction, together with in situ measurements and other airborne observations, will help evaluate the L2A (cloud and aerosol optical properties) and L2B (radial winds) ADM-Aeolus products over distances of several hundreds of kilometres.

5 EUREC^{4}A^{++}: An Opportunity for Complementary Investigations

The intensive observations of the atmosphere and of the surface that will be collected during EUREC^{4}A campaign will provide an opportunity to address additional scientific issues. A few of them are mentioned below in the context of what we call EUREC^{4}A^{++}, but more are almost certain to arise in the next years.

5.1 Rectification of Large-Scale Vertical Motions by the Diurnal Cycle of Shallow Clouds

In addition to providing large-scale context for the aircraft missions, measurements by the large-scale sounding array will supplement (and test) meteorological analyses to help answer scientific questions such as: What is the role of transient disturbances and their influence over large-scale vertical motion in modulating convective mass flux and large-scale diabatic heating? What drives the diurnal cycle of vertical motion and clouds in the trades?

Mounting evidence indicates that the Hadley cell over the remote oceans is characterized by a pronounced diurnal cycle in overturning motion, quite distinct from the influences of land (Nitta and Esbensen 1974a; Gille et al. 2003; Wood et al. 2009). It is possible that this diurnal cycle owes fundamentally to the response of deep convective clouds in the ITCZ to the diurnal cycle of direct shortwave absorption (Nitta and Esbensen 1974a), although this topic is unresolved. Observations from suppressed regimes in the Indian Ocean warm pool region reveal that the diurnal cycle in large-scale vertical motion is intimately tied to a diurnal cycle in the shallow convective-cloud population: clouds deepen each afternoon as subsidence relaxes, while the afternoon increase in more active, precipitating clouds leads to more cold pools that in turn augment cloud area fraction (Ruppert and Johnson 2015, 2016). Experiments conducted with a LES framework suggest that this diurnal cloud feedback between large-scale vertical motion and macroscopic cloud properties augments diabatic heating, thus impacting large-scale circulation, on longer timescales through nonlinear rectification (Ruppert 2016).

An additional objective of EUREC^{4}A will therefore be to diagnose the relationships between radiation, clouds, and large-scale vertical motion on the diurnal timescale, and to relate this timescale to other modes of variability. The target hypothesis pertaining to this process is that the diurnal shortwave heating cycle drives a diurnal cycle of deep convection in the ITCZ, which in turn drives a diurnal cycle of large-scale overturning motion in the greater Hadley cell.

5.2 Cloud Microphysics

Cloud macrophysical properties such as cloud fraction or cloud water content are very much influenced by the large-scale environment (e.g., the strength of large-scale vertical motion, tropospheric subsidence, surface temperature, tropospheric humidity). In this context, unravelling the impact of microphysical processes on cloud macrophysical properties or radiative fluxes is challenging because microphysics, cloud macrophysics and large-scale environmental conditions vary in concert. Given that the large-scale dynamical and thermodynamical conditions, and cloud macrophysical properties, will be quantified by EUREC^{4}A, it will be easier to assess the impact of microphysics on cloud macrophysics for given large-scale conditions, and to explore the dependence of microphysics on large-scale environmental conditions or convective organization. For instance, we might investigate whether in observations, precipitation affects the growth of shallow cumuli as LES studies suggest (Vogel et al. 2016).

EUREC^{4}A will offer several opportunities for microphysical measurements: besides in situ liquid and total water content, cloud droplet size distribution and particle imaging on-board the ATR-42 (Table 1), aerosol measurements at the BCO may also be supplemented by measurements from research vessels. In situ data could be collected by small autonomous vehicles launched from the ships or the island (e.g., drones) as well as a tethered HeliKite capable of carrying payloads of up to 100 kg to heights of 3 km. The possible deployment of additional aircraft (from the UK and US), that would focus on microphysical measurements, is also being considered. Such measurements would greatly help advance understanding of cloud-aerosol interactions, entrainment and mixing processes and the onset of precipitation within shallow clouds.

5.3 Shallow Clouds and Convective Momentum Transport

Concurrent measurements of clouds and the large-scale horizontal wind profile during EUREC^{4}A will also help address open and long-standing questions regarding the two-way interaction of clouds (convection) and winds. Past studies have investigated which parameters that represent the large-scale atmospheric state may best predict low-level cloud amount (sometimes referred to as “cloud-controlling” factors). Although this is challenging within the trade-wind cumulus regime—there is no single strong predictor, and correlations on timescales less than a month are small—the correlation between low cloud amount and the near-surface wind speed appears one of the stronger correlations (Brueck et al. 2015; Nuijens et al. 2015b). Daily surface wind speeds have also been found to correlate well with daily averaged rain cover.

This relationship may reflect the influence of surface wind speed on the surface enthalpy fluxes, and hence the depth of convection and the depth of the trade-wind layer. But the relationship may not just represent a one-way interaction: clouds themselves also influence the wind profile through convective momentum transport. This transport would alter winds across a much deeper layer than dry convection and turbulence in the subcloud layer can do. Depending on the wind profile beyond the subcloud layer, convective momentum transport may therefore slow down or accelerate winds near the surface through processes which may also depend on the degree of mesoscale organization of the cloud field. The wind measurements made during EUREC^{4}A will help investigate how winds influence clouds and vice-versa. The opportunity to enhance these measurements by using two different wind lidar systems as part of a funded ADM-Aeolus validation mission using DLR's Falcon aircraft is being considered.

5.4 Improving Climate and Weather Forecast Models

In addition to helping evaluate the processes that control trade-cumuli and cloud feedbacks in LES and GCM models (Sect. 4.1), EUREC^{4}A will help evaluate more generally the physics of large-scale climate and weather models. Indeed these models still exhibit significant biases in the representation of clouds and circulation in the trades. For instance, most of them overestimate the reflection of solar radiation by trade-cumuli despite an underestimate of the cloud fraction and/or the cloud water (the so-called too few, too bright problem, Karlsson et al. 2008; Nam et al. 2012). Models also exhibit persistent biases in their simulation of the surface wind stress (Wang and Carton 2003; Simpson et al. 2014) which, as discussed above, can relate to wrong representations of the surface drag and/or of the momentum transport by shallow convective clouds (Polichtchouk and Shepherd 2016; Schlemmer et al. 2017). The comparison with EUREC^{4}A observations of short-term forecasts run with such models will help disentangle sources of model errors in the representation of physical processes and their interaction with the large-scale circulation. Moreover, as discussed in Sect. 3.3, a better understanding and assessment of the different contributions to the subcloud layer energy budget will help assess the hypotheses underlying the cumulus mass flux closures used in convective parameterizations.

More specifically, EUREC^{4}A will serve as a testbed for high-resolution modelling approaches developed by the Caribbean Institute of Meteorology and Hydrology, Meteo-France, KNMI and ECMWF to deliver operational forecasts to the Caribbean countries. Improving the quality of these forecasts is critical, especially regarding high-impact weather. In Barbados for instance, heavy precipitation produced by severe weather frequently produce significant flash flooding. Landslides, particularly on neighbouring islands with greater orographic relief, result in significant social and economic losses including loss of property and livelihoods and on occasion loss of life. Losses from such events can range from 25–200% of national Gross Domestic Product setting back national development by more than a decade in some instances. In order to reduce losses, in recent years there has been a significant effort to improve early warning hydro-meteorological forecasts over the Caribbean through the development of high-resolution (4 km) numerical weather forecasts using the Advance Research Weather Research and Forecasting (WRF) model. By providing a reference data set for the evaluation of cloud, turbulence and convective parameterizations, EUREC^{4}A will help improve the physics of the model, and eventually the quality of the rainfall forecasts.

5.5 Improving Model Convective Parameterizations

Another important objective of the experiment is to provide field experimental data by which new and existing convective parameterizations can be tested. Measurements of mass fluxes during EUREC^{4}A can be used to test closure assumptions, as discussed, but measurements of convective fluxes would provide an opportunity to test how the convective scheme distributes the energy it carries across cloud base.

We know from many observations that the vertical enthalpy flux establishes a temperature lapse rate that is very close to moist adiabatic, and most existing schemes are designed to accomplish this. For this reason, comparing the parameterized enthalpy flux against observations is a weak test of performance. On the other hand, there is no corresponding universal water vapour profile, and it can be shown that the vertical flux of water depends on such processes as entrainment and cloud microphysics. For this reason, the vertical subgrid-scale water flux by a convection scheme also presents an opportunity for a strong test of convective schemes.

Measuring the vertical flux of water in a field programme is extremely challenging, as the decisive component is the small residual between large, but opposed, vertical turbulence fluxes and those by precipitation. During EUREC^{4}A, the density of the sondes, and the availability of water vapour lidar profiling, and the variety of radar products for estimating precipitation, should provide an excellent opportunity to constrain the water budget, perhaps to a degree that it can help test convective parameterizations. Fortunately, in the Tropics, fluctuations of moist static energy are strongly dominated by fluctuations in water vapour, because temperature perturbations are usually very small above the boundary layer. Thus, the moist static energy budget also offers and opportunity for deducing the convective flux of water vapour, principally by estimating the terms in the budget, or partially integrating measurements of the budget from the top of the atmosphere (or the surface) to the flight level, and then estimating the convective flux as a residual.

Note that it will be possible to assess the overall quality of the field measurements needed to test convective schemes by verifying that:

$$\frac{1}{g}\int_0^{P_s}\left(\frac{\partial h}{\partial t}+\mathbf{V}\cdot\nabla h\right)\mathrm{d}P+F_{\mathrm{TOA}}-F_{\mathrm{SFC}}=0 \tag{6}$$

where h is the moist static energy, $\mathbf{V}$ the three-dimensional wind, F_{TOA} the net radiative flux at the top of the atmosphere and F_{SFC} the total (radiative plus turbulent) surface energy flux. In previous field experiments, errors in some of the terms in (6) resulted in a nonzero sum, which had to be corrected by making adjustments to the vertical velocity used in the advection term (Emanuel and Zivkovic-Rothman 1999). Hopefully, these undesirable errors can be largely avoided in EUREC^{4}A by obtaining more accurate estimates of vertical velocity as well as improved estimates of radiative and surface energy fluxes. By providing more accurate field data sets—ones satisfying column water and moist static energy budgets—more rigorous tests of cumulus parameterizations should be possible. During EUREC^{4}A, these methods will be tested for regimes of shallow convection. If successful they could later be applied in regions of deep convection.

5.6 Ocean Eddies

EUREC^{4}A will also be an opportunity to study ocean-atmosphere interactions in the Atlantic, especially the role of ocean mesoscale eddies. The ocean is a fundamentally turbulent fluid full of fine-scale structures such as eddies, fronts, jets and filaments (McWilliams 2016). These oceanic structures, grouped as mesoscale (10–500 km, 10–100 days) and submesoscale (hundreds of metres to kilometres, daily timescales) dynamics, are recognized as key contributors to the ocean circulation (e.g., Zhang et al. 2013b). There is also increasing evidence that they impact air–sea interactions and influence the winds and clouds of the overlying atmosphere (Chelton et al. 2004; Ferreira and Frankignoul 2008; Frenger et al. 2013; Byrne et al. 2015). However, few observations are available to quantify the role of ocean eddies in the transport of water properties and in air–sea interactions, especially in the tropics.

Automatic eddy detection from satellite observations show the presence of mesoscale eddies (Fig. 10) not only in the mid- and higher latitudes but also in the Tropics (Laxenaire

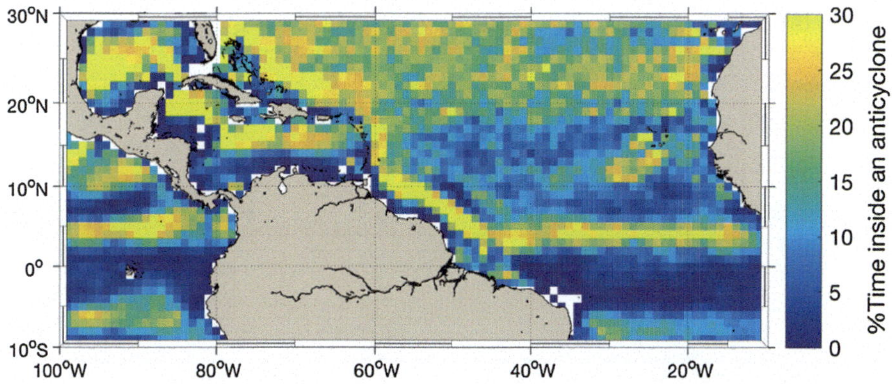

Fig. 10 The route of ocean eddies. Statistics of ocean mesoscale eddies derived from satellite altimetry (shown is the fraction of the time inside an anticyclonic eddy)

et al. 2017, accepted). In particular, intense warm ocean eddies (i.e. anticyclonic eddies) converge in the western tropical Atlantic, offshore of Barbados. These eddies come from the south (tropical Atlantic and South Atlantic) and from the East (from Cape Verde and the western Africa margin). Eddies such as the anticyclonic features associated with North Brazil Current Rings eventually carry freshwater, originating from the Amazon/Orinoco river, into the region. Linkages between the freshwater surplus by the rivers and intensification of storms and cyclones have been reported for the large scale (Reul et al. 2014) and for the region east of the Antilles, in the Caribbean, for individual eddies (Rudzin et al. 2017).

Inspecting historical ship data in the EUREC^{4}A campaign area (RV SONNE SO172; Fig. 11) shows the impact of low salinity water in an anticyclonic North Brazil Current Ring that is associated with increased air temperature and the occurrence of very local air/sea interactions. During the EUREC^{4}A campaign observations by aircraft, research vessels and autonomous platforms will make it possible to observe the coupled system and assess the role of mesoscale eddies on the lower tropical atmosphere in a comprehensive manner. In particular, the association of air and ship measurements will allow us to characterize contrasted ocean mesoscale eddies (e.g., cyclonic/anticyclonic, deep/shallow mixing) estimate the role of these eddies in modifying atmosphere-ocean coupling and in influencing the mesoscale organization of the atmosphere and shallow clouds in particular.

Moreover, a synoptical study from different research vessels measuring different mesoscale eddies across the experimental area will provide new information on water-mass characteristics advected by the regional eddies. These are of twofold importance, as they inform studies of mesoscale eddies in general, but also in a particularly interesting region where over a narrow zone of longitudes the upper and lower limbs of the Atlantic Meridional Overturning Circulation flow over one another. A relatively wide range of mesoscale eddy observations will provide quantitative assessments of their role in transport of mass, heat, freshwater, carbon and other biogeochemical variables enabling a measure of the role of the oceanic mesoscale in this overturning circulation.

The observing strategy will make use of near real-time satellite altimetry data (AVISO Ssalto/Duacs) to identify ocean mesoscale eddies and then one of the ships will be used to survey the eddy and to deploy autonomous oceanographic observing platforms such as underwater electric gliders and wave-gliders. A second ship will characterize the surrounding background field in parallel. Vertical profiling of the water column (temperature,

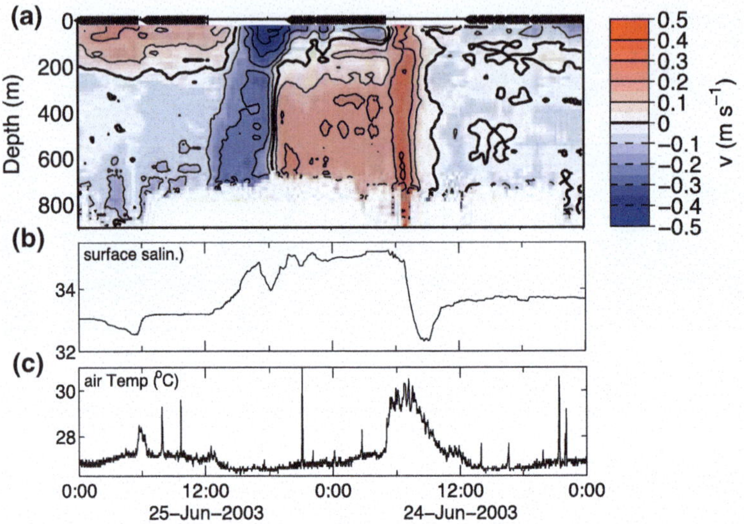

Fig. 11 RV SONNE (SO172) ship survey through an anticyclonic eddy (North Brazil Current Ring) west of Barbados (ship was going east to west, time axis is reversed for clarity). **a** Meridional current section (triangles at 0 m denote ship was stationary), **b** sea surface salinity (psu), and **c** air temperature

salinity, currents, oxygen, and other properties e.g., carbon related quantities) will be measured. Argo profiling floats will be deployed within anticyclonic eddies within the Guiana and North Equatorial currents in the year preceding the EUREC^{4}A campaign. More information about the instrumentation and the measurements envisioned is given in Appendix 2. This strategy will be tested and refined before the campaign by undertaking a set of preliminary studies based on the analysis of available data from satellites, ships, Argo profiling floats, and eddy-resolving numerical simulations.

Together, and perhaps the biggest plus, is that oceanic and atmospheric data collected during EUREC^{4}A will help build a data set capable of evaluating a wholly new generation of coupled ocean-atmosphere models, ones capable of resolving (rather than parameterizing) both convective eddies (cumulus convection) in the atmosphere, and the mesoscale dynamics of the ocean.

5.7 Capacity Building

Building national and regional resilience to increasing climate variability, climate change and extreme weather events in the Caribbean includes increasing national and regional weather and climate data, related knowledge platforms and human capacity. By involving university students and other young scientists of the region in the field campaign and the long-term research activities related to it, EUREC^{4}A will help train the next generation of regional climate scientists and operational forecasters, develop databases that will facilitate the dissemination and use of the data collected in the area during field studies, and will promote international partnership and collaboration networks.

6 Conclusions

By characterizing, for the first time, both the macrophysical properties of shallow cumuli and the large-scale environment in which convection and clouds are embedded, the EUREC^{4}A campaign will test developing ideas about what controls the cloud amount in the trades. It should elucidate the role of convection and large-scale circulations in low-cloud feedbacks and thus address one of the central questions of the World Climate Research Programme's Grand Challenge on *Clouds, Circulation and Climate Sensitivity* (Bony et al. 2015). Through its alignment with two flagship missions of the European Space Agency and the cutting-edge of modelling, EUREC^{4}A should also provide a new reference data set which can be used to assess the modelling and the remote sensing for the years to come. The experimental strategy proposed for the campaign is ambitious. However, it builds on a legacy of ongoing field studies, particularly ground measurements at the Barbados Cloud Observatory and field measurements as part of the NARVAL (December 2013) and NARVAL2 (August 2016) campaigns, as well as extensive experience from process models. Ongoing analysis of these measurements and simulations are being used to test and refine the experimental strategy of EUREC^{4}A so as to maximize the scientific gains from the planned measurements.

A compact and well-defined experimental strategy opens up the mission to other partners with complementary interests. This campaign should therefore be considered as an opportunity to nucleate larger international efforts and to underpin additional investigations ranging from factors influencing cloud microstructure and warm rain formation, to the role of mesoscale eddies in the ocean.

Acknowledgements The authors acknowledge Aurélien Bourdon, the CNRS-Météo-France-CNES SAFIRE facility for the scientific airborne operations (http://www.safire.fr), Didier Bruneau and Jacques Pelon for technical discussions, and the professional ULA pilot Franck Toussaint and the Air Creation company for having made the ULA-borne lidar tests possible. The paper benefited from stimulating discussions at the International Space Science Institute (ISSI) workshop on "Shallow clouds and water vapour, circulation and climate sensitivity". The EUREC^{4}A project is supported by the European Research Council (ERC), under the European Union's Horizon 2020 research and innovation programme (Grant Agreement No. 694768), by the Max Planck Society and by DFG (Deutsche Forschungsgemeinschaft, German Research Foundation) Priority Program SPP 1294.

Appendix 1: Aircraft Instrumentation

Airborne platforms are one means to probe the thermodynamic, dynamic and cloud properties of the atmosphere. EUREC^{4}A will be centred around measurements from two aircraft, namely the French ATR-42 and the German HALO, carrying complementary payloads. The HALO measurements aloft will characterize with downward-looking instruments the large-scale (several thousands of km^2) environment in which clouds form, and the radiative properties of the clouds therein. The ATR-42 measurements in the shallow cloud layer will constrain cloud macrophysical and microphysical properties, turbulent mixing processes and shallow circulations.

The airborne and in situ measurements of EUREC^{4}A will also advance the evaluation and calibration of space-borne measurements, and their interpretation in terms of geophysical variables (Wendisch and Brenguier 2013). It will be the case in particular for observations from EarthCare and ADM-Aeolus, which are two flagship satellite missions of the European Space Agency's Living Planet Programme devoted to the observation of clouds and circulation, and whose measurements will be mimicked by several instruments on board both aircraft.

The ATR-42 Aircraft and Its Instrumentation

The French ATR-42 is a bi turbo-prop aircraft from SAFIRE that has the capability of flying in the lower troposphere (ceiling at about 8 km) with a maximum range of about 1800 km. It will fly a series of low-level legs just above cloud base (around 1 km), about 100 km long and spaced by about 20 km, as a way to sample the cloud field within the area encompassed by HALO circles. A particularity of the aircraft instrumentation (summarized in Table 1) is that it will include sideways and vertically pointing lidar and radars that will probe the atmosphere horizontally and vertically, aiming at measuring the cloud fraction at cloud base. The last leg before refuelling will be flown either below cloud level, to measure surface turbulent fluxes, temperature at the sea surface and in the subcloud layer, and near-surface radiation, or near the trade-inversion level (around 2 km) to characterize cloud microphysics, measure radiative fluxes. Given the mean science speed of the aircraft (about $100\,\mathrm{m\,s^{-1}}$) and the endurance expected for the envisioned payload, the ATR-42 will make two four-hr flights per day bracketed by the daily nine-hr flight of HALO.

The ATR-42 will be equipped with advanced instrumentation including the multi-wavelength Leandre New Generation (LNG) HSR backscatter lidar (Bruneau et al. 2015), the 95 GHz Doppler radar RASTA (RAdar SysTem Airborne, Delanoë et al. 2013), the mini-cloud radar BASTA (Delanoë et al. 2016), and the mini-lidar ALiAS developed at LSCE (Chazette et al. 2007; Chazette 2016). The first three instruments have been developed at LATMOS[1] to characterize clouds, aerosol particles and hydrometeor particle velocities. Although RASTA and the HSR lidar will be observing the atmosphere above and below the aircraft, the mini-lidar ALiaS and the BASTA Doppler radar will be staring sideways from the aircraft windows in order to measure the cloud fraction and cloud optical properties just above cloud base (Sect. 3.2). In addition, the payload will include a thermal infrared radiometer (CLIMAT-AV, Conveyable Low-Noise Infrared Radiometer for Measurements of Atmosphere and Ground Surface Targets -Airborne Version) developed at the Laboratoire d'Optique Atmosphérique (Brogniez et al. 2003) to measure sea surface temperature, and hemispheric broadband pyrgeometer (Kipp and Zonen CGR4) and pyranometer (Kipp and Zonen CMP22) to measure upwelling and downwelling radiative fluxes in the longwave and shortwave, respectively. Finally, the aircraft will measure temperature, moisture and wind along its trajectory at a high frequency ($25\,\mathrm{s^{-1}}$) for turbulence statistics calculations.

The LNG airborne Lidar system is a three-wavelength (1064, 532, and 355 nm) backscatter lidar with polarization and high-spectral-resolution capability at 355 nm[2]. It operates in a direct detection mode (measurement of the backscattered light intensity), which has the advantage of relying on both particulate and molecular scattering, and allows

[1] http://rali.projet.latmos.ipsl.fr, http://basta.projet.latmos.ipsl.fr

[2] Based on the emitter characteristics given in Table 1 of (Bruneau et al. 2015), the nominal ocular hazard distance at each wavelength IS: 0 m at 355 nm, 333 m at 532 nm and 184 m at 1064 nm. These values are the same for the horizontal and vertically pointing mode. They also comply with norms NF EN 60825-1 and CEI 6025-1.

extended ranges and capabilities. Thanks to its two-wave interferometry system (Mach–Zehnder Interferometer) and its capability of high-spectral-resolution UV analysis, it can determine both optical parameters of aerosol and clouds and measurements of along-line-of-sight wind velocity, based on the Doppler effect on particles (Bruneau and Pelon 2003; Bruneau et al. 2015). When scattering particles are moving, the wavelength of the scattered light is shifted by a small amount as a function of speed. The Doppler wind lidar measures this change of wavelength to determine the velocity of the wind in the direction of the light pulse. For a vertically pointing lidar, vertical velocity measurements are thus possible. The capability of a lidar system using a Mach–Zehnder interferometer is described in Bruneau and Pelon (2003) and is not be detailed here. A simplified equation describing the precision of LNG-derived wind velocity measurement along the line of sight can be obtained from Bruneau and Pelon (2003) (see their equation 23) as: $\sigma(V) \approx \frac{100}{SNR} \frac{R_\beta}{R_\beta - 1}$ (in $m.s^{-1}$), where SNR is the signal-to-noise ratio on the 355-nm parallel channel and R_β the backscatter ratio (i.e. the total backscatter divided by the molecular backscatter) measured on the same channel. The accuracy of the LNG system along the line of sight after aircraft motions are removed has been assessed for a case of cirrus clouds by Bruneau et al. (2015) based on the apparent aircraft speed derived from LNG ground-echo, and based on results from a recent field campaign (NAWDEX—North Atlantic Waveguide and Downstream Impact Experiment, fall 2016): the accuracy of LNG wind speed measurements in these conditions was expected to be around $1\ m\ s^{-1}$ (D. Bruneau, personal communication).

The Airborne Lidar for Atmospheric Studies (ALiAS) was built at LSCE following a precursor instrument (Chazette et al. 2007; Chazette 2016) flown on-board an ultra-light aircraft (ULA). It is based on a frequency-tripled Nd:YAG laser (ULTRA) manufactured by QUANTEL[3] emitting in the near ultraviolet (355 nm), thus satisfying eye safety requirements at the output window. The UV pulse energy is 30 mJ, and the pulse repetition rate is 20 Hz. The acquisition system is based on a PXI (PCI eXtensions for Instrumentation) technology with a sampling frequency of 200 MHz (initial resolution along the line of sight equal to 0.75 m). The receiver includes two channels for the detection of the elastic backscatter from the atmosphere in the parallel and perpendicular polarization planes relative to the linear polarization of the emitted radiation. It was designed to monitor both the aerosol and hydrometeor distributions and dispersions in the low and middle troposphere. After specific signal analysis, including laser shot accumulation and low pass filtering, the final resolution along the line of sight is between 15 and 30 m. With a 15-cm diameter telescope, the lidar is compact ($\sim 70 \times 45 \times 18$ cm), lightweight (<50 kg for both optics and electronics), robust to vibration, requires moderate power (< 500 W) and can thus be easily mounted aboard an aircraft for horizontal shooting. Its wide field of view (FOV) of 2.3 mrad ensures a full-overlap of the transmitter and receiver paths beyond 100 to 200 m.

The BASTA (Bistatic Radar System for Atmospheric Studies) radar is a mini-cloud W-band radar (weighting only 32 kg) that measures the Doppler velocity and the reflectivity at 95 GHz (Delanoë et al. 2016). Its specificity, compared to traditional pulsed radars, is that instead of transmitting a large amount of energy for a very short time period (as a pulse), a lower amount of energy is transmitted continuously. The radar can be used in several modes depending on application, including 12.5- and 25-m-vertical resolution modes. The sensitivity of the instrument is about -40 dBZ at 1 km for 3-s integration and a

[3] http://www.quantel.com

range-gate of 25 m. Its unambiguous range is 12 and 18 km for the 12.5- and 25-m range-gate modes, respectively. The high mobility of the system allows to install the system with an horizontal pointing configuration. The most will be made of the bistatic nature of the system, with emitting and receiving radar signals through two side-by-side windows. Calculations suggest that a cloud radar having a sensitivity of −35 dBZ at 1 km in an atmosphere with 80 % relative humidity will detect clouds with a reflectivity below −30 dBZ over 1400 m, and will not detect any liquid cloud further 3250 m. Considering a liquid cloud with a constant LWC of 0.6 g/m3 (which is a large upper bound) next to the aircraft, the radar should be able to detect clouds below −30 dBZ until 1000 m and liquid cloud until 2000 m.

The RASTA (RAdar SysTem Airborne) radar measures the Doppler velocity and the reflectivity at 95 GHz (W-band) along a radial defined by the pointing direction of the antenna (Delanoë et al. 2013). During EUREC^{4}A, a 4-antenna configuration will be used, that will include 3 upward-looking beams (zenith, 28 degrees off-nadir perpendicular to the aircraft motion, and 20 degrees off-zenith and opposite the aircraft motion) and one nadir pointing antenna. This unique configuration allows for the retrieval of the three-dimensional wind field, i.e. the three components of the wind on vertical plan above the aircraft, by combining the independent measurements of the projected wind vector on radar line of sights. The independent Doppler radial velocities are provided by the multi-beam antenna system. The radar range is 15 km, its range resolution is 60 m and its horizontal resolution ranges from 100 to 150 m depending on aircraft speed. RASTA measurements of vertical velocities within clouds, combined with lidar-radar estimates of the cloud fraction at cloud base, will help develop an estimate of the convective mass flux at cloud base that will be completely independent of the one derived from the analysis of the subcloud layer mass budget (Sect. 3.3).

The CLIMAT-AV thermal infrared radiometer (Brogniez et al. 2003) measures (at nadir) radiances simultaneously in three narrowband channels centred at 8.7, 10.8, and 12.0 micron, with about 1 mm of full width at half maximum. It uses a 7-Hz sampling frequency and performs measurements within a 50-mrad field of view, which corresponds to a footprint of about 50 m at a 1-km range. CLIMAT-AV is very similar to the CALIPSO IIR system. The absolute accuracy of brightness temperature measurements is about of 0.1 K, and its sensitivity is of the order of 0.05 K. The radiances measured by CLIMAT-AV will be used to estimate the sea surface temperature.

The HALO Aircraft and Its Instrumentation

The German high-altitude and long-range research aircraft HALO is a modified Gulfstream G550 business jet with a long endurance (more than 10 flight hours), a long range (about 8000 km), and a high ceiling (15.5 km) (Wendisch et al. 2016). In cooperation with the DLR and the Universities of Cologne, Hamburg, Leipzig and Munich, it will be equipped with an extensive set of remote sensing instrumentation (summarized in Table 2) including: the differential absorption and high-spectral-resolution lidar system (WALES, Water vApour Lidar Experiment in Space), HAMP (the HALO Microwave Package) which includes the cloud radar MIRA36 (36 GHz) and a microwave radiometer, the spectral imager specMACS, and an instrument system that measures spectrally resolved upward and downward solar radiances and irradiances (SMART). The payload also includes in situ measurements of the meteorological properties along the flight track (BAHAMAS), and the ability to launch dropsondes using the AVAPS system. To measure broadband upward and downward longwave radiances and remotely sensed surface temperatures, it is planned to add instruments such as the hemispheric broadband pyranometer (solar) and pyrgeometer (thermal infrared) and the CIMEL/CLIMAT-AV instruments used on the ATR-42. A cooled infrared imaging spectral camera will also be integrated.

MIRA36 is a commercially available METEK Ka-band (36 GHz) cloud research radar with polarization and Doppler capability to determine vertical velocity in clouds and precipitation. Together with microwave radiometers in the K, V, W, F, and G-band the MIRA36 is part of HAMP (Mech et al. 2014).

The lidar system WALES is a combined differential absorption and high-spectral-resolution lidar (HSRL) system developed and built at the Deutsches Zentrum für Luft- und Raumfahrt (Wirth et al. 2009). WALES is capable of nearly simultaneously emitting four wavelengths, three online and one offline, in the water vapour absorption band between 935 and 936 nm. The three online wavelengths achieve the necessary sensitivity needed for measurements over the whole range of tropospheric water vapour concentration. The vertical resolution of the raw data is 15 m. In addition to the 935-nm channel, the receiver is equipped with polarization-sensitive aerosol channels at 532 and 1064 nm, the first one with high-spectral-resolution capabilities using an iodine filter in the detection path (Esselborn et al. 2008). This allows for collocated measurements of humidity, optical depth, clouds and aerosol optical properties.

SpecMACS is an imaging cloud spectrometer developed at LMU (Ewald et al. 2016) consisting of two commercial spectral camera systems in the visible near-infrared (VNIR: 400–1000 nm) and in the shortwave infrared (SWIR: 1000–2500 nm). The nominal spectral resolution is 3 nm and 10 nm for the VNIR and for the SWIR, respectively. SpecMACS produces a spectrally resolved line image. For a flight level of about 10 km, a spatial resolution in the order of 10 m for cloud objects at a distance.

SMART (Spectral Modular Airborne Radiation Measurement System) consists of a set of spectral solar radiation sensors including radiances and irradiances (Wendisch et al. 2001; Ehrlich et al. 2008). All quantities are obtained for the wavelength range of 0.3–2.2 μm with spectral resolution of 2–16 nm full width of half maximum (FWHM), which is sufficient to analyse the spectral characteristics of spectral absorption bands of ice and liquid water. While the irradiance sensors provide spectral albedo at flight-level representative for a specific area, the measurement frequency of 2 Hz and the 2.1° field of view of the radiance sensor allows identifying cloud inhomogeneities (about 200 m footprint for cloud top at 5 km and flight altitude of 10 km).

A similar payload was used during the NARVAL2 campaign and during the NAWDEX campaign of September–October 2016 over the North Atlantic.

Appendix 2: Oceanographic Instrumentation

The thermal structure of the upper ocean has been historically observed from oceanographic ships and from expendable bathythermograph (XBT). Since the early 2000s, the advent of the Argo array of autonomous profiling floats has significantly increased the ocean sampling to achieve near-global coverage for the first time over the upper 1800 m and with a nominal resolution of 3 degrees in latitude/longitude. However, these new global observations are still very sparse and do not provide adequate measurements along boundaries of the oceans and within mesoscale eddies. This represents an acute weakness in our present understanding of ocean and atmosphere dynamics and their role in shaping the Earth's climate variability and changes. To qualify and quantify the role of ocean eddies in the transport of water properties and in air–sea interactions, a number of oceanographic measurements will be necessary.

Ideally, in the year preceding the operational phase of EUREC^{4}A we aim to deploy Argo profiling floats within anticyclonic eddies within the Guiana and North Equatorial currents. By associating these data with satellite observations of the atmosphere (clouds, water vapour, winds, etc), it will be possible to follow the joint evolution of eddies and lower atmospheric properties. Then, during the operational phase of EUREC^{4}A, it will be appropriate to work with two ships measuring both air–sea fluxes, surface atmospheric properties and vertical profiles of temperature and salinity in the upper 2000 m of the ocean.

Ocean vertical profiles of temperature, salinity, currents, oxygen and other biogeochemical properties to also assess carbon related quantities will be acquired by a deep-reaching classical CTD rosette, equipped with sampling bottles and Acoustic Doppler Current profilers (ADCP; 150 or 300 kHz; one upward looking and one downward looking). To increase the sampling resolution, it will be very important to implement between CTD stations, at least for temperature, salinity and pressure, a very manageable and easy-to-use vertical profiler such as the Teledyne OceanScience Underway CTD/RapidCast (UCTD). On both ships, a microstructure vertical profiler would help infer turbulence linked with eddies and air–sea interactions.

The ships are equipped with an underway Thermosalinograph system (TSG) measuring near-surface temperature, salinity and, in many case, fluorescence. The vertical structure of ocean currents down to $>$ 1000m depth along the ship track is recorded by ADCPs (38 and 75 kHz) mounted in the ships hull. Further instrumentations on the ships include standard marine atmospheric observing devices such as LIDARs (for vertical profiles of temperature, humidity and wind) and radiometers and precipitation gauges that allow to derive local ocean-atmosphere heat and freshwater fluxes. Operations of autonomous observing platforms such as underwater electric gliders and wave-gliders in regions of particular interest (e.g., frontal systems) will complement the observing efforts. The autonomous observing effort will start before the operational phase of EUREC^{4}A and allow to observe the upper ocean conditions and their temporal and spatial evolution before, during and after the EUREC^{4}A core field phase.

References

Albrecht BA, Betts AK, Schubert WH, Cox SK (1979) Model of the thermodynamic structure of the trade-wind boundary layer: part I. theoretical formulation and sensitivity tests. J Atmos Sci 36(1):73–89. doi:10.1175/1520-0469(1979)036<0073:MOTTSO>2.0.CO;2

Albrecht BA, Bretherton CS, Johnson D, Scubert WH, Frisch AS (1995) The Atlantic stratocumulus transition experiment ASTEX. Bull Am Meteorol Soc 76(6):889–904. doi:10.1175/1520-0477(1995)076<0889:TASTE>2.0.CO;2

Alkasem A, Szczap F, Cornet C, Shcherbakov V, Gour Y, Jourdan O, Labonnote LC, Mioche G (2017) Effects of cirrus heterogeneity on lidar CALIOP/CALIPSO data. J Quant Spectrosc Radiat Transf 202:38–49. doi:10.1016/j.jqsrt.2017.07.005

Asrar G, Bony S, Boucher O, Busalacchi A, Cazenave A, Dowell M, Flato G, Hegerl G, Kallen E, Nakajima T, Ratier A, Saunders R, Slingo J, Sohn BJ, Schmetz J, Stevens B, Zhang P, Zwiers F (2015) Climate symposium 2014: findings and recommendations. Bull Am Meteorol Soc 96(9):ES145–ES147. doi:10.1175/BAMS-D-15-00003.1

Augstein E, Riehl H, Ostapoff F, Wagner V (1973) Mass and energy transports in an undisturbed Atlantic trade-wind flow. Mon Weather Rev 101(2):101–111. doi:10.1175/1520-0493(1973)101<0101:MAETIA>2.3.CO;2

Augstein E, Schmidt H, Ostapoff F (1974) The vertical structure of the atmospheric planetary boundary layer in undisturbed trade winds over the Atlantic ocean. Bound Layer Meteorol 6(1):129–150. doi:10.1007/BF00232480

Bony S, Dufresne JL (2005) Marine boundary layer clouds at the heart of tropical cloud feedback uncertainties in climate models. Geophys Res Lett 32(L20):806. doi:10.1029/2005GL023,851

Bony S, Dufresne JL, LeTreut H, Morcrette JJ, Senior C (2004) On dynamic and thermodynamic components of cloud changes. Clim Dyn 22:71–86

Bony S, Stevens B, Frierson DMW, Jakob C, Kageyama M, Pincus R, Shepherd TG, Sherwood SC, Siebesma AP, Sobel AH, Watanabe M, Webb MJ (2015) Clouds, circulation and climate sensitivity. Nat Geosci 8:261–268. doi:10.1038/ngeo2398

Borg LA, Bennartz R (2007) Vertical structure of stratiform marine boundary layer clouds and its impact on cloud albedo. Geophy Res Lett 34(5):l05807. doi:10.1029/2006GL028713

Boucher O, Randall D, Artaxo P, Bretherton C, Feingold G, Forster P, Kerminen VM, Kondo Y, Liao H, Lohmann U, Rasch P, Satheesh S, Sherwood S, Stevens B, Zhang X (2013) Clouds and aerosols. Cambridge University Press, Cambridge. book section 7:571–658. doi:10.1017/CBO9781107415324.016

Bretherton CS (2015) Insights into low-latitude cloud feedbacks from high-resolution models. Philos Trans R Soc Lond A Math Phys Eng Sci. doi:10.1098/rsta.2014.0415

Brient F, Bony S (2012) How may low-cloud radiative properties simulated in the current climate influence low-cloud feedbacks under global warming? Geophys Res Lett 39(20):l20807. doi:10.1029/2012GL053265

Brient F, Bony S (2013) Interpretation of the positive low-cloud feedback predicted by a climate model under global warming. Clim Dyn 40(9):2415–2431. doi:10.1007/s00382-011-1279-7

Brient F, Schneider T, Tan Z, Bony S, Qu X, Hall A (2016) Shallowness of tropical low clouds as a predictor of climate models' response to warming. Clim Dyn 47(1):433–449. doi:10.1007/s00382-015-2846-0

Brogniez G, Pietras C, Legrand M, Dubuisson P, Haeffelin M (2003) A high-accuracy multiwavelength radiometer for in situ measurements in the thermal infrared. Part II: behavior in field experiments. J Atmos Ocean Technol 20(7):1023–1033. doi:10.1175/1520-0426(2003)20<1023:AHMRFI>2.0.CO;2

Brueck M, Nuijens L, Stevens B (2015) On the seasonal and synoptic time-scale variability of the north atlantic trade wind region and its low-level clouds. J Atmos Sci 72(4):1428–1446. doi:10.1175/JAS-D-14-0054.1

Bruneau D, Pelon J (2003) Simultaneous measurements of particle backscattering and extinction coefficients and wind velocity by lidar with a mach-zehnder interferometer: principle of operation and performance assessment. Appl Opt 42(6):1101–1114. doi:10.1364/AO.42.001101

Bruneau D, Pelon J, Blouzon F, Spatazza J, Genau P, Buchholtz G, Amarouche N, Abchiche A, Aouji O (2015) 355-nm high spectral resolution airborne lidar lng: system description and first results. Appl Opt 54(29):8776–8785. doi:10.1364/AO.54.008776

Byrne D, Papritz L, Frenger I, Münnich M, Gruber N (2015) Atmospheric response to mesoscale sea surface temperature anomalies: Assessment of mechanisms and coupling strength in a high-resolution coupled model over the south atlantic. J Atmos Sci 72(5):1872–1890. doi:10.1175/JAS-D-14-0195.1

Chazette P (2016) Exceptional aerosol pollution plume observed using a new ULA-lidar approach. Atmosph Environ 141:470–480. doi:10.1016/j.atmosenv.2016.07.015

Chazette P, Sanak J, Dulac F (2007) New approach for aerosol profiling with a lidar onboard an ultralight aircraft: application to the african monsoon multidisciplinary analysis. Environ Sci Technol 41:8335–8341

Chazette P, Marnas F, Totems J, Shang X (2014) Comparison of iasi water vapor retrieval with h_2o-raman lidar in the framework of the mediterranean hymex and charmex programs. Atmos Chem Phys 14(18):9583–9596. doi:10.5194/acp-14-9583-2014

Chelton DB, Schlax MG, Freilich MH, Milliff RF (2004) Satellite measurements reveal persistent small-scale features in ocean winds. Science 303(5660):978–983. doi:10.1126/science.1091901

Ciesielski PE, Schubert WH, Johnson RH (1999) Large-scale heat and moisture budgets over the astex region. J Atmos Sci 56(18):3241–3261. doi:10.1175/1520-0469(1999)056<3241:LSHAMB>2.0.CO;2

Cornet C, Labonnote L, Szczap F (2010) Three-dimensional polarized Monte Carlo atmospheric radiative transfer model (3DMCPOL): 3D effects on polarized visible reflectances of a cirrus cloud. J Quant Spectrosc Radiat Transf 111(1):174–186. doi:10.1016/j.jqsrt.2009.06.013

Dal Gesso S, Siebesma AP, de Roode SR (2015) Evaluation of low-cloud climate feedback through single-column model equilibrium states. Q J R Meteorol Soc 141(688):819–832. doi:10.1002/qj.2398

Delanoë J, Protat A, Jourdan O, Pelon J, Papazzoni M, Dupuy R, Gayet JF, Jouan C (2013) Comparison of airborne in situ, airborne radar–lidar, and spaceborne radar-lidar retrievals of polar ice cloud properties sampled during the polarcat campaign. J Atmos Ocean Technol 30(1):57–73. doi:10.1175/JTECH-D-11-00200.1

Delanoë J, Protat A, Vinson JP, Brett W, Caudoux C, Bertrand F, du Chatelet JP, Hallali R, Barthes L, Haeffelin M, Dupont JC (2016) Basta: a 95-GHz fmcw doppler radar for cloud and fog studies. J Atmos Ocean Technol 33(5):1023–1038. doi:10.1175/JTECH-D-15-0104.1

Ehrlich A, Bierwirth E, Wendisch M, Gayet JF, Mioche G, Lampert A, Heintzenberg J (2008) Cloud phase identification of arctic boundary-layer clouds from airborne spectral reflection measurements: test of three approaches. Atmos Chem Phys 8(24):7493–7505. doi:10.5194/acp-8-7493-2008

Emanuel KA, Zivkovic-Rothman M (1999) Development and evaluation of a convection scheme for use in climate models. J Atmos Sci 56:1766–1782

Esselborn M, Wirth M, Fix A, Tesche M, Ehret G (2008) Airborne high spectral resolution lidar for measuring aerosol extinction and backscatter coefficients. Appl Opt 47(3):346–358. doi:10.1364/AO.47.000346

Ewald F, Kölling T, Baumgartner A, Zinner T, Mayer B (2016) Design and characterization of specMACS, a multipurpose hyperspectral cloud and sky imager. Atmos Meas Tech 9(5):2015–2042. doi:10.5194/amt-9-2015-2016

Ferreira D, Frankignoul C (2008) Transient atmospheric response to interactive sst anomalies. J Clim 21(3):576–583. doi:10.1175/2007JCLI1704.1

Frenger I, Gruber N, Knutti R, Münnich M (2013) Imprint of southern ocean eddies on winds, clouds and rainfall. Nat Geosci 6(8):608–612

Garcia JR, Mellado JP (2014) The two-layer structure of the entrainment zone in the convective boundary layer. J Atmos Sci 71(6):1935–1955

Gettelman A, Kay JE, Shell KM (2012) The evolution of climate sensitivity and climate feedbacks in the community atmosphere model. J Clim 25(5):1453–1469. doi:10.1175/JCLI-D-11-00197.1

Ghate VP, Miller MA, DiPretore L (2011) Vertical velocity structure of marine boundary layer trade wind cumulus clouds. J Geophys Res Atmos 116(D16):d16206. doi:10.1029/2010JD015344

Ghate VP, Miller MA, Zhu P (2016) Differences between nonprecipitating tropical and trade wind marine shallow cumuli. Mon Weather Rev 144(2):681–701. doi:10.1175/MWR-D-15-0110.1

Gille ST, Llewellyn Smith SG, Lee SM (2003) Measuring the sea breeze from quikscat scatterometry. Geophys Res Lett 30(3):1114. doi:10.1029/2002GL016230

Holland JZ (1970) Preliminary report on the bomex sea–air interaction program. Bull Am Meteorol Soc 51(9):809–820. doi:10.1175/1520-0477(1970)051<0809:PROTBS>2.0.CO;2

Holland JZ, Rasmusson EM (1973) Measurements of the atmospheric mass, energy, and momentum budgets over a 500-kilometer square of tropical ocean. Mon Weather Rev 101(1):44–55. doi:10.1175/1520-0493(1973)101<0044:MOTAME>2.3.CO;2

Horváth A, Gentemann C (2007) Cloud-fraction-dependent bias in satellite liquid water path retrievals of shallow, non-precipitating marine clouds. Geophys Res Lett 34(22):l22806. doi:10.1029/2007GL030625

Illingworth AJ, Barker HW, Beljaars A, Ceccaldi M, Chepfer H, Clerbaux N, Cole J, Delanoë J, Domenech C, Donovan DP, Fukuda S, Hirakata M, Hogan RJ, Huenerbein A, Kollias P, Kubota T, Nakajima T, Nakajima TY, Nishizawa T, Ohno Y, Okamoto H, Oki R, Sato K, Satoh M, Shephard MW, Velàzquez-Blàzquez A, Wandinger U, Wehr T, van Zadelhoff GJ (2015) The earthcare satellite: the next step forward in global measurements of clouds, aerosols, precipitation, and radiation. Bull Am Meteorol Soc 96(8):1311–1332. doi:10.1175/BAMS-D-12-00227.1

Karlsson J, Svensson G, Rodhe H (2008) Cloud radiative forcing of subtropical low level clouds in global models. Clim Dyn 30(7):779–788. doi:10.1007/s00382-007-0322-1

Kollias P, Albrecht B (2010) Vertical velocity statistics in fair-weather cumuli at the arm twp nauru climate research facility. J Clim 23(24):6590–6604. doi:10.1175/2010JCLI3449.1

Lamer K, Kollias P, Nuijens L (2015) Observations of the variability of shallow trade wind cumulus cloudiness and mass flux. J Geophys Res Atmos 120(12):6161–6178. doi:10.1002/2014JD022950

Laxenaire R, Speich S, Blanke B, Chaigneau A, Pegliasco C (2017) Extended agulhas ring trajectories connecting the Indo-Atlantic western boundaries as inferred from altimetry. Geophys Res Lett (accepted)

LeMone MA, Pennell WT (1976) The relationship of trade wind cumulus distribution to subcloud layer fluxes and structure. Mon Weather Rev 104(5):524–539. doi:10.1175/1520-0493(1976)104<0524:TROTWC>2.0.CO;2

Lenschow DH, Krummel PB, Siems ST (1999) Measuring entrainment, divergence, and vorticity on the mesoscale from aircraft. J Atmos Ocean Technol 16(10):1384–1400. doi:10.1175/1520-0426(1999)016<1384:MEDAVO>2.0.CO;2

Lenschow DH, Savic-Jovcic V, Stevens B (2007) Divergence and vorticity from aircraft air motion measurements. J Atmos Ocean Technol 24(12):2062–2072. doi:10.1175/2007JTECHA940.1

Lilly DK (1968) Models of cloud-topped mixed layers under a strong inversion. Q J R Meteorol Soc 94:292–309

Mayer B (2009) Radiative transfer in the cloudy atmosphere. EPJ Web Conf 1:75–99. doi:10.1140/epjconf/e2009-00912-1

McWilliams JC (2016) Submesoscale currents in the ocean. Proc R Soc Lond A Math Phys Eng Sci. doi:10.1098/rspa.2016.0117

Mech M, Orlandi E, Crewell S, Ament F, Hirsch L, Hagen M, Peters G, Stevens B (2014) Hamp—the microwave package on the high altitude and long range research aircraft (halo). Atmos Meas Tech 7(12):4539–4553. doi:10.5194/amt-7-4539-2014

Medeiros B, Nuijens L (2016) Clouds at barbados are representative of clouds across the trade wind regions in observations and climate models. Proc Nat Acad Sci 113(22):E3062–E3070. doi:10.1073/pnas.1521494113

Medeiros B, Stevens B, Held I, Zhao M, Williamson DL, Olson J, Bretherton C (2008) Aquaplanets, climate sensitivity, and low clouds. J Clim 21:4974–4991

Medeiros B, Stevens B, Bony S (2015) Using aquaplanets to understand the robust responses of comprehensive climate models to forcing. Clim Dyn 44(7):1957–1977. doi:10.1007/s00382-014-2138-0

Muller CJ, Held IM (2012) Detailed investigation of the self-aggregation of convection in cloud-resolving simulations. J Atmos Sci 69:2551–2565. doi:10.1175/JAS-D-11-0257.1

Nam C, Bony S, Dufresne JL, Chepfer H (2012) The 'too few, too bright' tropical low-cloud problem in CMIP5 models. Geophys Res Lett. doi:10.1029/2012GL053421

Naumann AK, Stevens B, Hohenegger C, Mellado JP (2017) A conceptual model of a shallow circulation induced by prescribed low-level radiative cooling. J Atmos Sci (in press)

Neggers R, Stevens B, Neelin JD (2006) A simple equilibrium model for shallow-cumulus-topped mixed layers. Theor Comput Fluid Dyn 20(5):305–322. doi:10.1007/s00162-006-0030-1

Neggers RAJ (2015) Exploring bin-macrophysics models for moist convective transport and clouds. J Adv Model Earth Syst 7(4):2079–2104. doi:10.1002/2015MS000502

Nitta T, Esbensen S (1974a) Diurnal variations in the western atlantic trades during the bomex. J Meteorol Soc Jpn Ser II 52(2):254–257

Nitta T, Esbensen S (1974b) Heat and moisture budget analyses using bomex data. Mon Weather Rev 102(1):17–28. doi:10.1175/1520-0493(1974)102<0017:HAMBAU>2.0.CO;2

Norris JR (1998) Low cloud type over the ocean from surface observations. Part II: geographical and seasonal variations. J Clim 11:383–403

Nuijens L, Stevens B, Siebesma AP (2009) The environment of precipitating shallow cumulus convection. J Atmos Sci 66(7):1962–1979. doi:10.1175/2008JAS2841.1

Nuijens L, Serikov I, Hirsch L, Lonitz K, Stevens B (2014) The distribution and variability of low-level cloud in the north Atlantic trades. Q J R Meteorol Soc 140(684):2364–2374. doi:10.1002/qj.2307

Nuijens L, Medeiros B, Sandu I, Ahlgrimm M (2015a) The behavior of trade-wind cloudiness in observations and models: the major cloud components and their variability. J Adv Model Earth Syst 7(2):600–616. doi:10.1002/2014MS000390

Nuijens L, Medeiros B, Sandu I, Ahlgrimm M (2015b) Observed and modeled patterns of covariability between low-level cloudiness and the structure of the trade-wind layer. J Adv Model Earth Syst 7(4):1741–1764. doi:10.1002/2015MS000483

Oue M, Kollias P, North KW, Tatarevic A, Endo S, Vogelmann AM, Gustafson WI (2016) Estimation of cloud fraction profile in shallow convection using a scanning cloud radar. Geophys Res Lett 43(20):10998–11006. doi:10.1002/2016GL070776

Polichtchouk I, Shepherd TG (2016) Zonal-mean circulation response to reduced air–sea momentum roughness. Q J R Meteorol Soci 142(700):2611–2622. doi:10.1002/qj.2850

Rauber RM, Stevens B, Ochs HT III, Knight C, Albrecht BA, Blyth AM, Fairall CW, Jensen JB, Lasher-Trapp SG, Mayol-Bracero OL, Vali G, Anderson JR, Baker BA, Bandy AR, Burnet E, Brenguier JL, Brewer WA, Brown PRA, Chuang P, Cotton WR, Di Girolamo L, Geerts B, Gerber H, Göke S, Gomes L, Heikes BG, Hudson JG, Kollias P, Lawson RP, Krueger SK, Lenschow DH, Nuijens L, O'Sullivan DW, Rilling RA, Rogers DC, Siebesma AP, Snodgrass E, Stith JL, Thornton DC, Tucker S, Twohy CH, Zuidema P (2007) Rain in shallow cumulus over the ocean: the RICO campaign. Bull Am Meteorol Soc 88(12):1912–1928. doi:10.1175/BAMS-88-12-1912

Raymond DJ (1995) Regulation of moist convection over the west pacific warm pool. J Atmos Sci 52(22):3945–3959. doi:10.1175/1520-0469(1995)052<3945:ROMCOT>2.0.CO;2

Reul N, Chapron B, Lee T, Donlon C, Boutin J, Alory G (2014) Sea surface salinity structure of the meandering gulf stream revealed by smos sensor. Geophys Res Lett 41(9):3141–3148. doi:10.1002/2014GL059215

Rieck M, Nuijens L, Stevens B (2012) Marine boundary layer cloud feedbacks in a constant relative humidity atmosphere. J Atmos Sci 69:2538–2550. doi:10.1175/JAS-D-11-0203.1

Roca R, Brogniez H, Chambon P, Chomette O, Cloché S, Gosset ME, Mahfouf JF, Raberanto P, Viltard N (2015) The Megha–Tropiques mission: a review after three years in orbit. Front Earth Sci 3:17. doi:10.3389/feart.2015.00017

Rodts SMA, Duynkerke PG, Jonker HJJ (2003) Size distributions and dynamical properties of shallow cumulus clouds from aircraft observations and satellite data. J Atmos Sci 60(16):1895–1912. doi:10.1175/1520-0469(2003)060<1895:SDADPO>2.0.CO;2

Rudzin JE, Shay LK, Jaimes B, Brewster JK (2017) Upper ocean observations in eastern caribbean sea reveal barrier layer within a warm core eddy. J Geophys Res Oceans 122(2):1057–1071. doi:10.1002/2016JC012339

Ruppert JH (2016) Diurnal timescale feedbacks in the tropical cumulus regime. J Adv Model Earth Syst 8(3):1483–1500. doi:10.1002/2016MS000713

Ruppert JH, Johnson RH (2015) Diurnally modulated cumulus moistening in the preonset stage of the Madden–Julian oscillation during dynamo. J Atmos Sci 72(4):1622–1647. doi:10.1175/JAS-D-14-0218.1

Ruppert JH, Johnson RH (2016) On the cumulus diurnal cycle over the tropical warm pool. J Adv Model Earth Syst 8(2):669–690. doi:10.1002/2015MS000610

Schlemmer L, Bechtold P, Sandu I, Ahlgrimm M (2017) Uncertainties related to the representation of momentum transport in shallow convection. J Adv Model Earth Syst 9:1269–1291. doi:10.1002/2017MS000915

Seifert A, Heus T (2013) Large-eddy simulation of organized precipitating trade wind cumulus clouds. Atmos Chem Phys 13(11):5631–5645. doi:10.5194/acp-13-5631-2013

Sherwood SC, Bony S, Dufresne JL (2014) Spread in model climate sensitivity traced to atmospheric convective mixing. Nature 505:37–42. doi:10.1038/nature12829

Siebesma AP, Cuijpers JWM (1995) Evaluation of parametric assumptions for shallow cumulus convection. J Atmos Sci 52(6):650–666. doi:10.1175/1520-0469(1995)052<0650:EOPAFS>2.0.CO;2

Siebesma AP, Bretherton CS, Brown A, Chlond A, Cuxart J, Duynkerke PG, Jiang H, Khairoutdinov M, Lewellen D, Moeng CH, Sanchez E, Stevens B, Stevens DE (2003) A large eddy simulation intercomparison study of shallow cumulus convection. J Atmos Sci 60(10):1201–1219. doi:10.1175/1520-0469(2003)60<1201:ALESIS>2.0.CO;2

Simpson IR, Shaw TA, Seager R (2014) A diagnosis of the seasonally and longitudinally varying midlatitude circulation response to global warming. J Atmos Sci 71(7):2489–2515. doi:10.1175/JAS-D-13-0325.1

Stevens B (2006) Bulk boundary-layer concepts for simplified models of tropical dynamics. Theor Comput Fluid Dyn 20(5):279–304. doi:10.1007/s00162-006-0032-z

Stevens B, Ackerman AS, Albrecht BA, Brown AR, Chlond A, Cuxart J, Duynkerke PG, Lewellen DC, Macvean MK, Neggers RAJ, Sánchez E, Siebesma AP, Stevens DE (2001) Simulations of trade wind cumuli under a strong inversion. J Atmos Sci 58(14):1870–1891. doi:10.1175/1520-0469(2001)058<1870:SOTWCU>2.0.CO;2

Stevens B, Farrell D, Hirsch L, Jansen F, Nuijens L, Serikov I, Brügmann B, Forde M, Linne H, Lonitz K, Prospero JM (2016) The Barbados Cloud Observatory: Anchoring investigations of clouds and circulation on the edge of the ITCZ. Bull Am Meteorol Soc 97(5):787–801. doi:10.1175/BAMS-D-14-00247.1

Stubenrauch CJ, Rossow WB, Kinne S, Ackerman S, Cesana G, Chepfer H, Girolamo LD, Getzewich B, Guignard A, Heidinger A, Maddux BC, Menzel WP, Minnis P, Pearl C, Platnick S, Poulsen C, Riedi J, Sun-Mack S, Walther A, Winker D, Zeng S, Zhao G (2013) Assessment of global cloud datasets from satellites: Project and database initiated by the GEWEX radiation panel. Bull Am Meteorol Soc 94(7):1031–1049. doi:10.1175/BAMS-D-12-00117.1

Szczap F, Cornet C, Alkasem A, Gour Y, Labonnote L, Jourdan O (2013) A 3D polarized Monte Carlo LIDAR System simulator for studying effects of cirrus inhomogeneities on CALIOP/CALIPSO measurements. In: IRS2012 proceedings

Tomassini L, Voigt A, Stevens B (2015) On the connection between tropical circulation, convective mixing, and climate sensitivity. Q J R Meteorol Soc 141(689):1404–1416. doi:10.1002/qj.2450

van Zanten MC, Stevens B, Nuijens L, Siebesma AP, Ackerman AS, Burnet F, Cheng A, Couvreux F, Jiang H, Khairoutdinov M, Kogan Y, Lewellen DC, Mechem D, Nakamura K, Noda A, Shipway BJ, Slawinska J, Wang S, Wyszogrodzki A (2011) Controls on precipitation and cloudiness in simulations of trade-wind cumulus as observed during RICO. J Adv Model Earth Syst. doi:10.1029/2011MS000056

Vial J, Dufresne JL, Bony S (2013) On the interpretation of inter-model spread in cmip5 climate sensitivity estimates. Clim Dyn 41(11):3339–3362. doi:10.1007/s00382-013-1725-9

Vial J, Bony S, Dufresne JL, Roehrig R (2016) Coupling between lower-tropospheric convective mixing and low-level clouds: physical mechanisms and dependence on convection scheme. J Adv Model Earth Syst. doi:10.1002/2016MS000740

Vial J, Bony S, Stevens B, Vogel R (2017) Mechanisms and model diversity of trade-wind shallow cumulus cloud feedbacks : a review. Surv Geophys. doi:10.1007/s10712-017-9418-2

Vogel R, Nuijens L, Stevens B (2016) The role of precipitation and spatial organization in the response of trade-wind clouds to warming. J Adv Model Earth Syst 8(2):843–862. doi:10.1002/2015MS000568

Wang J, Carton JA (2003) Modeling climate variability in the tropical Atlantic atmosphere. J Clim 16(23):3858–3876. doi:10.1175/1520-0442(2003)016<3858:MCVITT>2.0.CO;2

Wang JJ, Young K, Hock T, Lauritsen D, Behringer D, Black M, Black PG, Franklin J, Halverson J, Molinari J, Nguyen L, Reale T, Smith J, Sun B, Wang Q, Zhang JA (2015) A long-term, high-quality, high-vertical-resolution gps dropsonde dataset for hurricane and other studies. Bull Am Meteorol Soc 96(6):961–973. doi:10.1175/BAMS-D-13-00203.1

Webb M, Senior C, Sexton D, Ingram W, Williams K, Ringer M, McAvaney B, Colman R, Soden B, Gudgel R et al (2006) On the contribution of local feedback mechanisms to the range of climate sensitivity in two GCM ensembles. Clim Dyn 27(1):17–38

Webb MJ, Lock AP, Bretherton CS, Bony S, Cole JNS, Idelkadi A, Kang SM, Koshiro T, Kawai H, Ogura T, Roehrig R, Shin Y, Mauritsen T, Sherwood SC, Vial J, Watanabe M, Woelfle MD, Zhao M (2015) The impact of parametrized convection on cloud feedback. Philos Trans R Soc Lond A Math Phys Eng Sci. doi:10.1098/rsta.2014.0414

Wendisch M, Brenguier JL (2013) Airborne measurements for environmental research: methods and instruments. Wiley-VCH Verlag GmbH & Co. KGaA, Weinheim. doi:10.1002/9783527653218

Wendisch M, Müller D, Schell D, Heintzenberg J (2001) An airborne spectral albedometer with active horizontal stabilization. J Atmos OceanTechnol 18(11):1856–1866. doi:10.1175/1520-0426(2001)018<1856:AASAWA>2.0.CO;2

Wendisch M, Pöschl U, Andreae MO, Machado LAT, Albrecht R, Schlager H, Rosenfeld D, Martin ST, Abdelmonem A, Afchine A, Araùjo AC, Artaxo P, Aufmhoff H, Barbosa HMJ, Borrmann S, Braga R, Buchholz B, Cecchini MA, Costa A, Curtius J, Dollner M, Dorf M, Dreiling V, Ebert V, Ehrlich A, Ewald F, Fisch G, Fix A, Frank F, Fütterer D, Heckl C, Heidelberg F, Hüneke T, Jäkel E, Järvinen E, Jurkat T, Kanter S, Kästner U, Kenntner M, Kesselmeier J, Klimach T, Knecht M, Kohl R, Kölling T, Krämer M, Krüger M, Krisna TC, Lavric JV, Longo K, Mahnke C, Manzi AO, Mayer B, Mertes S, Minikin A, Molleker S, Mnch S, Nillius B, Pfeilsticker K, Pöhlker C, Roiger A, Rose D, Rosenow D, Sauer D, Schnaiter M, Schneider J, Schulz C, de Souza RAF, Spanu A, Stock P, Vila D, Voigt C, Walser A, Walter D, Weigel R, Weinzierl B, Werner F, Yamasoe MA, Ziereis H, Zinner T, Zöger M (2016) Acridicon-chuva campaign: Studying tropical deep convective clouds and precipitation over amazonia using the new german research aircraft halo. Bull Am Meteorol Soc 97(10):1885–1908. doi:10.1175/BAMS-D-14-00255.1

Wentz FJ, Spencer RW (1998) SSM/I rain retrievals within a unified all-weather ocean algorithm. J Atmos Sci 55(9):1613–1627. doi:10.1175/1520-0469(1998)055<1613:SIRRWA>2.0.CO;2

Winker DM, Pelon J, McCormick MP (2003) The CALIPSO mission: Spaceborne lidar for observation of aerosols and clouds. Proc SPIE 4893:1–11

Wirth M, Fix A, Mahnke P, Schwarzer H, Schrandt F, Ehret G (2009) The airborne multi-wavelength water vapor differential absorption lidar wales: system design and performance. Appl Phys B 96(1):201. doi:10.1007/s00340-009-3365-7

Wood R, Köhler M, Bennartz R, O'Dell C (2009) The diurnal cycle of surface divergence over the global oceans. Q J R Meteorol Soc 135(643):1484–1493. doi:10.1002/qj.451

Yanai M, Esbensen S, Chu JH (1973) Determination of bulk properties of tropical cloud clusters from large-scale heat and moisture budgets. J Atmos Sci 30(4):611–627. doi:10.1175/1520-0469(1973)030<0611:DOBPOT>2.0.CO;2

Zhang M, Bretherton C (2008) Mechanisms of low cloud-climate feedback in idealized single-column simulations with the community atmospheric model, version 3 (cam3). J Clim 21(18):4859–4878. doi:10.1175/2008JCLI2237.1

Zhang M, Bretherton CS, Blossey PN, Austin PH, Bacmeister JT, Bony S, Brient F, Cheedela SK, Cheng A, Genio AD, Roode SR, Endo S, Franklin CN, Golaz JC, Hannay C, Heus T, Isotta FA, Dufresne JL, Kang IS, Kawai H, Köhler M, Larson VE, Liu Y, Lock AP, Lohmann U, Khairoutdinov MF, Molod AM, Neggers RAJ, Rasch P, Sandu I, Senkbeil R, Siebesma AP, Siegenthaler-Le Drian C, Stevens B,

Suarez MJ, Xu KM, Salzen K, Webb MJ, Wolf A, Zhao M (2013a) CGILS: Results from the first phase of an international project to understand the physical mechanisms of low cloud feedbacks in single column models. J Adv Model Earth Syst 5:826–842. doi:10.1002/2013MS000246

Zhang Z, Zhang Y, Wang W, Huang RX (2013b) Universal structure of mesoscale eddies in the ocean. Geophys Res Lett 40(14):3677–3681. doi:10.1002/grl.50736

Zhao G, Di Girolamo L (2007) Statistics on the macrophysical properties of trade wind cumuli over the tropical western Atlantic. J Geophys Res 112(D10):d10204. doi:10.1029/2006JD007371

Zinner T, Mayer B, Schröder M (2006) Determination of three-dimensional cloud structures from high-resolution radiance data. J Geophys Res Atmos 111(D8):d08204. doi:10.1029/2005JD006062

The manufacturer's authorised representative in the EU is Springer Nature Customer Service Centre GmbH, Europaplatz 3, 69115 Heidelberg, Germany. If you have any concerns regarding our products, please contact ProductSafety@springernature.com

Printed and bound by CPI Group (UK) Ltd, Croydon, CR0 4YY
05/10/2026
02228132-0006